Environmental Compliance and Sustainability

Environmental Compliance and Sustainability

Global Challenges and Perspectives

Daniel T. Rogers

CRC Press
Taylor & Francis Group
Boca Raton London New York

CRC Press is an imprint of the
Taylor & Francis Group, an **informa** business

CRC Press
Taylor & Francis Group
6000 Broken Sound Parkway NW, Suite 300
Boca Raton, FL 33487-2742

© 2020 by Taylor & Francis Group, LLC
CRC Press is an imprint of Taylor & Francis Group, an Informa business

No claim to original U.S. Government works

Printed on acid-free paper

International Standard Book Number-13: 978-0-367-00230-5 (Hardback)

Visit the Taylor & Francis Web site at
http://www.taylorandfrancis.com

and the CRC Press Web site at
http://www.crcpress.com

To Danielle, my wife, my friend, my companion

Contents

PART TWO   *Environmental Regulations*

PART THREE   *Compliance and Sustainability*

Preface

After studying, conducting research, and working at cleaning up the environment at thousands of sites in more than 20 countries, I have come to certain conclusions that are different than I imagined when I first set out to dedicate my professional life in doing my part to help protect and restore the environment from the harmful and cruel effects of pollution. For instance, I was woefully unaware of the extent and harm that pollution inflicts on all life on Earth. I did not fully appreciate the fact that pollution does not respect the boundaries between countries or between different media, such as the land, atmosphere, and water. Pollution migrates, sometimes quickly, and may even change form, often poisoning everything in its path. I was disappointed and alarmed when I realized how little Science knew about the natural world just a few centimeters beneath our feet in urban environments. I kept telling myself that this is where earth and environmental science should focus, this is the front line, this is where our environmental story begins, because this is the contact between civilization and the natural world.

It has taken me more than 40 years to:

(1) decipher and begin to understand the complex web of interrelated human activities that contribute to pollution of the Earth
(2) begin to understand the behavior of pollution once released into the environment
(3) learn how and why certain pollutants are more dangerous than others
(4) understand how each country chooses to protect the environment and how those actions are effective, or not, at actually protecting the environment
(5) evaluate what sustainability measures are required to reduce harmful impacts
(6) evaluate who should be responsible for correcting the wrongs that humans have done to the Earth

From the study of the items listed above, I have reached a couple of rather simple conclusions, they are: (a) wherever you are, it's all the same and (b) everyone must do their fair share to protect the air, protect the water, and protect the land. I have learned from cleaning up sites of environmental contamination that prevention is key and that it starts with awareness and environmental regulation. In addition, I have learned a hard truth that sometimes we can't make things better no matter what degree of financial resources or advanced technology we employ once pollution is released. Therefore, awareness and environmental regulation are simply not enough. Many other actions are needed to reduce, prevent, and eliminate the human impact on the Earth that require action at the global, national, local, and individual level, which will require coordination and commitment of every person at all levels.

Climate change is just one of many examples of how humans have negatively impacted Earth through the introduction of pollution in its many forms into the natural environment on a global scale. Other impacts caused by pollution that may not be in the public eye but are just as significant include impacts to the land, oceans, surface water, groundwater, and the biosphere. Each are related and affects the others and involves the physical and chemical composition and dynamics of pollutant behavior within the natural world.

Quantifying anthropogenic impacts on a global scale has only recently been available through new technologies and research but the data clearly demonstrates that anthropogenic impacts are substantial and are getting worse. However, now that the many aspects that have caused much of the environmental degradation have been identified, we can now move toward action to reduce those impacts. Those actions must include many aspects of how humans behave and treat the Earth and involve changing our lifestyle through modifications of transportation, energy needs, shelter, and agriculture. We must remember that we all need to work together because as stated earlier, pollution does not respect boundaries. Pollution from one country often migrates and affects all environmental media of other countries and sometimes the entire planet.

Examining environmental regulations of more than 50 countries has revealed a wide spectrum of effective and non-effective methods at environmental protection. It has also unearthed reasons why seemingly robust and comprehensive regulations sometimes do not work. Examining and comparing regulations and other sociological, political, and economic conditions of one country with another has also revealed numerous methods where significant improvements are possible.

To summarize, we must now accept the reality that humans have adversely impacted the entire Earth and that our efforts to improve our environment since the enactment of environmental regulations in the United States and worldwide have been inadequate. To underscore this point, earth scientists are now convinced that we have now moved the needle of geologic time into a new period called the Anthropocene. This is significant because the definition of a Geologic Age is a time period that affects the entire Earth and will be recorded in the Earth's history and cannot be reversed. In order to support a human population on Earth we must not consume and pollute at current levels. In order to achieve sustainable living, vibrant, and diverse world instead of a world without humans, we need to diligently work cooperatively toward living in productive harmony with nature instead of trying to change nature. This means that we must protect the air, protect the water, and protect the land. By protecting these three media, we can also protect the biosphere.

In conclusion, after more than 40 years of trying to make things better and repair and restore what we have polluted, I am convinced that we as humans do not desire to purposely pollute and cause harm to our planet. The majority of harmful actions that have occurred throughout our brief history have been unintended or have been the result of ignorance and not of malice. It is with this belief that I have hope.

Acknowledgments

I wish to thank many regulatory agencies and scientific research organizations who assisted over the years providing research, opinions, and advice to many tough questions concerning protecting human health and the environment that have contributed to the development of this book. Some of the more significant include the International Union of Geological Sciences, Geological Society of America, International Association of Hydrogeologists, United States Environmental Protection Agency, United Nations, World Health Organization, European Environmental Agency, Environment Canada, United States Geological Survey, British Geological Survey, National Oceanic and Atmospheric Administration, Illinois State Geological Survey, California Environmental Protection Agency, California Department of Toxic Substances and Control, Delaware Department of Natural Resources and Environmental Control, Illinois Environmental Protection Agency, Indiana Department of Environmental Management, Iowa Department of Natural Resources, Kansas Department of Health and Environment, Maryland Department of Environment, Michigan Department of Environmental Quality, New Jersey Department of Environmental Protection, Ohio Environmental Protection Agency, Oregon Department of Environmental Quality, and the Wisconsin Department of Natural Resources.

Individuals who have provided guidance and advice from the research and academic field over the years include my colleagues at the University of Michigan, Dr Martin Kaufman and Dr Kent Murray and Dr Ken Howard at the University of Toronto, Dr Jack Sharp at the University of Texas, Dr Richard Berg at the Illinois State Geological Survey, Dr Robert Bobrowsky at the University of British Columbia, Dr Peter Kolesar at Utah State University, Dr Derek Wong at Bureau of Veritas, Dr Colin Booth at Northern Illinois University, Dr Mike Barcelona at Western Michigan University, Dr Mary Ann Thomas at USGS, Dr Garth van der Kamp at Environment Canada, Dr Vladimir Kovalevsky at the Russian Academy of Sciences, Dr John Moore with IAH, Dr Hugo Loaiciga with the University of California, and Dr John Chilton with the British Geological Survey.

When dealing with environmental regulations of the United States and the world, one must also have guidance from the legal field. I wish to thank those special individuals who have been a source of inspiration, counsel, and debate concerning environmental regulations who include Chris Athas, Esq, Ed Brosius, Esq, James Enright, Esq, Rick Glick, Esq, Geneva Halliday, Esq, Brian Houghton, Esq, Michael Krautner, Esq, Stephen Lewis, Esq, Michael Maher, Esq, Chris McNevin, Esq, Granta Y. Nakayama, P. C., Esq, Jeryl Olsen, Esq, Ron Paterson, Esq, Tom Petermann, Esq, Doug Schleicher, Esq, Stephen Smith, Esq, and Tom Wilczak, Esq.

The author wishes to thank the following professional colleagues for their assistance, cooperation, and ideas over the many years that it took to prepare this book: Lauren Alkadis, Cathee Andrews, Tony Anthony, Ann Barry, Gary Blinkiewicz, Tom Buggey, Kristine Casper, Chris Christensen, Russ Chadwick, Bob Cigale, Tom Cok, Marc DeLoecker, Sheryl Doxtader, Ian Drost, Rob Ellis, Rose Ellison, Rob Ferree, Jennifer Formoso, Orin Gelderloos, Steve Hoin, Heather Hopkins, George Karalus, Steve Kitler, Rick Linnville, Daniel J. Lombardi, Bill Looney, Kim Myers, Ernie Nimister, David O'Donnell, Tom O'Hara, Ray Ostrowski, Paul Owens, Bruce Patterson, Fred Payne, Sara Pearson, Mark Penzkover, Robert Ribbing, Eric Ross, Jeanne Schlaufman, Dave Slayton, Dave Smith, Rebecca Spearot, Ed Stewart, Patricia Thornton, Mary Vanderlaan, Chris Venezia, Nick Welty, Cheryl Wilson, and Tom Wenzel.

Lastly, I wish to thank my wife, Danielle, for her constant and unwavering support and for being the main sounding board for many of the ideas and approaches in preparing this manuscript.

Author

Daniel T. Rogers is the Director of Environmental Affairs at Amsted Industries Incorporated in Chicago, Illinois. Amsted Industries is a diversified manufacturing company of industrial components serving railroad, vehicular, construction, and building markets. Amsted Industries has more than 60 manufacturing locations in 15 countries. Mr Rogers participates in environmental due diligence for acquisitions and divestitures, investigation and remediation of contamination, creating and evaluating environmental compliance and sustainability programs, and providing environmental advice, oversight, training, and negotiation strategies at various levels within the organization.

Over the last few decades, Mr Rogers has published nearly 100 research papers in professional and academic publications and peer-reviewed journals on subjects including environmental geology, hydrogeology, geologic vulnerability and mapping, contaminant fate and transport, urban geology, environmental site investigations, contaminant risk, brownfield re-development, remediation, pollution prevention, environmental compliance, and sustainability. He has co-authored *Urban Watersheds: Geology, Contamination and Sustainable Development* (Taylor & Francis 2011) and is also the author of *Environmental Geology of Metropolitan Detroit* (1996) and has published surficial geologic and contaminant vulnerability maps of the Rouge River watershed in southeastern Michigan. In addition, he is a contributing author of an encyclopedia of global social concerns.

He has taught geology and environmental chemistry at Eastern Michigan University and The University of Michigan and has presented guest lectures at numerous colleges and universities both in the United States and internationally.

Photograph of the author by Danielle Rogers on the summit of Whistler Peak in Jasper National Park, Canada, with Mount Robson in the background.

Acronyms, Elements, Symbols, Molecules, and Units of Measure

ACRONYMS

ACGIH	American Conference of Governmental Industrial Hygienists
ADP	Atmospheric Decontamination Plan
APCP	Air Prevention and Control of Pollution Act
APP	Atmospheric Prevention Plan
AST	Aboveground storage tank
ASTM	American Society for Testing Materials
ATSDR	Agency for Toxic Substances and Disease Registry
BACT	Best Available Control Technology
BLM	Bureau of Land Management
BMP	Best management practice
BPA	Bisphenol A
BTEX	Benzene, toluene, ethyl benzene, and xylenes
CAA	Clean Air Act
CAAC	Clean Air Alliance of China
CAS	Chemical Abstract Service
CalEPA	California Environmental Protection Agency
CEPA	Canadian Environmental Protection Act
CESQG	Conditionally exempt small quantity generator
CFC	Chlorofluorocarbons
CMEE	China Ministry of Ecology and Environment
CONAGUA	Mexico National Water Commission
CERCLA	Comprehensive Environmental Response, Compensation and Liability Act
CFR	Code of Federal Regulation
CRF	Contaminant or pollution risk factor
CRF$_{air}$	Contaminant or pollution risk factor for air
CRF$_{gw}$	Contaminant or pollution risk factor for groundwater
CRF$_{soil}$	Contaminant or pollution risk factor for soil
CSO	Combined sewer overflow
CVOC	Chlorinated volatile organic hydrocarbon
CWA	Clean Water Act
DBP	Di-n-butyl phthalate
DCE	dichloroethylene, dichloroethane
DEHP	Bi or Bis(2-ethylhexyl) phthalate
DEP	Diethyl phthalate
DDT	dichlorodiphenyltrichloroethane
DIVs	Dutch Intervention Values
DNAPL	Dense nonaqueous phase liquid
DOT	Department of Transportation
EEA	European Environment Agency
EEC	European Economic Community
EIA	Environmental Impact Assessment
EIS	Environmental Impact Statement
EHS	Extremely hazardous substance

EH&S	Environmental health and safety
EINETICS	European Information and Observation Network
EPCRA	Emergency Planning and Community Right-to-Know Act
EPA	Environmental Protection Agency
ESA	Environmental site assessment
ESPM	Elimination, Substitution, Prevention, and Minimization
ETS	Emission Trading System
EU	European Union
FDA	Food and Drug Administration
FEPCA	Federal Environmental Pesticide Control Act
FFDCA	Federal Feed, Drug, and Cosmetic Act
FIFRA	Federal Insecticide, Fungicide, and Rodenticide Act
GACT	Generally Available Control Technology
GHG	Greenhouse gas
GWP	Global warming potential
HAP	Hazardous Air Pollutant
HAZMAT	Hazardous materials
HMTA	Hazardous Materials Transportation Act
HCFC	Hydrochlorofluorocarbons
HFC	Hydrofluorocarbon
HMTA	Hazardous Materials and Transportation Act
HRS	Hazard Ranking System
IED	Industrial Emissions Directive
KDNREP	Kentucky Department of Natural Resources and Environmental Protection
LEPC	Local Emergency Planning Committee
LNAPL	Light nonaqueous phase liquid
LOAEL	Lowest-observed-adverse-effect level
LQG	Large quantity generator
MACT	Maximum Achievable Control Technology
MCL	Maximum contaminant level
MCLG	Maximum contaminant level goal
MDEP	Massachusetts Department of Environmental Protection
MDEQ	Michigan Department of Environmental Quality
MEK	Methyl ethyl ketone
MEP	China Ministry of Environmental Protection
MNR	Russian Ministry of Natural Resources
MS4S	Large Municipal Separate Storm Sewer Systems
MSDS	Material safety data sheet
MTBE	Methyl tertiary butyl ether
NAAQS	National Ambient Air Quality Standards
NAFTA	North American Free Trade Agreement
NASA	National Aeronautics and Space Administration
NATO	North Atlantic Treaty Organization
NCZMP	National Coastal Zone Management Program
NERRS	National Estuarine Research Reserve System
NEA	Norwegian Environment Agency
NEPA	National Environmental Protection Act
NEPC	Australian National Environmental Protection Council
NESHAPS	National Emission Standards for Hazardous Air Pollutants
NGS	National Geographic Society
NHTSA	National Highway Traffic Safety Administration

NJDEP	New Jersey Department of Environmental Protection
NOAA	National Oceanic and Atmospheric Administration
NOM	Mexico Standards or Norms
NOV	Notice of violation
NPDES	National Pollutant Discharge Elimination System
NPDWS	National Primary Drinking Water Standards
NPL	National Priority List
NPS	Nonpoint source
NRC	National Response Center
NRC	Nuclear Regulatory Commission
ODS	Ozone depleting substance
OECD	Organization for the Economic Co-operation and Development
OPP	Office of Pesticide Programs (USEPA)
OSHA	Occupational Safety and Health Administration
OSPA	Oil Spill Prevention Act
PAH	Polycyclic aromatic hydrocarbon
PFC	Perfluorocarbon
PBT	Persistent bio-accumulative toxic
PCB	Polychlorinated biphenyl
PCE	Tetrachloroethylene, tetrachloroethene, perchloroethylene, perchloroethene
PCP	Pentachlorophenol
PERC	Tetrachloroethylene, tetrachloroethene, perchloroethylene, perchloroethene
PEL	Permissible exposure limit
PFOA or PFAS	Per- and polyfluoroalkyl substances
PM	Particulate matter
PNA	Polynuclear aromatic hydrocarbon
POTW	Publicly owned treatment works
PPA	Pollution Prevention Act
PPE	Personal protective equipment
PREFEPA	Mexico Federal Attorney Generalship of Environmental Protection
PRP	Potential Responsible Party
PS	Point source
PSD	Prevention of Significant Deterioration
PTE	Potential to emit
PVC	Polyvinyl chloride
RCI	Reactivity, corrosivity, and ignitability
RCRA	Resource, Conservation, and Recovery Act
REC	Recognized environmental condition
RQ	Reportable quantity
SARA	Superfund Amendments and Reauthorization Act
SDWA	Safe Drinking Water Act
SDS	Safety data sheet
SEMARNAT	Mexico Secretariat of the Environment and Natural Resources
SEP	Supplemental environmental project
SERC	State Emergency Response Commission
SMG	Small Quantity generator
SPCC	Spill Prevention Control and Countermeasures Plan
SPLP	Synthetic precipitant leaching procedure
SRP	Spill response plan
SVOC	Semi-volatile organic compound
SWPPP	Stormwater pollution prevention plan

TCA	Trichloroethane
TCE	Trichloroethylene, trichloroethene
TCLP	Toxic characteristic leaching procedure
THM	Trihalomethane
TPQ	Threshold Planning Quantity
TSCA	Toxic Substance and Control Act
TSP	Total suspended particulates
UK	United Kingdom
UN	United Nations
UNFCCC	United Nations Framework Convention on Climate Change
UNESCO	United Nations World Heritage Site
USCB	United States Census Bureau
USCIA	United States Central Intelligence Agency
USDA	United States Department of Agriculture
USDC	United States Department of Commerce
USDOE	United States Department of Energy
USDHH	United States Department of Health and Human Services
USDOT	United States Department of Transportation
USEPA	United States Environmental Protection Agency
USFS	United States Forest Service
USGS	United States Geological Survey
USNRC	United States Nuclear Regulatory Commission
UST	Underground storage tank
VC	Vinyl chloride
VOC	Volatile organic compound
VSQG	Very small quantity generator
WDNR	Wisconsin Department of Natural Resources
WHO	World Health Organization

ELEMENTS

Ag	Silver
As	Arsenic
Ba	Barium
Br	Bromine
C	Carbon
Cd	Cadmium
Cl	Chlorine
Cr	Chromium
Cr III	Trivalent chromium
Cr VI	Hexavalent chromium
Cr+^{3}	Trivalent chromium
Cr+^{6}	Hexavalent chromium
Cu	Copper
F	Fluorine
H	Hydrogen
Hg	Mercury
I	Iodine
K	Potassium
N	Nitrogen
O	Oxygen
P	Phosphorus

Pb   Lead
S   Sulfur
Se   Selenium
Zn   Zinc

SYMBOLS

η Effective porosity
ρb bulk density
W_s Water solubility
ATM   Atmosphere
C°   degrees Celsius
F°   degrees Fahrenheit
Foc   Organic carbon partition coefficient
H   Henry's Law constant
Kd   Distribution coefficient
Koc   Fraction of total organic carbon
M   Mobility
MW   Molecular weight
P   Persistence
R   Retardation factor
T   Toxicity
VP   Vapor pressure

MOLECULES

CO   Carbon monoxide
CO$_2$ Carbon dioxide
CH$_4$ Methane
H$_2$O   Water
NOx   Nitrogen oxide
N$_2$O   Nitrous oxide
O$_3$ Ozone
SO$_2$ Sulfur dioxide
SF$_6$ Sulfur hexafluoride

UNITS OF MEASURE

mg/kg   milligram per kilogram
ug/kg   microgram per kilogram
mg/l   milligram per liter
ug/l   microgram per liter
g/l   gram per liter
mg/m^3   milligram per cubic meter
ug/m^3   microgram per cubic meter
PM2.5   Particulate matter less than 2.5 microns
PM10   Particulate matter less than 10 microns
ppm   part per million
ppmv   part per million by volume
ppb   part per billion
TPY   tons per year
BTU   British Thermal Unit

1 Compliance and Sustainability Overview and Themes

1.1 INTRODUCTION

The purpose of environmental regulation is to protect human health and the environment. To achieve this goal, nations of the world have learned to concentrate their efforts on environmental regulations that protect the air, protect the water, and protect the land. This forms the framework of environmental regulations throughout the world and is one of several central rhetorical themes of this book, which is "no matter where you are, it's all the same." Protect the air, protect the water, and protect the land. This approach is critical if humans are to restore a sustainable balance with the Earth because, as we shall discover, pollution does not respect political boundaries between countries. Pollution caused by one country often pollutes neighboring countries and sometimes the entire planet when the pollutants released are at the right place, at the right time, and in the right amount.

To understand environmental regulations, one must first understand pollution in its many types and forms and understand the behavior of pollution once released into the environment. Pollution and its behavior in nature is also a central theme of this book and is addressed in the first section.

The second section of this book examines environmental regulations and pollution throughout the world. Most often, environmental regulations are reactive and are born out of an incident that results in an environmental disaster that all too often causes enormous harm to humans and the environment. This book explores the many aspects of how humans have caused harm to the environment and examines how more than 50 countries choose to protect the air, water, and land everywhere on Earth including Antarctica and the oceans.

The third and final section of the book examines compliance with existing environmental regulations and sustainability. A model is presented for improving and tracking compliance and sustainability efforts to work toward the ultimate goal of environmental stewardship and achieving balance with nature. Based on scientific information presented in the first two sections of the book, numerous and extensive options are presented that provide a recipe of what must be done on a global, national, local, and individual level to modify environmentally destructive human behavior to environmentally constructive human behavior.

In order to restore the Earth to a living, vibrant, and diverse world, humans need to diligently pursue global cooperation and take immediate proactive sustainability measures so that we may once again live in productive harmony with nature instead of trying to change nature.

1.2 MAJOR THEMES OF THIS BOOK

The following sections provide a brief introduction into the central themes that are covered in this book and include the following:

1. Pollution and its behavior in nature
2. Land use and where cities are built
3. History of environmental regulations
4. Complexity and effectiveness of environmental regulations
5. Environmental regulations and pollution of the world
6. Maintaining compliance with environmental regulations
7. Sustainability at the global, local, national, and individual level
8. A model for sustainability

1.2.1   THEME 1: POLLUTION AND ITS BEHAVIOR IN NATURE

Air, water, and land are essential for life on Earth. Life becomes threatened when any one of these becomes contaminated. As the human population continues to increase, we continue to modify and impact our environment. In many instances these actions result in the introduction of pollutants into the environment that degrade not just the local environment but can now be observed on a global scale and affects all life on Earth. Therefore, we must evaluate how these actions are impacting our planet and either eliminate, modify, or reduce the harmful impacts if we and other organisms are to survive. Protection of these three essentials for life on Earth are the fundamentals for environmental management.

There would be no need for environmental laws and regulations, or the United States Environmental Protection Agency (USEPA) if there was no pollution. Therefore, it seems logical that we begin our journey by examining pollution in its many forms. However, it does not just begin and end with pollution. There are other activities that impact our environment that go beyond what is commonly termed pollution. Many of these other forms do not introduce chemicals into the natural world. For instance, any number of different land uses where humans inflict enormous harm on our environment include mining, petroleum development, urban land development, wetland destruction, deforestation, forest management techniques, and agriculture. As we are discovering, these activities do not just affect a centralized area where the activity took place. These types of activities often have a rippling effect that negatively impact surrounding areas and in some cases spreads like a disease that affects the entire planet. As we will discover in the course of this book, this is where sustainability then plays an important and central role. Sustainability goes beyond addressing pollution and examines how humans, albeit unintentional, adversely impact our planet as though we ourselves are a form of pollution.

In order to fully understand pollution and its effects, we need to address four aspects of that:

- Physical chemistry of pollution
- Behavior of pollution in nature
- Nature's response to pollution
- The effects of pollution on humans and the environment

There are thousands of different types of pollution. To understand the behavior of pollution in nature, one must evaluate the physical chemistry of pollution, and how pollution behaves in the geologic realms beneath the surface, in water, and in the atmosphere. The following is a shortened list of chemical and organic pollutants commonly present in the environment (Kaufman et al. 2011):

- Volatile organic compounds (VOCs)
  - Dense non-aqueous phase liquids
  - Light non-aqueous phase liquids
- Polynuclear aromatic hydrocarbons (PNAs or PAHs)
- Semi-volatile organic compounds (SVOCs)
- Polychlorinated biphenyls (PCBs)
- Pesticides and herbicides
- Heavy metals
- Others including:
  - Common fertilizers including nitrates, phosphorus, and potassium
  - Greenhouse gases
  - Chlorofluorocarbons
  - Carbon dioxide
  - Carbon monoxide
  - Particulates (dust)

- Ozone
- Bacteria such as *E. coli*
- Viruses
- Invasive species
- Pharmaceuticals
- Cyanide
- Asbestos
- Acids
- Bases
- Radioactive compounds
- Dioxins
- Polyfluoroalkyl substances
- Emerging contaminants

Sources of pollution include many human activities performed primarily at the surface resulting in the release of toxic substances into the environment. These toxic substances may be transported over time, or remain relatively close to their source before they are degraded, transformed, or destroyed. Pollutants released into the environment are often mixtures, each contaminant is unique chemically, and the geologic environment in which the pollutants are released is also unique. Understanding the physical chemistry of each chemical and the geologic environment into which they are released becomes necessary for characterizing their fate and transport in nature (Rogers et al. 2007).

During their transport and before reaching their final sink, certain pollutants may reside at multiple intermediate sinks for different periods of time. Intermediate sinks include surface water, groundwater, and the atmosphere, because the contaminants held within these water-containing sinks will flow and ultimately reach the oceans. Aquifers with a very low hydraulic conductivity are for practical purposes considered final sinks, as are inland bogs and some wetlands. Sediment and soil can function as intermediate or as final sinks, since erosion may move both of these unconsolidated materials. The oceans are almost always a final sink of pollution, but wave action occasionally brings pollutants onshore.

To a large extent, the level of human health and/or environmental risk is a function of two fundamental concepts: *mobility* and *persistence*. Mobility is a measure of a substance's potential to migrate. Persistence is a measure of a substance's ability to remain in the environment before being degraded, transformed, or destroyed (Rogers et al. 2007). Substances with higher toxicity pose greater health risks, but the risk to humans and the environment grows exponentially if the chemical is both mobile and persistent. For example, a highly toxic but immobile chemical may affect a few people in a warehouse through inhalation, whereas a mobile and persistent chemical of moderate toxicity can contaminate a public water supply, or migrate to a different sink where the potential for widespread human exposures and ecosystem damages is much higher.

We have control over the chemicals we use, and where and how we use them. Control over the geologic environment, however, is beyond our means. Therefore, we must understand the geologic environment where our urban areas are located and develop methods to minimize or eliminate the potentially harmful effects of contaminants upon human health and the environment. A logical first step is through an *understanding* of urban geology, followed by an *evaluation* of the extent that a given urban area's geology influences the migration of pollutants. In addition, since water plays a critical role in assessing a region's vulnerability to pollution, the analyses performed require an understanding of water occurring at the Earth's surface and beneath.

Throughout the world, the largest cities share a geologic environment dominated by unconsolidated sedimentary deposits and are also located near water (see Table 1.1). Most of those sedimentary deposits are saturated with water very near the surface and function as sources of drinking water and/or as hydraulic connections to surface water and ecosystems. Moreover, water is considered the universal solvent, so any pollution released into the environment from anthropogenic

TABLE 1.1

Water Bodies and Near-Surface Geology Near the Major Cities of the World

City	Location	Estimated Metropolitan Population (millions)	Geology	Water Body
Tokyo	Japan	37.8	Unconsolidated	Pacific Ocean
Shanghai	China	34.8	Unconsolidated	Pacific Ocean
Jakarta	Indonesia	31.7	Unconsolidated	Pacific Ocean
Delhi	India	26.4	Unconsolidated	Yamuna River
Seoul	Korea	25.5	Unconsolidated	Han River
Beijing	China	24.9	Unconsolidated	Yongding River
New York City	USA	23.8	Unconsolidated	Atlantic Ocean, Hudson River
Mexico City	Mexico	21.6	Unconsolidated	Lerma River, Santiago River
Sao Paulo	Brazil	21.2	Unconsolidated	Atlantic Ocean
Cairo	Egypt	20.5	Unconsolidated	Nile River
Los Angeles	USA	18.7	Unconsolidated	Pacific Ocean
Moscow	Russia	16.9	Unconsolidated	Moskve River
Istanbul	Turkey	15.2	Unconsolidated	Turkish Straits
London	England	14.2	Unconsolidated	Thames River
Buenos Aires	Argentina	13.1	Unconsolidated	Atlantic Ocean
Paris	France	12.6	Unconsolidated	Seine River
Rio de Janeiro	Brazil	12.3	Unconsolidated	Atlantic Ocean
Chicago	USA	9.8	Unconsolidated	Lake Michigan
Johannesburg	South Africa	9.6	Unconsolidated	Jukskei River
Riyadh	Saudi Arabia	7.7	Unconsolidated	Simbacom River
Santiago	Chile	6.7	Unconsolidated	Pacific Ocean
Berlin	Germany	6.1	Unconsolidated	Spree River
Toronto	Canada	5.9	Unconsolidated	Lake Ontario
Sydney	Australia	5.0	Unconsolidated	Pacific Ocean

Source: United Nations. World Population Prospects. United Nations Department of Economic and Social Affairs. Population Division. 2017. https://esa.un.org/unpd/wpp/data. (accessed December 14, 2017.)

or natural sources has the potential to migrate and cause harm. Scientific factors that control the severity of harm to the environment from pollutant releases are: (1) the geologic and hydrogeologic environment, (2) the physical chemistry of the contaminants and amounts released, and (3) the mechanism in which the release occurs (Rogers 1996; Murray and Rogers 1999; Kaufman et al. 2005; Rogers 2018). There are numerous but costly techniques available for investigating, managing, and reducing the harmful effects of urban water pollution (Rogers 2016a;2016b).

The largest polluting activity of land world-wide is agriculture. The largest polluter of air world-wide are motor vehicles, which include automobiles, diesel trucks, trains, buses, marine vessels, and aircraft. The largest polluter of water world-wide are biological pollutants largely from human waste and from agricultural activities that include pesticides and herbicides, fertilizers, and erosion (United Nations 2016a).

The final resting place for much of the pollution of the world are the oceans.

With respect to pollution, we will learn that:

- No country, continent, or ocean is pollution free
- Pollution caused by humans is everywhere
- Significant environmental degradation caused by humans has touched every country

- Pollution does not respect country borders
- Pollution will continue to migrate in the air, water, and on land
- The oceans have been treated as the human privy and garbage disposal of the world

1.2.2 Theme 2: Land Use and Where Cities Are Built

For nearly 10,000 years, humans have applied themselves toward changing their environment for their own benefit and enjoyment. Over the past few decades there has been a realization that perhaps changing our environment may not be a wise choice. John Muir once said that "changing nature cannot be attained by force but by understanding." This phrase forms one of the focal points of this book as it relates to environmental regulations and sustainability. Smart land use in concert with nature can play a significant role in shaping a sustainable future in preventing pollution and minimizing its harm on the environment.

The surface of the Earth is approximately 510,100,000 square kilometers (Coble et al. 1987). Approximately 70% is covered by water or 361,800,000 square kilometers and 30% is land or 148,300,000 square kilometers. The amount of land on Earth that is farmed is estimated at 40% of 59,320,000 square kilometers (United Nations 2016a). This is a huge number, especially since the amount of urbanized land on Earth is estimated to be only 2.7% and is home to roughly half the current human population (United Nations 2016a). The amount of land currently used for farming becomes an even larger value since 10.4% of Earth's land or 15,600,000 square kilometers is covered by ice and is not farmable, 20% or 29,660,000 square kilometers is mountainous, and another 20% or 29,660,000 square kilometers is desert and has little or no topsoil (Coble et al. 1987). When factoring out those land areas not suitable for farming, only 7% of land remains available on Earth.

Despite the availability of specific methods and procedures, the natural environment of most urban areas is not well understood. A common thread of all major cities of the world is that they are all located near water and on unconsolidated sediments (see Table 1.1). Modifications to land use can have a significant impact on improving the environment and greatly increasing sustainability measures. Repurposing the land we already have developed can potentially significantly reduce the amount of developed land, even with an expanding human population, as we shall see when employing several techniques and options for smart land use that include numerous available options we shall explore in this book.

1.2.3 Theme 3: History of Environmental Regulations

In order to know where we are heading into the future as a species we often must reflect back at our history for perspective. From a historical point of view, our treatment of the environment has not been kind. As recently as the early portion of the 20th century, environmental awareness and legislation were generally lacking not only in the United States but world-wide. There were a few international agreements that primarily focused on boundary waters, navigation, and fishing rights along shared waterways between countries. However, they ignored pollution and ecological issues. In fact, the most convenient and least costly method of waste disposal up until the middle of the 20th century was "up the stack or down the river" (Haynes 1954).

The publishing of the Rachel Carson book *Silent Spring* in 1962 described the environmental effects of polychlorinated biphenyls (PCBs) on birds and catapulted the environmental movement in the United States into the public eye (Carson 1962). The 1960s and 1970s are generally described as the age of environmentalism in the United States. This is when the United States realized that environmental degradation had significantly affected the air and water quality of many urban areas, especially the air in Los Angeles and water in Lake Erie, one of the Great Lakes. Starting on Earth Day in 1971, the Ad Council and Keep America Beautiful campaign aired public awareness commercials on television depicting a Native American shedding a tear when looking out over the polluted landscape of the United States. The commercial stated "people start pollution, people can stop pollution" (United States Advisory Council on Historic Preservation 2017).

Although laws focused on protecting human health or the environment can be traced back more than 2,000 years, it took the United States in the 1970s with the passage of the most significant environmental regulations and laws to set the example for the rest of the world (Hahn 1994).

Environmental regulations of the United States and the world are largely based on reactions to an incident that caused enormous harm to human health and the environment. A few of the many examples of incidents that resulted in enactment of environmental regulations include solvents in groundwater at Love Canal in New York, dioxins in shallow soil at Times Beach in Missouri, a release of toxic gas at Bhopal in India, the fire on the Cuyahoga River near Cleveland, the Exxon Valdez oil spill in Alaska, and the recent incident on the Deepwater Horizon oil drilling platform in the Gulf of Mexico.

1.2.4　THEME 4: COMPLEXITY AND EFFECTIVENESS OF ENVIRONMENTAL REGULATIONS

Factors that influence the effectiveness of environmental regulations vary from country to country and within each country as well. The United States is no different. Factors that influence the degree to which a country has effective environmental regulations include (Bates and Ciment 2013):

- Political will
- Environmental tragedies, harm to the environment, and lessons learned
- Cost
- Geography and climate
- Enforcement
- Incentives
- Lack of overriding social factors such as war, political structure, greed, poverty, hunger, lack of infrastructure, educational awareness, and corruption

Environmental laws and regulations in the United States and the world have continually become more numerous and complex to the point that even many scientists and other professionals need assistance at interpreting, understanding, and applying many environmental laws and regulations.

The growth of the number and complexity of environmental laws and regulations can be more easily understood by simply applying the principle of the scientific term called entropy. Entropy is a principle that states that as time progresses forward there is more disorder which increases complexity. This can be applied to our human civilization as well. Our civilization has become more complex with time as a result of our technological advances, population increase, and other factors. This in turn has put pressure on the natural world and our response has been the passage of environmental laws and regulations as an attempt to establish order in a disordered and out-of-equilibrium environment that humans caused. To make this point more clear, a list of laws that are commonly attributed to protecting human health and the environment just in the United States include (USEPA 2017):

- Rivers and Harbors Act of 1899
- Antiquities Act of 1906
- Atomic Energy Act of 1946
- Atomic Energy Act of 1954
- Brownfield Revitalization Act of 2002
- Clean Air Act (CAA) of 1970
- Clean Water Act (CWA) of 1972
- Coastal Zone Management Act of 1972

- Comprehensive Environmental Response, Compensation, and Liability Act (CERCLA) of 1980
- Emergency Planning and Community Right-to-Know Act of 1986
- Endangered Species Act of 1973
- Energy Policy Act of 1992
- Energy Policy Act of 2005
- Federal Power Act of 1935
- Federal Feed, Drug, and Cosmetic Act (FFDCA) of 1938
- Federal Insecticide, Fungicide, and Rodenticide Act (FIFRA) of 1947
- Fish and Wildlife Coordination Act of 1968
- Fisheries Conservation and Management Act of 1976
- Food Quality Protection Act of 1996
- Global Climate Protection Act of 1987
- Hazardous Materials Transportation Act (HMTA) of 1975
- Insecticide Act of 1910
- Lacey Act of 1900
- Marine Mammal Protection Act of 2015
- Marine Protection Act of 1972
- Migratory Bird Treaty Act of 1916
- Mineral Leasing Act of 1920
- National Environmental Policy Act of 1969
- National Forest Management Act of 1976
- National Historic Preservation Act of 1966
- National Parks Act of 1980
- Noise Control Act of 1974
- Nuclear Waste Policy Act of 1982
- Occupational Safety and Health Act (OSHA) of 1970
- Ocean Dumping Act of 1988
- Oil Spill Prevention Act of 1990
- Pollution Prevention Act (PPA) of 1990
- Refuse Act of 1899
- Resource, Conservation, and Recovery Act (RCRA) of 1976
- Rivers and Harbors Act of 1899
- Safe Drinking Water Act (SDWA) of 1974
- Superfund Amendments and Reauthorization Act (SARA) of 1986
- Surface Mining Control and Reclamation Act of 1977
- Toxic Substance Control Act of 1976
- Wild and Scenic Rivers Act of 1968

1.2.5 THEME 5: ENVIRONMENTAL REGULATIONS AND POLLUTION OF THE WORLD

A few of some of the most important conclusions reached in this book concerning environmental regulations of the world include:

- Each country has environmental regulations
- No two countries are exactly the same
- Each country has a unique set of attributes that influence the effectiveness of environmental protection that includes climate, geography, geology, social and economic concerns, and politics
- The platform of environmental regulations of each country are similar to that of the United States or the European Union

- Although some countries are doing a better job than others, each country struggles with protecting human health and the environment for various reasons, two of which are universal and include a growing population and urban expansion
- Every country has been significantly affected by pollution of the air, water, and land

On Earth, the largest polluting activity is agriculture. The largest polluter of air are motor vehicles that include automobiles, diesel trucks, trains, buses, marine vessels, and aircraft. The largest polluter of water world-wide are biological pollutants largely from human waste and from agricultural activities that include pesticides and herbicides, fertilizers, and erosion (United Nations 2016a).

The final resting place for much of the pollution of the world are the oceans.

International organizations such as the WHO and the United Nations have begun to evaluate on a global scale, the quality of fresh water and the air on Earth. According to the WHO (2013 and 2015) and the United Nations (2016b and 2016c), approximately 2.8 billion humans lack access to basic sanitation and improved drinking water. While this number is improving, it still represents nearly 30% of the planet's human population. In addition, according to the WHO (2018), nine out of ten people on Earth breathe air containing high levels of pollution. The United Nations estimates that most land pollution is caused by agricultural activities, such as grazing, use of pesticides, fertilizers, irrigation, plowing, and confined animal facilities (United Nations 2016a).

Some of these actions have affected an individual site or location and others have affected the entire planet. This book explores how these actions have molded and influenced the development of environmental regulations world-wide and identifies and explores many new approaches that can be undertaken to improve our environment. These improvements and modifications range from the individual to global level through several avenues that include education, science-based assessments, improving existing environmental regulations, and enacting new environmental regulations with a goal of achieving a human population that is sustainable with Earth.

1.2.6 THEME 6: MAINTAINING COMPLIANCE WITH ENVIRONMENTAL REGULATIONS

Environmental compliance is sometimes perceived as just another set of rules that is easy to accomplish. Just follow the rules. This premise fails on several levels. First, the human population is ever increasing and our thirst for land and modern conveniences increases as our standard of living is raised. Second, our technology is ever changing with new products and chemicals. Third, the Earth is dynamic. There is change, getting more complex as more information and science is discovered, learned, and shared. The weather changes constantly and is perhaps the best example of how to imagine environmental management and risk. It also changes, as does everything. Everything obeys the second law of thermodynamics, namely entropy, in that nature tends to become more complex as time marches on.

Understanding the natural setting of our urban and industrial areas. Knowing the chemicals that are used and how they might cause harm to the environment and humans if they are released. What alternative chemicals are available. How to reduce chemical and energy use. Preventing chemical releases to the environment. Quantifying the costs involved with cleaning up the environment once a release occurs. Developing environmental stewardship. These are all subjects that should be taken into account when evaluating the environmental health of a country, province, state, city, manufacturing facility, or an individual household, and it all starts with conducting an environmental audit.

The significant lessons and themes to remember when evaluating and working with maintaining compliance with environmental laws include:

- It starts with conducting an environmental audit
- Being in compliance with environmental regulations has little to do with sustainability
- Work yourself out of environmental regulations

- There is little we can control, but those things we do control are meaningful and can make a big difference
- Limit chemical use
- Lastly, we must remember that "no matter where you are, it's all the same"

We are learning that we live on one planet with no environmental boundaries and that contamination does not respect boundaries. Another point is that environmental regulations are built on the USEPA or EU platforms and those platforms rely on three basic principles:

1. Protect the air
2. Protect the water
3. Protect the land

1.2.7 THEME 7: SUSTAINABILITY AT THE GLOBAL, NATIONAL, LOCAL, AND INDIVIDUAL LEVEL

The United Nations has framed sustainability goals, termed "Global Goals," which include (United Nations 2019):

- No poverty
- Zero Hunger
- Good health
- Well being
- Quality Education
- Gender Equality
- Clean Water
- Effective Sanitation
- Affordable Energy
- Clean Energy
- Economic Growth
- Satisfactory Employment
- Industry
- Innovation
- Infrastructure
- Reduced Inequalities
- Sustainable Cities
- Responsible Consumption
- Climate Action
- Life Below Water
- Life on Land
- Peace and Justice
- Partnerships
- Life in the Air

At least five actions at the global level should be undertaken to improve sustainability at every level and include:

1. Cohesive environmental policy and enforcement
2. Controlling the human population
3. Modifying the human diet
4. Financial fairness and equity distribution
5. Modifying business models

The majority of subject areas where environmental regulations at national levels are either in need of improvement or where new regulations are needed to assist sustainability initiatives include the following subject areas:

- Modifying environmental enforcement emphasis and policy
- Science-based improvements to environmental regulations to account for regional risks and climate change
- Regulate the farming industry and farmland no different than other industries and regulate the farm industry through the USEPA
- Reducing and regulating pesticides and herbicides more strictly
- Reducing or eliminating some fertilizers
- Septic tank legislation
- Additional invasive species controls
- Urban planning and land use
- Residential and household waste
- Urban air improvements
- Further or new restrictions on harmful chemicals, especially those that are persistent and mobile in the environment
- Emerging contaminants, such as those described in Chapter 2
- Improving pollution prevention techniques and awareness
- Sustainability legislation
- Noise controls
- Plastic reduction or elimination

The individual level is perhaps where the most improvement is needed in both action and education of sustainability. Some individual sustainability action items include (USEPA 2018):

- Bike to work or to school
- Walk
- Carpool
- Consider living closer to work
- Install solar house panels
- Compost yard waste and food waste
- Plant a vegetable garden
- Collect rain water from roof drains to water plants and for other outdoor needs
- Drive a more fuel efficient automobile, hybrid, or electric automobile
- Consume less meat
- Consume more plant food
- Use mass transit
- Dry clothes by hanging them outside
- Consider an electric lawn mower and snow blower
- Lower water heater temperature
- Do not consume water from plastic bottles
- Recycle plastic, paper, glass, and aluminum (contact for municipality for additional information and recycling opportunities)
- Properly dispose of electronic equipment (contact your local municipality for more information concerning other items such as light bulbs, batteries, mercury switches)
- Properly dispose of other household items including cleaners, solvents, paints, and other items (contact your local municipality for a comprehensive local list)
- Utilize reusable shopping bags
- End the use of plastic bags and plastic bottles

- Choose to purchase items with minimal packaging material (remember if you don't consume or use it, it becomes a waste)
- Upgrade appliances with improved energy efficiency
- Install a heat blanket on the hot water heater and lower its temperature 10 to 20%
- Use low-flow shower heads
- Take shorter showers
- Wash clothes in cold water
- Purchase organically grown and eco-friendly food
- Become more educated
- Encourage and inspire others
- Turn your computer off rather than using sleep mode or screen saver
- Print documents only when required
- Unplug electrical devices such as toasters, printers, chargers, computers, and other items when not in use
- Use recycled paper
- Drink more tap water
- Don't drink from disposable cups
- Replace your vehicle's air filter every 3,000 miles
- Keep vehicle tires at recommended air pressure
- Do full loads of laundry
- Run the dishwasher only when full
- Limit dry cleaning clothes and other items
- Turn off the water while brushing your teeth
- Receive and pay bills online
- Call companies that participate in junk mail advertisements and request that they no longer participate in junk mail advertisements
- Avoid purchasing food in individual serving sizes
- Avoid purchasing items that are considered single-use
- Turn down the heat or turn air conditioning up when not at home
- Lower the thermostat during winter and especially at night and increase the thermostat during the summer
- Contact your local and state-elected officials and encourage them to support sustainable legislation
- Participate in volunteer outdoor cleanup efforts in your community
- Plant trees
- Consider planting native vegetation rather than grass
- Limit fertilizer and pesticide and herbicide use
- Use citrus-based cleaners rather than volatile organic compounds
- Consider downsizing your home
- Improve insulation of your home
- Purchase more items from garage sales, thrift stores, resale, and consignment shops
- Donate more items
- Sell items at a garage sale, resale, or consignment shop
- Purchase bleach-free paper products
- Reuse wrapping paper and envelopes
- Purchase only items that you need
- Collect and donate packing peanuts to local shipping store
- Purchase rechargeable batteries
- Recycle motor oil, ink cartridges, and tires (contact your local municipality for more information)
- Conserve water (contact your local municipality for more information)

Embracing some or all of the above listed items into the everyday life of individuals will have a positive impact on the environment. In addition, incorporating the sustainability measures listed above will result in significant cost savings for each individual.

1.2.8 THEME 8: BUILDING A MODEL FOR SUSTAINABILITY

A sustainability model for any location in the world may seem impossible because of the sheer number of potential variables that greatly increase the complexity and because of the many differences between countries. However, this is just not the case. To build a model for compliance and sustainability we must consider all the major themes presented above and then combine them into a single comprehensive and understandable format, namely an equation. To follow the themes in this book and to achieve a model for sustainability we must understand the environment and also understand the aspects of how humans impact the environment. At the local or site level or even the individual level, sustainability is the outcome of analyzing the aspects of facility operations or habits with the natural environment and developing engineering methods to eliminate or reduce the potential for realization of harmful impacts.

Two required inputs that must be calculated to build a sustainability model are the contaminant risk factor (CRF) and the geologic vulnerability factor.

Geologic vulnerability is a characterization of the natural environment. Essentially, geologic vulnerability is a measure of nature's response when it becomes polluted.

The CRF is a significant aspect of operations. If it weren't for chemicals, environmental risk would be greatly reduced but would not be eliminated. To understand the risk of chemicals and their behavior when released into the environment, one must understand each chemical's toxicity, persistence, and mobility. Each of these factors are critically important when combined with geologic vulnerability because it can be used as a predictor to evaluate whether a release of any particular chemical will cause harm and at what scale.

The natural environment reacts differently and uniquely to each chemical. However, there are geologic and hydrologic areas that are especially sensitive to contamination where a synergistic effect is realized that greatly amplifies potential harmful effects. Each chemical has a different associated risk factor for each environmental media that are air, water, and soil that is dependent upon its toxicity and physical chemistry (mobility and persistence).

To combine our understanding of nature and pollution with aspects of facility operations and risk reduction actions, a logical place to start is with the environmental audit. The environmental audit is convenient because it examines facility operations that have the potential to emit pollutants into the air, the water, and to the land. In addition, an environmental audit also examines management practices and engineering controls undertaken to eliminate or manage the risks involved with minimizing the potential for a release of contaminants to the environment. It is also a convenient place to begin because most every industrial facility on the planet should conduct an environmental audit at some scale. Therefore, much of the required information already exists.

Facility risks can then be characterized with two measured variables that include aspect risks of operations and risk reduction actions. Much of this information can be obtained and scored from information in an environmental audit.

Now all the elements to build a model for sustainability are in place and include three basic components:

1. Understanding the natural environment
2. A detailed analysis of the aspects of operations
3. Evaluating the effectiveness or ineffectiveness of risk reduction actions

Combining these three elements we arrive at an equation with an output that is termed the sustainability index (Rogers 2018).

To evaluate the sustainability model, it was tested for a period of 10 years that included five 2-year evaluation periods, at as many as 67 manufacturing facilities located in 12 countries. Each facility was tracked using the sustainability model outlined above and is presented in detail in Chapter 10. Over the 10-year period, improvements in risk reduction measures have been realized by an average of 80% resulting in a significant reduction in environmental non-compliance and releases of hazardous substances to the environment and has greatly improved the sustainability of each facility.

Efforts to reduce operational aspects were largely offset by new or more strict environmental regulations, facility expansion, new equipment, or increased production that required additional permitting and regulatory limitations. However, even with the offsets, slightly more than 50% of facilities reduced operational aspects. These opportunities were most often realized through elimination, substitution, prevention, or minimization (ESPM) efforts and include:

- Eliminating chemical use with high CRFs (i.e., chlorinated VOC, PCBs, other VOCs, and hexavalent chrome)
- Reducing hazardous waste generation that resulted in a lower hazardous waste classification (i.e., large quantity generator to small quantity generator)
- Eliminating the requirement for a stormwater permit by increasing onsite surface water infiltration and eliminating any offsite discharge of stormwater
- Changing wastewater discharge from surface water to a Publicly Owned Treatment Works
- Upgrading wastewater treatment systems to reduce the amount of pollutant load and reduce the amount of water discharged
- Reducing wastewater discharges through operational modifications
- Reducing wastewater discharge through recycling efforts
- Reclassification of solid waste to beneficial reuse
- Assisting other companies in networking and increasing recycling opportunities
- Upgrading operations with new technologies that produce less waste
- Working with suppliers and purchasers to reduce packaging wastes and increase recycling efforts

The most significant challenge over the 10-year implementation and evaluation period was changing cultural attitude and behavior. Up until the 1970s, there generally was a disregard for the environment that perhaps had persisted for centuries. The 1970s saw the birth of the environmental movement. The 1980s and 1990s saw the passage and implementation of most of the major environmental laws of the United States and in many other countries of the world. The United States and many other developed countries have made significant progress at cleaning up the environment and making many places a better place to live for us and all the other animals and plants on Earth. However, much remains to be done even in the United States. The most difficult actions that have yet to be taken but must be taken is changing our cultural attitude and behavior toward Earth.

1.3 SUMMARY AND IMPLEMENTATION

The success of the sustainability program outlined in this book could not have been realized without acknowledging and overcoming a cultural attitude and behavior ingrained in most capitalist organizations, which is a constant drive for growth and profit first and everything else second. The issues and challenges surrounding and weaved within the solutions for overcoming pollution will take commitment and sacrifice from each of us. One person may not make much of a difference but collectively we can and must.

Changing attitude and culture required several steps and involved all organization levels. Those steps included:

- A corporate culture of respect and commitment of establishing and maintaining environmental stewardship with the Earth

- Education and training
- A clear plan with a stated purpose and goals
- Incentives, rewards, and penalties
- Measuring and recording performance
- Data analysis
- Plan adaptations including additions or subtractions and adjusting timelines

The sustainability model introduced in this book can be implemented at the global, national, local, or individual level to create a proactive, cohesive, and effective platform to attain the ultimate goal of sustainability and environmental stewardship.

REFERENCES

Bates, C.G. and Ciment, J. 2013. *Encyclopedia of Global Social Issues.* M.E. Sharpe Publishers. New York. 1450p.

Carson, R. 1962. *Silent Spring.* Houghton Mifflin. Boston, MA.

Coble, C.R., Murray, E.G., and Rice, D.R. 1987. *Earth Science.* Prentice-Hall Publishers. Englewood Cliffs, NJ. 502p.

Hahn, R.W. 1994. United States Environmental Policy: Past, Present and Future. *Natural Resource Journal.* John F. Kennedy School of Government, Harvard University. Cambridge, Massachusetts. Vol. 34. No 1. pp. 306–348.

Haynes, W. 1954. *American Chemical Industry – A History.* I–IV. Van Nostrand Publishers. New York.

Kaufman, M.M., Rogers, D.T., and Murray, K.S. 2005. An Empirical Model for Estimating Remediation Costs at Contaminated Sites. *Journal of Water, Air and Soil Pollution.* Vol. 167. pp. 365–386.

Kaufman, M.M., Rogers, D.T., and Murray, K.S. 2011. *Urban Watersheds.* CRC Press. Boca Raton, FL. 583p.

Murray, K.S. and Rogers, D.T. 1999. Groundwater Vulnerability, Brownfield Redevelopment, and Land Use Planning. *Journal of Environmental Planning and Management.* Vol. 42. No. 6. pp. 801–810.

Rogers, D.T. 1996. *Environmental Geology of Metropolitan.* Clayton Environmental Consultants. Novi, MI.

Rogers, D.T., Murray, K.S., and Kaufman, M.M. 2007. Assessment of Groundwater Contaminant Vulnerability in an Urban Watershed in Southeast Michigan, USA. *In:* Howard, K.W.F. editor. *Urban Groundwater – Meeting the Challenge.* Taylor & Francis. London, England.

Rogers, D.T. 2016a. Next Generation of Urban Hydrogeologic Investigations. *Journal of the Italian Geological Society.* Rome, Italy. Vol. 39. No. 1. pp. 349–353.

Rogers, D.T. 2016b. Scientific Advancements That Improve the Conceptual Site Model in Urban Hydrogeological Site Investigations. *35th International Geological Congress.* Paper 3019. Cape Town, South Africa.

Rogers, D.T. 2018. Derivation of a Comprehensive Environmental Risk Model for Urban Groundwater Protection. *International Association of Hydrogeologists Congress.* Vol. 1. Daejeon, Korea.

United Nations. 2016a. *Human Development Report.* United Nations Development Programme. http://www.hdr.undr.org/sites/2017. (accessed September 16, 2017).

United Nations. 2016b. *Global Drinking Water Quality Index and Development and Sensitivity Analysis Report.* UNEP Water Programme Office. Burlington, Ontario, Canada. 58p.

United Nations. 2016c. *World Air Pollution Status.* United Nations News Center. New York. https://www.un.org/sustainabledevelopment/2016/09. (accessed August 29, 2017).

United Nations. 2017. *World Population Prospects.* United Nations Department of Economic and Social Affairs. Population Division. https://esa.un.org/unpd/wpp/data. (accessed December 14, 2017).

United Nations. 2019. *Sustainability Goals.* https://www.un.org/sustainabledevelopment/. (accessed April 20, 2019).

United States Advisory Council on Historic Preservation. 2017. *National History Preservation Act.* https://www.achp.gov. (accessed April 3, 2017).

United States Environmental Protection Agency. 2017. *Summary of Environmental Regulations of the United States.* https://epa.gov/summary-environmental-laws. (accessed February 24, 2017).

United States Environmental Protection Agency (USEPA). 2018. *What Is Sustainability?* https://www.epa.gov/sustainability. (accessed December 1, 2018).

World Health Organization (WHO). 2013. *Water Quality and Health Strategy 2013–2020.* http://www.who.int/water_sanitation_health/dwq/en/. (accessed October 18, 2018).

WHO. 2015. *Progress on Drinking Water and Sanitation.* World Health Organization. New York. 90p.

WHO. 2018. *Nine Out of Ten People Breathe Unhealthy Air.* Geneva. Switzerland. https://www.who.int/news-room/detail/02-02-2018. (accessed May 30, 2018).

Part One

Pollution and Its Behavior

2 Pollution

2.1 INTRODUCTION

There would be no need for environmental laws and regulations or the United States Environmental Protection Agency (USEPA) if there was no pollution. Therefore, it seems logical that we begin our journey with examining pollution in its many forms. However, it does not just begin and end with pollution. There are other activities that impact our environment that go beyond what is commonly termed pollution. Many of these other forms do not introduce chemicals into the natural world. For instance, any number of different land uses where humans inflict enormous harm on our environment include mining, petroleum development, urban land development, wetland destruction, deforestation, forest management techniques, and agriculture. As we are discovering, these activities do not just affect a centralized area where the activity took place. These types of activities often have a rippling effect that negatively impact surrounding areas and in some cases spreads like a disease that affects the entire planet. As we will discover in the course of this book, this is where sustainability then plays an important and central role. Sustainability goes beyond addressing pollution and examines how humans, albeit unintentional, adversely impact our planet as though we ourselves are a form of pollution.

Pollution and **contamination** are used synonymously to mean the introduction into the environment by humans of substances that are harmful or poisonous to human health and ecosystems (van der Perk 2006). Over time the term pollution evolved to include not only substances, but energy wastes such as heat, light, and noise. Pollution now also includes many types of land use that inflicts harm on the environment as stated above. For our purposes in this chapter, we will concentrate on chemical pollutants. We will discuss other forms of pollution in subsequent chapters and by the end of the book we will perhaps suggest that the term pollution be redefined.

A chemical or substance becomes pollution when it is released into the environment either inadvertently or improperly – at the wrong place or in the wrong amounts. For example, milk becomes a contaminant when large quantities are released into a stream. In urban areas, contaminants are everywhere – in the air, soil, water, inside buildings, and in our homes. Most households contain chemicals that would be considered contaminants and hazardous if they were released into the environment or disposed of improperly (see Figure 2.1). These chemicals include:

• Cleaners	• Solvents	• Gasoline
• Pesticides	• Herbicides	• Lawn fertilizers
• Some paints	• Plastic	• Oil
• Grease	• Dirt	• Electronic equipment

In addition, the list of common household products shown below contain contaminants if they are improperly disposed and include:

- Computer equipment
- Televisions
- Some electrical equipment
- Most batteries
- Some building products
- Wood with certain applied preservatives or coatings

FIGURE 2.1   Household items that can become contaminants (From USGS. Volatile Organic Compounds in the Nation's Ground Water and Drinking-Water Supply Wells. USGS Circular 1292. Reston, VA, 2006.)

Thousands of environmental pollutants exist, with the following categories of chemical and organic contaminants commonly present in the environment (Kaufman et al. 2011):

- Volatile organic compounds (VOCs)
  - Dense non-aqueous phase liquids
  - Light non-aqueous phase liquids
- Polynuclear aromatic hydrocarbons (PNAs or PAHs)
- Semi-volatile organic compounds (SVOCs)
- Polychlorinated biphenyls (PCBs)
- Pesticides and herbicides
- Heavy metals
- Others including:
  - Common fertilizers including nitrates, phosphorus, and potassium
  - Greenhouse gases
  - Chlorofluorocarbons
  - Carbon dioxide
  - Carbon monoxide
  - Particulates (dust)
  - Ozone
  - Bacteria such as *E. coli*
  - Viruses
  - Invasive species
  - Pharmaceuticals
  - Cyanide
  - Asbestos
  - Acids
  - Bases
  - Radioactive compounds
  - Dioxins

- Polyfluoroalkyl substances
- Emerging contaminants

Before addressing individual contaminant groups, we must introduce and discuss the general concepts of toxicity by examining the types, differences, and potential effects of exposure to pollutants. The chapter will conclude with an examination of each contaminant category listed above.

2.2 TOXICITY OF POLLUTION

Toxicity or **potency** is the degree to which a chemical or substance is able to inflict damage to an exposed organism (USEPA 1989a). Note that toxicity does not equal risk. The difference between toxicity and risk is based primarily upon the length of exposure to a chemical or substance, and whether the dosage received from this exposure is enough to cause harm. Toxic substances surround us and are present at most all locations on Earth. However, and as stated above, there is only risk if we are exposed to a substance or chemical long enough and at a high enough dose to cause harm. Toxicity does not assess risk – this determination is reserved for a risk assessment. A **risk assessment** is an evaluation that combines toxicity of a contaminant to possible exposure pathways. A risk assessment is, therefore, a measure of the nature and magnitude of health risks and ecological receptors from chemical contaminants that may be present in the environment (USEPA 2019a).

There are three basic types of toxic categories:

1. Chemical or substance, including inorganic and organic substances such as acids and bases, flammable liquids, metals, etc.
2. Biological, including bacteria and viruses
3. Physical, including sound and vibration, heat and cold, light, radiation, etc.

Toxicity is measured by the effects on a whole organism, an individual organ, tissue, or even a cell. Populations are most often used to measure toxicity since any one individual typically may have a different level of response to a toxin at a certain dose or concentration. The most common measure of chemical or substance toxicity is termed the LD_{50}; defined as the concentration or dose that is lethal to 50% of the population being tested (USEPA 1989a). When direct data are not available the LD_{50} is estimated by comparing the substance to other similar chemicals and organisms.

Another important measure of toxicity is the **lowest-observed-adverse-effect level** (LOAEL) or **threshold effect value.** The LOAEL is the lowest tested dose of a chemical or substance causing a harmful or adverse health effect (ATSDR 2019a). Factors influencing the LOAEL include:

- The chemical's solubility in body fluids
- The particle size and state of the chemical
- Route of exposure
- Residence time of the chemical in the body
- Individual susceptibility

Units of dose are expressed as the mass of chemical per unit mass of body weight in milligrams per kilogram (mg/kg). These units are employed in order to evaluate the relative toxicities between animals of different species and size. To develop the most accurate human exposure limits, extensive animal studies are initially used to establish extrapolated human dosage limits for a specified chemical, and these estimates are refined by human health studies conducted on individuals known to have been exposed to the same chemical. The Occupational Safety, and Health Administration (OSHA) has established a list of exposure limits for over 600 individual chemicals and refers to exposure limits as Permissible Exposure Limits (PELs) (ACGIH 2019). Most chemicals or substances have the potential to exhibit some adverse health effect on humans or other organisms given a certain set of circumstances. **Adverse health effect** is defined as a change in body function or cell structure potentially leading to disease or health problems (ATSDR 2019a). The effect that a specific chemical or substance may exhibit on a living organism is dependent upon the following factors (USEPA 1989a; 2005):

- Nature of the chemical or substance
- Concentration
- Route of exposure
- Length of time of exposure
- Individual susceptibility

Chemicals and substances enter the human body through three **routes of exposure:** inhalation, ingestion, and dermal adsorption (USEPA 1989a). When exposure occurs, chemicals may produce one or more of these symptoms:

• Tissue irritation	• Eye irritation	• Rash
• Dizziness	• Bleeding	• Hair loss
• Vision loss	• Hearing loss	• Nausea
• Anxiety	• Narcosis	• Headache
• Vomiting	• Diarrhea	• Pain
• Fever	• Tremors	• Difficulty breathing
• Psychotic behavior	• Euphoria	• Cancer
• Death	• Burning sensations	• Others

Human response to exposure of a chemical is described as acute or chronic. An **acute** response is generally characterized as a single high dose with rapid onset and disappearance of symptoms. A **chronic** response involves a stimulus lingering for a period of time (ATSDR 2019a). Exposure to a contaminant may not trigger an immediate response. Instead, there is a **latency period** – a duration of time without observable effects. Certain chemicals or substances, such as asbestos and forms of mercury, have a latency period of up to 10 to 20 years (USEPA 2005). Exposure to some chemicals or substances may result in the development of cancer, and any chemical, substance, radionuclide, or radiation contributing to the development or propagation of cancer is termed a carcinogen (United States Department of Health and Human Services 2005). **Cancer** exists when cells in the body become abnormal and grow or multiply out of control (ATSDR 2019a). There is also often some selectivity involved with chemical exposure, as some chemicals may target certain organs, parts of the body, or reproduction. For example, **teragens** may have an adverse effect on a developing fetus, **mutagens** may induce genetic changes which could affect future generations through reproduction, and **heptotoxins** are chemicals posing a risk of liver damage (ATSDR 2019a).

Carcinogens are grouped into five general categories (ACGIH 2019):

- Group A1: Confirmed human carcinogen
- Group A2: Suspected human carcinogen
- Group A3: Confirmed animal carcinogen with unknown relevance to humans
- Group A4: Not classified as a human carcinogen
- Group A5: Not suspected as a human carcinogen

It is much more difficult to evaluate the toxicity of a mixture of contaminants compared to a single chemical compound or substance. This is because the interaction of a mixture of contaminants may produce enhanced or diminished effects. Common mixtures include gasoline, often containing more than 250 individual chemical compounds, industrial wastes, or a malfunctioning sewage treatment plant, which may result in discharging many chemical and biological contaminants (USEPA 2019b).

The following sections describe the major contaminant categories. For each contaminant category, specific contaminants, their common uses, physical attributes, and toxicities are presented.

2.3 VOLATILE ORGANIC COMPOUNDS (VOCS)

Volatile organic compounds (VOCs) are organic compounds that generally volatilize or evaporate readily under normal atmospheric pressure and temperatures (USGS 2006; USEPA 2019b). Table 2.1 is a list of common VOCs and includes those sought out and analyzed when conducting a subsurface environmental investigation (USGS 2006; USEPA 2019b). Included in Table 2.1 is the Chemical Abstracts Service (CAS) registry number for each chemical compound. The CAS registry identifies each compound with a unique numerical identifier and includes over 143 million substances (American Chemical Society 2019). In general, VOCs have a high vapor pressure (evaporate quickly), low-to-medium solubility, and low molecular weight (USGS 2006). Most VOCs are considered toxic or harmful to humans and other organisms (USEPA 2019c). In addition, many VOCs are flammable and must be handled with extreme care.

The use of VOCs in the United States continues to increase (Storrow 2019; USEPA 2019b). Some VOCs have had, and continue to have, very heavy usage. An example is gasoline, which contains numerous VOC compounds and its production and use continues to increase. Current gasoline consumption in the United States is approximately 1.6 billion liters per day (400 million gallons per day) (United States Energy Information Service 2019). Table 2.1 lists common uses of VOCs. As noted, many VOCs are flammable and are components of gasoline and other fuels such as diesel fuel, kerosene, and fuel oil. In addition, VOCs are commonly used as solvents, ingredients in paints, paint thinners, in the manufacturing process of pharmaceuticals, for caffeine extraction, nail polish removers, dry cleaning chemicals, mothballs, pesticides and fumigants, adhesives, refrigerants, and as a by-product of chlorination for water purification.

For hydrogeological purposes, VOCs are separated into two general categories: dense non-aqueous phase liquids (DNAPLs) and light non-aqueous phase liquids (LNAPLs) (USGS 2009a; 2009b). An additional category of VOCs are the trihalomethanes. Each is discussed separately in the following sections.

2.3.1 LIGHT NON-AQUEOUS PHASE LIQUIDS (LNAPLS)

Light non-aqueous phase liquids (LNAPLs) are liquids lighter than water and do not mix or dissolve in water readily. Common LNAPL compounds include the VOCs; benzene, ethyl benzene, toluene, and xylenes. These compounds are also often symbolized by the acronym BTEX. BTEX compounds are common components of gasoline and are often used as indicator analytes when evaluating whether there has been a release of gasoline or other similar fuels to the environment (USGS 2009b). LNAPLs also include other compounds lighter than water such as polynuclear aromatic hydrocarbons (PAHs), which are discussed in Section 2.4.

BTEX compounds are aromatic hydrocarbons with a **benzene ring** forming the backbone of its molecular structure (Jensen 2009). As depicted within the compound benzene shown in Figure 2.2, the benzene ring consists of a hexagonal arrangement of six carbon atoms located at the vertices. Each atom is bonded to its adjacent atoms by a single covalent bond and by an unusual ring bond of electrons shared by all six carbon atoms. Benzene and ethyl benzene are considered A1 carcinogens (USEPA 2019c).

Health effects from exposure to benzene include headaches, dizziness, drowsiness, rapid heart rate, tremors, and unconsciousness. Exposure to high levels can result in death (ATSDR 2007a). Health effects from exposure to toluene include tiredness, confusion, loss of appetite, memory loss,

TABLE 2.1
Common Uses of VOCs

Compound (listed alphabetically)	CAS Registry Number	Common Uses
Acetone[a]	67-64-1	Nail polish remover, paint thinner, laboratory chemical
Benzene[b,c]	71-43-2	Gasoline component, solvent, pharmaceutical, dyes, and plastics
Bromodichloromethane	75-27-4	Trihalomethane (by-product of water purification)
Bromoform	75-25-2	Trihalomethane (by-product of water purification)
Bromomethane	75-83-9	Fumigant
Carbon disulfide[d]	75-15-0	Cellulose manufacturing, soil fumigant
Carbon tetrachloride[e]	56-23-5	Solvent, dry cleaning, agriculture, formerly used in fire Extinguishers
Chlorobenzene	108-90-7	Solvent, insecticides
Chloroethane	75-00-3	Solvent, pharmaceutical manufacturing
Chloroform	67-66-3	Trihalomethane (by-product of water purification)
Chloromethane	74-87-3	Solvent, agricultural chemicals, and cellulose
Dibromochloromethane	124-48-1	Trihalomethane (by-product of water purification)
1,1-Dichloroethane	75-34-3	Solvent, manufacture of plastic wrap and adhesives
1,2-Dichloroethane	107-06-2	Solvent, paint ingredient, fumigant, vinyl chloride production
1,1-Dichloroethene	75-35-4	Manufacturing of plastics, adhesives, refrigerants
cis-1.2-Dichloroethene	156-59-2	Fat extraction from meat & fish, refrigerants, pharmaceuticals
trans-1,2-Dichloroethene	156-60-5	Fat extraction from meat & fish, refrigerants, pharmaceuticals
Dichloromethane	79-09-2	Paint and stain remover, aerosol propellant, caffeine extraction
1,2-Dichloropropane	78-87-5	Solvent, stain remover, former soil fumigant
cis-1,3-Dichloropropene	10061-01-5	Solvent
trans-1,3-Dichloropropene	10061-02-06	Solvent
Ethyl benzene[f]	100-41-4	Gasoline component, solvent, styrene manufacturing
2-Hexanone[g]	591-78-6	Paint and paint thinner, used to dissolve oil and waxes
Methyl ethyl ketone (MEK)	78-93-3	Solvent
Methyl isobutyl ketone (MIBK)	108-10-1	Solvent, used in metal extraction
Methyl-tert-butyl ether (MTBE)	1634-04-4	Former gasoline additive, formerly used to dissolve gallstones
Naphthalene[h]	91-20-3	Gasoline component, mothballs, insecticide
Styrene	100-42-5	Gasoline component, coating, paint, rubber, adhesives
1,1,2,2-Tetrachloroethane	79-34-5	Solvent, ingredient in paints and pesticides
Tetrachloroethene (PCE)[i]	127-18-4	Solvent, dry cleaning, textile processing
Toluene[j]	108-88-3	Gasoline component, solvent
1,1,1-Trichloroethane[j]	71-55-6	Solvent, cosmetic ingredient, aerosol products, textile processes
1,1,2-Trichloroethane	79-00-5	Solvent, manufacturing of plastic wraps
Trichloroethene (TCE)[j]	79-01-6	Solvent, caffeine extraction, dry cleaning, paint and ink, rubber
Vinyl acetate	108-05-4	Polymer

(Continued)

TABLE 2.1 (CONTINUED)
Common Uses of VOCs

Compound (listed alphabetically)	CAS Registry Number	Common Uses
Vinyl chloride[j]	75-01-4	Rubber, paper, and glass industries, PVC manufacturing
Xylenes[k]	1330-20-7	Gasoline component, paint thinner ingredient, plasticizer

Source: USGS. 2006. Volatile Organic Compounds in the Nation's Ground Water and Drinking-Water Supply Wells. USGS Circular 1292. Reston, VA. 2006: USEPA. 2019b. EPA's Report on the Environment (ROE). 2019b. https://www.epa.gov/report-environment. (accessed March 24, 2019.)

[a] Levy (2009).
[b] Lide (2008).
[c] ATSDR (2007a).
[d] Holleman and Wiberg (2001).
[e] ATSDR (2007b).
[f] ATSDR (1995).
[g] ATSDR (2005a).
[h] ATSDR (2001a).
[i] Doherty, R.E. (2000).
[j] ATSDR (2006a).
[k] ATSDR (2007c).

loss of color vision, and nausea (ATSDR 2001a), and exposure to ethyl benzene may cause irreversible damage to the inner ear and hearing loss, dizziness, and kidney damage (ATSDR 2007b). Exposure to xylenes can result in headaches, lack of muscle coordination, dizziness, and confusion. Exposure to very high levels of xylenes can cause unconsciousness (ATSDR 2007c).

Methyl-tert-butyl ether (MTBE) is a LNAPL compound that was produced exclusively as a gasoline additive. MTBE belongs to a group of chemicals known as "oxygenates" because they raise the oxygen content of gasoline and thereby raise the octane level. MTBE is a colorless liquid at room temperature, and is very volatile and flammable (USEPA 2013). The purpose of adding MTBE to gasoline was to increase the efficiency of combustion in automobiles enabling them to run cleaner and emit fewer pollutants. However, the use of MTBE has declined significantly because of (1) health concerns and (2) MTBE has been detected in many groundwater aquifers used as drinking-water sources in the United States (USEPA 2013). The United States has now banned the use of MTBE because of its propensity to contaminate groundwater and the high costs incurred to remove it from groundwater (USEPA 2013). MTBE is also used to dissolve gallstones. Patients treated

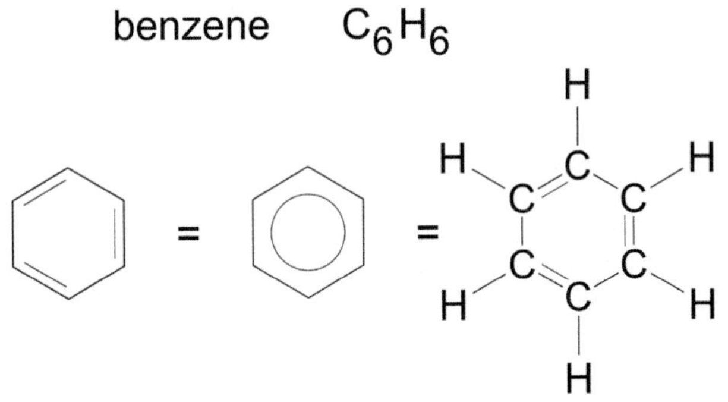

FIGURE 2.2   Benzene and the benzene ring.

for gallstones using MTBE have MTBE delivered directly to the gallbladder through surgically inserted tubes. Health effects from exposure to MTBE may include nose and throat irritation, headaches, nausea, dizziness, and mental confusion. Currently, evidence suggesting that MTBE may cause cancer is inadequate (ATSDR 2019b).

2.3.2 DENSE NON-AQUEOUS PHASE LIQUIDS (DNAPLs)

Dense non-aqueous phase liquids (DNAPLs) are liquids denser than water, and do not mix or dissolve readily in water (USGS 2006). DNAPL compounds include many common solvents and coal tar (Suthersan and Payne 2005). They are also commonly referred to as chlorinated solvents, halogenated VOCs, or chlorinated VOCs because chlorine is in the atomic structure, and the most common uses of these compounds are for cleaning and degreasing (USGS 2006). Chlorinated VOCs have been in use for nearly 100 years, have been widely used by industry, and many household products contain them. Figure 2.3 shows some household products containing chlorinated solvents (USGS 2006).

Halogenated VOCs are a group of organic compounds with a halogen atom as part of its molecular structure. Halogens include the elements Fluorine (F), Chlorine (Cl), Bromine (Br), or Iodine (I). Part of the uniqueness of halogenated VOCs is they tend to have a very weak tendency to form hydrogen bonds with water. This lack of affinity for water means the halogenated VOCs – especially those with fluorine or chlorine – tend to be hydrophobic and have low solubility (Suthersan and Payne 2005).

Common halogenated VOCs include:

• Tetrachloroethene (PCE)	• Trichloroethene (TCE)
• *cis*-1,2-Dichloroethene (DCE)	• Trans-1,2-Dichloroethene (DCE)
• Vinyl chloride	• 1,1,1-Trichloroethane (TCA)
• 1,1-Dichloroethene (DCE)	• Methylene chloride
• Carbon tetrachloride	• Chloroform
• Chlorobenzene	• 1,2-Dichlorobenzene

FIGURE 2.3 Household products containing chlorinated solvents. (From USGS. Volatile Organic Compounds in the Nation's Ground Water and Drinking-Water Supply Wells. USGS Circular 1292. Reston, VA, 2006.)

Many VOCs halogenated with chlorine have high electro negativities and form rather strong bonds with the carbon atoms in their structure. In addition, the substitution of a hydrogen atom by a chlorine atom enhances the inertness of the molecule. This inertness results in many halogenated VOCs being rather persistent when released into the environment. A degradation sequence, however, does exist for a group of the most commonly used halogenated VOCs, and is shown in Figure 2.4. This degradation sequence becomes very important when we discuss the fate and migration of these compounds in the next chapter. Vinyl chloride is an A1 carcinogen and many of the others are currently under further review (ASTDR 2006a; 2019c; USEPA 2019c). Health effects from overexposure to most chlorinated solvents include lung irritation, difficulty walking and speaking, poor coordination, dizziness, headache, nausea, sleepiness, unconsciousness, and even death (ATSDR 2019c; 2019d; 2019e).

2.3.3 Trihalomethanes (THMs)

Trihalomethanes (THMs) are a group of VOCs in which three of the four atoms of methane (CH_4) are replaced by halogen atoms. Many THMs are used as refrigerants, solvents, and are also by-products produced during water purification and chlorination. THMs can form as a by-product when chlorine and bromine are used to disinfect water for drinking or recreational use; they form from a reaction between chlorine and/or bromine with organic matter in the water being treated (USEPA 2019d). Chloroform has also been used as an anesthetic and is often used in swimming pools as a disinfectant (ATSDR 1997a). Trihalomethanes are currently considered a suspected human carcinogen (Group: A2) (USEPA 2019c). According to the USEPA (2019d), THMs may cause adverse health effects at high concentrations. Exposure to high concentrations of bromoform may interfere with normal brain function and cause sleepiness (ATSDR 2005b). USEPA has established maximum allowable concentration of THMs in drinking water at 80 parts per billion for the combined concentration of chloroform, bromoform, bromodichloromethane, and dibromochloromethane (USEPA 2019d).

2.4 POLYNUCLEAR AROMATIC HYDROCARBONS

Polynuclear or **polycyclic aromatic hydrocarbons** (PNAs or PAHs) are a group of compounds formed synthetically or naturally during the incomplete combustion of coal, oil, tar, gas, wood, and

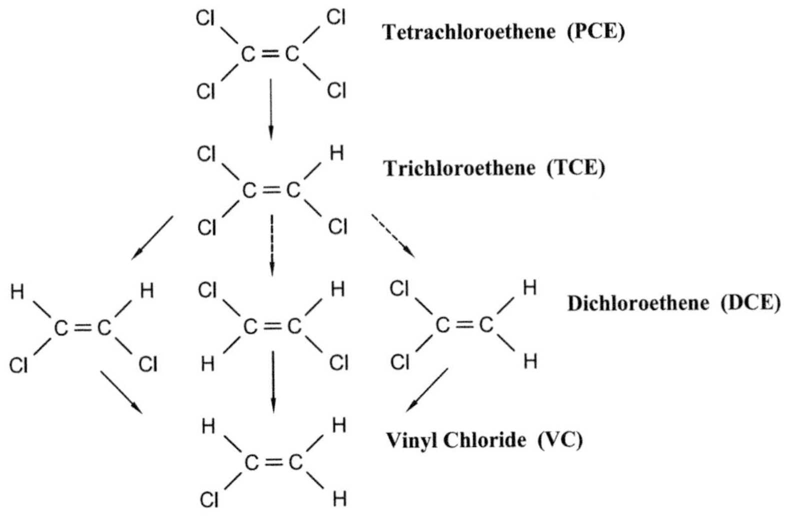

FIGURE 2.4 DEGRADATION SEQUENCE OF PCE.

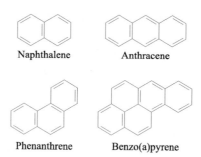

FIGURE 2.5 Molecular structure of the select PAH compounds; naphthalene, anthracene, phenanthrene, and benzo(a)pyrene.

garbage, and may also be present in tobacco, medicines, dyes, plastics, pesticides, and charbroiled meat (United States Department of Health and Human Services [USDHH] 1995). Other sources of PAHs include heavy petroleum, including diesel fuels, kerosene, aviation fuels, heavy home-heating oils, oils, used oil, and many lubricants (Missouri Department of Natural Resources 2006). Overall, there are more than 100 distinct PAH compounds. Chemically, PAHs are composed of multiple benzene rings (Figure 2.5), are lighter than water, and are generally not very soluble in water. In addition, PAHs strongly attach themselves or sorb to soil grains, especially organic soils and clay (USEPA 1989b). Because PAHs are not very soluble and are lighter than water, they are considered LNAPLs.

With two rings, naphthalene is the smallest of the PAHs. Naphthalene is a common PAH compound, and is characterized by its distinctive odor and common use in mothballs. Anthracene and phenanthrene are composed of three rings. Phenanthrene is an **isomer** of anthracene and they share many physical properties except boiling point and water solubility (Fetzer 2000). With its five rings, benzo(a)pyrene is the largest PAH molecule, and is the only known human carcinogen (Group: A1) of the four compounds shown in Figure 2.5 (USEPA 2019c). Common PAH compounds are listed in Table 2.2 along with their CAS registry number (USEPA 2019e; ATSDR 2019f).

Animal studies have indicated PAHs can cause harmful effects on the skin, body fluids, and ability to fight disease (ATSDR 2019e). Health effects from exposure to naphthalene may cause damage to red blood cells. Exposure to high levels of naphthalene may cause nausea, vomiting, diarrhea, dizziness, blood in the urine, and a yellow color to the skin (ATSDR 2019e). Currently, only benzo(a)pyrene and fluoranthene are considered A1 carcinogens (USEPA 2019c).

2.5 POLYCHLORINATED BIPHENYLS

Polychlorinated biphenyls (PCBs) are a group of synthetically produced compounds (USEPA 2019f). PCBs were produced in the United States from 1929 until 1979 when they were banned because of human and environmental health effects (USEPA 2019f). Some of the physical properties of PCBs include (Phillips 1986; Barbalace 2009):

- Very low solubility in water
- High relative solubility in organic solvents, fats, and oil
- Low vapor pressures
- Strongly sorb to soil grains, especially organic-rich soils and clays
- Stable compounds
- Do not readily degrade
- Not flammable
- Resistant to oxidation, reduction, addition, elimination, and electrophilic substitution
- High boiling point

TABLE 2.2
Common PAH Compounds

Compound (alphabetically)	CAS Registry Number	Common Uses
Acenaphthene	83-32-9	Dyes, pesticides, pharmaceuticals, oil component
Acenaphthylene	208-96-8	Automobile exhaust, oil and lubricant component
Anthracene	120-12-7	Dyes, stains, wood preservatives, insecticides, oil
Benz[a]anthracene	56-55-3	Automobile exhaust, oil and lubricant component
Benzo[b]fluoranthene	205-99-2	Automobile exhaust, oil and lubricant component
Benzo[k]fluoranthene	92-24-0	Organic semiconductor, oil component
Benzo[ghi]perylene	198-55-0	Photoconductor, pitch, coal tar, tobacco smoke
Benzo[a]pyrene	50-32-8	Pitch, coal tar, automobile exhaust, tobacco smoke
Chrysene	218-01-9	Wood preservative, oil component, dyes
Dibenzo[a,h]anthracene	53-70-3	Wood preservative, insecticides, oil component
Fluoranthene	206-44-0	Automobile exhaust, oil component
Fluorene	86-73-7	Automobile exhaust, dyes, plastics, pesticides
Indeno[1,2,3-cd]pyrene	193-39-5	Pitch, coal tar, oil component
Naphthalene	91-20-3	Moth balls, oil and gasoline component
Phenanthrene	85-01-8	Cigarette smoke, oil and lubricant component
Pyrene	129-00-0	Dyes, coal tar and pitch, oil component

Source: Agency for Toxic Substances and Disease Registry (ATSDR). 2019f. *Factsheet for PAHs.* http://www. atsdr.cdc.gov/tfacts69.html. (accessed March 24, 2019). 2019f.); USEPA. *Polycyclic Aromatic Hydrocarbons (PAHs).* 2019e. https://www.epa.gov/sites/production/files/2014-03/documents/pahs_ factsheet_cdc_2013.pdf. (accessed March 24, 2019.)

Due to the physical properties listed above, PCBs were widely used in many different industrial and commercial products, including (USEPA 2019f):

- Transformer and capacitors
- Other electrical equipment including voltage regulators, switches, reclosers, bushings, and electromagnets
- Oil used in motors and hydraulic systems
- Older electrical devices that contain capacitors
- Fluorescent light ballasts
- Cable insulation
- Thermal insulation material including fiberglass, felt, foam, and cork
- Adhesives and tape
- Oil-based paint
- Caulking
- Plastics
- Carbonless copy paper
- Floor finish

PCBs are mixtures of chlorinated organic compounds called congeners (ATSDR 2019g). A **congener** is a related compound or compounds in a specific chemical family. There are a large number of congeners, with approximately 209 present in PCB mixtures (USEPA 2019f). PCBs were produced synthetically through electrophilic chlorination of a biphenyl molecule with chlorine gas. In the United States, PCBs are commonly known under the trade name aroclors. **Aroclors** are mixtures of PCBs distinguished by a four-digit numbering system. The first two digits refer to the number of carbon atoms in a PCB molecule, and the second two digits indicate the percentage of chlorine by mass in the mixture (USEPA 2019f). For example, PCB aroclor 1254 means the mixture contains 12 carbon atoms and is 54% chlorine by weight (ATSDR 2019g). As the degree of chlorination increases, the melting point increases and the vapor pressure and solubility decrease. Common PCB aroclors are listed in Table 2.3 (USEPA 2019f). Figure 2.6 shows the basic structure of a PCB molecule.

Currently, there is inadequate information to establish whether PCBs are a human carcinogen (USEPA 2019c), but the World Health Organization (2003) suspects they are (Group: A2). Notwithstanding cancer, they pose significant other dangers to human health. PCBs have been shown to cause cancer in laboratory animals, and have also been shown to cause a number of serious non-cancer health effects in animals, including the immune system, reproductive system, nervous system, endocrine system, and others (USEPA 2019c).

2.6 SEMI-VOLATILE ORGANIC COMPOUNDS (SVOCS)

Semi-volatile organic compounds (SVOCs) are a group of organic-based compounds much less volatile than VOCs. A variety of SVOCs are used in clothing and building materials to provide flexibility, water resistance, or stain repellence, as well as to inhibit ignition or flame retardants (ATSDR 2002a; USEPA 2010). Common groups of organic compounds associated with SVOCs are phthalates, phenols, amines, and esters.

TABLE 2.3
Common PCB Aroclors

PCB Aroclor	CAS Registry Number	Percent Chlorine	Average Chlorine Atoms per Molecule	Average Molecular Weight
Aroclor 1016	12674-11-2	15.5 to 16.5	1.05	160
Aroclor 1221	11104-28-2	20.5 to 21.5	1.15	192
Aroclor 1232	11141-16-5	31.5 to 32.5	2.04	221
Aroclor 1242	53469-21-9	42	3.10	261
Aroclor 1248	12672-29-6	48	3.90	288
Aroclor 1254	11097-69-1	54	4.96	327
Aroclor 1260	11096-82-5	60	6.30	372
Aroclor 1262	37323-23-5	62	6.80	389
Aroclor 1268	11100-14-4	68	8.70	453

Source: USEPA. Polychlorinated Biphenyls Fact Sheet: Basic Information. 2019f. http://www.epa.gov/e pawaste/hazard/tsd/pcbs/about.htm. (accessed March 24, 2019.)

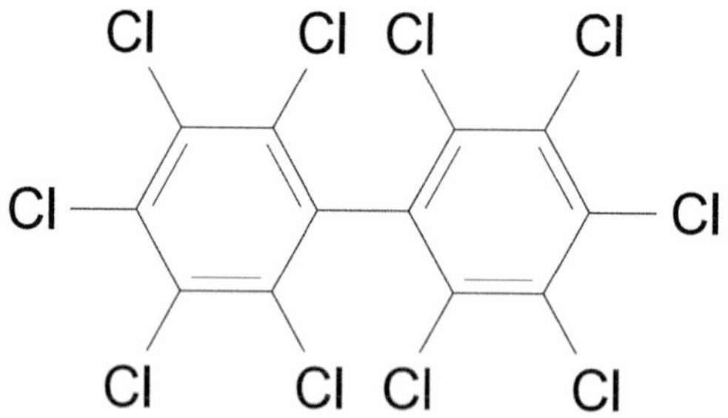

FIGURE 2.6 Basic PCB structure.

2.6.1 Phthalates

The name phthalates is derived from phthalic acid – itself derived from naphthalene and previously discussed earlier in this chapter. Phthalates exhibit low water solubility, high oil solubility, and low volatility (Fetzer 2000). There are approximately 25 distinctive phthalate compounds, including the common ones that include:

- Diethyl phthalate (DEP)
- Di-n-butyl phthalate (DBP)
- Bi or Bis(2-ethylhexyl) phthalate (DEHP)

The structural formula and CAS registration number for each of the three common phthalate compounds are listed in Table 2.4 (USEPA 2019b). The molecular structure of Di(2-exthylhexyl) phthalate is shown in Figure 2.7.

Phthalate compounds are primarily used as plasticizers. Phthalates are added to plastics to increase their flexibility, transparency, durability, and longevity, and are also used to soften polyvinyl chloride (PVC) – a common pipe material. More than a billion pounds of phthalates are used in a variety of products every year (ATSDR 2019h). The variety of products using phthalates includes (ATSDR 2002a):

• Coatings	• Plastics	• Pharmaceuticals	• Gelling agents
• Stabilizers	• Binders	• Dispersants	• Eye shadow
• Moisturizer	• Perfume	• Children's toys	• Detergents
• Nail polish	• Hair spray	• Modeling clay	• Packaging
• Waxes	• Printing inks	• Fishing lures	• Caulk
• Paints	• Curtains	• Vinyl upholstery	• Auto interiors
• Toothpaste	• Candles	• Food containers	• Many others

TABLE 2.4

Structural Formula and CAS Registration Numbers for Three Common Phthalates (USEPA 2019g)

Name	CAS Registration Number	Chemical Formula
Diethyl phthalate (DEP)	84-66-2	$C_6H_4 (COOC_2H_5)_2$
Di-n-butyl phthalate (DBP)	84-74-2	$C_6H_4 [COO(CH_2)_3CH_3]_2$
Di or Bis(2-ethylhexyl) phthalate (DEHP)	117-81-7	$C_6H_4[COOCH_2CH(C_2H_5)(CH_2)_3CH_3]_2$

Source: USEPA. Integrated Risk Information System. 2019c. https://www.epa.gov/iris. (accessed March 24, 2019.)

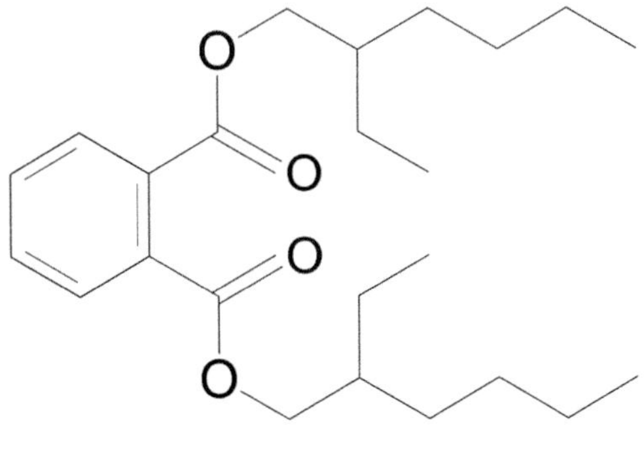

FIGURE 2.7 Molecular structure of Di(2-ethylhexyl) phthalate.

According to ATSDR (2019h) and USEPA (2019g), DEHP is a suspected human carcinogen (Group: A2), while DEP and DBP are not classified as human carcinogens (Group: A4). According to ATSDR, DEHP is not toxic at the low levels usually present in the environment. In animals, DEHP has been found to damage the liver and kidney and has affected the ability to reproduce.

2.6.2 PHENOL

Phenol is a group of organic compounds with a hydroxyl group (-OH) attached to a carbon atom in a benzene ring. Phenol compounds do occur naturally, and their presence in plant foliage discourages herbivores from consuming the plant material (Fetzer 2000; McMurry 2009). Figure 2.8 is a diagram showing the basic structure of a phenol molecule. The simplest phenol compound is carbolic acid (C_6H_5OH); also called phenol.

Phenol combined with formaldehyde forms a widely used polymer typically referred to as a phenolic resin. Phenolic resins have a wide range of industrial and commercial applications including (USEPA 2019h):

• Billiard balls	• Countertops	• Plastics
• Electronic components	• Coating	• Composites
• Abrasives	• Adhesives	• Felt bonding
• Foundry applications	• Friction products	• Refractory products
• Rubber additives	• Auto brakes	• Many others

A compound called pentachlorophenol (PCP) is a phenol compound with the addition of chlorine into its molecular structure (Figure 2.9). PCP has been used as an herbicide, insecticide, fungicide, algaecide, and disinfectant. Other uses include leather, masonry products, and wood preservatives, and utility poles often employ PCP as a wood preservative (USEPA 2006).

Bisphenol A or BPA is an organic phenolic compound with two phenol functioning groups. BPA is a very common compound and is used heavily in the production of plastic products including bottles for drinking water and numerous other consumer products (USDHH 2008). Some concerns

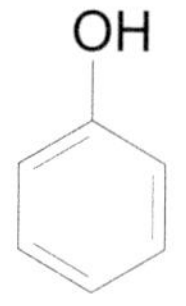

FIGURE 2.8   Basic structure of phenol.

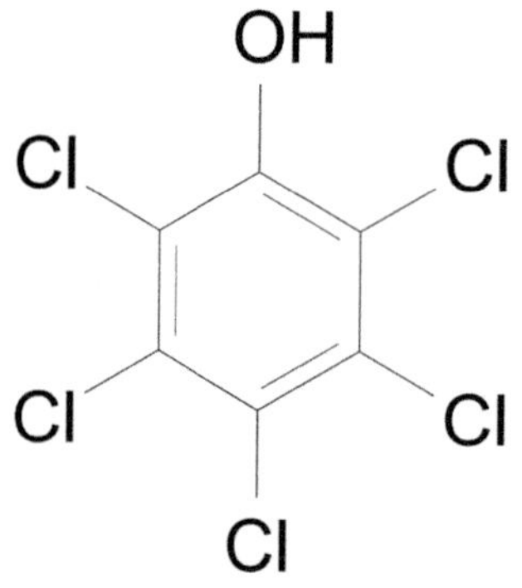

FIGURE 2.9   Basic molecular structure of pentachlorophenol.

about health effects from exposure to BPA have been expressed by the USDHH (2008), as type 7 plastics (polycarbonates) may leach PBA into the liquids they hold. In 2009, Canada banned the use of PBA in polycarbonates in baby bottles (Carwile et al. 2009).

2.6.3 AMINES

Amines are organic compounds that contain nitrogen and are **basic**. Three widely used amine compounds include methylamine, dimethylamine, and trimethylamine, and are prepared by the reaction of ammonia with methanol in the presence of a silicoaluminate catalyst (McMurry 2009). Amines are used in the manufacturing of dyes, plant growth regulators, resins, as a precursor in the manufacturing of tires, pharmaceuticals (ephedrine), pesticides, and surfactants (ATSDR 2019i). Trimethylamine also forms naturally from decomposing plant and animal matter (USEPA 2019i). Amines are often gas at room temperature and pressure and are very soluble in water (USEPA 2019i).

2.6.4 ESTERS

Esters are a group of more than 50 compounds known collectively as acid derivatives. Ester compounds contain a modified carboxylic acid group (– COOH), in which the acidic hydrogen atom has been replaced by a different organic functional group. Common uses of esters include (ATSDR 2019j; USEPA 2019j):

- Flame retardants
- Nail polish removers
- Glue
- Fragrances
- Clothing

Polyester is the most common and most widely produced ester compound. Esters are also naturally occurring and are found in flowers, in fats as triesters derived from glycerol and fatty acids, and in wine (ATSDR 2019j). At high concentrations, ester compounds can irritate the skin and mucous membranes and cause breathing difficulties (USEPA 2019c).

2.7 HEAVY METALS

Heavy metals naturally exist in the environment at amounts termed natural background levels.

When human activities introduce additional heavy metals into the environment and the background levels are exceeded, they then may be considered contaminants. In terms of their characteristics, heavy metals are generally not soluble in water, except for arsenic and chromium VI at a neutral pH (Murray et al. 2004; Kaufman et al. 2011). Heavy metals are not volatile but are often released into the atmosphere through the combustion of coal, automobile exhaust, metal production, and other methods (Murray et al. 2004). The most common heavy metals in urban areas include (Murray et al. 2004):

• Arsenic	• Barium	• Chromium III	• Chromium VI
• Cadmium	• Copper	• Lead	• Mercury
• Nickel	• Selenium	• Silver	• Zinc

Industrial watersheds are especially prone to heavy metal contamination. In an extensive study of the Rouge River watershed in southeastern Michigan, Murray et al. (2004) discovered elevated

TABLE 2.5

Common Uses for Each of the Select Heavy Metals (Pradyot 2003; Krebs 2006)

Heavy Metal	Atomic Number	Uses
Arsenic	33	Wood preservative, poison, insecticide, pigments, chemical weapons
Barium	56	Superconductors, pigments, fireworks, lubricants, optics
Cadmium	48	Batteries, plastic stabilizer, pigments, metal plating, coatings, alloys
Chromium III	24	Pigments, inks, glass, steel additive, dyes, leather tanning, refractory, alloys
Chromium VI	24	Metal plating, corrosion resistance additive, wood preservative
Copper	29	Wire, building products, piping, jewelry, electromagnetics, brass, and other alloys
Lead	82	Batteries, paint, ceramics, firearms, industrial coolant, electrodes, solder, construction materials, alloys, formerly a gasoline additive
Mercury	80	Electrical switches, thermometers, manometers, medical and dental applications, cosmetics, mercury-vapor lamps, formerly used in hat making. Mercury is the only metal evaluated that is a liquid and has a measurable vapor pressure at standard pressure and temperature
Nickel	28	Batteries, steel additive, magnets, coins, alloys
Selenium	34	Photocells, electronics, semiconductors, steel additive, alloys, copier and printing drums, glass manufacturing, pigments
Silver	47	Coins, electronics, circuit boards, alloys, mirrors, decorative items, jewelry, photographic films, batteries
Zinc	30	Metal plating, rust inhibitor (galvanization), brass alloy, batteries, cathodic protection, paint pigment, fire retardant, propellant, photocopying products, medical applications

Source: Pradyot, P. Handbook of Inorganic Chemical Compounds. McGraw Hill. New York, NY. 2003: Krebs, R. E. The History and Use of Earth's Chemical Elements: A Reference Guide. Greenwood Publishing Group. Oxford, England. 2006.

levels of barium, cadmium, chromium, copper, nickel, lead, and zinc in surface soil. Table 2.5 lists the common uses for each of the heavy metals listed above. The toxicity and carcinogenicity of heavy metals vary widely. Brief explanations of their general toxicities are given below:
Arsenic is considered a human carcinogen (USEPA 2019c).

- Barium exposure can cause gastrointestinal disturbances and muscular weakness if it is ingested in a soluble form (ATSDR 2013).
- Cadmium is considered a suspected or probable human carcinogen (ATSDR 2012).
- Chromium VI is considered a human carcinogen (USEPA 2019c). Chromium III is an essential nutrient that helps the body use sugar, protein, and fat (ATSDR 2019k).
- Copper is essential for good health in small amounts. High levels, however, can be harmful and cause nausea, vomiting, and diarrhea. Very high amounts can damage the liver and kidneys (ATSDR 2018a).
- Lead is considered a suspected or probable human carcinogen (USEPA 2019c). Lead can damage the nervous system and the brain (ATSDR 2019l).
- Mercury exposure may affect the brain and central nervous system (ATSDR 2019m).
- Nickel exposure most often results in an allergic reaction. Nickel may also affect the lungs (ATSDR 2018b).
- Selenium has beneficial and potentially harmful effects. Low doses of selenium are necessary to maintain good health. High doses can cause harmful health effects such as nausea, vomiting, and diarrhea. Prolonged exposure to selenium can cause a disease called selenosis (ATSDR 2018c).

- Silver exposure may result in a condition called arygia if exposure is prolonged. Arygia is a condition that causes a blue-gray discoloration of the skin and other body tissue (ATSDR 2018d).
- Zinc is essential for good health in small amounts. Too little zinc can cause hair and weight loss. Harmful effects do not usually occur until levels of ingestion exceed 10 to 15 times the recommended amount. Effects of overexposure to zinc include stomach cramps, nausea, and vomiting (ATSDR 2018e).

2.8   OTHER CONTAMINANTS

The remaining contaminants encountered in urban areas span a wide variety of organic and inorganic compounds and substances. These contaminants include:

- Pesticides and herbicides
- Fertilizers such as nitrates, phosphorus, and potassium
- Pharmaceuticals
- Bacteria such as coliform bacteria
- Viruses
- Cyanide
- Asbestos
- Acids
- Bases
- Radioactive compounds
- Dioxins
- Invasive species
- Contaminants predominantly present only in air include:
  - Greenhouse gases including carbon dioxide, methane, nitrous oxides, and others
  - Carbon monoxide
  - Particulates (dust)
  - Radon
  - Ozone
  - Many others
- Emerging contaminants
- And others

2.8.1   PESTICIDES AND HERBICIDES

Many of the compounds previously discussed, including several VOCs, SVOCs, and arsenic are/were used or are/were ingredients contained in pesticides and herbicides. Simply put, pesticides and herbicides are manufactured to kill things. USEPA (2019k) defines a **pesticide** as preventing, destroying, repelling, or mitigating any pest. An **herbicide** is defined as a substance that is used to kill unwanted plants commonly referred to as weeds. **Pests** are defined as living organisms occurring where they are not wanted or causing damage to crops or humans or other animals (USEPA 2019k). Examples include:

- Insects
- Mice or other animals
- Unwanted plants (weeds)
- Fungi
- Microorganisms such as bacteria and viruses
- Invasive species that include many plants and animals

Perhaps the most famous of all pesticides is the now banned substance called **DDT** (dichlorodiphenyltrichloroethane). It became famous as an environmental contaminant after the book *Silent Spring* written by Rachel Carson was published in 1962 (Carson 1962). The book cataloged the environmental impacts of the indiscriminate spraying of DDT in the United States and questioned the logic of releasing large amounts of chemicals into the environment without fully understanding their effects on ecology or human health. DDT was widely used in agriculture and by consumers in the 1940s and 1950s. DDT (CAS Registry Number 50-29-3) is now considered a probable carcinogen (Group: A2) by USEPA (2019c) and is no longer used as a pesticide in the United States after it was banned in 1972. The 2001 United Nations Environmental Program meeting held in Stockholm, Sweden (and put into effect in 2004) permitted its use for "vector control" – organisms that produce pathogens, such as mosquitoes) (USEPA 2019l). The structure of DDT is shown in Figure 2.10.

Common pesticides and herbicides, including DDT, are listed in Table 2.6 along with the respective CAS Registry Numbers and carcinogenicity ratings (USEPA 2019k).

USEPA banned the use of the pesticide **chlordane** in 1983 because of potential environmental and human health concerns for all applications except termite control (ATSDR 2011; USEPA 2016).

Malathion is an organophosphate insecticide widely used in agriculture, residential landscaping, and public recreation areas. Malathion is used widely to control mosquitoes, the West Nile virus, and was used in the 1980s in southern California to combat the Mediterranean Fruit Fly (ATSDR 2003). Exposure to high amounts of malathion can cause difficulty breathing, chest tightness, vomiting, cramps, diarrhea, blurred vision, sweating, headaches, dizziness, loss of consciousness, and possibly death (ASTDR 2003). If appropriate treatment is provided rapidly, there may be no long-term harmful effects.

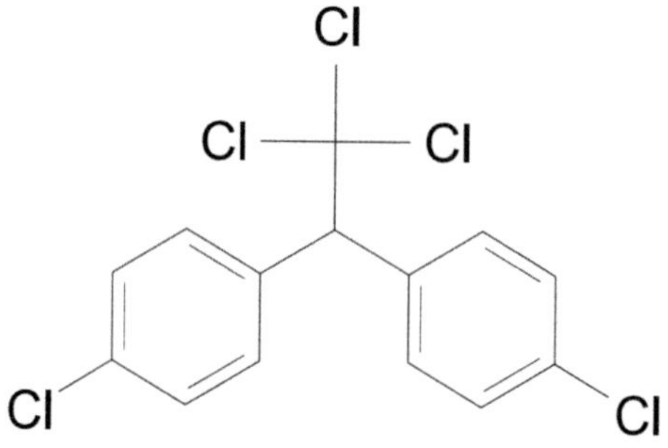

FIGURE 2.10 Basic molecular structure of DDT.

TABLE 2.6

Common Pesticides and Herbicides (2019k)

Pesticide	CAS Registry Number	Carcinogenicity Rating	Chemical Formula
DDT	50-29-3	Group: A2	$C_{14}H_9Cl_5$
Chlordane	57-74-9	Group: A2	$C_{10}H_6Cl_8$
Malathion	121-75-5	Group: A4	$C_{10}H_{19}O_6\ PS_2$
Permethrin	52645-53-1	Group: A2	$C_{21}H_{20}Cl_2O_3$
Toxaphene	8001-35-2	Group: A2	$C_{10}H_{10}Cl_8$[a]
Herbicide	**CAS Registry Number**	**Carcinogenicity Rating**	**Chemical Formula**
Glyphosate	1071-83-6	Group: A4	$C_3H_8NO_5\ P$
2,4-D	94-75-7	Group: A4	$C_8H_6Cl_2O_3$
Pentachlorophenol	87-86-5	Group: A2	$C_6H_4Cl_5O$

[a] Represents a mean chemical formula since toxaphene is composed of a mixture of compounds.

Source: USEPA. Pesticides. 2019k. https://www.epa.gov/pesticides. (accessed March 24, 2019.)

Permethrin is a widely used synthetic insecticide and insect repellent used on cotton, wheat, maize, and alfalfa crops. It is also used to kill parasites on chickens and other poultry, and as a flea treatment for dogs. Permethrin is considered a neurotoxin and is highly toxic to both freshwater and estuarine aquatic organisms (ATSDR 2005c).

Toxaphene is an insecticide that is composed of a mixture of over 670 chemicals (ATSDR 1997b). Toxaphene was one of the most widely used insecticides in the United States until 1982 when use dropped significantly and then was banned in 1990. It was primarily used in the southern states where it was applied to cotton to control pests. However, it was also used elsewhere to control pests on livestock and to kill unwanted fish in lakes (ATSDR 1997b). Exposure to toxaphene may cause damage to the lungs, nervous system, and kidneys and can even cause death if exposure is extreme (ATSDR 1997b).

Glyphosate and **2,4D** are widely used herbicides in the United States. A common name or trade name for glyphosate is *Roundup*. USEPA (2019c) does not currently classify glyphosate and 2,4D as carcinogenic. However, these two compounds may affect the immune system, kidneys, and the liver (USEPA 2019c). A review of the chemical formulas of the compounds listed in Table 2.6 indicates many of these pesticides and herbicides are **organochlorines** – an organic compound containing at least one covalently bonded chlorine atom. Chlorine released into the environment presents special challenges because: 1) it is highly toxic; 2) tends to be persistent due to the strength of the chlorine-carbon bonds; and 3) has an affinity for fatty tissues in vertebrates (fish and mammals), thus enabling bioaccumulation, a process where concentrations of a toxin increase at higher trophic levels in a food chain. Numerous studies have noted the presence of trace amounts of organochlorines in human breast milk (Calle et al. 2002).

2.8.2 DIOXINS

Compounds generally referred to as **dioxins** represent a diverse set of halogenated substances and include other substances called furans. There are 75 dioxin isomers and 135 furan isomers. Dioxins are not intentionally produced and have no known use (ATSDR 2008a); they form unintentionally as a by-product of many industrial processes involving chlorine such as waste incineration and combustion. Dioxin compounds may also form as a by-product during the manufacture of chlorinated compounds and during paper bleaching (ATSDR 2008a). A dioxin compound consists of two benzene molecules joined with two oxygen bridges. Figure 2.11 shows the basic structure of dioxin.

According to ATSDR (2008a), the most toxic dioxin compound is 2,3,7,8-tetrachlorodibenzo-p-dioxin or TCDD (CAS Registry Number 1746-01-6). Exposure to TCDD may lead to a condition known as chloroacne resulting in severe acne-like skin lesions occurring mainly on the face and upper body. USEPA (2019c) lists many of the dioxin compounds as either a known or suspected carcinogen. Dioxin compounds have been shown to bioaccumulate in humans and wildlife and they behave as teratogens and mutagens (USEPA 2019m).

2.8.3 FERTILIZERS

Fertilizers are chemical compounds designed to promote plant and fruit growth when applied (USEPA 2019b). The most common fertilizers are nitrogen (N), phosphorus (P), and potassium (K).

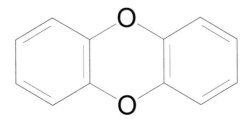

FIGURE 2.11   Basic structure of a dioxin molecule.

Nitrogen is found primarily in an organic form in soils, but can also occur as nitrate. Because nitrate is very soluble and mobile, it is transported by surface water to rivers, lakes, and streams where it can promote algal growth. In many cases the algal growth is extensive. Nitrate can also contaminate drinking water. Phosphorus occurs in soil in organic and inorganic forms, but as it is more soluble than nitrate, can also be depleted in soil through surface water runoff. Phosphorus can also promote algal growth in rivers, lakes, and streams, because it is a limiting nutrient in fresh water. Potassium (K) in fertilizers is commonly incorporated as potash – an oxide of potassium that includes the compounds potassium chloride, potassium sulfate, potassium nitrate, and potassium carbonate. The term "potash" comes from the historical practice of extracting potassium carbonate (K_2CO_3) by leaching wood ashes and evaporating the solution in large iron pots. Fertilizers containing potassium generally do not promote algal growth (USEPA 2019b).

The United States Department of Agriculture (USDA) has tracked fertilizer use in the United States since 1960. According to the USDA, approximately 63.5 kilograms (140 pounds) of fertilizer containing N-P-K are applied each year per acre of land farmed and this amount has increased more than 200% since 1960 (USDA 2019). In 2018, USDA estimated that 1.41 million square kilometers (350 million acres) were planted for crops, which means that approximately 22.2 trillion kilograms (49 trillion pounds) of fertilizers were applied to farmland in the United States (USDA 2019). In urban areas, fertilizers containing N-P-K are common and routinely applied to residential lawns and golf courses to help maintain green grass and gardens (USEPA 2019n). According to USEPA (2019c), nitrogen, phosphorus, and potassium fertilizers are not currently known to cause cancer. Exposure to high concentrations may cause nausea and vomiting (USEPA 2019c).

2.8.4 Cyanide

Cyanide is any chemical compound containing the cyano group – a carbon atom triple-bonded to a nitrogen atom. Common cyanide compounds include hydrogen cyanide, potassium cyanide, and sodium cyanide. Certain bacteria, fungi, and algae can produce cyanide, and cyanide is present in a number of foods and plants (ATSDR 2006b).

Cyanide compounds occur as gases, liquids, and solids. Inorganic cyanides are commonly salts of the cyano anion CN^-. Organic compounds containing the cyano group are called nitriles. Those compounds that are able to release the cyano group CN^- ion are highly toxic to humans and animals (ATSDR 2006b). Hydrogen cyanide is a colorless gas with a faint, bitter, and almond-like odor. Sodium cyanide and potassium cyanide are both white solids with a bitter, almond-like odor when volatilization occurs. Cyanide and hydrogen cyanide have industrial applications in metal plating, metallurgy, some mining applications for extraction of precious metals, and in the manufacturing of plastics (ATSDR 2006b). According to USEPA (2019c), cyanide and related compounds are not classified as carcinogens (Group: A4), but they are highly toxic. According to USEPA (2019c) and ATSDR (2006b), exposure to cyanide can cause rapid breathing, low blood pressure, headaches, coma, and death.

2.8.5 Asbestos

Asbestos (CAS Registry Number 1332-21-4) is the name given to a group of six different fibrous minerals that include (ATSDR 2001b):

- Amosite
- Chrysotile
- Crocidolite
- And the fibrous forms of tremolite, actinolite, and anthrophyllite

The term asbestos describes a variety of fibrous, nonflammable minerals with flexibility and high tensile strength. Their unique properties were used mostly between the 1940s and 1970s in fireproof insulation, vinyl flooring, pipe insulation, ceiling tiles, brake linings, and roof coatings. Chrysotile, a serpentine mineral, is also known as "white asbestos" and makes up about 95% of asbestos found in buildings in the United States. The other asbestos minerals – crocidolite, amosite, anthophyllite, tremolite, and actinolite – are amphiboles that aren't commonly used in commercial products (USEPA 2019o). Asbestos usually occurs in the form of very thin elongated fibers. If asbestos is disturbed, the fibers may create an airborne dust, which when inhaled can penetrate tissues deep within the lungs and cause asbestosis, mesothelioma, and lung cancer. The United States OSHA and Environmental Protection Agency (EPA) make no distinction between the two kinds of asbestos – chrysotile and amphibole. OSHA began regulating workplace asbestos in 1970, around the time when miners and construction workers, who worked with the fibrous material, began reporting serious lung disease. At that time, the United States used about 800,000 metric tons of asbestos per year. OSHA published the Asbestos Standard for the Construction Industry, which outlined four categories of asbestos contamination, and specified precautions and disposal techniques for each class. For instance, if asbestos insulation is removed (Class 1), contractors and supervisors trained in asbestos removal must be onsite wearing respirators and protective clothing (USEPA 2019o).

Starting in 1973, USEPA began a process of banning certain uses of asbestos and designated it a Class A1 human carcinogen (USEPA 2019c). USEPA's response to asbestos has changed over the years. In 1983, the agency's asbestos handbook stated that removing the material is always appropriate, while their 1990 handbook acknowledged that asbestos removal may cause more contamination than leaving it in place. USEPA's advice on asbestos is neither to rip it all out in a panic nor to ignore the problem under a false presumption that asbestos is risk free. Asbestos material in buildings should be identified and managed appropriately. The main commercial and industrial value of asbestos lies in its ability to be heat resistant. For this reason, asbestos was widely applied in the United States in manufactured goods and building construction as a heat insulator for these products (USEPA 2019o):

- Roofing, ceiling, and floor tiles
- Window caulking
- Pipe wrap and pipe insulation
- Paper products
- Friction products, such as automobile brakes, clutches, and transmissions
- Heat-resistant fabrics
- Packaging
- Gaskets
- Coatings
- Some vermiculite and talc containing products
- Fire resistant doors

Exposure to asbestos is almost always through the inhalation route. Asbestos affects the lungs and may lead to a condition called asbestosis. USEPA classifies asbestos as a human carcinogen (Group: A1) (USEPA 2009c). According to ATSDR (2001b), we are all exposed to some asbestos in the air we breathe, especially in urban areas. Fortunately, the concentrations in air are generally very low, being on the order of 0.00001 to 0.0001 fibers of asbestos per milliliter of air (USEPA 2019o).

2.8.6 Acids and Bases

An **acid** increases the concentration of the hydrogen ion H^+ when dissolved in water and lowers the pH (potential hydrogen) of the solution. The bare hydrogen ion, H^+, is short for the hydronium ion, H_3O^+, since a bare H^+ does not exist in a solution. Conversely, a **base** increases the concentration of the hydroxide ion OH^- when dissolved in water, and raises the pH of the solution (Meyers 2003). Common acids and bases are listed in Table 2.7 (Meyers 2003). We recognize acids and bases by

TABLE 2.7

Common Acids and Bases

Common Acids		Common Bases	
Acid	**Chemical Formula**	**Base**	**Chemical Formula**
Acetic acid	$HC_2H_3O_2$	Ammonia	NH_3
Benzoic acid	$HC_7H_5O_2$	Aniline	$C_6H_5NH_2$
Boric acid	H_3BO_3	Dimethylamine	$(CH_3)_3NH$
Carbonic acid	H_2CO_3	Ethylamine	$C_2H_5NH_2$
Cyanic acid	$HCNO$	Hydrazine	H_2H_4
Formic acid	$HCNO_2$	Hydroxylamine	NH_2OH
Hydrocyanic acid	HCN	Methylamine	CH_3NH_2
Hydrofluoric acid	HF	Pyridine	C_5H_5N
Hydrogen sulfide	H_2S	Urea	NH_2CONH_2
Hydrochloric acid	HCl	Potassium hydroxide	KOH
Nitric acid	HNO_3	Sodium bicarbonate	$NaHCO_3$
Phosphoric acid	H_3PO_4	Sodium hydroxide	$NaOH$
Pyuvic acid	$HC_3H_3O_3$	Calcium hydroxide	$Ca(OH)_2$
Sulfuric acid	H_2SO_4		

Source: Meyers, R. The Basics of Chemistry. Greenwood Press. Westport, CT. 2003.

their simple properties, such as taste, and conclude the sour taste of a lemon indicates it must be acidic. Bases tend to taste bitter. On the pH scale, any substance with a pH less than 7 (the neutral point) is acidic and any substance having a pH greater than seven is basic. Acids and bases are widely used in industry, and are present in many widely consumed foods and drinks (see Table 2.7). Stronger acids and bases are used in household cleaners and detergents, especially those used on glassware and in ovens (Meyers 2003).

Acids or bases are only toxic if they are strong, meaning they are of relatively low or high pH. Exposure to strong acids and strong bases causes respiratory irritation and burning, and causes skin burns. Significant exposure may cause severe burns and even death (USEPA 2019c). Currently, adequate information is not available to evaluate the potential carcinogenic effects of common acids and bases (USEPA 2019c).

Ammonia is a common basic chemical widely used as a household cleaning agent and in many industrial applications (ATSDR 2018f). Ammonia is present naturally throughout the environment in air, soil, and water. Exposure to high levels of ammonia may cause lung, skin, and throat irritation. Some people with asthma may react more negatively to the inhalation of ammonia (ATSDR 2018f).

Hydrochloric acid (also referred to as hydrogen chloride) is a common acid widely used in industry as a cleaning agent, in the manufacturing of PVC, in making steel, and making leather. Hydrochloric acid is also present in humans and other organisms as a gastric acid (ATSDR 2002b), and sometimes exists as an acid mist. This mist may cause skin and lung irritation, and skin burns can occur if you are exposed to a highly concentrated mist for a prolonged period (ATSDR 2002b).

2.8.7 Radioactive Compounds

Radioactive decay occurs when an unstable atomic nucleus spontaneously loses energy by emitting ionizing particles and radiation. This decay, or loss of energy, results in an atom of one type (parent nuclide) transforming into an atom of a different type (daughter nuclide). All elements with an atomic number greater than 80 possess radioactive isotopes, and all isotopes of elements with an atomic number greater than 83 are radioactive (Kathren 1991).

Some radioactive compounds deserving special attention include (Kathren 1991):

• Beryllium	• Calcium	• Carbon	• Potassium
• Cadmium	• Cesium	• Iodine	• Strontium
• Palladium	• Tin	• Radon	• Radium
• Thorium	• Uranium	• Plutonium	

Radon (CAS Registry Number 10043-92-2) is the most common radioactive compound present in urban areas, and it has the potential for adverse human health effects (ATSDR 2000). USEPA (2019c) classifies radon as a human carcinogen (Group: A1), and exposure to radon for a long period of time at elevated concentrations may cause cancer. Radon is a decay product of uranium, found naturally in the Earth's crust. It is one of the heaviest substances existing as a gas under normal conditions of pressure and temperature.

The highest average radon concentrations in the United States are found in Iowa, southeastern Pennsylvania, and Appalachian Mountain areas. During the decay process, alpha, beta, and gamma radiation are released. Alpha particles can travel only short distances and cannot travel through your skin. Beta particles can penetrate through your skin but not your whole body. Gamma particles can penetrate your whole body. Radon is normally present at very low levels in outdoor air but may be present at higher levels in indoor air, especially in basements and buildings with poor ventilation and in well water (ATSDR 2000).

2.8.8 GREENHOUSE GASES

Greenhouse gases are gases in the atmosphere capable of absorbing and emitting radiation within the thermal infrared range (Karl and Trenberth 2003; USEPA 2017; 2018a). Greenhouse gases cause the Earth to warm up and since the start of the Industrial Revolution in the early 18th century, levels of greenhouse gases in the Earth's atmosphere have increased (USEPA 2018a). Figure 2.12 shows the greenhouse gas effect. The primary effects of greenhouse gases are climate change related and not directly related to toxicity.

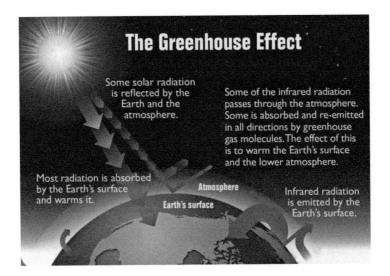

FIGURE 2.12 Greenhouse gas effect. (From USEPA 2018a. Global Greenhouse Gas Emissions Data. 2018a. https://epa.gov/ghgemissions/global-greenhouse-gas-emissions-data. (accessed December 31, 2018.))

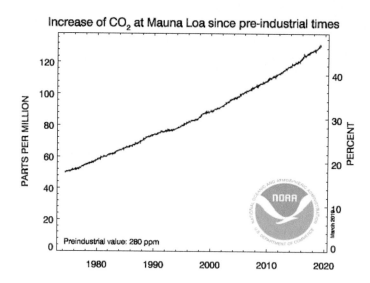

FIGURE 2.13 Carbon dioxide concentrations from 1960 to 2018 at Mauna Loa, Hawaii. (From National Oceanic and Atmospheric Administration (NOAA). 2018. Atmospheric CO2 at Mauna Loa Observatory. 2018. http://www.esrl.noaa.gov/gmd/ccgg/trends. (accessed December 31, 2018.))

Greenhouse gases are not generally investigated at the specific site of environmental contamination. It is important, however, to discuss greenhouse gases because of their potential impacts on climate change. Figure 2.13 shows the increase of carbon dioxide in the atmosphere since 1960 (NOAA 2018). The yearly variation shown in Figure 2.13 is seasonal and is attributed to extraction of carbon dioxide from plant matter during photosynthesis (NOAA 2018). The greenhouse gases include the following compounds (USEPA 2018a):
Carbon dioxide (CO_2)

- Methane (CH_4)
- Nitrous oxide (N_2O)
- Fluorinated gases including:
  - Hydrofluorocarbons (HFCs)
  - Perfluorocarbons (PFCs)
  - Sulfur hexafluoride (SF_6) – an industrially produced gas with a global warming potential (GWP) almost 24,000 times that of CO_2. The GWP measures the
  - relative contribution of a gas to global warming based on its ability to absorb infrared radiation, its residence time in the atmosphere, and wavelengths of energy absorbed. For comparison purposes, CO_2 has a GWP = 1.

Greenhouse gases originate from a variety of sources, such as (USEPA 2018a):

- Fossil fuel combustion:
  - Coal
  - Gasoline and diesel fuel in automobiles and trucks
  - Aviation fuels
  - Home-heating fuels such as home-heating oil and kerosene
- Industrial processes – Refineries and cement making. Cement making is an often overlooked source; it accounts for 5–7% of anthropogenic CO_2
- Waste disposal facilities
- Electrical generation
- Mining

- Residential and commercial sources
- Agriculture

Figure 2.14 shows a breakdown of yearly greenhouse gas emissions by source.

Increasing carbon dioxide levels in the atmosphere have been linked to the burning of fossil fuels (USEPA 2018a), and the evidence is clear. Figure 2.15 shows the percent contribution of each type of greenhouse gas on a global scale and Figure 2.16 shows the contribution by economic sector (USEPA 2018a). The current total estimated emissions of greenhouse gases world-wide exceed 6,780 million metric tons (USEPA 2018a).

Some negative effects attributed to climate change caused by rising carbon dioxide levels in the atmosphere can be observed in the western portion of North America in Canada and the United States. Millions of acres of forest are being killed by the pine bark beetle moving into areas that were once too cold for the beetle to thrive (United States Forest Service 2019). Figure 2.17 shows the Yosemite National Park in California where the National Park Service is actively cutting down thousands of trees infected with the beetle. Figure 2.18 is a photograph from Jasper National Park in Canada, more than 2,000 kilometers north of Yosemite, where an estimated 80% of the trees are infected and dying from the pine bark beetle.

2.8.9 CARBON DIOXIDE

Carbon dioxide is a greenhouse gas which we covered above, but that is just half the story. We must also discuss the buildup of carbon dioxide in the world's oceans. The oceans were once believed to be so vast and deep that the effects of pollutions would never directly threaten the oceans. This has been proven false. The rate of accumulation of pollution is proportional to the human population. The more people, the more pollution ends up in the oceans.

Pollution in the oceans does not all originate from direct discharge but also is delivered through the atmosphere from increased concentrations of carbon dioxide in the atmosphere that is partially

Share of global carbon dioxide emissions from fuel combustion (2015)

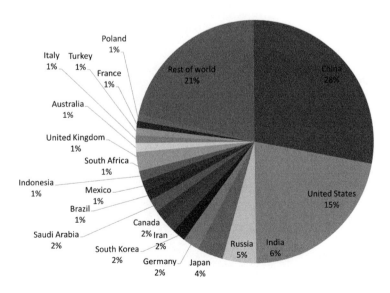

FIGURE 2.14 2015 Greenhouse gas contribution by country. (From Union of Concerned Scientists. Each Country's share of CO2 Emissions. 2018. https://www.ucsusa.org/global-warming/science-and-impacts/science/each-countrys-share-of-co2.html. (accessed March 24, 2019.))

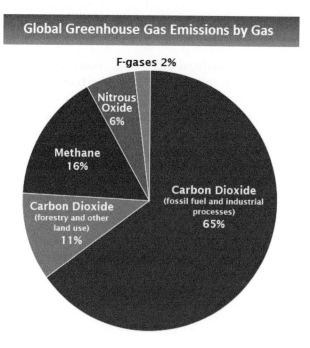

FIGURE 2.15 Greenhouse gas contribution in the United States in 2015 by type. (From USEPA. Global Greenhouse Gas Emissions Data. 2018a. https://epa.gov/ghgemissions/global-greenhouse-gas-emissions-data. (accessed December 31, 2018.))

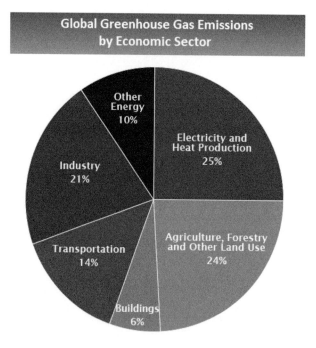

FIGURE 2.16 Greenhouse gas contribution in the United States by economic sector. (From USEPA 2018a. Global Greenhouse Gas Emissions Data. 2018a. https://epa.gov/ghgemissions/global-greenhouse-gas-emissions-data. (accessed December 31, 2018.))

FIGURE 2.17 Yosemite National Park in the United States showing infected trees with the bark beetle (Photograph by Daniel T. Rogers).

FIGURE 2.18 Jasper National Park in Canada showing forested areas infected with the bark beetle (Photograph by Daniel T. Rogers).

absorbed by the oceans and causes acidification of the surface layers. Absorption of carbon dioxide by the oceans has been estimated to be as high as 25%, which represents millions of tons of carbon dioxide absorbed by the oceans every year. The additional carbon dioxide absorbed by the oceans reacts with ocean water and forms an acid called carbonic acid and is lowering the pH of the oceans world-wide. In fact, the oceans are acidifying faster than they have in some 300 million years. The effect of the acidification has been linked to bleaching of coral reefs, including the Great Barrier Reef off the northeastern coast of Australia (Natural Resource Defense Council 2018).

2.8.10 Carbon Monoxide

Carbon monoxide (CO) is a colorless, odorless, nonirritating gas that is very toxic to humans and animal life (USDHH 2009). Carbon monoxide is formed naturally and anthropogenically. The human body produces a small amount of CO when red blood cells convert protoporphyrin into bilirubin. Synthetically, most carbon monoxide is created by the incomplete combustion of fossil fuels, with the highest percentage coming from automobile exhaust. Inside homes, significant sources of carbon dioxide emissions can be natural gas and home-heating oil furnaces, hot water heaters, appliances, wood burning stoves, and fireplaces. Carbon monoxide is also created as a by-product of several industrial processes including metal melting and chemical synthesis (USDHH 2009).

Carbon monoxide is one of the most common causes of fatal air poisoning (USDHH 2009). When inhaled, carbon monoxide combines with hemoglobin to produce carboxyhemoglobin. A condition known as anoxemia arises because carboxyhemoglobin cannot deliver oxygen to body tissues efficiently. Exposure to high levels of carbon monoxide may cause headache, nausea, vomiting, dizziness, lethargy, seizures, coma, and even death (USDHH 2009). Concentrations of less than 700 parts per million of CO may cause up to 50% of the body's hemoglobin to convert to carboxyhemoglobin. In the United States, OSHA has established a long-term exposure standard for carbon monoxide at 50 parts per million (United States Department of Labor 2019).

2.8.11 Ozone

Ozone (O_3) is a gas present in Earth's upper atmosphere and at ground level. The ozone occurring at ground level is considered an air pollutant by USEPA (2019p), whereas the ozone occurring in the upper atmosphere is not considered a pollutant, because it is beneficial to life on Earth by acting as a filter of ultraviolet (UV) radiation produced by the sun (USEPA 2019q). Ozone from the ground level to a height of approximately 6 miles in the troposphere is the main constituent of urban smog (USEPA 2019q).

Until the mid-1990s, ozone in the stratosphere had been gradually being destroyed by anthropogenically produced chemicals referred to as ozone-depleting substances (ODS), including (USEPA 2019q):

- Chlorofluorocarbons (CFCs)
- Hydrochlorofluorocarbons (HCFCs)
- Halons
- Methyl bromide
- Carbon tetrachloride
- Methyl chloroform

These substances were commonly used in coolants, foaming agents, fire extinguishers, solvents, pesticides, and aerosol propellants. Once released into the environment they

tend to degrade very slowly (USEPA 2019q). It should also be noted that hydrofluorocarbons (HFCs) have been substituted widely for CFCs. While this substitution has slowed the harmful effects brought on by free chlorine in the upper atmosphere, HFCs are potent greenhouse gases. Ground-level ozone is not emitted directly into the air, but is created by chemical reactions between oxides of nitrogen (NOx) and VOCs in the presence of sunlight. Emissions from automobile exhaust, gasoline vapors, chemical solvents, electrical generating facilities, and certain factories are some of the major sources that emit compounds leading to the generation of ozone (USEPA 2019p). Table 2.8 shows the percent contributions of these major sources.

Ground-level ozone is of a particular concern in urban regions of the United States. During the summer, strong sunlight and hot weather produces the conditions necessary for producing harmful levels of ozone. Overexposure to ground-level ozone can result in difficulty breathing and other respiratory effects (USEPA 2019p), especially for the elderly, very young, and those with existing respiratory ailments.

2.8.12 Sulfur Dioxide

Sulfur dioxide (SO_2) is released naturally through volcanic eruptions. Anthropogenically, significant sources of sulfur dioxide include automobile exhaust, the burning of coal, and some industrial processes (USEPA 2018b). Sulfur dioxide is a component of smog. Other components of smog include ozone, carbon monoxide, particulate matter, VOCs, and nitrous oxides (USEPA 2018b).

Figure 2.19 shows an example of smog. When released into the atmosphere, sulfur dioxide often reacts with water vapor and eventually forms sulfuric acid (H_2SO_4). When precipitation occurs with

TABLE 2.8

Sources and Contribution of NOx and VOCs (USEPA 2019p)

Compound	Source	Percent Contribution
NOx	Motor vehicles	56
	Utilities	22
	Industrial, commercial, and residential fuel consumption	17
	Other	5
VOCs	Industrial and commercial processes	50
	Motor vehicles	45
	Consumer solvents	5

Source: USEPA. Ground Level Ozone. 2019p. https://www.epa.gov/ground-level-ozone-pollution. (accessed March 24 2019.)

FIGURE 2.19 Layer of smog in Los Angeles, California (Photograph by Daniel T. Rogers).

a pH lower than that of natural rain, which is considered to be a pH of 5.6, it is considered **acid rain** (USEPA 2018b). Figure 2.20 is a diagram showing acid rain development.

According to USEPA (2018b), acid rain and smog are serious environmental problems affecting large parts of the United States, especially heavily urbanized areas. Figure 2.21 shows the areas of the United States having levels of smog capable of causing an adverse health effect in humans (USEPA 2018c).

2.8.13 PARTICULATE MATTER

Particulate matter (PM) is also known as particle pollution. PM is a complex mixture of very small particles and liquid droplets. Particle pollution consists of a number of components, including (USEPA 2019r):

- Acids
- Sulfates
- Metals
- Nitrates
- Organic compounds
- Soil or dust particles

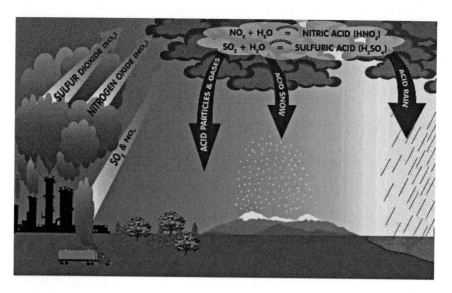

FIGURE 2.20 Development of acid rain (From New York State Department of Environmental Conservation. Acid Deposition Large Graphic. 2010. http://www.dec.ny/chemical/41293.html (accessed April 14, 2019.))

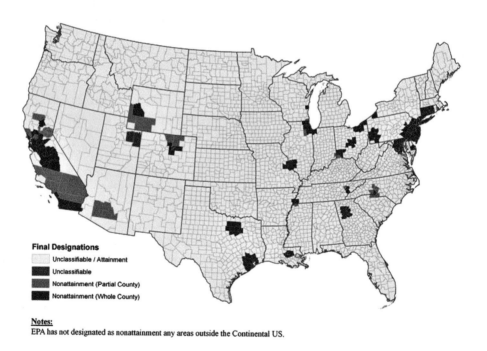

FIGURE 2.21 Non-attainment areas of the United States for smog-forming air pollutants in 2018. (From USEPA. Non-attainment Areas of the United States. 2018c. https://archive.epa.gov/ozonedesignations/web/jpg/desfinal.jpg. (accessed March 24, 2019.))

The size of the particles is directly related to their potential for causing adverse health effects. Particles less than ten micrometers (μm) or less are small enough to pass through the nose and throat and enter the lungs while breathing (USEPA 2019r). Once inhaled, these particles can affect the heart and lungs and cause substantially adverse health effects. Figure 2.22 is a diagram of particle size relative to a human hair (USEPA 2019r).

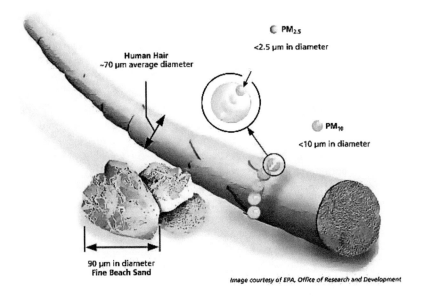

FIGURE 2.22   Particle size representation. (From USEPA. Particulate Matter (PM) Pollution. 2019r. https://www.epa.gov/pm-pollution. (accessed March 25, 2019.))

USEPA groups PM into two categories (USEPA 2019r):

- Particles ranging in size from 2.5 μm to 10 μm are considered inhalable coarse particles; most likely to be present near roadways and dust-producing industries
- Particles less than 2.5 μm are considered fine particles and are present in smoke and haze. These particles can be emitted from forest fires or power plants, certain industries, and automobiles.

Adverse health effects caused by inhalation of PM include (USEPA 2019r):

- Increased respiratory irritation, coughing, and difficulty breathing
- Decreased lung function
- Aggravated asthma
- Development of chronic bronchitis
- Irregular heartbeat
- Nonfatal heart attacks
- Premature death in people with heart or lung disease

An additional concern is the potential exposure to other contaminants sorbed onto PM, which are then ingested or inhaled. This topic will be covered in greater detail in Chapter 4.

2.8.14   Bacteria, Parasites, and Viruses

Although we do not always think of bacteria, parasites, and viruses as everyday environmental contaminants, they are common and can cause many diseases and adverse health effects. Technically, bacteria, parasites, and viruses can contaminate water supplies and are also present in the air, in soil, on and within food, and most other surface surfaces.

   Bacteria are organisms made up of just one cell. They are capable of multiplying themselves through a process called "binary fission," whereby a single bacterium grows to approximately twice its normal size and then splits into two daughter cells that are copies of the original bacterium.

Bacteria live everywhere, even inside most organisms. Most bacteria are harmless and some are beneficial by destroying other harmful bacteria within our bodies (Madigan et al. 2008). However, some may cause disease such as tuberculosis. One of the more common harmful bacteria is a group called *E. coli*, which is short for Escherichia coli. Most *E. coli* are harmless but a strain called sero-type O157:H7 can cause food poisoning in humans. Three basic shapes of bacteria exist: rounded, rod shaped, and spiral. *E. coli* bacteria is rod shaped. The ability for *E. coli* to survive for a brief period outside the body creates the potential for the bacteria to spread and infect other people. The spread of *E. coli* usually occurs when there is poor sanitation or when untreated sewerage is discharged from municipal wastewater treatment plants, which occasionally occurs during flood events (USEPA 2009). Adverse health effects from exposure to *E. coli* bacteria typically include gastroenteritis, urinary tract infections, skin rashes, and neonatal meningitis (USEPA 2009).

A **virus** is a sub-cellular infectious agent capable of replicating itself inside the cells of another organism (Madigan et al. 2008). They are typically 100 times smaller than a bacterium. Viruses consist of two parts consisting of: 1) DNA or RNA molecules carrying genetic information; and 2) a protein coat protecting the genes. Some viruses may also have an outside layer of fat surrounding them while they are outside a cell. Viruses spread in many ways, including aerosol routes (coughing and sneezing), through infected or contaminated water, and exchange of body fluids. Viruses cause diseases such as the common cold, influenza, chickenpox, mumps, Ebola, and HIV (Madigan et al. 2008).

A **parasite** is an organism living on or within a different organism (the host) at the expense of the host organism. Common examples of parasites causing adverse health effects in humans are cryptosporidium and giardia; both may cause severe intestinal disorder (United States Center for Disease Control 2009). Exposure to these two parasites occurs by consuming affected water (United States Center for Disease Control 2009). In humans, giardia creates uncomfortable but curable gastrointestinal symptoms, whereas cryptosporidium can create life-long gastrointestinal symptoms in persons with weak immune systems and may result in death.

2.8.15 INVASIVE SPECIES

We will consider invasive species as a form of pollution. This is because the definition of invasive species tracks closely with other forms of pollution we have discussed. An **invasive species** is any kind of living organism that is not native to an ecosystem and causes harm that is directly or indirectly introduced by humans or is the result of human activity (USDA 2018; USEPA 2018d).

Invasive species' mode of invasion is commonly as a stowaway during the transportation of goods. USEPA (2018d) estimated that 30% of invasive species to the Great Lakes was as ballast water in ships carrying cargo. Ballast water is taken into a ship or discharged from a ship as it loads or unloads cargo to accommodate the ship's weight changes. Invasive species is a world-wide problem and has been for hundreds of years. Most wildlife and environmental agencies are losing the battle with invasive species for one simple and disturbing reason; this type of pollution breeds.

To make this point clear, the United Nations established the Global Invasive Species Program (GISP) in 1997 to develop a global strategy to deal with invasive species (United Nations 2018). World-wide, there are thousands of invasive species (United Nations 2018). Invasive species are characterized as aggressive when they are transported into areas where they are not naturally occurring usually because they do not have a natural predator or other organisms have not evolved defenses (United Nations 2018). The top 13 invasive species currently causing the most harm in the United States includes (USEPA 2018d):

1. Burmese Python. Introduced in the Florida Everglades, they were considered established in 2002 and have reduced deer, raccoon, marsh rabbits, bobcats, possum, and alligators by as much as 99%.
2. Emerald Ash Borer. Originally from Russia, China, and Japan, they were accidentally introduced in southeastern Michigan in 2002 in infected wood crates. The Emerald

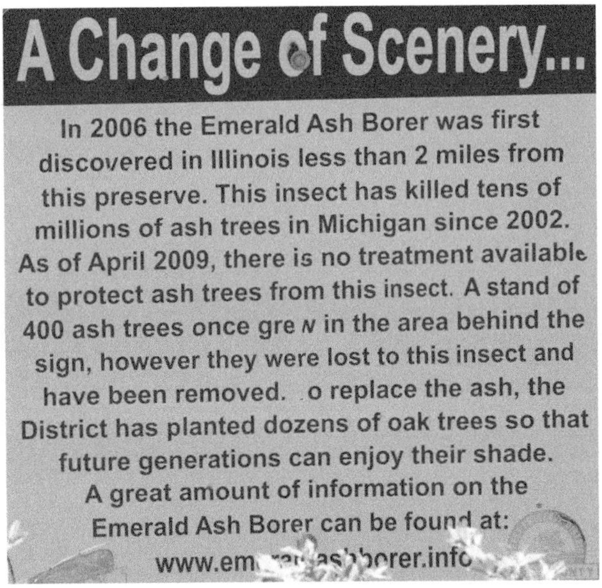

FIGURE 2.23   Sign describing damage inflicted by the Emerald Ash Borer (Photograph by Daniel T. Rogers).

Ash has infected and killed hundreds of millions of Ash trees in the United States (see Figure 2.23).

3. Nutrua. Commonly called swamp rats. They were imported from South America into the southern United States for their fur but now have grown out of control and are in the wild. Their endless digging along the banks of rivers and streams kill plants and fish and causes soil erosion.

4. European Starling. Since their introduction into the United States in the 1890s, they have become the most successful foraging bird in the United States, with a population now estimated at 200 million.

5. Northern Snakehead. The most feared fish in the Chesapeake Watershed. A sharp-toothed fish from Korea and China that originally infested the Chesapeake Watershed has now spread to other parts of Virginia and Delaware.

6. Brown Marmorated Stink Bug. Stink Bugs destroy fruits and vegetables to such an extent that they have been known to increase the price of produce. They first showed up in Allentown, Pennsylvania in 1998 after crawling out of a cargo ship. Originally they are from China.

7. Feral Hogs. Essentially farm pigs gone wild. They are established in 47 states with massive populations in Texas, Florida, Georgia, Virginia, and North Carolina.

8. Lionfish. Lionfish are native to the Pacific Ocean. They were first spotted off the Miami coast in the mid-1980s. They are characterized as eating everything that they can stuff into their mouths. That's how they obtained their name. They are destroying life on the coral reefs of the Atlantic Ocean.

9. Norway Rat. First introduced in the United States in 1775 and now they are everywhere, including Hawaii and Alaska.

10. Tegu. Tegus are brown lizards that are muscular, fast, and love to eat eggs. They are known to harass pets and invade homes where they were released in Florida. Game officials have largely given up on eradicating them from the wild.

11. Asian Citrus Psyllid. The Asian Citrus Psyllid carries bacteria that attacks and kills fruit trees, especially orange trees. Estimates are that 80% of the orange trees in Florida are infected.

FIGURE 2.24   The invasive plant Kudzu in North Carolina (Photograph by Daniel T. Rogers).

12. Brown Tree Snake. The Brown Tree Snake has infected the island of Guam and has deci-
    mated the local bird population. They are so out of control on the island that they are even
    responsible for power outages.
13. Kudzu. First introduced into the United States in 1877. Kudzu is an invasive plant species
    that grows very fast and smothers other plants in the southeastern United States as far
    north as Indiana. Kudzu is now growing and killing other plant species in the United States
    at a rate of up to 1,000 square kilometers per year (See Figure 2.24).

The number of different types of invasive species in North America alone total the following
(USDA 2018):

- Plants – 1,911 species
- Insects – 6,647 species
- Pathogens – 1,278 identified
- Animals – 220 species

The number of invasive species and diseases in North America alone totals nearly 10,000
(USDA 2018).

2.8.16   POLYFLUOROALKYL SUBSTANCES

Per- and Polyfluoroalkyl substances are a group of synthetic chemicals that have been manufactured and used in a variety of industries and consumer products world-wide since the 1940s.
They are commonly referred to as PFOA or PFAS. These chemicals are very persistent and
mobile in the environment and tend to bioaccumulate. Although these chemicals are no longer
manufactured in the United States, they can be present in many consumer goods that include
carpet, leather and apparel, textiles, paper, packaging, coatings, rubber, and plastic (USEPA
2018e).

  Toxicological studies indicate that these chemicals can cause reproductive and developmental
abnormalities, liver and kidney damage, immunological effects in laboratory animals, and cause
tumors in animals. The most consistent findings are increased cholesterol levels among exposed

FIGURE 2.25   PFAS HEALTH ADVISORY IN MICHIGAN, USA. (PHOTOGRAPH BY DANIEL T. ROGERS).

populations, with more limited findings related to low infant birth weights, effects on the immune system, cancer, and thyroid hormone disruption (USEPA 2018e).

In certain parts of the United States, these chemicals have contaminated water supplies and streams and lakes, prompting health advisories (see Figure 2.25).

2.8.17   EMERGING CONTAMINANTS

Through research, we are now learning about the presence in the environment of many chemicals and microbes that historically were not considered contaminants (USGS 2019s). Emerging contaminants originate from urban and agricultural sources and impact soil and groundwater at many urban locations. The potential health risks posed by emerging contaminants are not fully known. Many emerging contaminants enter the environment from residential waste products and agriculture, and this fact has prompted a shift in traditional thinking that most releases of contaminants into the environment were from industrial sources (Barnes et al. 2008).

Emerging contaminants include a wide variety of compounds consisting of (Bell et al. 2019; Barnes et al. 2008; USEPA 2019s):

- Pharmaceuticals and drugs including:
  - Antibiotics
  - Steroids
  - Antibacterial chemicals
  - Hormones
  - Narcotics
  - Many other legal and illegal drugs
- Insect repellants
- Solvents
- Detergents
- Plasticizers
- Fire retardants
- Veterinary antibiotics
- Pesticides
- Others

Common pharmaceuticals and drugs with the capability to become contaminants if not properly disposed of include (USEPA 2019s):

- Hormones, such as testosterone
- Antibiotics, such as penicillin
- Sildenafil citrate, commonly known as Viagra
- Benzoylmethylecgonine, commonly known as cocaine

Benzoylmethylecgonine or **cocaine** is a stimulant affecting the central nervous system and also acts as an appetite suppressant. Antibiotics inhibit the growth of bacteria. Sildenafil citrate is an arterial stimulant that was originally intended to treat high blood pressure (Barnes et al. 2008). Testosterone is a male sex hormone, an anabolic steroid, and affects the growth of muscle mass. Figures 2.26 and 2.27 show the structures of penicillin and benzoylmehylecgonine, respectively. Table 2.9 lists the CAS Number and molecular formula for benzoylmethylecgonine, penicillin, sildenafil citrate, and testosterone.

Two other emerging contaminants of note include a group of compounds called perchlorates and the compound 1,4-Dioxane. **Perchlorates** are colorless and odorless salts. They are a group of compounds including:

- Magnesium perchlorate ($MgClO_4$)
- Potassium perchlorate ($KClO_4$)
- Ammonium perchlorate (NH_4ClO_4)
- Sodium perchlorate ($NaClO_4$)
- Lithium perchlorate ($LiClO_4$)

Perchlorates are very reactive and are commonly used in explosives, fireworks, road flares, and rocket motors (USEPA 2019t; ATSDR 2008b). Perchlorate may also be present in bleach as an impurity. Adverse health effects of exposure to perchlorates include the ability of the thyroid

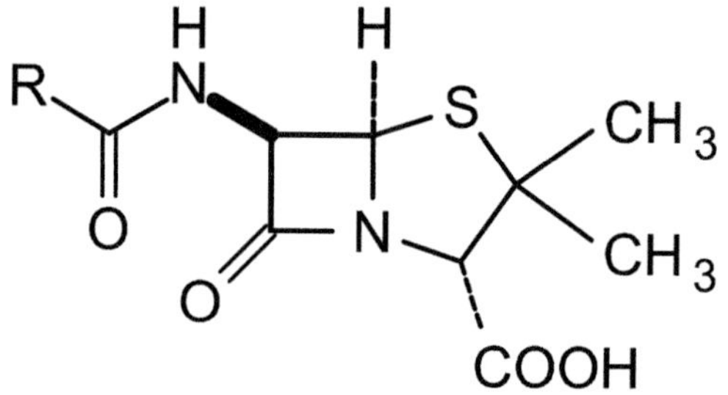

FIGURE 2.26 Structure of the antibiotic penicillin.

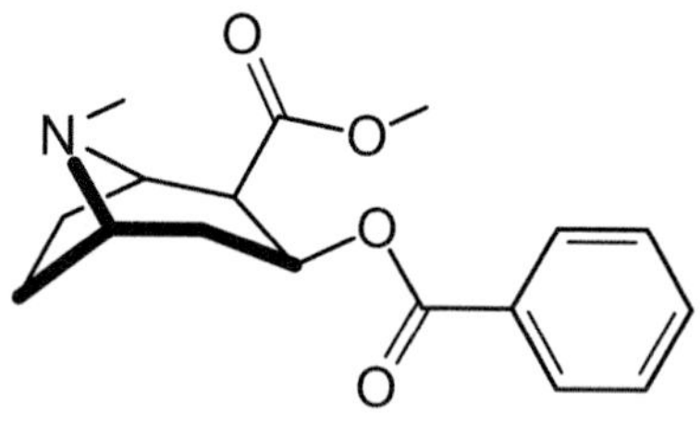

FIGURE 2.27 Structure of benzoylmethylecgonine.

TABLE 2.9
CAS Registration Numbers and Molecular Formulas for Select Emerging Pharmaceuticals and Drugs

Compound	CAS Registration Number	Molecular Formula
Benzoylmethylecgonine	50-36-2	$C_{17}H_{21}NO_4$
Sildenafil citrate	139755-83-2	$C_{22}H_3N_8O_4S$
Testosterone	58-22-0	$C_{19}H_{28}O_2$
Penicillin	Not listed	$*R\text{-}C_9H_{11}N_2O_4S$

* R = Indicates a variable group attached to the molecule

Source: USEPA. Contaminants of Emerging Concern. 2019s. https://www.epa
.gov/wqc/contaminants-emerging-concern-including-pharmaceuticals-
and-personal-care-products. (accessed March 25, 2019.)

gland to uptake iodine. Iodine is needed to produce hormones that regulate many body functions. USEPA does not currently list any of the perchlorate compounds as human carcinogens (USEPA 2019c).

1,4-Dioxane ($C_4H_8O_2$) is a clear liquid that easily dissolves in water. It is one of three isomer varieties of dioxanes, and is primarily used as an industrial solvent, with less widespread use in cosmetics, shampoos, and detergents (ATSDR 2007d). The ability of 1,4-Dioxane to dissolve so easily in water and its penchant for not being biodegradable results in 1,4-Dioxane easily contaminating surface water and groundwater. Inadequate information is available for classifying the carcinogenicity of 1,4-Dioxane (USEPA 2019c).

1,2,3,-Trichloropropane (TCP) is a chlorinated volatile organic compound with high chemical stability and relatively high solubility (Bell et al. 2019; California Environmental Protection Agency [CalEPA] 2019). TCP is considered a probable human carcinogen (USEPA 2019c). TCP is used as a cleaning solvent and as a pesticide, especially in California. CalEPA established a drinking-water notification concentration for TCP of 0.005 micrograms per liter (ug/l). TCP has been detected in groundwater at numerous locations in California, especially within agricultural areas (CalEPA 2019).

2.9 SUMMARY AND CONCLUSION

Now we know there are thousands of different types of pollutants existing everywhere. They are in the air we breathe, the water we drink, the food we eat, and in the dirt we play in. They are organic and inorganic. Many naturally occur – many do not. Some we know about and most we do not. Some are not so toxic and some are very toxic. Some may cause cancer and some may not. All of them have the ability to cause some adverse health effect in humans, to other organisms, or negatively impair or impact the environment, if the exposure and dose were just right. Otherwise, they would not be considered pollutants.

This now leads us to the next set of questions we need to explore. How do contaminants behave once they are released into the environment? Do they degrade? Where would we go to find them? How long do they last? Evaluating the behavior of pollutants in the environment is commonly referred to as fate and transport assessment. Analysis of the fate and transport of pollutants once released into the environment is crucial for accurately assessing the risk posed by a specific chemical. Just because a pollutant exists does not mean there will be a risk to human health or the environment. There must be a completed pathway – the pollutant must be transported from its point of release to a place where exposure can occur. Chapter 3 discusses this central concept, and describes the fate and transport of many pollutants introduced in this chapter.

REFERENCES

Agency for Toxic Substances and Disease Registry (ATSDR). 1995. 2-Hexanone. CAS Registry Number 591-78-6. ATSDR ToxFAQs. Atlanta, GA.

ATSDR. 1997a. Chloroform. CAS Registry Number 127-18-4. ATSDR ToxFAQs. Atlanta, GA.

ATSDR. 1997b. Toxaphene. CAS Registry Number 8001-35-2. ATSDR ToxFAQs. Atlanta, GA.

ATSDR. 2000. *Radon Toxicity: Who Is At Risk?* ATSDR. Atlanta, GA.

ATSDR. 2001a. Toluene. CAS Registry Number 108-88-3. ATSDR ToxFAQs. Atlanta, GA.

ATSDR. 2001b. Asbestos. CAS Registry Number 1332-21-4. ATSDR ToxFAQs. Atlanta, GA.

ATSDR. 2002a. Di(2-ethylhexyl) Phthalate. CAS Registry Number 117-81-7. ATSDR ToxFAQs. Atlanta, GA.

ATSDR. 2002b. Hydrogen Chloride. CAS Registry Number 7647-01-0. ATSDR ToxFAQs. Atlanta, GA.

ATSDR. 2003. Malathion. CAS Registry Number 121-75-5. ATSDR ToxFAQs. Atlanta, GA.

ATSDR. 2005a. Naphthalene. CAS Registry Number 91-20-3. ATSDR ToxFAQs. Atlanta, GA.

ATSDR. 2005b. Bromoform. CAS Registry Number 75-25-2. ATSDR ToxFAQs. Atlanta, GA.

ATSDR. 2005c. Permethrin: Toxicologic Information About Pesticides. CAS Registry Number 52645-53-1. ATSDR. Atlanta, GA.

ATSDR. 2006a. Vinyl Chloride: CAS Registry Number 75-01-4. ATSDR. Atlanta, GA.

ATSDR. 2006b. Cyanide: CAS Registry Numbers 74-90-8, 143-33-9, 151-50-8, 592-01-8, 544-92-3, 506-61-6, 460-19-5, and 506-77-4. ATSDR. Atlanta, GA.

ATSDR. 2007a. Benzene. CAS Registry Number 71-43-2. ATSDR ToxFAQs. Atlanta, GA.

ATSDR. 2007b. Ethybenzene. CAS Registry Number 100-41-4. ATSDR ToxFAQs. Atlanta, GA.

ATSDR. 2007c. Xylene. CAS Registry Number 1330-20-7. ATSDR ToxFAQs. Atlanta, GA.

ATSDR. 2007d. 1,4-Dioxane. CAS Registry Number 123-91-1. ATSDR ToxFAQs. Atlanta, GA.

ATSDR. 2008a. Dioxin. https://www.atsdr.cdc.gov/toxprofiles/TP.asp?id=366&tid=63. (accessed March 24, 2019).

ATSDR. 2008b. Perchlorates. CAS Registry Numbers 10034-81-8, 7778-74-7, 7790-98-9, 7601-89-0, and 7791-03-9. ATSDR ToxFAQs. Atlanta, GA.

ATSDR. 2011. Chlordane. CAS Registry Number 57-74-9. https://www.atsdr.cdc.gov/substances/toxsubstance.asp?toxid=62. (accessed March 24, 2019).

ATSDR. 2012. Cadmium. CAS Registry Number 7440-43-9. https://www.atsdr.cdc.gov/toxprofiles/TP.asp?id=48&tid=15. (accessed March 24, 2019).

ATSDR. 2013. Barium. CAS Registry Number 7440-39-3. https://www.atsdr.cdc.gov/toxfaqs/tf.asp?id=326&tid=57. (accessed March 24, 2019).

ATSDR. 2018a. Copper. Case Registry Number 7440-50-8. https://www.atsdr.cdc.gov/toxprofiles/TP.asp?id=206&tid=37. (accessed March 24, 2019).

ATSDR. 2018b. Nickel. Case Registry Number 7440-02-0. https://www.atsdr.cdc.gov/toxprofiles/tp.asp?id=245&tid=44. (accessed March 24, 2019).

ATSDR. 2018c. Selenium. Case Registry Number 7782-49-2. https://www.atsdr.cdc.gov/toxprofiles/tp.asp?id=153&tid=28. (accessed March 24, 2019).

ATSDR. 2018d. Silver. Case Registry Number 7440-22-4. https://www.atsdr.cdc.gov/toxprofiles/tp.asp?id=539&tid=97. (accessed March 24, 2019d).

ATSDR. 2018e. Zinc. Case Registry Number 7440-66-6. https://www.atsdr.cdc.gov/toxprofiles/tp.asp?id=302&tid=54. (accessed March 24, 2019).

ATSDR. 2018f. Ammonia. Case Registry Number 7664-41-7. https://www.atsdr.cdc.gov/toxprofiles/tp.asp?id=11&tid=2. (accessed March 24, 2019).

ATSDR. 2019a. ATSDR Glossary of Terms. Center for Disease Control. http://atsdr.cdc.gov/glossary.html. (accessed March 24, 2019a).

ATSDR. 2019b. Toxicological Profile for Methyl-tert-butyl-Ether. https://www.atsdr.cdc.gov/toxprofiles/tp.asp?id=228&tid=41. (accessed March 24, 2019).

ATSDR. 2019c. Toxicological Profile for Vinyl Chloride. https://www.atsdr.cdc.gov/substances/toxsubstance.asp?toxid=51. (accessed March 24, 2019).

ATSDR. 2019d. Toxicological Profile for Tetrachloroethylene. https://www.atsdr.cdc.gov/substances/toxsubstance.asp?toxid=48. (accessed March 24, 2019).

ATSDR. 2019e. Toxicological Profile for Trichloroethylene. https://www.atsdr.cdc.gov/substances/toxsubstance.asp?toxid=30. (accessed March 24, 2019).

ATSDR. 2019f. Factsheet for PAHs. http://www.atsdr.cdc.gov/tfacts69.html. (accessed March 24, 2019).

ATSDR. 2019g. Toxicological Profile for Polychlorinated Biphenyls. ATSDR ToxFAQs. Atlanta, GA. https://www.atsdr.cdc.gov/toxprofiles/tp.asp?id=142&tid=26. (accessed March 24, 2019).

ATSDR. 2019h. Phthalates Factsheet. https://www.cdc.gov/biomonitoring/Phthalates_FactSheet.html. (accessed March 24, 2019).

ATSDR. 2019i. Aromatic Amines. https://www.atsdr.cdc.gov/substances/toxchemicallisting.asp?sysid=36. (accessed March 24, 2019).

ATSDR. 2019j. Phosphate Ester Flame Retardants. https://www.atsdr.cdc.gov/phs/phs.asp?id=1118&tid=239. (accessed March 24, 2019).

ATSDR. 2019k. Chromium. Case Registry Number 7440-47-3. https://www.atsdr.cdc.gov/substances/toxsubstance.asp?toxid=17. (accessed March 24, 2019).

ATSDR. 2019l. Lead. Case Registry Number 7439-92-1. https://www.atsdr.cdc.gov/substances/toxsubstance.asp?toxid=22. (accessed March 24, 2019).

ATSDR. 2019m. Mercury. Case Registry Number 7439-97-6. https://www.atsdr.cdc.gov/substances/toxsubstance.asp?toxid=24. (accessed March 24, 2019).

American Chemical Society. 2019. CAS Registry Numbers. https://www.cas.org/support/documentation/chemical-substances. (accessed March 24, 2019).

American Conference of Governmental Industrial Hygienists (ACGIH). 2019. Guide to Occupational Exposure Values. ACGIH. Cincinnati, OH.

Barbalace, R.C. 2009. The Chemistry of Polychlorinated Biphenyls. EnvironmentalChemistry.com. September 2003. http://www.environmentalchemistry.com/ypgi/chemistry/pcb.html. (accessed October 26, 2009).

Barnes, K.K., Kolpin, D.W., Furlong, E.T., Zaugg, S.D., Meyer, M.T., and Barber, L.B. 2008. A National Reconnaissance of Pharmaceuticals and Other Organic Wastewater Contaminants in the United States. *Journal of Science in the Total Environment*. Vol. 402. No. 2–3. pp. 192–200.

Bell, C.A., Gentile, M., Kalve, E., Ross, I., Horst, J., and Suthersan, S. 2019. *Emerging Contaminants*. CRC Press. Boca Raton, FL. 439p.

California Environmental Protection Agency (CalEPA). 2019. *1,2,3,-Trichloropropane (TCP)*. https://www.waterboards.ca.gov/drinking_water/certlic/drinkingwater/123TCP.html. (accessed March 25, 2019).

Calle, E.E, Frumkin, H., Henley, S.J, Savitz, D.A., and Thun, M.J. 2002. Organochlorines and Breast Cancer Risk, CA. *A Cancer Journal for Clinicians*. Vol. 52. p. 301.

Carson, R. 1962. *Silent Spring*. Houghton Mifflin. Boston, MA.

Carwile, J.L., Luu, H.T., Bassett, L.S., Driscoll, D.A., Yuan, C., Chang, J.Y., Ye, X., Calafat, A.M., and Michels, K.B. 2009. Use of Polycarbonate Bottles and Urinary Bisphenol A Concentrations. *Environmental Health Perspectives*, online May 12, 2009. doi:10.1289/ehp.0900604.

Doherty, R.E. 2000. A History of the Production and Use of Carbon Tetrachloride, Tetrachloroethylene, Trichloroethylene, and 1,1,1-Trichloroethane in the United States: Part 1 – Historical Background; Carbon Tetrachloride and Tetrachloroethylene. *Journal of Environmental Forensics*. Vol. 1. No. 2. pp. 69–81.

Fetzer, J.C. 2000. *The Chemistry and Analysis of Large Polycyclic Aromatic Hydrocarbons*. John Wiley & Sons. New York.

Holleman, A.F. and Wiberg, E. 2001. *Inorganic Chemistry*. Academic Press. San Diego, CA.

Jensen, W.B. 2009. The Origin of the Circle Symbol for Aromaticity. *Journal of Chemical Education*. Vol. 86. No. 4. pp. 423–425.

Karl, T.R. and Trenberth, K.E. 2003. Modern Global Climate Change. *Journal of Science*. Vol. 302. pp. 1719–1723.

Kathren, R. 1991. *Radioactivity and the Environment*. Taylor & Francis Publishers. Lieden, The Netherlands.

Kaufman, M.M., Rogers, D.T., and Murray, K.S. 2011. *Urban Watersheds*. CRC Press. Boca Raton, FL. 583p.

Krebs, R.E. 2006. *The History and Use of Earth's Chemical Elements: A Reference Guide*. Greenwood Publishing Group. Oxford, England.

Levy, A.B. 2009. Acetone. In: *Ullmann's Encyclopedia of Industrial Chemistry*, 7th Edition. John Wiley & Sons – VCH. New York.

Lide, D.R. 2008. Physical Constants of Organic Compounds. In: *Handbook of Physics and Chemistry*, 89th Edition. CRC Press. Boca Rotan, FL.

Madigan, M.T., Martinko, J.M., Dunlap, P.V., and Clark, D.P. 2008. 12th Edition. *Brock Biology of Microorganisms*. Prentice-Hall, Inc. New York.

McMurry, J.E. 2009. *Fundamentals of Organic Chemistry*. Brook Cole Publishing Company. New York.

Meyers, R. 2003. *The Basics of Chemistry*. Greenwood Press. Westport, CT.

Missouri Department of Natural Resources. 2006. *Missouri Risk-Based Corrective Action for Petroleum Storage Tank Sites Sampling for Polynuclear Aromatic Hydrocarbons*. Hazardous Waste Fact Sheet. Jefferson City, MO.

Murray, K.S., Rogers, D.T., and Kaufman, M.M. 2004. Heavy Metals in an Urban Watershed in Southeastern Michigan. *Journal of Environmental Quality*. Vol. 33. pp. 163–172.

National Oceanic and Atmospheric Administration (NOAA). 2018. Atmospheric CO_2 at Mauna Loa Observatory. http://www.esrl.noaa.gov/gmd/ccgg/trends. (accessed December 31, 2018).

Natural Resources Defense Council (NRDC). 2018. Ocean Pollution: The Dirty Facts. https://nrdc.org/stories/ocean-pollution-dirty-facts. (accessed November 18, 2018).

New York State Department of Environmental Conservation. 2010. Acid Deposition Large Graphic. http://www.dec.ny/chemical/41293.html (accessed April 14, 2019).

Phillips, D.J. 1986. *PCBs and the Environment*: Volume 2. CRC Press. Boca Raton, FL.

Pradyot, P. 2003. *Handbook of Inorganic Chemical Compounds*. McGraw Hill. New York.

Storrow, B. 2019. Emissions Growth in the United States and Asia Fueled by Record Carbon Levels in 2018. *Science*. https://www.sciencemag.org/news/2019/03/emissions-growth-united-states-asia-fueled-record-carbon-levels-2018. (accessed March 25, 2019).

Suthersan, S.S. and Payne, F.C. 2005. *In Situ Remediation Engineering*. CRC Press. Boca Raton. FL. 430p.

Union of Concerned Scientists. 2018. Each Country's Share of CO2 Emissions. https://www.ucsusa.org/global-warming/science-and-impacts/science/each-countrys-share-of-co2.html. (accessed March 24, 2019).

United Nations. (2018). Biodiversity: Invasive Species. United Nations System-Wide Earthwatch. https://www.un.org/earthwatch/biodiversity/invasivespecies.html. (accessed December 29, 2018).

United States Center for Disease Control (CDC). 2009. *Cryptosporidium Fact Sheets*. Department of Health and Human Services. CDC. Atlanta, GA.

United States Department of Agriculture (USDA). 2018. National Invasive Species Information Center (NISIC). https://www.invasivespeciesinto.gov. (accessed December 29, 2018).

USDA. 2019. *United States Fertilizer Use and Cost*. USDA. Washington, DC.

United States Department of Health and Human Services (USDHH). 1995. Toxicological Profile for Polycyclic Aromatic Hydrocarbons. Agency for Toxic Substances and Disease Registry. Atlanta, GA.

USDHH . 2005. Report on Carcinogens, 11th Edition. Public Health Service. National Toxicology Program. Washington, DC.

USDHH. 2008. NTP – CERHR Monograph on the Potential Human Reproductive and Developmental Effects of Bisphenol A. National Institute of Health Publication No. 08-594. Washington, DC.

USDHH. 2009. *Draft Toxicological Profile for Carbon Monoxide*. ATSDR. Washington, DC.

United States Department of Labor. 2019. Occupational Safety and Health Administration (OSHA) Permissible Exposure Limit for Carbon Monoxide. https://www.osha.gov/dsg/annotated-pels/tablez-1.html. (accessed March 25, 2019).

United States Energy Information Service. 2019. United States Consumption of Gasoline. https://www.eia.gov/tools/faqs/faq.php?id=23&t=10. (accessed June 5, 2019).

United States Environmental Protection Agency (USEPA). 1989a. Risk Assessment Guidance for Superfund. Volume I. Human Health Evaluation Manual (Part A). EPA/540/1-89/002. Office of Emergency and Remedial Response. Washington, DC.

USEPA. 1989b. Transport and Fate of Contaminants in the Subsurface. USEPA Center for Environmental Research Information. EPA/625/4-89/019. Cincinnati, OH.

United States Environmental Protection Agency (USEPA). 2005. Guidelines for Carcinogenic Risk Assessment. EPA/630/P-03/001F. USEPA. Washington, DC.

United States Environmental Protection Agency (USEPA). 2006. Pentachlorophenol Consumer Fact Sheet. Washington, DC.

USEPA. 2009. *E. Coli Bacteria. Total Coliform Rule-Basic Information*. USEPA. Washington, DC.

USEPA. 2010. Semi-volatile Organic Compounds. https://cfpub.epa.gov/si/si_public_record_report.cfm?Lab=NRMRL&dirEntryId=226943. (accessed April 20, 2019).

USEPA. 2013. Overview of Methyl Tertiary Butyl Ether (MTBE). https://archive.epa.gov/mtbe/web/html/faq.html. (accessed March 24, 2019).

USEPA. 2016. Chlordane. https://www.epa.gov/sites/production/files/2016-09/documents/chlordane.pdf. (accessed March 24, 2019).

USEPA. 2017. Overview of Greenhouse Gases. https://19january2017snapshot.epa.gov/ghgemissions/overview-greenhouse-gases_.html. (accessed March 25, 2019).

USEPA. 2018a. Global Greenhouse Gas Emissions Data. https://epa.gov/ghgemissions/global-greenhouse-gas-emissions-data. (accessed December 31, 2018).

USEPA. 2018b. Sulfur Dioxide (SO2) Pollution. https://www.epa.gov/so2-pollution. (accessed March 25, 2019).

USEPA. 2018c. Non-attainment Areas of the United States. https://archive.epa.gov/ozonedesignations/web/jpg/desfinal.jpg. (accessed March 24, 2019).

USEPA. 2018d. Invasive Species. https://www.epa.gov/greatlakes/invasive-species. (accessed December 29, 2018).

USEPA. 2018e. Basic Information on PFAS Chemicals. https://www.epa.gov/pfas/basic-information-pfas. (accessed December 30, 2018).

USEPA. 2019a. Risk Assessment: What Is it? https://www.epa.gov/risk_.html. (accessed March 24, 2019).

USEPA. 2019b. EPA's Report on the Environment (ROE). https://www.epa.gov/report-environment. (accessed March 24, 2019).

USEPA. 2019c. Integrated Risk Information System. https://www.epa.gov/iris. (accessed March 24, 2019).

USEPA. 2019d. National Primary Drinking Water Regulations: Trihalomethanes. https://www.epa.gov/ground-water-and-drinking-water/national-primary-drinking-water-regulations. (accessed March 24, 2019).

USEPA. 2019e. Polycyclic Aromatic Hydrocarbons (PAHs). https://www.epa.gov/sites/production/files/2014-03/documents/pahs_factsheet_cdc_2013.pdf. (accessed March 24, 2019).

USEPA. 2019f. Polychlorinated Biphenyls Fact Sheet: Basic Information. http://www.epa.gov/epawaste/hazard/tsd/pcbs/about.htm. (accessed March 24, 2019).

USEPA. 2019g. Risk Management for Phthalates. https://www.epa.gov/assessing-and-managing-chemicals-under-tsca/risk-management-phthalates. (accessed March 24, 2019).

USEPA. 2019h. Factsheet for Phenols. https://www.epa.gov/sites/production/files/2016-09/documents/phenol.pdf. (accessed March 24, 2019).

USEPA. 2019i. Factsheet for Triethylamine. https://www.epa.gov/sites/production/files/2016-09/documents/triethylamine.pdf. (accessed March 24, 2019).

USEPA. 2019j. Organic Esters of Phosphoric Acid. https://archive.epa.gov/pesticides/reregistration/web/html/index-198.html. (accessed March 24, 2019).

USEPA. 2019k. Pesticides. https://www.epa.gov/pesticides. (accessed March 24, 2019).

United States Environmental Protection Agency (USEPA). 2019l. DDT – A Brief History and Status. https://www.epa.gov/ingredients-used-pesticide-products/ddt-brief-history-and-status. (accessed March 24, 2109).

USEPA. 2019m. Dioxin. https://www.epa.gov/dioxin. (accessed March 24, 2019).

USEPA. 2019n. Agriculture Nutrient Management and Fertilizer. https://www.epa.gov/agriculture/agriculture-nutrient-management-and-fertilizer. (accessed March 25, 2019).

USEPA. 2019o. Asbestos. https://www.epa.gov/asbestos. (accessed March 25, 2019).

USEPA. 2019p. Ground Level Ozone. https://www.epa.gov/ground-level-ozone-pollution. (accessed March 24 2019).

USEPA. 2019q. Upper Atmosphere Ozone. https://www3.epa.gov/region1/eco/uep/ozone.html. (accessed March 24, 2019).

USEPA. 2019r. Particulate Matter (PM) Pollution. https://www.epa.gov/pm-pollution. (accessed March 25, 2019).

USEPA. 2019s. Contaminants of Emerging Concern. https://www.epa.gov/wqc/contaminants-emerging-concern-including-pharmaceuticals-and-personal-care-products. (accessed March 25, 2019).

USEPA. 2019t. Technical Fact Sheet on Perchlorate. https://www.epa.gov/fedfac/technical-fact-sheet-perchlorate. (accessed March 25, 2019).

United States Geological Survey (USGS). 2006. Volatile Organic Compounds in the Nation's Ground Water and Drinking-Water Supply Wells. USGS Circular 1292. Reston, VA.

United States Geological Survey (USGS). 2009a. Toxic Substance Hydrology Program. *DNAPL*. http://www.usgs.gov/definitions/dnapl_def.html. (accessed October 21, 2009).

USGS. 2009b. Toxic Substance Hydrology Program. *BTEX*. http://www.usgs.gov/definitions/btex.html. (accessed October 21, 2009).

United States Forest Service. 2019. *Bark Beetles and Climate Change in the United States*. Climate Change Resource Center. https://www.fs.usda.gov/ccrc/topics/bark-beetles-and-climate-change-united-states. (accessed May 12, 2019).

Van der Perk, M. 2006. *Soil and Water Contamination: From Molecular to Catchment Scale*. Balkema: Proceedings and Monographs in Engineering, Water and Earth Sciences. Taylor & Francis. London, UK.

World Health Organization (WHO). 2003. *Polychlorinated Biphenyls: Human Health Aspects*. WHO. Geneva, Switzerland.

3 The Behavior of Pollution in Nature

3.1 INTRODUCTION

We now know what pollution is, how and where it originates, and what it can do to our bodies if we are exposed to most forms of pollution. We know that pollution only presents a risk to human health and the environment if there is a completed exposure pathway. So, where do the different types of pollution end up when released into the environment and how do humans and the environment become exposed to pollution?

The answer to these questions is obtained through an understanding of the process called **contaminant fate and transport,** defined as the sequence of anthropogenic and natural events involving a pollution source, its mobilization or transport, and its ultimate fate or resting place, termed a **sink** (Rogers et al. 2007). Sources of pollution include many human activities performed primarily at the surface resulting in the release of toxic substances into the environment. These toxic substances may be transported over time, or remain relatively close to their source before they are degraded, transformed, or destroyed. Because: (1) pollutants released into the environment are often mixtures; (2) each contaminant is unique chemically; and (3) the geologic environment in which the pollutants are released is also unique, understanding the physical chemistry of each, the microbiologic environment, and the geologic environment into which they are released becomes necessary for characterizing their fate and transport (Rogers et al. 2007).

During their transport and before reaching their final sink, certain pollutants may reside at multiple intermediate sinks for different periods of time. Intermediate sinks include surface water, groundwater, and the atmosphere, because the contaminants held within these water-containing sinks will flow and ultimately reach the oceans. Aquifers with a very low hydraulic conductivity are for practical purposes considered final sinks, as are inland bogs and some wetlands. Sediment and soil can function as intermediate or as final sinks, since erosion may move both of these unconsolidated materials. The oceans are almost always a final sink of pollution, but wave action occasionally brings pollutants onshore. To a large extent, the level of human health and/or environmental risk is a function of two fundamental concepts introduced in this chapter: *mobility* and *persistence*. **Mobility** is a measure of a substance's potential to migrate. **Persistence** is a measure of a substance's ability to remain in the environment before being degraded, transformed, or destroyed (Rogers et al. 2007). In Chapter 2, we noted substances with higher toxicity pose greater health risks, but the risk to humans and the environment grows exponentially if the chemical is both mobile and persistent. For example, a highly toxic but immobile chemical may affect a few people in a warehouse through inhalation, whereas a mobile and persistent chemical of moderate toxicity can contaminate a public water supply, or migrate to a different sink where the potential for widespread human exposures and ecosystem damages is much higher.

The following sections discuss how and where contaminant releases occur, how their migration through the soil, groundwater, and atmosphere proceeds, and how they end up in sinks. We conclude with a brief description of the fate and transport for each pollutant group.

3.2 RELEASES OF POLLUTION INTO THE ENVIRONMENT

The fate and transport of pollution begins with their release into the environment. Pollutant releases originate from numerous sources and under different circumstances; they vary in degree, concentration, duration, mass, volume, and whether a single contaminant is released or if a mixture of contaminants are released. Each of these factors influences their fate and transport. For instance, some releases may be very small and avoid detection. On the other hand, some releases are so large and sudden (e.g., a tanker spill) they become the leading story of the next newscast. Sudden and large releases increase the probability of severe environmental impairment or destruction, especially if they occur in or near a sensitive ecological area.

Historically, there was no regulation of the disposal of wastes potentially containing pollutants. Until the mid-20th century, the most convenient and least costly method of waste disposal was "up the stack or down the river" (Haynes 1954). We now know the perception of a pollutant leaving your immediate vicinity and being gone forever is not true. Today, many releases of pollutants are permitted but carefully monitored, such as wastewater discharges from industrial and municipal sources and those to the atmosphere from industrial and commercial sources and power plants. The legislation governing such releases (the Clean Water Act of 1972 and the Clean Air Act of 1990) permits a point source (such as a wastewater plant or a smokestack) to release specific compounds at low concentrations to ensure there are no adverse health and environmental effects detectable on the media they are released into (water or air). Some releases, as we shall discover and discuss in detail, are not regulated and include the majority of household wastes and releases of pollution from many agricultural sources. A majority of the other pollutant releases into the environment are unintentional or accidental and occur from numerous sources under a multitude of circumstances, and some of these include (Fetter 1993; Rogers 1996; USGS 2006a):

- Permitted releases of contaminants in wastewater and air emissions
- Spills and leaks from several types of containers or operations, including:
  - Drums of various sizes and shapes
  - Pipelines
  - Above ground storage tanks
  - Underground storage tanks
  - Tanker trucks and other transport vehicles
  - Tanker ships
  - Railroad tanker cars
  - Aircraft
- Accidents or collisions involving automobiles where gasoline and oil may be released, railroad derailments, etc.
- Septic systems
- Sewer leaks
- Landfills and dumps
- Automobile exhaust
- Sumps and dry wells
- Former disposal lagoons
- Animal feedlots
- Pesticide and herbicide application
- Fertilizer application
- Injection wells
- Application of pesticides, herbicides, and fertilizers
- Application of deicing compounds (i.e., road salts)

Exploring and evaluating the fate and transport of pollutants in the environment is a complex interplay between the environment and the contaminant. And, as Figure 3.1 shows, fate and transport are also influenced by the method and location of contaminant release, and the volume, mass, and duration of the release. We will break down and describe these factors influencing the migration of contaminants, degradation of contaminants, and then describe how the fate and transport of each group of substances are influenced by the group's specific chemistry.

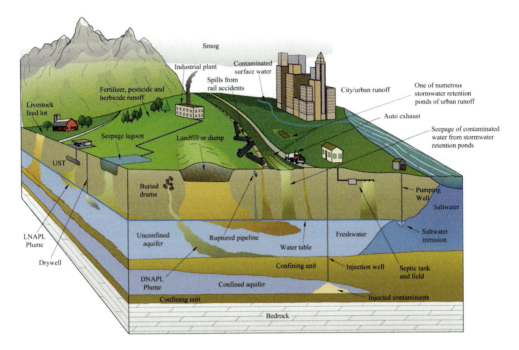

FIGURE 3.1   Sources and locations of many contaminant releases.

3.3   PRINCIPLES OF POLLUTANT FATE AND TRANSPORT

Interaction with the environment begins immediately after a pollutant has been released. Once released, a pollutant can do three things (Hemond and Fechner-Levy 2000):

- Stay put
- Migrate in soil, water, air, or a combination of media
- Degrade, transform, or get destroyed

The factors controlling these three outcomes depend upon the physical chemistry of the contaminant and the characteristics of the receiving environment (USGS 2006a; Rogers et al. 2007). For example, some pollutants may change physical states. An example is many of the more than 250 chemical compounds that comprise gasoline that begin to immediately evaporate after exposure to the atmosphere. Other pollutants may degrade in a manner of minutes after being released, while some may last for thousands of years or sometimes longer, as with certain radioactive compounds. Therefore, computing a mass balance should be the first action when assessing any particular release, as this will ensure the mass or volume of contaminant released is accurately measured (Equation 3.1). Conducting a mass balance also serves to validate or refute our current understanding of the behavior of the different environmental elements influencing the pollutant once it has been released.

A simple mass balance is expressed as:

$$\text{Amount released} = \text{amount recovered}$$
$$+ \text{amount lost to the environment} \qquad (3.1)$$
$$\times \left(\text{air} + \text{water} + \text{soil} \right)$$

Once in the environment, pollutants can and often move between soil, surface water, groundwater, and the atmosphere, and they can also degrade. Many pollutants degrade quickly if conditions are favorable, yet others can persist and last for years or decades if conditions are favorable (USGS 2006a). Factors degrading contaminants fall into two broad categories: biotic degradation and abiotic degradation. **Biotic** degradation involves microorganisms or fungi, and occurs when an organism, such as a bacterium, uses a contaminant as a source of food and either degrades or transforms the contaminant (USGS 2006a). **Abiotic** degradation involves other processes not including microorganisms. Examples of abiotic degradation include photolysis – degradation as a result of exposure to sunlight (USGS 2006a) – and hydrolysis. Hydrolysis involves cleaving a molecule into two parts by the addition of a molecule of water.

When examining pollutant degradation in the environment, it is important not to confuse dilution with degradation. If given enough time, contaminants may become diluted, and this process results in a decrease of the pollutant concentrations per unit volume of the media being measured. However, dilution is not degradation, since it does not involve a chemical transformation of the pollutant. We will discuss degradation in more detail in Section 3.3.2.

3.3.1 BASIC POLLUTANT TRANSPORT CONCEPTS

Pollutant transport in the environment is dominated by three physical transport mechanisms (USEPA 1996a; USGS 2006a): advection/convection, molecular diffusion, and dispersion. **Advection** is the horizontal transport of any property of the atmosphere and water. Common examples are the transfer of heat by wind and sediment transport within a flowing stream. **Convection** is the vertical advection of air, water, or other fluid as a result of thermal differences. We introduced the concept of convection in Chapter 2 as the driving force behind plate tectonics. **Molecular diffusion** is the movement of a chemical from an area of higher concentration to an area of low concentration due to the random motion of the chemical molecules. **Dispersion** (also referred to as hydrodynamic dispersion) is the tendency for pollutants to spread out from the path of the expected advective flow (USGS 2006a). Occasionally, the effects of diffusion and dispersion are treated together, but for the purposes of this book we treat them separately.

The rate of advective transport of a pollutant is often expressed in terms of flux density. **Flux density** is the mass of a chemical transported across an imaginary surface of a unit area per unit of time. Equation 3.2 shows this relationship (Hemond and Fechner-Levy 2000), which is independent of the media involved (soil, surface water, groundwater, or the atmosphere).

$$J = CV \qquad (3.2)$$

Where:

J = flux density = (Mass/[Length × width] × Time) or $[M/L^2 \times T]$
C = concentration of the chemical per cubic liter or meter of media $[M/L^3]$
V = velocity [Length/Time] or [L/T]

An example of molecular diffusion is shown in Figure 3.2 (Payne et al. 2008). From the release time to infinity, a pollutant released into a fluid such as air or water will diffuse throughout the fluid at random locations.

Fick's First Law of diffusion (Equation 3.3) can be used to predict the diffusive flux of a pollutant (solute) across an imaginary plane as a function of the rate of change in concentration with distance (Hemond and Fechner-Levy 2000).

$$J = -D(dC/dx) \quad (\text{one dimension}) \qquad (3.3)$$

Where:

J = the flux density $[M/L^2 \times T]$
D = the Fickian mass transport coefficient $[L^2/T]$
C = chemical concentration $[M/L^3]$
X = is the distance over which a concentration change is being considered [L]

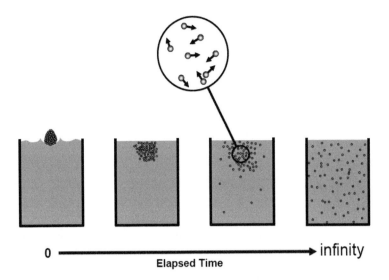

0 ——————————————————————→ infinity

Elapsed Time

FIGURE 3.2   Molecular diffusion. (From Payne, F.C., Quinnan, J.A. and Potter, S.T. Remediation Hydraulics. CRC Press. Boca Raton, FL. 2008.)

Note: In simple calculations the minus sign is often omitted if the direction of Fickian transport is clear.

An example of the effects of subsurface dispersion is presented in Figure 3.3 (Hemond and Fechner-Levy 2000).

General factors that influence or control the rate of migration of a contaminant include (Hornsby 1990):

- Physical and chemical properties of the contaminants themselves
- The geological environment where the release occurs
- Climatological factors
- Vegetation factors

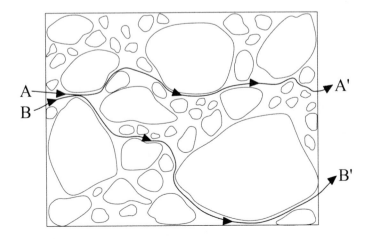

FIGURE 3.3   Effects of dispersion on subsurface migration in a porous media.

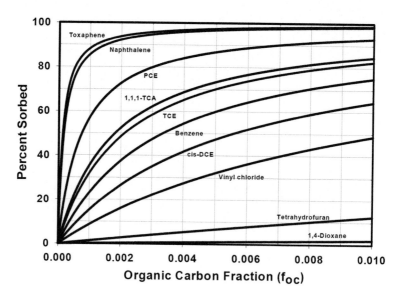

FIGURE 3.4 Sorbed percent of select VOC compounds and its relation to increasing amounts of organic carbon within geologic media. (Adapted from Payne, F.C. et al. *Remediation Hydraulics*, CRC Press. Boca Raton, FL. 2008. With permission.)

Specific physical properties affecting migration of contaminants in soil include (USGS 2006a; Rogers et al. 2007; Payne et al 2008):

- Solubility in water. The more soluble a contaminant is in water, the more mobile it will be in the subsurface environment.
- Vapor pressure. As vapor pressure increases, affinity to volatilize increases and the more likely a contaminant will be present in the gas phase.
- Molecular weight. The higher the molecular weight, the greater the energy required to transport the contaminant in the horizontal direction. The increased molecular weight may induce the contaminant to migrate downward in areas of steep slopes (see Figure 3.4).
- Chemical stability and persistence. Stable compounds have more time to migrate and may migrate further if conditions are favorable before they are degraded, transformed, or destroyed.
- Sorptive properties. Sorptive processes include adsorption and absorption. The lower the sorptive properties, the higher the migration potential. This relationship is shown in Figure 3.4, which graphs the percent sorbed (nonaqueous) of select volatile organic compounds (VOCs) in relation to organic carbon content within the subsurface. For this relationship to exist, the number of sorption sites within the geological materials has not been exceeded (Payne et al. 2008).

Specific geologic factors affecting the migration of pollutants in soil include (Rogers 1996; USGS 2006a; Rogers et al. 2007; 2012):

- Composition. Soil type has a significant influence on the migration of contaminants. In general, coarse-grained soils, such as sand and gravel, do not impede the migration of contaminants nearly as much as soils composed of clay.
- Porosity and permeability. Direct evidence of migration potential is hydraulic conductivity. Soils composed of sand and gravel have a hydraulic conductivity hundreds to thousands of times higher than soils composed of clay. Although soils or sediments composed of clay may have a high relative porosity, they are usually not as permeable because the porosity is generally not interconnected.

In some cases, however, as demonstrated in Chapter 3, secondary porosity such as root fragments and vertical fractures can make a seemingly impervious clay deposit much more permeable than expected.

- Organic carbon content. As just discussed, an increased total organic content in soil may impede the migration of certain types of contaminants as long as the contaminant mass does not exceed the holding capacity of the soil. Soils or sediments composed of sand and gravel generally have lower organic carbon content compared to soils composed of clay.
- Soil chemistry. The pH, redox potential, and other soil chemistry factors influence the contaminant migration of many different types of compounds. Many metals, for instance, are particularly sensitive to pH differences in soil, and these differences – along with the characteristics of each metal – influence their migration patterns in the environment.
- Stratigraphy. This is where heterogeneity and the anisotropic nature of the geologic sediments play a significant role at both a micro and macro scale. The geology beneath the surface can and does change dramatically in just a few feet in any direction. The result is differing sediment types and chemical composition, including pH and redox potential, of subsurface layers acting to impede or enhance contaminant migration.
- Unconformities. The presence of unconformities influences the migration of groundwater as well as the migration of contaminants. Hydrogeologically, the presence of an unconformity indicates there is a surface or plane in the subsurface geologic environment. This surface or plane often produces a significant difference in the hydraulic conductivities within the soils or sediments above and below the unconformity, especially if these units are fine-grained sediments such as silts or clays. As a result, contaminants released in this type of location use the unconformity as a sink and migrate much further than expected.

Specific climatological factors affecting migration of pollutants in soil include (Freeze and Cherry 1978; Kaufman et al. 2011):

- Freeze-thaw cycles. Freeze-thaw cycles can lead to the development of vertical fractures in the soil to depths approaching 10 meters. These vertical fractures are a type of secondary porosity, and if present, can greatly increase the migration potential of contaminants vertically through the soil column. In addition, the freezing of near-surface soils may trap contaminants at the surface and lead to increased contaminant loading during warmer periods when the ice melts.
- Liquid precipitation. Because water is the universal solvent, geographic locations receiving abundant rainfall play a significant role in enhancing the migration of contaminants, especially if they are soluble. Rainfall also enhances the migration of contamination through the physical transport of particles with sorbed contamination on their surfaces.
- Snowfall. Airborne deposition of contaminants may become temporarily trapped in seasonal snowpack and released when the snowpack melts. Increased contaminant loading to the environment may occur during warmer periods when the snowpack melts (Wania et al. 1998).
- Wind. Many locations within the United States contain significant amounts of wind-blown deposits, especially in the southwest. Contaminants with a high sorption potential may become attached to fine wind-blown sediment grains and transported over long distances. In addition, volatile contaminants released as a gas are routinely transported by wind.
- Water vapor. Humidity plays a significant role in the water cycle by affecting air movement, reacting with contaminants in the gas phase and contaminants released into the atmosphere attached or sorbed onto particulate matter. Contamination goes along for the ride when precipitation formed around contaminated condensation nuclei is transported from the atmosphere to the lithosphere and then entrained by surface runoff.
- Fog. Fog can be an effective agent for the transfer of acid rain by transferring it to vegetation or other surface materials through direct contact.

- Flood events and hurricanes. Due to their catastrophic nature and magnitudes, floods and hurricanes may not only increase contaminant migration, they can also cause significant releases. During the 1993 flood of the Mississippi River, numerous barrels containing hazardous waste were swept away and deposited in the Gulf of Mexico.
- Solar energy. Sunlight breaks down some contaminants through a process called photolysis. In toxic microorganisms, ultraviolet light passes easily through cell walls, cytoplasm, and nuclear membranes and prevents DNA replication.

Specific vegetative factors that affect migration of pollutants in soil include (Nudunuri et al. 1998; Kaufman et al. 2011):

- Roots. These pathways within shallow subsurface geological materials can enhance the migration of contaminants.
- Certain plants can uptake contaminants. Many types of plants and trees have the capability with their root systems to assist in the removal of contaminants from shallow subsurface soil. Contaminants may be stored in plant tissues or are transformed through biologic processes.
- Microorganisms. Microorganisms in the soil often biodegrade many different types of contaminants by using the contaminants themselves as a source of food.

3.3.2 BASIC POLLUTANT DEGRADATION CONCEPTS

The degradation of specific compounds in the environment is expressed in terms of their half-life. **Half-life** is the average amount of time required to degrade half or 50% of a specific pollutant population (USEPA 1998a; 1998b). Contaminants degrade through biotic or abiotic processes, and the processes controlling their rate of decay depends upon the following factors (USEPA 1996a; 1996b; USGS 2006a):

- The nature of the release. This group of factors includes:
  - Media receiving the release. Was it into the atmosphere, surface or subsurface soil, surface water, ocean, directly to groundwater, or a combination of media?
  - Amount (volume or mass) of the release
  - Number of contaminants released. Was it a single contaminant – or a mixture of several contaminants?
  - Physical state of the release (i.e., liquid, solid, or gas)
  - Time duration of the release
- Geologic environment. The most significant geologic factors include:
  - Soil composition and other physical characteristics such as permeability, porosity, moisture content, composition, extent and distribution, thickness, total organic carbon content, pH, redox potential, dissolved oxygen (if saturated), and other parameters
  - Depth to bedrock, type, composition, distribution, fractures, permeability, porosity of bedrock, and other parameters
  - Terrain and topography
  - Potential surface and subsurface migration pathways
- Climatic factors
  - Release location. Different climates – deserts, mountains, humid areas, or temperate regions can influence the type and rate of degradation.
- Surface water features
  - Distance to surface water bodies and their type. Immature streams, mature streams, rivers, lakes, wetlands, and bogs can differ in pH due to the rock composition of their channels and bottoms and the amount of organic matter they receive from outside; inputs termed allochthonous.

- Weather conditions at the time of the release. Weather conditions are often important and sometimes overlooked as potentially significant. Those conditions affecting degradation and migration include:
  - Temperature
  - Humidity
  - Precipitation
  - Wind speed and direction
- Biologic factors
  - The type, distribution, and amount of microorganisms will influence the rate and can determine if degradation even occurs
- Anthropogenic factors. Anthropogenic factors are often overlooked and frequently significant. These include:
  - Physical landscape alteration (i.e., buildings, roads, parking lots, etc.)
  - Surface water drainage modifications including stormwater control and wetland destruction
  - Alteration of native vegetation
  - Introduction of invasive vegetation
  - Regional contaminant loading, including sources, duration, type, release points, and physical state of contaminants (i.e., solid, liquid, or gas)
  - Developmental history of the area and region

3.3.2.1 Biotic Degradation

Microbes have the ability to oxidize a variety of organic contaminants including many VOCs, PAHs, and other compounds. This capability arises from their enormous variety, populations, rapid growth, and diversity of environmental niches. Soluble organic compounds with low molecular weights such as alcohols and organic acids are metabolized and degraded rapidly by microbes, perhaps because these compounds also occur naturally and microbes have evolved to degrade them more efficiently (Hemond and Fechner-Levy 2000). Halogenated synthetic or anthropogenic compounds, however, are not easily degraded by microbes (USEPA 1998b), and some contaminants that escape biotic degradation bioaccumulate in the bodies of organisms. The rate of biodegradation by microorganisms generally slows if the organic contaminants possess the following (Hemond and Fechner-Levy 2000):

- Increasing molecular weight
- Decreasing water solubility
- Presence of benzene or aromatic rings
- A large amount of branching within the molecular structure
- Presence of halogen atoms in the structure (chlorine, fluorine, bromine, or iodine)

3.3.2.2 Abiotic Degradation

Abiotic degradation applies to degradation processes accomplished without microorganisms. Common abiotic degradation processes include:

- Photolysis
- Hydrolysis
- Reduction - oxidation
- Radioactive decay

Photolysis (sometimes referred to as photodegradation or photochemical degradation), occurs in the presence of sunlight. Common examples of this process include the fading of colored and dyed objects and the transformation of plastic object textures from pliable to brittle. Photolysis is most common in the atmosphere, surface water, and at the Earth's surface. The degree of degradation caused by sunlight depends on the wavelength spectrum of the light, the intensity of light exposure, and duration (Hemond and Fechner-Levy 2000). If the energy per photon is sufficient to break a specific chemical bond photolysis can be initiated. Once begun, increased light intensity will result in a faster rate of degradation. Failure to achieve the energy level required to break a bond means degradation will not occur. Ultraviolet light is especially effective at degrading many organic contaminants (USEPA 1996a).

Degradation by photolysis occurs within compounds capable of absorbing light energy and is often observed in organic compounds with double bonds between their carbon atoms. Many VOCs, polycyclic aromatic hydrocarbons (PAHs), and semi-volatile organic compound (SVOCs) characterized by a benzene ring fit this pattern. Under favorable conditions, the degradation of many organic compounds by photolysis may occur in a short period of time – from a few hours to a few days (Lyman et al. 1990).

The process of **hydrolysis** occurs when a water molecule breaks. Contaminant degradation by hydrolysis also destroys a molecule of contaminant. Two types of chemical compounds are susceptible to degradation by hydrolysis (Schwartzenbach et al. 1993):

- Alkyl halides, straight-chained or branch-chained hydrocarbons where one or more hydrogen atoms have been replaced by a chlorine, fluorine, bromine, or iodine atom. Using an "X" to represent a halogen atom and an "R" to represent the hydrocarbon group, the basic hydrolysis reaction is shown in Equation 3.4:

$$H_2O + R - X \rightarrow R - OH + H^+ + X^-$$
(3.4)

- Esters, compounds containing a modified carboxylic acid group (– COOH), where the acid hydrogen atom has been replaced by a different organic functional group. The process of hydrolysis within this group converts the ester compound into the "parent" organic acid and alcohol (Equation 3.5):

$$H_2O + CH_3COOC_2H_5 \rightarrow CH_3COOH + C_2H_5OH$$
(3.5)

In reduction–oxidation degradation reactions (redox), electrons are transferred from one atom to another. **Chemical reduction** is defined as the addition of electrons and **chemical oxidation** is defined as the loss of electrons (Hemond and Fechner-Levy 2000). In a reaction involving Atoms A and B, if Atom A gains an electron, it is reduced, and Atom B, having donated an electron, is the **reductant**. Because Atom B loses an electron, B is oxidized and Atom A is the **oxidant**. Each reaction involving the loss or gain of an electron is termed a **half-reaction**. The oxidation of contaminants can occur very rapidly through combustion or incineration. Here, fire transforms the contaminants through oxidation at greatly elevated temperatures and uses the cooking, heating, and transportation applications (Hemond and Fechner-Levy 2000).

3.3.3 Transport and Fate of Pollutants in Soil

Folklore holds that the presence of soil protects groundwater quality by filtering contaminants before they reach and impact groundwater (Hornsby 1990). Soil does have a limited ability to filter contamination; however, it does not do a perfect job of holding, filtering, degrading, transforming, or destroying contaminants. These capabilities also depend upon a number of factors related to the chemistry of the contaminant and the geological environment where the contaminant is released.

Soil is defined as the unconsolidated mineral matter on the immediate surface of the Earth (Soil Science Society of America 1987). Basic to an understanding of soil are the factors affecting its development and ultimate physical structure. The composition, texture, and thickness of soil are influenced by its source material, plant growth, micro- and macro-organisms, climate, topography, the process of formation (e.g., alluvial, fluvial, and glacial), and physical and chemical weathering since original formation (Brady and Well 1999). Structurally, soil is composed of three phases: soil gases, soil water, and organic and inorganic solids (Sawhney and Brown 1989). The gas and water phases may comprise 25–50% of the total volume of surface soil, especially at shallow depths (USEPA 1999).

Pollutants released into the soil can migrate within all three phases. Once a pollutant is resident in soil, these factors determine its migration rate (Schnoor 1996; USEPA 1999; Kaufman et al. 2011):

- Contaminant mass released
- Duration of the release
- Physical chemistry of the contaminant
- Physical chemistry of the soil (e.g., pH, redox potential and mineralogy)
- Amount of water present
- Permeability of the soil
- Retention capacity of the soil
- Distribution of plant matter
- Biological interaction between the contaminant and indigenous microorganisms

All of these factors must be well understood before an accurate assessment of the fate and transport of a contaminant in soil can be made.

Contaminants migrate through the soil by two basic processes: diffusion and mass flow. The rates of diffusion and mass flow greatly depend upon the local geology and the physical chemistry of the contaminant. Diffusion of substances through soil and aquifer materials occurs in response to differences in energy from one point to another. These energy gradients may be caused by differences in temperature or chemical concentrations within the contaminated area. In most cases, however, the principal process moving a contaminant through soil is mass flow or advection, because contaminants generally want to move downward through the soil under the force of gravity (USGS 2006a). Contaminant-specific physical and chemical attributes affecting the migration of contaminants in soil include (USEPA 1996a; Wiedemeier et al. 1999):

- Solubility
- Vapor pressure
- Density
- Chemical stability
- Persistence
- Adsorption potential

In soil, solid-phase contaminants migrate much more slowly than liquid phase contaminants, and tend to remain relatively close to their point of release or deposition (USEPA 1999). Before they can migrate a significant distance, solid-phase contaminants must change phase or undergo a transformation process. For example, heavy metals – a solid-phase contaminant – typically remain at their point of release or deposition. If they undergo oxidation, however, their solubility and other properties enabling migration may increase (Lindsay 1979). And once a contaminant begins to dissolve in water, it may also be subject to further transformation reactions induced by indigenous bacteria present in the surface soil (Sutherson and Payne 2005).

If a source continues to emit contaminant in solid or liquid form that dissolves in water, the underlying soil will eventually become saturated. The leading edge of contamination will migrate either horizontally or vertically or both as long as the retention capacity of the soil is exceeded (USEPA 1999). When the contaminant release stops, the migration of the liquid contaminant will significantly decrease as the soil regains its retention capacity (Kaufman et al. 2011). Since soil is also composed of gas, contaminant migration through the vapor phase is often observed with contaminants having higher relative vapor pressures. VOCs are frequent participants in this type of migration. Capillary forces can also induce the migration of liquid phase contaminants.

3.3.4 Transport and Fate of Pollutants in Surface Water

The transport of contaminants in surface water is dominated by turbulent advective flow. Because the rate of flow in a river or stream varies significantly by location and over time, estimating the contaminant flow involves averaging the streamflow variations and contaminant concentrations over a specified time interval. Conducting measurements at multiple locations also provides a more accurate measure of the rates of streamflow and contaminant transport.

Molecular diffusion also influences contaminant migration in surface water. Turbulent flow is characterized by water moving in with constantly changing and unpredictable patterns. The swirls resulting from turbulent flow are called **eddies,** and they appear in many sizes, volumes, and velocities. Random mixing of the water within eddies creates turbulent diffusion and also influences mass transport. Wave action can create similar eddying effects in lakes and other non-flowing water bodies. Analytically, the transport rates for chemicals in surface water are expressed in terms of flux density. Flux density is the mass of a chemical transported across an imaginary surface of a unit area per unit of time (Equation 3.2) (Hemond and Fechner-Levy 2000). Fick's First Law (see Equation 3.3) is also used to describe the flux density of mass transport by turbulent dispersion (Hemond and Fechner-Levy 2000).

Figure 3.5 depicts a municipal or industrial wastewater plant discharging to surface water and many of the ensuing contaminant fate and transport processes, including (Kaufman et al. 2011): Transport of discharged wastewater solute downstream

- Mixing due to turbulent advection and turbulent diffusion
- Photolysis
- Hydrolysis
- Biodegradation
- Transformation
- Volatilization to the atmosphere
- Sorption of contaminants in sediment
- Bioaccumulation of contaminants by animal and plant life
- Dilution

Molecular diffusion and groundwater discharge and recharge are not shown in the figure.

Since most urban areas of the United States obtain their potable water from surface sources, wastewater discharges are a concern. Treatment costs rise when the source of supply is contaminated, and the risks of biological contamination also increase. Moreover, a majority of urban areas are located along rivers and streams and other surface water bodies, and many of these water bodies have already been degraded – some significantly (USGS 1995a; 1995b; Kaufman et al. 2011).

When certain contaminants are released to surface waters through overland flow, stormwater runoff, or wastewater discharge, they may accumulate in sediments. Compounds with a higher likelihood of accumulating have the following physical characteristics:

- Low solubility
- High molecular weight
- Low potential to degrade
- High sorption potential

Contaminants having these physical chemistry attributes typically do not sustain themselves in surface water unless the rate of flow is substantial. Even then, they may be carried along the bottom of

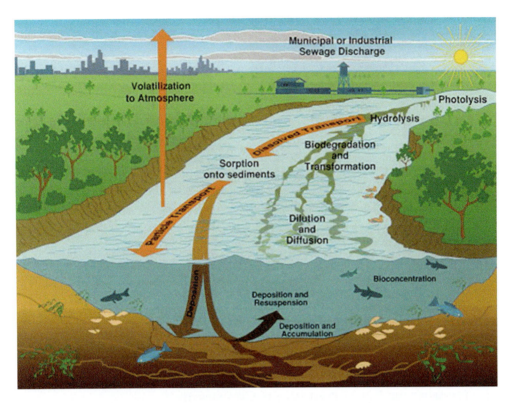

FIGURE 3.5 Fate and transport effects in surface water. (From USGS, Contaminants in the Mississippi River, USGS Circular 1133, Washington, DC, 1995b.)

the stream or river until the carrying capacity of surface water is insufficient and the contaminants settle to the bottom. As shown in Figure 3.6, certain locations in the stream bottom become a sink for these contaminants as they accumulate (Kaufman et al. 2011). If the source of contamination persists, greater amounts of the contaminant will be deposited. The accumulation of contaminants in sediments increases the exposure risk to aquatic and terrestrial plant and animal life. If any of the contaminants exhibit bioaccumulation properties, they may proceed up the food chain from bottom-dwelling macro-invertebrates to small fish and eventually to larger fish, predatory birds, and other organisms. Humans are situated at the top of the food chain, and the potential risks to human health must be considered when evaluating the fate and transport of contaminants in surface water (USEPA 2019a).

Contaminants considered to be bioaccumulative include (USEPA 2019a):

1. Mercury
2. PCBs
3. Chlordane
4. Dioxins
5. DDT

The number of surface water bodies under advisory include (See Figure 3.6) (USEPA 2019a):

- 44% of the nation's total lake acres (excluding the Great Lakes), representing approximately 18 million acres of surface water
- 35% of the nation's total river miles, or approximately 1.3 million miles
- 45% of the nation's contiguous coastal waters
- 100% of the Great Lakes and their connecting waters

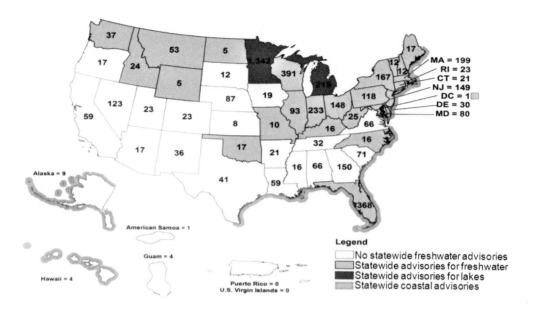

FIGURE 3.6 Health advisories for bioaccumulative contaminants in surface water by state. (From USEPA. Advisories and Technical Resources for Fish and Shellfish Consumption. 2019b. https://www.epa.gov/fish-tech.) (Accessed March 25, 2019.)

3.3.5 Transport and Fate of Pollutants in Groundwater

Transport of contaminants in groundwater is dominated by three factors: advection, dispersion, and molecular diffusion. When applied to groundwater, **advection** is the movement of contaminants by the bulk motion of groundwater flow, dispersion is the tendency for contaminants to spread out from the path of the expected advective flow, and diffusion is the action of spreading of molecules from areas of high concentration to areas of low concentration at the molecular level. Groundwater flow lacks turbulent diffusion because velocities are typically much slower. In some instances, however, groundwater does display turbulent dispersion, especially in karst topography where water flowing beneath the surface flows and behaves much like a stream at the surface. Figure 3.7 shows a spill from an underground storage tank (USGS 2006a). Here, advective transport of contaminants in groundwater is occurring at the water table boundary. Diffusion, biodegradation, volatilization, and recharge from surface precipitation affecting the contaminant migration are also shown. The effects of dispersion and diffusion are represented by the spreading of the contaminant plume as it migrates from a hole or rupture at the bottom of the tank (USGS 1998).

The representation of dispersion in Figure 3.8 is overly simplistic, because the geology of unconsolidated sediments is very complex and typically displays a high degree of heterogeneity and anisotropic distribution patterns. As a result, contaminants migrating in unconsolidated deposits do not migrate uniformly, but migrate within the physical parameters of advection dictated by the particular subsurface geology. This concept is represented in Figure 3.9, where a contaminant (solute) is shown migrating in the more permeable layers. More highly permeable layers have a higher hydraulic conductivity and behave as preferred groundwater and contaminant migration pathways. These layers are essentially super highways for groundwater and contaminant transport, and in some instances the hydraulic conductivity is from 100 to sometimes 1,000 times greater over distances of just a few centimeters.

Zones with higher permeability move water more quickly and have a higher flux density. If more water moves through these higher permeability zones, then a potentially greater contaminant mass also moves through. Figure 3.8 shows this relationship at the right of the diagram, downgradient from the source. As contamination continues to migrate along these flow paths of higher

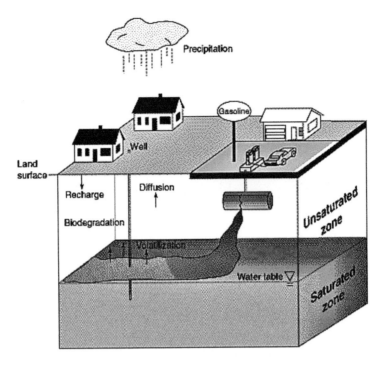

FIGURE 3.7 Advective transport and other processes effecting the migration of contaminants in groundwater. (From USGS, Simulating transport of volatile organic compounds in the unsaturated zone using the computer model R-UNSAT, USGS Fact Sheet 019-08, Washington, DC, 1998.)

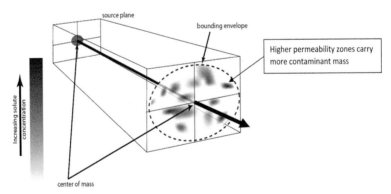

FIGURE 3.8 Effects of diffusion over time within a contaminant plume. (Adapted from Payne, F.C. et al. *Remediation Hydraulics*, CRC Press, Boca Raton, FL. 2008. With permission.)

permeability dictated by the subsurface geology, diffusion of contaminants into less permeable zones occurs (Figure 3.8). The top portion of the figure represents the flow paths of contaminants in the early stages of migration, and the later stages of migration are shown in the figure's bottom portion. Over time, the contaminant (solute) has diffused into the less permeable, lower hydraulic conductivity geologic materials (Payne et al. 2008).

The transport of contaminants in groundwater is also influenced by many of the same factors affecting the migration of contaminants in unsaturated soil or the vadose zone:

- Physical chemistry of the contaminants:
  - Solubility
  - Molecular weight

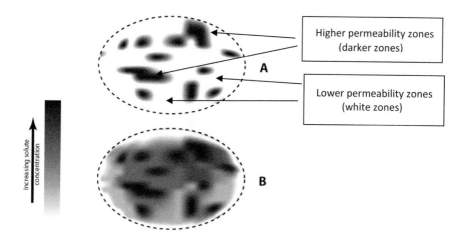

FIGURE 3.9 Contaminant migration in heterogeneous and anisotropic geologic media. (Adapted from Payne, F.C. et al. *Remediation Hydraulics*, CRC Press, Boca Raton, FL. 2008. With permission.)

- • Vapor pressure
- • Stability and persistence
- Sorption potential
- Type, distribution, and amount of microorganisms
- Tendency to biodegrade
- Dissolved oxygen content of groundwater
- Geological factors:
  - Stratigraphy (including thickness and distribution of geological units down to micro-stratigraphic scales at the centimeter or even millimeter)
  - Presence of unconformities
  - Soil or sediment chemistry
  - Organic carbon content
  - Porosity and permeability
  - Composition
- Climate factors:
  - Freeze and thaw cycles
  - Recharge from surface precipitation
  - Flood events
  - Seasonal climatic variations
- Vegetative factors including types and distribution of surface vegetation, root networks, and water requirements

Sorption potential has a significant effect on the migration of contaminants in groundwater because it slows the migration of contaminants even as the flow rate of the transporting groundwater remains constant. This effect is termed **retardation** (USGS 2006a), and the degree of retardation present depends upon the specific contaminants' sorptive affinity and the amount of total organic carbon in the aquifer matrix (see Figure 3.5).

Many contaminants are captured by pumping wells or migrate to surface water if the travel times and/or distances are short enough before they degrade. Figure 3.10 shows an example of travel times and capture zone in groundwater beneath urban areas. Any contaminant reaching groundwater within the area marked *capture zone* has the potential to enter the public water supply if the contaminant does not degrade before reaching a public water supply well. Several sources of contamination listed in Section 3.2 are also shown: septic tanks, underground storage tank (USTs),

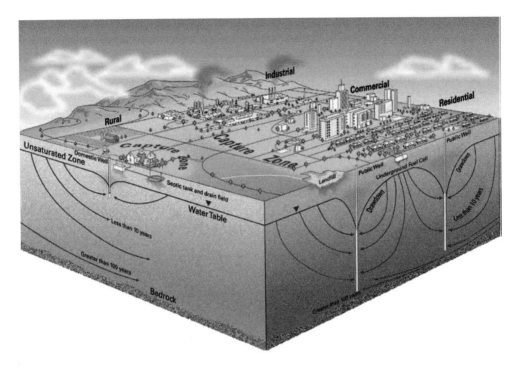

FIGURE 3.10 Potential groundwater travel time and capture zone beneath an urban area. (From USGS. Volatile Organic Compounds in Nation's Ground-Water and Drinking-Water Supply Wells. National Water-Quality Assessment Program. USGS Circular 1292. Washington, D.C. 2006a.)

landfills, industrial facilities, and power plants. These contaminant sources are typical for any urban area within the United States, and pose distinct threats to contaminate a public or private water supply.

3.3.5.1 Karst Topography

In karst topographical settings, the transport of groundwater contamination may behave similarly to surface water. Some karst formations may exhibit turbulent advective flow because they have flow rates approaching the velocities observed in surface water flow (Heath 1983). Figure 3.11 shows an example of contaminant flow in a karst aquifer (USGS 1995a).

3.3.6 Transport and Fate of Pollutants in the Atmosphere

Different contaminants affect different portions of the atmosphere. For instance, chlorofluorocarbons (CFCs) affect the protective ozone layer in the upper portions of the atmosphere. Figure 3.12 shows the layers of the atmosphere and the location of the ozone layer within the stratosphere.

Contaminant behavior in the atmosphere is very similar to the behavior of contaminants observed in surface water. Advective transport, turbulent diffusion, and molecular diffusion also influence contaminant migration in the atmosphere (Hemond and Fechner-Levy 2000). Figure 3.13 shows smoke from a fire billowing up (turbulent diffusion) into the atmosphere and the horizontal movement of the smoke by advective transport. We also see a type of advective transport called convection; in this process air rises due to thermal differences in the atmosphere. Turbulent advective mixing by wind and convection of the atmosphere is most significant within the troposphere (Schlatter 2009). These forces are very effective at transporting contaminants in the gas phase, and also do a good job of moving solid particulate matter in the atmosphere (USEPA 2008).

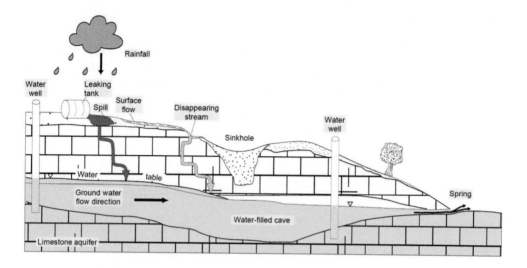

FIGURE 3.11 Contaminant migration in a karst aquifer. (From USGS 1995a. Ground-Water Quality Protection. Open-File Report 95-376. Nashville, Tennessee. 1995a. http://www.pubs.usgs.gov/of/1995/ofr-953 76.) (Accessed March 25, 2019.)

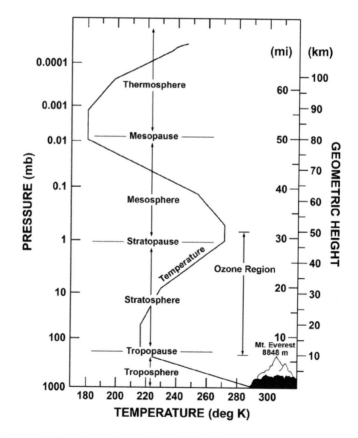

FIGURE 3.12 Layers of the atmosphere. (From United States Standard Atmosphere, *The Standard Atmosphere of the United States.* National Oceanic and Atmospheric Administration, National Aeronautics and Space Administration, and the United States Air Force. Washington, DC: United States Government Printing Office, 1976.)

Rising smoke due to
advective convection

Advective transport of
smoke by wind and
dispersion

Billowing
smokeindicative of
turbulent diffusion

FIGURE 3.13   Smoke rising into the atmosphere from a fire. (Photograph by Daniel T. Rogers.)

Anthropogenic sources of atmospheric contaminants are significant and present themselves as an array of different contaminants released in high volumes annually (USEPA 1998a). Most contaminants are released into the atmosphere from anthropogenic sources at or near the surface, with most of the impacts occurring in the troposphere, and to a lesser degree the stratosphere. Contaminants released near the ground surface can mix throughout the troposphere in a few weeks, but it can take years or decades for them to reach the stratosphere (Hemond and Fechner-Levy 2000). Records of atmospheric contaminants and their effects can be traced to the 13th century when King Edward I banned the burning of kiln coal in London due to its impacts on air quality (Wilson and Sprengler 1996).

Temperature and pressure are two important factors affecting the migration of contaminants in the atmosphere. The reason temperature and pressure play a much more significant role in atmospheric fate and transport than in soil and water is because of the ideal gas laws – a combination of Boyle's and Charles' gas laws describing the relationships between temperature, density, and pressure. Temperature and pressure typically decrease with increasing altitude in the troposphere. The temperature in the lower portion of the stratosphere is relatively constant, and helped give rise to its name, which means "stratified" (see Figure 3.12). Table 3.1 lists the standard temperatures and atmospheric pressure within the atmosphere (U.S. Standard Atmosphere 1976).

Oxygen is a recent addition to the atmosphere in geological terms. The origin of oxygen began with algae production approximately 2.45 billion years ago (Farquhar et al. 2000; Raub and Kirschvink 2008). We do not have a definitive explanation of oxygen's atmospheric origins, but one hypothesis states oxygen levels rose when the volcanism providing large amounts of hydrogen to the atmosphere declined. Methane and carbon dioxide were also gradually displaced by oxygen until levels rose rapidly to about 2.3 bYBP. The presence of oxygen in the atmosphere plays a significant role and affects contamination in the environment through oxidation reactions and rates of combustion (USEPA 1991). Table 3.2 lists the basic composition of the atmosphere (NASA 1976).

Contaminants initially released into the atmosphere often do not remain in the air; they settle out and contaminant the soil or surface water. Some contaminants, however, remain in the atmosphere for long periods of time, and other contaminants initially released into soil or water may volatilize and contaminate the air. In some cases, air contaminants may settle out of the atmosphere and

TABLE 3.1

Standard Atmospheric Temperature and Pressure with Increasing Altitude

Altitude		Pressure	Temperature	
Feet	**Meters**	**(atm)**	**°F**	**°C**
0	0	1.000	59.0	15.0
2,000	610	0.943	51.9	11.0
4,000	1,219	0.888	44.7	7.0
6,000	1,826	0.836	37.6	3.1
8,000	2,438	0.786	30.5	−0.8
10,000	3,048	0.738	23.3	−5.0
15,000	4,572	0.564	5.5	−14.7
20,000	6,096	0.459	−12	−24.4
30,000	9,144	0.297	−48	−44.4
40,000	13,123	0.185	−67	−55
60,000	18,288	7.1×10^{-2}	−67	−55
80,000	24,384	2.7×10^{-2}	−67	−55
100,000	30,480	1.0×10^{-2}	−67	−55
140,000	42,672	2.0×10^{-3}	74	23.3
180,000	54,864	5.7×10^{-4}	170	76.7
220,000	67.056	1.7×10^{-4}	92	33.3
300,000	91,440	1.5×10^{-5}	27	−2.8
380,000	115,824	7.7×10^{-7}	188	86.7

Source: United States Standard Atmosphere. The Standard Atmosphere of the United States. National Oceanic and Atmospheric Administration, National Aeronautics and Space Administration, and the United States Air Force. United States Government Printing Office, Washington, D.C. 277 p. 1976.

adsorb onto a soil grain on the land surface only to be picked up later by the wind and be sent airborne again. Factors controlling whether a contaminant remains in the atmosphere include:

- Physical and chemical factors of the contaminants including vapor pressure, molecular weight, solubility, reactivity
- Geography and local topography
- Climate and weather conditions

The average person inhales approximately 20,000 liters of air per day (USEPA 2008a). Each year, the World Health Organization (2019) estimates 7 million people die from causes directly attributable to air pollution, with the elderly and young children at the most risk. Specific diseases caused by prolonged exposure to air contaminants are chronic and often do not immediately appear after exposure. These diseases include heart disease, lung cancer, and bronchitis. The burning of fossil fuels in power plants and automobile, truck, and bus exhaust account for 90% of all air pollution in the United States (USEPA 2008a). Figure 3.14 shows some of the significant sources, methods of transport, and removal of air pollutants in the atmosphere (USEPA 2008a).

TABLE 3.2
Composition of the Atmosphere

Gas	Chemical Symbol	Mean Molecular Weight (m mol^{-1})	Concentration Parts per Million by Volume [PPMv])
Nitrogen	N_2	28.013	780,840
Oxygen	O_2	31.999	209,460
Argon	Ar	39.948	9,340
Carbon dioxide	CO_2	44.010	384
Neon	Ne	20.180	18.18
Helium	He	4.003	5.24
Methane	CH_4	16.043	1.774
Krypton	Kr	83.798	1.14
Hydrogen	H_2	2.106	0.56
Nitrous oxide	N_2O	44.012	0.32
Xenon	Xe	131.293	0.09
Ozone	O_3	47.998	0.01 to 0.10

Source: United States Standard Atmosphere. The Standard Atmosphere of the United States. National Oceanic and Atmospheric Administration, National Aeronautics and Space Administration, and the United States Air Force. United States Government Printing Office, Washington, D.C. 277 p. 1976.

Deposition of contaminants onto the land from the atmosphere occurs in two different ways:

1. Dry deposition. Dry deposition is typically dust or particulate matter settling out of the air. The amount of dry deposition depends upon the amount of suspended particles, wind speed and duration, and particle size. Figure 3.15 shows a dust storm potentially depositing a significant amount of dry material. Contaminants are often present within dry deposition events, especially in urban areas where they may be sorbed onto the surfaces of particulate matter in the air (USEPA 2008).
2. Wet deposition. Occurs when snow, fog, or a rain droplet forms and then dissolves or carries a contaminant to the surface. Acid rain is a good example of wet deposition. Figure 3.16 shows the wet deposition of contaminants (USEPA 2019c).

3.4 FATE AND TRANSPORT OF POLLUTANTS

We now briefly discuss the fate and transport behavior of each contaminant group discussed in Chapter 2.

3.4.1 VOCs

VOCs are organic compounds that generally volatilize or evaporate readily under normal atmospheric pressure and temperatures. They usually have a high vapor pressure, low-to-medium solubility, and low molecular weight. As a result of these chemical characteristics, VOCs are common air, soil, and water contaminants (USGS 2006a; USEPA 2008a). Automobile exhaust contains VOCs. When combined with other common air pollutants and sunlight, urban smog will form if atmospheric conditions are favorable – that is, there is an ample supply of the combined sources of

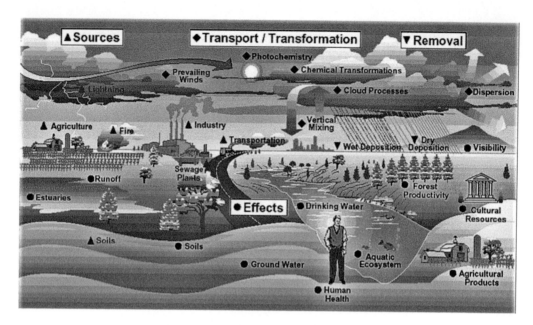

FIGURE 3.14   Sources, transport methods, and removal of air pollution. (From USEPA. *Taking Toxics Out of the Air.* EPA451/K-98-001. Washington, DC. 1998a.)

FIGURE 3.15   Dust storm that demonstrates transport and dry deposition of particulates. (From National Oceanic and Atmospheric Administration (NOAA). 2010. The May 29th Dust Storm. 2019. http://www.crh. noaa.gov/ddc/?n=dust.) (Accessed March 25, 2019.)

VOCs and other smog-forming contaminants. This type of smog formation produces **photochemical smog** and is shown in Figure 3.17.

VOCs are also released directly onto the ground surface through leaks or spills at or near the surface. Sources of these leaks include underground storage tanks, service stations, refineries, and pipelines. Because of these surface and shallow subsurface releases, VOCs are common groundwater contaminants and have been detected in the groundwater of numerous aquifers in the United States (USGS 2006a). A study of groundwater in the United States detected VOCs at a concentration of 0.02 ug/L in more than 50% of approximately 3,500 samples collected from 100 different groundwater aquifers across the country (USGS 2006a). The VOCs detected most often included bromoform, bromodichloromethane, chloroform, chloromethane, 1,1-dichloroethane,

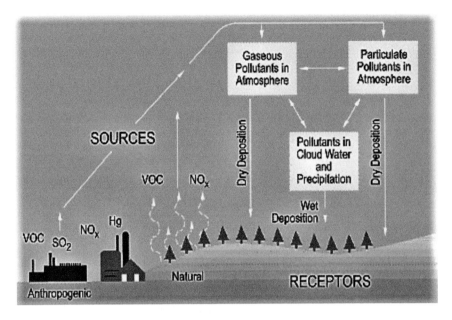

FIGURE 3.16   Process of wet and dry deposition of air contaminants. (From USEPA. *What Is Acid Rain.* 2019c. http://www.epa.gov/acidrain/what/index.html.) (Accessed March 25, 2019.)

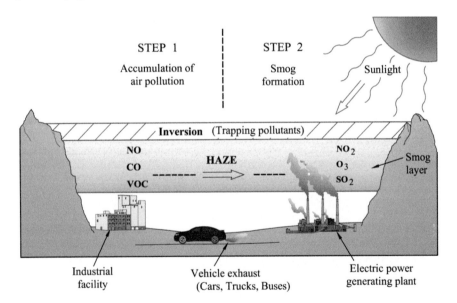

FIGURE 3.17   Formation of photochemical smog.

dichlorodifluoromethane, methylene chloride, dibromodichloromethane, Methyl-tert butyl ether, Trichloroethene, Tetrachloroethene, 1,1,1-Trichloroethane trans-1,2-dichloroethene, toluene, and trichlorofluoromethane.

This same study indicates the vulnerable nature of many aquifers of the United States, and their location corresponds with many urban areas. In fact, of the 28 major urban areas of the United States, 27 have detectable concentrations of VOCs (only Kansas City is missing) and represent a population of 125.4 million – over 41% of the entire US population. Figure 3.18 shows the locations where VOCs were detected (USGS 2006a).

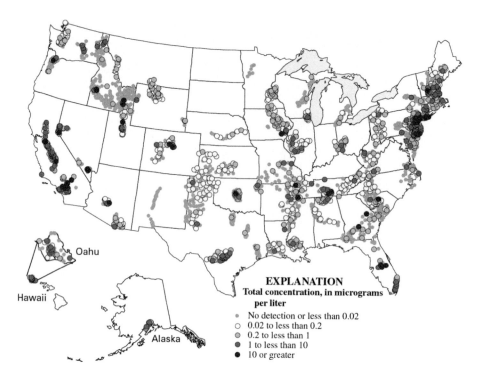

FIGURE 3.18 Occurrence of VOCs in groundwater aquifers of the United States. (From USGS. Volatile Organic Compounds in Nation's Ground-Water and Drinking-Water Supply Wells. National Water-Quality Assessment Program. USGS Circular 1292. Washington, D.C. 2006a.)

VOCs exist as light nonaqueous phase liquid (LNAPLs) and dense nonaqueous phase liquid (DNAPLs). Because LNAPLs are lighter than water, they tend to float on top of groundwater, whereas the heavier than water DNAPLs tend to sink through the water column in an aquifer if conditions are favorable, as depicted in Figure 3.19 (USGS 2006a).

Contaminant degradation rates vary widely and depend on many factors including: 1) the nature of the release; 2) physical chemistry of the contaminants themselves; 3) the geological environment

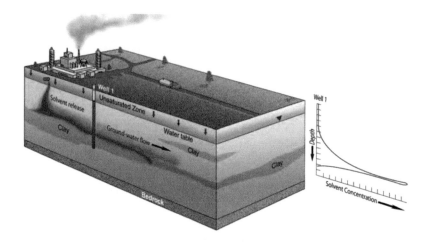

FIGURE 3.19 Migration of DNAPL compound. (From USGS. Volatile Organic Compounds in Nation's Ground-Water and Drinking-Water Supply Wells. National Water-Quality Assessment Program. USGS Circular 1292. Washington, D.C. 2006a.)

where the contaminants are released; and 4) the presence, type, and distribution of microorganisms (Howard et al. 1997; USEPA 1998b; McKone and Enoch 2002; USGS 2006a).

Degradation rates also vary by media. In general, organic compounds in the atmosphere, including VOCs, degrade more quickly than the same organic compounds released and migrating to subsurface soil and groundwater (USEPA 1998a; 1998b).

VOCs not degrading very easily include many of the chlorinated solvents, or DNAPL compounds (USEPA 1996a; 1996b). Many chlorinated VOC compounds including PCE, TCE, 1,1,1-TCA, DCE, and vinyl chloride are considered very persistent in the environment once released and can remain present for decades (USEPA 1996a; Suthersan et al. 2017). Figure 3.20 shows an example of the degradation of tetrachloroethene. The degradation sequence follows a modified Dominico (1987) analytical solution for mass transport in a consecutive and irreversible fourth-order differential equation with a default persistence of 2, 4, 6, and 8 years for each compound (see Figure 3.21). As shown in the figure, the maximum vinyl chloride concentration is achieved approximately 20 years following the release. The example in the figure uses an initial concentration of tetrachloroethene of 1,000 ug/L (Payne and Rogers 1997).Other VOC compounds, such as benzene, toluene, ethyl benzene, and xylenes (BTEX) are not typically as persistent in the environment and have been known to biodegrade in a few months to years if conditions are favorable (Rogers 1995; USEPA 1996a). The VOC compound MTBE is persistent in the environment. Its relatively high solubility in water and low sorptive properties compared to other common VOC contaminants has resulted in significant MTBE-contaminated groundwater supplies at many urban locations throughout the United States (USGS 2006a).

The trihalomethane VOCs include chloroform, bromoform, bromodichloro-methane, and dibromochloromethane. These compounds have been detected in the groundwater of many US aquifers (USGS 2006a). Trihalomethanes have high relative vapor pressures and commonly evaporate quickly when in contact with the atmosphere. Therefore, exposure to trihalomethanes is of special concern during showering and washing (ATSDR 1997a; 2005a). They degrade by photolysis when exposed to direct sunlight and can also be degraded by microorganisms (ATSDR 1997a; 2005a).

3.4.2 PAHs

PAHs are LNAPL compounds, and being lighter than water they float on surface water and groundwater. They do not readily dissolve in water and have low vapor pressures compared to most VOCs. PAHs are common constituents of automobile exhaust, especially from diesel fuel (USGS 2006a),

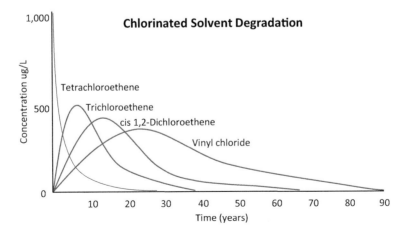

FIGURE 3.20 Chlorinated solvent degradation. (Modified from Payne, F.C. and Rogers, D.T. Chlorinated Solvent Degradation – A Consecutive, Irreversible Reaction Sequence. *Remediation of Recalcitrant Chlorinated Volatile Organic Compounds in Groundwater*. National Groundwater Association, Dublin, Ohio. Groundwater Management Book 21. P. 97-109. 1997.)

$$\frac{C_x}{C_0} = \exp\left\{\frac{x}{2\alpha_x}\left[1 - \left(1 + \frac{4\lambda\alpha_x}{v}\right)^{\frac{1}{2}}\right]\right\} erf\left[\frac{Y}{4(\alpha_y x)^{\frac{1}{2}}}\right] erf\left[\frac{Z}{4(\alpha_z x)^{\frac{1}{2}}}\right]$$

Where,

C_x - contaminant concentration in a downgradient well along the plume centerline
    at a distance x (mg/L),
C_0 - contaminant concentration in the source well (mg/L),
x - centerline distance between the downgradient well and source well (ft),
α_x, α_y, and α_z - longitudinal, transverse, and vertical dispersivity (ft), respectively,
    $D_x = \alpha_x \times v$, $D_y = \alpha_y \times v$, $D_z = \alpha_z \times v$,
λ - degradation rate constant (1/day),
    $\lambda = 0.693/t_{1/2}$ (where $t_{1/2}$ is the degradation half-life of the compound).
v - groundwater velocity (ft/day),
Y - source width (ft),
Z - source depth (ft),
erf - error function,
exp - exponential function.

FIGURE 3.21 Dominico steady-state model for multidimensional transport of decaying contaminant species analytical transport equation. (From Domenico, P.A. An Analytical Model for Multidimensional Transport of Decaying Contaminant Species. *Journal of Hydrology.* 91:49–58. 1987.)

and are common air contaminants in urban areas. In addition, PAHs have a high sorptive affinity and can therefore attach to particulate matter (USGS 1996a; ATSDR 1996). Many PAH compounds biodegrade under favorable conditions. The half-life of PAHs is shortest in the atmosphere due to photochemical degradation and lasts just a few days or weeks (ATSDR 1996). The half-life of PAH compounds in soil and groundwater is longer and may last for several years or decades. When released in soil, PAHs tend to migrate more slowly than VOCs because of their higher molecular weight and sorption to soils with high organic content.

3.4.3 PCBs

Once in the environment, PCBs do not readily degrade – they remain in the environment for long periods of time and often cycle between air water and soil. PCBs can be carried long distances attached to particulate matter and have been detected in snow and sea water far away from any known point of release (USEPA 1996a). This transport capability is confirmed by their world-wide detection. The lighter the PCB compound (fewer number of chlorine atoms in its structure) (see Figure 2.6), the farther it can be transported from its release point.

PCBs are not very soluble, have high sorptive potential and low vapor pressures. This combination of physical properties largely determines their environmental distribution, as they are only present at high concentrations in water or air when sorbed to particulate matter. Sinks where PCBs are frequently detected include soil near release points and sediments in rivers, streams, and lakes (USEPA 1996a). PCBs, bioaccumulate in the leaves and aboveground parts of plants and food crops, and in aquatic organisms and fish where PCBs are present in sediments (Section 3.3.4). As a result, humans and other organisms ingesting impacted plant material, food crops, or organisms containing PCBs may bioaccumulate PCBs in their body tissues (ATSDR 2001a; USEPA 2009a).

3.4.4 SVOCs

SVOCs are much less volatile than VOCs, but notable exceptions here are the amine compounds that exist as a gas at room temperature and standard pressure (ATSDR 2002a). In terms of solubility, phthalates and phenols do not readily dissolve in water, whereas amines and esters may dissolve, become mobile, and reach groundwater (ATSDR 2002a; 2002b; 2002c; 2002d). When released into the environment, SVOCs are commonly detected in soil because they have high sorptive potentials. Many SVOCs, including pentachlorophenol, are degraded by microorganisms under favorable conditions and are also susceptible to photolysis and hydrolysis (ATSDR 2001b; 2002a). Amines and esters degrade in minutes when exposed to direct sunlight (ATSDR 1999a; ATSDR 2002a; 2002b; 2002c; 2002d).

3.4.5   HEAVY METALS

Heavy metals are released directly to air, water, and soil. In most cases, these contaminants do not remain in the atmosphere for long periods of time because they have high specific gravity and become deposited onto the land surface shortly after being emitted. Mercury, however, has been detected as far away as 50 miles from its source after being released into the atmosphere (USEPA 1997). Lead is also considered a common air pollutant (USEPA 2008). Major sources of lead include metal melting facilities, battery manufacturing, and leaded gasoline and fuels. The good news is there has been a 92% decrease in atmospheric lead concentrations over the period from 1980 through 2018. Contributing to the observed decrease has been the removal of lead from gasoline and fuels, increased efficiency in air pollution control equipment, and the regulation of lead emissions sources (USEPA 2019a).

The solubility of heavy metals in water is very low, except for arsenic and chromium VI at a neutral pH. Due to their inability to form a solution, the preferred sinks for heavy metals are soil and sediments. Many metals undergo some transformation such as oxidation after being released into the environment, but are not destroyed and remain in the environment (ATSDR 1999b; 1999c; 2003; 2004a; 2005b; 2005c; 2007a; 2007b; 2007c; 2008a; 2008b). Due to their low solubility and high specific gravities, they tend to remain near release points. This is why increased concentrations of heavy metals are present in the near-surface soil of urban areas as a result of releases from anthropogenic sources (Murray et al. 2004). Some heavy metals such as mercury accumulate in sediment of lakes, rivers, and streams. Mercury may undergo a transformation through a process known as methylation after being released into the environment, typically when it reaches a surface water body (USEPA 1997). The process of methylation transforms elemental mercury to methyl mercury (CH_3Hg). Methyl mercury is the most toxic form of mercury and has the potential to bioaccumulate in aquatic organisms, including fish (USGS 2000).

On land and in sediments, other heavy metals such as barium, cadmium, chromium, lead, copper, and zinc can accumulate in plant matter if they are present at sufficient concentrations in soil within the root zone of plants (Nudunuri et al. 1998). Removal of contaminants from near-surface soil is not uncommon and sometimes is the preferred alternative. Suitable locations for removal include closed landfill sites requiring a vegetative cap to minimize erosion potential, or sites where contamination is shallow, relatively static, and has become a chronic problem (Singer et al. 2003). **Phytoremediation** – the removal of contamination using plants – has been applied to other contaminants including organic compounds, but with limited success.

3.4.6   OTHER POLLUTANTS

The remaining contaminants encountered in urban areas span a wide variety of organic and inorganic compounds. Their fate and transport are briefly discussed here.

3.4.6.1   Pesticides and Herbicides

Pesticides and herbicides are released into the environment for specific purposes, and are most commonly detected in near-surface soils and surface water rather than groundwater (USGS 2006b). The highest concentrations of pesticides exist in the nation's streams and sediments within urban areas where they have been detected in 83% of streams and 70% of sediments (USGS 2006b). The total tonnage of pesticide and herbicide use in the US has remained constant during the 1980s and early

1990s but has increased by nearly 60% in the last 20 years, mostly for agriculture and home and garden purposes (USEPA 2017).

Permethrin and toxaphene are two other widely-used compounds in insecticides and pesticides, respectively. The fate and transport concern with permethrin centers around its use as an insect repellent, application to crops, and flea treatments for pets. All of these activities involve human exposure (ATSDR 2005d), and with a half-life of approximately 28 days, there is often adequate time for human contact. Toxaphene strongly sorbs to soil particles and is not very soluble in water (SRC 2019). Common sinks where toxaphene may be present include sediments and soil, from where it bioaccumulates in fish and mammals. Toxaphene is also in the atmosphere since it evaporates when in a solid form or is dissolved (ATSDR 1997b). It is estimated the half-life of toxaphere is more than 10 years in soil (SRC 2019), so there is a good chance it is still present at appreciable concentrations in the environment.

3.4.6.2   Dioxins

Since one method of dioxin formation is through incineration and combustion, dioxin compounds are present in the atmosphere and have been detected around the globe (ATSDR 2006a). When dioxins are released at the surface, they typically sorb to soil particles and are often detected in sediments in lakes, rivers, and streams acting as sinks for dioxins (ATSDR 2006a). Dioxin compounds are considered bioaccumulative contaminants (Section 3.3.4), with the potential to build up in the food chain and yield detectable concentrations in the tissues of many animals (ATSDR 2006a; USEPA 2019d). Due to the presence of chlorine in their atomic structure, dioxins do not readily degrade once they are formed and released into the environment (USEPA 2006).

3.4.6.3   Fertilizers

The most common fertilizers include nitrogen (N), phosphorus (P), and potassium (K). A major sink for fertilizers is surface water because they are applied to the soil surface and are considered soluble in water and mobile – especially in the case of nitrates and phosphorus. Once in surface water, nitrate and phosphorus can promote excessive algal growth. Significant algal growth can deplete the dissolved oxygen in surface water and cause suffocation and death to aquatic organisms. The solubility of some fertilizers, combined with the relationship between surface water and groundwater, may lead to groundwater contamination. The natural process of enrichment of surface waters with plant nutrients is termed **eutrophication**. When anthropogenic activities such as fertilization or sewage discharges accelerate this natural process, **cultural eutrophication** occurs (McGucken 2000).

3.4.6.4   Cyanide

In the atmosphere, cyanide is most often present as hydrogen cyanide. When present in surface water, cyanide compounds usually will form hydrogen cyanide and then enter the atmosphere through evaporation. When released to soil, cyanide compounds are considered fairly mobile when the retention capacity of the soil is exceeded and may migrate and contaminate groundwater. Cyanide compounds are degraded by microorganisms when present at low concentrations. When concentrations of cyanide compounds are elevated, they tend to be toxic to microorganisms and resist degradation (ATSDR 2006b).

The half-life of cyanide in the atmosphere ranges from between one to three years (ATSDR 2006b). In soil and water, the half-life of cyanide compounds is much more difficult to estimate because the concentration, distribution, and presence of microorganisms available to degrade the cyanide compounds varies.

3.4.6.5   Asbestos

Asbestos fibers do not degrade, evaporate, or dissolve in water and remain virtually unchanged in the environment (ATSDR 2001c). Asbestos originates from naturally occurring minerals and is therefore present in the environment. Average background concentrations of asbestos in air range from 0.00001 to 0.0001 fibers per milliliter of air and are highest in urban areas (ATSDR 2001c).

Small diameter asbestos fibers can remain suspended in the atmosphere for a long period of time compared to larger fibers (those larger than ten microns) (ATSDR 2001c). Since asbestos was widely used in building materials, it is most common in urban areas and where natural deposits are present. Asbestos can become airborne through the disturbance of asbestos-containing materials during demolition or remodeling activities.

3.4.6.6 Acids and Bases

When released into soil, acids and bases neutralize rapidly. They are diluted when they come into contact with water if a difference in pH levels exists. Therefore, if environmental impairment occurs, it must be realized rapidly before the acid or base becomes neutralized. This impairment occurs with the majority of sudden and accidental releases, but does not hold true for acid rain, which generates effects with slower onsets.

Acids and bases may migrate a significant distance – sometimes more than 1 mile – when released in the atmosphere, and may cause significant impairment to living organisms exposed to their vapors (ATSDR 2004b; 2004c).

3.4.6.7 Radioactive Compounds

Radioactive compounds occur naturally, with the most common being radon. Radon is produced from the decay of uranium (ATSDR 2008c) and is present in the air, water, and soil. Radon may build up in basements, especially if cracks exist, or other subsurface structures located above natural deposits having higher relative uranium levels. The half-life of radon is approximately 4 days (ATSDR 2008c). Most of the human exposure attributed to other radioactive compounds results from medical devices, diagnostic treatments, testing equipment such as x-ray machines, and cancer therapy (Kathren 1991; ATSDR 2000).

3.4.6.8 Greenhouse Gases

Greenhouse gases decay very slowly and are primarily atmospheric contaminants. Some quantities of these gases are naturally removed from the atmosphere, such as the removal of carbon dioxide during photosynthesis (see Figure 3.22). However, the anthropogenic addition of carbon dioxide and other greenhouse gases into the atmosphere has greatly exceeded the capacity of the natural environment to remove them (USEPA 2009b). As a result, greenhouse gas concentrations have been increasing (USEPA 2018).

3.4.6.9 Carbon Dioxide

Carbon dioxide is a greenhouse gas and we discuss above its fate in the atmosphere, but there is another important method in which carbon dioxide is absorbed and that is by the oceans. The rate of accumulation of pollution is proportional to the human population. The more people, the more carbon dioxide ends up in the oceans.

Pollution in the oceans does not all originate from direct discharge but also is delivered through the atmosphere from increased concentrations of carbon dioxide in the atmosphere that is partially absorbed by the oceans and causes acidification of the surface layers. Absorption of carbon dioxide by the oceans has been estimated to be as high as 25%, which represents millions of tons of carbon dioxide absorbed by the oceans every year. The additional carbon dioxide absorbed by the oceans

$$6H_2O + \quad 6CO_2 \quad \xrightarrow{\text{sunlight}} \quad C_6H_{12}O_6 + \quad 6O_2$$

$$\text{Water} + \text{Carbon dioxide} \longrightarrow \text{Glucose} + \text{Oxygen}$$
$$\text{(a carbohydrate)}$$

FIGURE 3.22 Photosynthesis and the removal of carbon dioxide.

reacts with ocean water and forms an acid called carbonic acid and is lowering the pH of the oceans world-wide. In fact, the oceans are acidifying faster than they have in some 300 million years. The effect of the acidification has been linked to bleaching of coral reefs, including the Great Barrier Reef off the northeastern coast of Australia (Natural Resource Defense Council 2018).

3.4.6.10   Carbon Monoxide

Carbon monoxide is created when fuel is not burned completely. Carbon monoxide is formed naturally and anthropogenically. The most significant source of carbon monoxide is from automobile exhaust (USEPA 2019a). Inside homes, significant sources of carbon dioxide emissions are natural gas and oil furnaces, hot water heaters, appliances, wood-burning stoves, and fireplaces. Carbon monoxide is also created as a by-product of several industrial processes including metal melting and chemical synthesis. From 1980 to 2016, there was a decrease of greater than 80% of carbon dioxide in the United States. This decrease was attributed to improved air pollution control equipment for stationary and mobile sources of air pollution (USEPA 2019a).

3.4.6.11   Ozone

Ozone is a gas occurring in the Earth's upper atmosphere and at ground level. As noted in Chapter 7, ozone in the upper atmosphere is greatly beneficial to life on Earth because it filters UV radiation, but ozone occurring at ground level is considered an air pollutant (USEPA 2019a). Ground-level ozone is not emitted directly into the air – it is created by chemical reactions between oxides of nitrogen (NOx) and VOCs in the presence of sunlight. Emissions from automobile exhaust, gasoline vapors, chemical solvents, electrical generating facilities, and some factories trigger the production of ground-level ozone (USEPA 2009b). This variety of ozone is a concern in urban regions of the United States during the summer because strong sunlight and hot weather can generate higher levels (USEPA 2009b). Since the Clean Air Act of 1990, atmospheric ozone concentration in the United States has declined by 25% (USEPA 2019a). Better control of stationary and mobile sources of air pollution such as automobile exhaust is behind this improvement (USEPA 2019a).

3.4.6.12   Sulfur Dioxide

Sulfur dioxide is a component of smog and also combines with nitrous oxide compounds to eventually form sulfuric acid, which is commonly referred to as acid rain (USEPA 2019a). It is removed from the atmosphere during precipitation and is typically neutralized quickly in the soil if the pH of the soil is greater than seven. Some areas of the northeastern United States have soils lacking the ability to effectively neutralize the effects of acid rain, and there have been adverse effects on aquatic life and vegetation in the region. Efforts to reduce sulfur dioxide emissions from stationary and mobile sources have resulted in a decrease of 75% since 1980 in the United States (USEPA 2019a). Nevertheless, the pH of rain in the eastern United States remains acidic.

3.4.6.13   Particulate Matter

Urban areas have the highest concentrations of particulate matter, which is a significant distributor of contaminants in the atmosphere. Contaminants such as SVOCs, some VOCs, PCBs, and many pesticides and herbicides may sorb to a soil particle and travel a significant distance through wind action (USEPA 1998a). The size of particulate matter is significant because the largest sizes tend to settle to the ground surface first. Smaller particles can travel around the globe and remain suspended for years if favorable conditions exist (USEPA 2019a). Here is some good news: there was a 20 to 30% decline in atmospheric particulate matter within the urban areas of the United States between 2001 and 2016 (USEPA 2019a).

3.4.6.14   Bacteria, Parasites, and Viruses

Bacteria, parasites, and viruses are present in large numbers everywhere in the environment. They are in and on the food we eat, in and on our bodies, in the air we breathe and the water we drink, in soil, and at depths within the Earth. Many are beneficial, but some have the potential to adversely affect our health and well-being (Madigan et al. 2008).

3.4.6.15 Invasive Species

Invasive species have been around for a few hundred years and now number in the tens of thousands. Invasive species are difficult to impossible to contain once released, so prevention is key. The main reason invasive species are difficult to contain is that the mode of migration is heavily influenced by breeding. With an increasing human population, increased standard of living, a warming climate, and increased human travel, the prospects of increased invasive species issues are almost certain.

3.4.6.16 Polyfluoroalkyl Substances (PFAS)

PFAS chemicals are persistent, mobile, and toxic. They have been widely produced and used in numerous consumer products for decades and have just recently been identified as a contaminant. Due to its soluble nature, PFAS chemicals are most often detected in surface and groundwater and have also been detected in fish tissues. Reliable estimates on their degradation have not yet been quantified.

3.4.6.17 Emerging Contaminants

Emerging contaminants including many pharmaceuticals, 1,4-dioxane, 1,2,3-trichloropropane, PFAs compounds, and perchlorates are resistant to degradation and can remain in the environment for long periods of time (Bell et al. 2019). Emerging contaminants have been detected in groundwater where they can migrate long distances due to their relatively high solubility and resistance to degradation. Emerging contaminants are difficult to investigate because many have entered the environment from non-traditional sources such as residential septic systems and agricultural locations, as opposed to industrial sources (Bell et al. 2019).

3.5 SUMMARY AND CONCLUSION

There are tens of thousands of pollutants existing everywhere. After being released into the environment, they migrate in air, soil, and water. Some are persistent, while others are not. Some dissolve in water and some do not. Some are transported around the globe in the atmosphere while others do not. The geography, geology, hydrogeology, and atmospheric conditions all play a significant role in affecting the fate and transport of pollutants and determines their final disposition.

Halogenated contaminants (those containing chlorine, fluorine, bromine, or iodine within their structure) generally remain in the environment for long periods of time because, in large part, they are molecularly stable and synthetic compounds and the natural environment has difficulty in degrading them. Several of these compounds have the ability to accumulate in tissues of living organisms (e.g., PCBs), and through bioaccumulation may expose humans after they work their way up the food chain. Some contaminants can change form, such as a gas to a liquid or a liquid to a gas, and cycle between the soil, air, and water if they last long enough (e.g., some VOCs). Other contaminants not changing form can be found in sinks or areas where they accumulate (e.g., river and lake sediment). Risk only occurs when an exposure pathway exists. Toxicity, therefore, is not the only factor to consider when evaluating risk. Mobility and persistence of a contaminant must also be considered.

The next chapter examines geologic vulnerability to pollution, especially in urban regions of the world.

REFERENCES

Agency for Toxic Substances and Disease Registry (ATSDR). 1996. *Polycyclic Aromatic Hydrocarbons*. General Contaminant Class. ATSDR ToxFAQs. Atlanta, GA.

ATSDR. 1997a. *Chloroform*. CAS Registry Number 127-18-4. ATSDR ToxFAQs. Atlanta, GA.

ATSDR. 1997b. *Toxaphene*. CAS Registry Number 8001-35-2. ATSDR ToxFAQs. Atlanta, GA.

ATSDR. 1999a. *Dimethylamine*. CAS Registry Number 124-40-3. ATSDR ToxFAQs. Atlanta, GA.

ATSDR. 1999b. *Mercury*. CAS Registry Number 7439-97-6. ATSDR ToxFAQs. Atlanta, GA.

ATSDR. 1999c. *Silver*. CAS Registry Number 7440-22-4. ATSDR ToxFAQs. Atlanta, GA.

ATSDR. 2000. *Radon Toxicity: Who Is At Risk?* ATSDR. Atlanta, GA.

ATSDR. 2001a. *Polychlorinated Biphenyls*. ATSDR ToxFAQs. Atlanta, GA.

ATSDR. 2001b. *Pentachlorophenol*. CAS Registry Number 87-86-5. ATSDR ToxFAQs. Atlanta, GA.

ATSDR. 2001c. *Asbestos*. CAS Registry Number 1332-21-4. ATSDR ToxFAQs. Atlanta, GA.

ATSDR. 2002a. *Di(2-ethylhexyl) Phthalate*. CAS Registry Number 117-81-7. ATSDR ToxFAQs. Atlanta, GA.

ATSDR. 2002b. *Toxicological Profile for Flame Retardant Ester Compounds*. ATSDR ToxFAQs. Atlanta, GA.

ATSDR. 2002c. *Benzyl Acetate*. CAS Registry Number 140-11-4. ATSDR ToxFAQs. Atlanta, GA.

ATSDR. 2002d. *Ethyl Acetate*. CAS Registry Number 141-78-6. ATSDR ToxFAQs. Atlanta, GA.

ATSDR. 2003. *Selenium*. CAS Registry Number 7782-49-2. ATSDR ToxFAQs. Atlanta, GA.

ATSDR. 2004a. *Copper*. CAS Registry Number 7440-50-8. ATSDR ToxFAQs. Atlanta, GA.

ATSDR. 2004b. *Ammonia*. CAS Registry Number 7664-41-7. ATSDR ToxFAQs. Atlanta, GA.

ATSDR. 2004c. *Hydrochloric Acid*. CAS Registry Number 7647-01-0. ATSDR ToxFAQs. Atlanta, GA.

ATSDR. 2005a. *Bromoform*. CAS Registry Number 75-25-2. ATSDR ToxFAQs. Atlanta, GA.

ATSDR. 2005b. *Nickel*. CAS Registry Number 7440-02-0. ATSDR ToxFAQs. Atlanta, GA.

ATSDR. 2005c. *Zinc*. CAS Registry Number 7440-66-6. ATSDR ToxFAQs. Atlanta, GA.

ATSDR. 2005d. *Permethrin: Toxicologic Information about Pesticides*. CAS Registry Number 52645-53-1. ATSDR. Atlanta, GA.

ATSDR. 2006a. *Dioxins: Chemical Agent Briefing Sheet*. ATSDR. Atlanta, GA.

ATSDR. 2006b. *Cyanide*. CAS Registry Number 74-90-8, 143-33-9, 151-50-8, 592-01-8, 544-92-3, 506-61-6, 460-19-5, 506-77-4. ATSDR ToxFAQs. Atlanta, GA.

ATSDR. 2007a. *Lead*. CAS Registry Number 7439-92-1. ATSDR ToxFAQs. Atlanta, GA.

ATSDR. 2007b. *Barium*. CAS Registry Number 9440-39-3. ATSDR ToxFAQs. Atlanta, GA.

ATSDR. 2007c. *Arsenic*. CAS Registry Number 7440-38-2. ATSDR ToxFAQs. Atlanta, GA.

ATSDR. 2008a. *Cadmium*. CAS Registry Number 7440-43-9. ATSDR ToxFAQs. Atlanta, GA.

ATSDR. 2008b. *Chromium*. CAS Registry Number 7440-47-3. ATSDR ToxFAQs. Atlanta, GA.

ATSDR. 2008c. *Radon*. CAS Registry Number 14859-67-7. ATSDR ToxFAQs. Atlanta, GA.

Bell, C.A., Gentile, M., Kalve, E., Ross, I., Horst, J., and Suthersan, S. 2019. *Emerging Contaminants*. CRC Press. Boca Raton, FL. 439p.

Brady, N.C. and Well, R.R. 1999. *The Nature and Properties of Soils*, 12th Edition. Prentice- Hall. Upper Saddle River, NJ.

Domenico, P.A. 1987. An Analytical Model for Multidimensional Transport of Decaying Contaminant Species. *Journal of Hydrology*. Vol. 91. pp. 49–58.

Farquhar, J., Huiming, B., and Thiemens, M. 2000. Atmospheric Influence of Earth's Earliest Sulfur Cycle. *Science*. Vol. 289. No. 5480. pp. 756–758.

Fetter, C. 1993. *Contaminant Hydrogeology*. 2nd Edition. Prentice Hall. Upper Saddle River, NJ.

Freeze, R.A. and Cherry, J.A. 1978. *Groundwater*. Prentice Hall. Upper Saddle River, NJ.

Haynes, W. 1954. *American Chemical Industry – A History*. Vols. I–IV. Van Nostrand Publishers. New York.

Heath, R.C. 1983. *Basic Ground-Water Hydrology*. United States Geological Survey. Water Supply Paper 2220. United States Government Printing Office. Alexandria, VA.

Hemond, H.F. and Fechner-Levy, E.J. 2000. *Chemical Fate and Transport in the Environment*. Academic Press. London, England.

Hornsby, A.G. 1990. *How Contaminants Reach Groundwater*. University of Florida Institute of Food and Agriculture. Gainesville, FL.

Howard, P.H., Michalenko, E.M., Basu, D.K., and Aronson, D. 1997. *Handbook of Environmental Fate and Exposure Data for Organic Chemicals*. Vol. 5. Solvents 3. Lewis Publishers. Chelsea, MI.

Kathren, R. 1991. *Radioactivity and the Environment*. Taylor & Francis Publishers. Lieden, The Netherlands.

Kaufman, M.M., Rogers, D.T., and Murray, K.S. 2011. *Urban Watersheds*. CRC Press. Boca Raton, FL. 583p.

Lindsay, W.L. 1979. *Chemical Equilibria in Soils*. John Wiley & Sons, New York.

Lyman, W.J., Reehl, W.F., and Rosenblatt, D.M. 1990. *Handbook of Chemical Property Estimation Methods*. American Chemical Society. Washington, DC.

Madigan, M.T., Martinko, J.M., Dunlap, P.V., and Clark, D.P. 2008. *Brock Biology of Microorganisms*. 12th Edition. Prentice Hall. New York.

McGucken, W. 2000. *Lake Erie Rehabilitated: Controlling Cultural Eutrophication, 1960s–1990s*. University of Akron Press. Akron, OH.

McKone, T.E. and Enoch, K.G. 2002. *CalToxTM, A Multimedia Total Exposure Model Spreadsheet Users Guide*. Lawrence Berkeley National Laboratory. LBNL 47399. University of California, Berkeley, CA.

Murray, K.S., Rogers, D.T., and Kaufman, M.M. 2004. Heavy Metals in an Urban Watershed in Michigan. *Journal of Environmental Quality*. Vol. 33. pp. 163–172.

National Oceanic and Atmospheric Administration (NOAA). 2019. The May 29th Dust Storm. http://www.crh.noaa.gov/ddc/?n=dust. (accessed March 25, 2019).

Natural Resources Defense Council (NRDC). 2018. Ocean Pollution: The Dirty Facts. https://nrdc.org/stories/ocean-pollution-dirty-facts. (accessed November 18, 2018).

Nudunuri, K.V., Erickson, L.E., and Govindaraju, R.S. 1998. Modeling the Role of Active Biomass on the Fate and Transport of Heavy Metals in the Presence of Root Exudates. *Journal of Hazardous Waste Research*. Vol. 1. No. 9. pp. 1–25.

Payne, F.C. and Rogers, D.T. 1997. Chlorinated Solvent Degradation – A Consecutive, Irreversible Reaction Sequence. *Remediation of Recalcitrant Chlorinated Volatile Organic Compounds in Groundwater*. National Groundwater Association, Dublin, OH. Groundwater Management Book 21. pp. 97–109.

Payne, F.C., Quinnan, J.A., and Potter, S.T. 2008. *Remediation Hydraulics*. CRC Press. Boca Raton, FL.

Raub, T.D. and Kirschvink, J.L. 2008. A Pan-Precambrian Link between Deglaciation and Environmental Oxidation. In: Cooper, A.K., Barrett, P.J., Stagg, H., Stump, E. and Wise, W. editors. *Antarctica: A Keystone in a Changing World*. The National Academic Press. Washington, DC.

Rogers, D.T. 1995. *Intrinsic Bioremediation of Gasoline-Contaminated Groundwater – A Case Study*. Air and Waste Management Association. Annual Meeting. San Antonio, TX.

Rogers, D.T. 1996. *Environmental Geology of Metropolitan Detroit*. Clayton Environmental Consultants, Novi, MI.

Rogers, D.T., Murray, K.S., and Kaufman, M.M. 2007. Assessment of Groundwater Contaminant Vulnerability in an Urban Watershed in Southeast Michigan, USA. In: Howard, K.W.F. editor. *Urban Groundwater – Meeting the Challenge*. Taylor & Francis. London, England.

Rogers, D.T., Murray, K.S., and Kaufman, M.M. 2012. Environmental Risk Analysis through Integration of Geologic Vulnerability and Air, Water, and Soil Contaminant Risk Factor Derivation. International Geological Congress. p. 3003. Brisbane, Australia.

Sawhney, B.L. and Brown, K. (editors). 1989. Reactions and Movement of Organic Chemicals in Soils. Special Publication No 22. Soil Science of America. Madison, WI.

Schlatter, T.W. 2009. *Atmospheric Composition and Vertical Structure*. National Oceanic and Atmospheric Administration (NOAA). Boulder, CO.

Schnoor, J.L. 1996. *Environmental Modeling: Fate and Transport of Pollutants in Water and Soil*. John Wiley & Sons. New York.

Schwarzenbach, R.P., Gschwend, P.M., and Imboden, D.M. 1993. *Environmental Organic Chemistry*. John Wiley & Sons. New York.

Singer, A.C., Crowley, D.E., and Thompson, I.P. 2003. Secondary Plant Metabolites in Phytoremediation and Biotransformation. *Trends in Biotechnology*. Vol. 21. No. 3. pp. 123–130.

Soil Science Society of America. 1987. *Glossary of Soil Science Terms*. Madison, WI.

SRC. 2019. Environmental Fate Data Base (EFDB). CHEMFATE Chemical Search. http://srcinc.com/what-we-do/efbd.aspx. (accessed March 25, 2019).

Suthersan, S.S. and Payne, F.C. 2005. *In Situ Remediation Engineering*. CRC Press. Boca Raton. FL. 430p.

Suthersan, S.S., Horst, J., Schnobrich, M., Welty, N., and McDonough, J. 2017. *Remediation Engineering: Design Concepts*. 2nd edition. CRC Press. Boca Raton, FL. 603p.

United States Environmental Protection Agency (USEPA). 1991. *Air Pollution and Health Risk*. USEPA. EPA/450/3-90-022. Washington, DC.

USEPA. 1996a. *Transport and Fate of Contaminants in the Subsurface*. USEPA. EPA/625/4-89/019. Washington, DC.

USEPA. 1996b. *Bioscreen*. Natural Attenuation Decision Support System. Office of Research and Development. Washington, DC.

USEPA. 1997. *Mercury Study Report to Congress: Volume III: Fate and Transport of Mercury in the Environment*. USEPA. EPA-454/R-97-005. Washington, DC.

USEPA. 1998a. *Taking Toxics Out of the Air*. EPA451/K-98-001. Washington, DC.

USEPA. 1998b. *Chemical Fate Half-Lives for Toxics Release Inventory (TRI) Chemicals*. USEPA. Washington, DC.

USEPA. 1999. *Fundamentals of Soil Science as Applicable to Management of Hazardous Wastes*. Office of Research and Development. EPA/540/S-98/500. Washington, DC.

USEPA. 2006. *An Inventory of Sources and Environmental Releases of Dioxin-Like Compounds in the United States for the Years 1987, 1995, and 2000.* USEPA. EPA/600/P-03/002f. Washington, DC.

USEPA. 2008. *Findings on National Air Quality: Status and Trends Through 2006.* USEPA. EPA454/R-07-007. Research Triangle Park. NC.

USEPA. 2009a. *Biennial National Listing of Fish Advisors for 2008.* USEPA. EPA-823-F-09-007. Washington, DC.

USEPA. 2009b. *Ozone – Good Up High Bad Nearby. Air Quality Planning and Standards.* USEPA. Washington, DC.

USEPA. 2017. *Pesticide Industry Usage.* USEPA Biological and Economic Analysis Division. Office of Pesticide Programs. Washington, DC. 24p.

USEPA. 2018. Global Greenhouse Gas Emissions Data. https://epa.gov/ghgemissions/global-greenhouse-gas-emissions-data. (accessed December 31, 2018).

USEPA. 2019a. EPA's Report on the Environment (ROE). https://www.epa.gov/report-environment. (accessed March 24, 2019).

USEPA. 2019b. Advisories for Fish and Shellfish Consumption. https://www.epa.gov/fish-tech. (accessed March 25, 2019).

USEPA. 2019c. What Is Acid Rain. http://www.epa.gov/acidrain/what/index.html. (accessed March 25, 2019).

USEPA. 2019d. Integrate Risk Information System (IRIS). http://www.epa.gov/ncea/iris/intro.htm. (accessed March 25, 2019).

United States Geological Survey (USGS). 1995a. Ground-Water Quality Protection. Open-File Report 95–376. Nashville, Tennessee. http://www.pubs.usgs.gov/of/1995/ofr-95376. (accessed December 2009).

USGS. 1995b. *Contaminants in the Mississippi River.* USGS Circular 1133. Washington, DC.

USGS. 1998. *Simulating Transport of Volatile Organic Compounds in the Unsaturated Zone Using the Computer Model R-UNSAT,* USGS Fact Sheet 019–08, Washington, DC.

USGS. 2000. *Mercury in the Environment.* Fact Sheet 146–00. USGS. Washington, DC.

USGS. 2006a. *Volatile Organic Compounds in the Nation's Ground Water and Drinking-Water Supply Wells. National Water-Quality Assessment Program.* USGS Circular 1292. Washington, DC.

USGS. 2006b. *Pesticides in the Nation's Streams and Groundwater. 1992–2001 – A Summary.* USGS Fact Sheet 2006–3028. Washington, DC.

United States Standard Atmosphere. 1976. *The Standard Atmosphere of the United States. National Oceanic and Atmospheric Administration, National Aeronautics and Space Administration, and the United States Air Force.* United States Government Printing Office. Washington, DC. 277p.

Wania, F., Hoff, J.T., Jia, C.Q., and Mackey, D. 1998. The Effects of Snow and Ice on the Environmental Behavior of Hydrophobic Organic Chemicals. *Journal of Environmental Pollution.* Vol. 102. pp. 79–95.

Wiedemeier, T.H., Rifai, H.S., Newell, C.J., and Wilson, T.J. 1999. *Natural Attenuation of Fuels and Chlorinated Solvents in the Subsurface.* John Wiley & Sons. New York.

Wilson, R. and Spengler, J.D. editors. 1996. *Particles in Our Air: Concentration and Health Affects.* Harvard University Press. Cambridge, MA.

World Health Organization (WHO). 2019. *Health Statistics and Health Information Systems.* Mortality Database Tables. http://www.who.int/healthinfo/morttables/en/index/html. (accessed March 25, 2019).

4 Geologic Vulnerability

4.1 INTRODUCTION

Each day when we walk outside we are walking on a history book, yet most of us are unaware of its significance. The arrangement, thickness, and composition of the soil and sediment layers just centimeters deep have a profound influence on our lives and all life on Earth. These soil and sediment layers dictate where cities are built, where roads are built, and how buildings are constructed. Perhaps most significantly to our civilization at this juncture in our collective history, these soil and sediment layers are the meeting place between human civilization and the natural world. This contact between human civilization and the natural world is the proving ground and point at which contaminants released into the environment begin their destructive journey and eventually detrimentally impact on all living things on Earth. This is why we must understand our natural world in at least the same scientific detail that we need to understand pollution, because they are interactive with one other.

Up to this point we have defined and examined the many types of pollution and have also learned how pollutants behave in the environment once they are released. However, we have not discussed how the environment reacts to pollutants and how the environment influences pollutants. That is the subject of this chapter. We shall now look at our natural environment and explore how it influences pollution. We have control over the chemicals we use, and where and how we use them. Control over the geologic environment, however, is beyond our means. Therefore, we must understand the geologic environment where our urban areas are located and develop methods to minimize or eliminate the potential harmful effects of contaminants upon human health and the environment. A logical first step is through an *understanding* of urban geology, followed by an *evaluation* of the extent that a given urban area's geology influences the migration of pollutants. In addition, since water plays a critical role in assessing a region's vulnerability to pollution, the analyses performed require an understanding of water occurring at the Earth's surface and beneath.

Throughout the world, the largest cities share a geologic environment dominated by unconsolidated sedimentary deposits and are located near water (see Table 4.1). Most of those sedimentary deposits are saturated with water very near the surface, and function as sources of drinking water and/or as hydraulic connections to surface water, ecosystems, and ultimately to the oceans. Moreover, water is considered the universal solvent, so any pollution released into the environment from anthropogenic or natural sources has the potential to migrate and cause harm. Scientific factors that control the severity of harm to the environment from pollutant releases are: (1) the geologic and hydrogeologic environment, (2) the physical chemistry of the contaminants and amounts released, and (3) the mechanism in which the release occurs (Rogers 1996; Murray and Rogers 1999a; Kaufman et al. 2005; Rogers 2018). There are several techniques available for investigating and managing the complexity of urban water pollution (Rogers 2016a; 2016b). One important and critical concept to keep in mind when evaluating groundwater in urban areas is that the hydrologic cycle is modified and altered by human development (Wong et al. 2012).

Despite the availability of specific methods and procedures, the environmental assessment of many urban areas can become a daunting task. This situation arises because the near-surface geologic deposits in urban areas are poorly understood because they are difficult to study, require in depth study, are complex, have been anthropogenically disturbed, and exhibit high variability over short distances. Therefore, to achieve any level of success in mitigating environmental contamination, it becomes a prerequisite to understand the geology, hydrology, and the fate and migration of contaminants within its specific geology (Rogers 2014).

TABLE 4.1

Water Bodies and Near-Surface Geology Near the Major Cities of the World

City	Location	Estimated Metropolitan Population (millions)	Geology	Water Body
Tokyo	Japan	37.8	Unconsolidated	Pacific Ocean
Shanghai	China	34.8	Unconsolidated	Pacific Ocean
Jakarta	Indonesia	31.7	Unconsolidated	Pacific Ocean
Delhi	India	26.4	Unconsolidated	Yamuna River
Seoul	Korea	25.5	Unconsolidated	Han River
Beijing	China	24.9	Unconsolidated	Yongding River
New York City	USA	23.8	Unconsolidated	Atlantic Ocean, Hudson River
Mexico City	Mexico	21.6	Unconsolidated	Lerma River, Santiago River
Sao Paulo	Brazil	21.2	Unconsolidated	Atlantic Ocean
Cairo	Egypt	20.5	Unconsolidated	Nile River
Los Angeles	USA	18.7	Unconsolidated	Pacific Ocean
Moscow	Russia	16.9	Unconsolidated	Moskve River
Istanbul	Turkey	15.2	Unconsolidated	Turkist Straits
London	England	14.2	Unconsolidated	Thames River
Buenos Aires	Argentina	13.1	Unconsolidated	Atlantic Ocean
Paris	France	12.6	Unconsolidated	Seine River
Rio de Janeiro	Brazil	12.3	Unconsolidated	Atlantic Ocean
Chicago	USA	9.8	Unconsolidated	Lake Michigan
Johannesburg	South Africa	9.6	Unconsolidated	Jukskei River
Riyadh	Saudi Arabia	7.7	Unconsolidated	Simbacom River
Santiago	Chile	6.7	Unconsolidated	Pacific Ocean
Berlin	Germany	6.1	Unconsolidated	Spree River
Toronto	Canada	5.9	Unconsolidated	Lake Ontario
Sydney	Australia	5.0	Unconsolidated	Pacific Ocean

Source: United Nations. World Population Prospects. United Nations Department of Economic and Social Affairs. Population Division. https://esa.un.org/unpd/wpp/data. (accessed December 14, 2017.)

The focus of this chapter is to identify urban regions vulnerable to contamination and those areas where widespread contamination is less likely. To accomplish this task, an additional level of interpretation consisting of a comprehensive vulnerability analysis is added to the near-surface geologic maps of urban areas presented in Chapter 5. This chapter also explains why certain types of geology may be especially susceptible to contamination – a topic explored and discussed in greater detail later in this book.

4.2   SUBSURFACE VULNERABILITY AND VULNERABILITY MAP DEVELOPMENT

The concept of vulnerability of the subsurface to contamination originated in France during the 1960s and was introduced into the scientific literature by Albinet and Marget (1970). Since then, the concept of subsurface vulnerability has evolved to include both a distinction between and combination of vulnerability and risk assessment. Groundwater vulnerability is currently interpreted as a function of the natural properties of the overlying soil or sediments of the unsaturated zone, aquifer properties (e.g., effective porosity and recharge area), and aquifer material (Foster and Hirata 1988; Robins et al. 1994; Rogers et al. 2007).

Geologic vulnerability mapping can be divided into two groups: subjective rating methods and statistical and process-based methods (Focazio et al. 2001). The subjective rating methods are characterized by numerical scales representing low to high vulnerabilities. Typically the results are applied to large areas and used for policy and management objectives. By contrast, the statistical process-based methods produce finite values, such as areas exceeding specific water quality values. With these methods the results are usually not applied to large areas due to data gaps and variable geology. In addition, the results are generally obtained under more detailed site-specific assessments and used for purely scientific purposes (Focazio et al. 2001). In practice, the subjective rating methods are preferred for conducting vulnerability assessments on a watershed scale (Murray and Rogers 1999a).

The concept of geologic vulnerability relies on the assessment and representation of various hydrogeologic parameters such as vadose zone characteristics (e.g., thickness and infiltration capacity), depth to water, and amount of recharge (Zaporozec and Eaton 1996; Eaton and Zaporozec 1997). The utility of this concept, however, becomes more important when the geologic data are supplemented with environmental, economical, and political insight gained through past environmental cleanup efforts (Foster et al. 1994; Loague et al. 1998). A specific example of this data augmentation is provided at the end of the chapter.

Successful development of geologic vulnerability maps can be difficult to achieve in areas experiencing rapid growth. Urbanization and the artificial infrastructure it produces (e.g., sewers and detention ponds) can have a profound influence on the regional hydrogeology (Vuono and Hallenbeck 1995; Zaporozec and Eaton 1996; Kibel 1998). Basic processes affecting surface water and groundwater are modified, including surface water drainage patterns and velocities, evaporation rates, infiltration, and aquifer recharge (Burn et al. 2007; Fresca 2007; Howard et al. 2007; Mohorlok et al. 2007). The difficulties in vulnerability map development are also compounded by the differences in the amount and type of geologic and hydrogeologic information available in urban areas and rural settings. For these reasons, a uniform assessment of data while conducting vulnerability studies in urbanizing areas is difficult to achieve.

Geologic vulnerability mapping provides a starting point for quantifying anticipated environmental risk at a particular site and also highlight locations where additional information is required. In lieu of specific site information, this risk assessment can serve, if necessary, as a proxy for anticipated future cleanup costs. Additionally, this method can be used by other interested parties during the recycling of industrial sites to estimate the liability of sites (Stiber et al. 1995; Murray and Rogers 1999b). With respect to water resource allocation, since surface and groundwater interact, mapping groundwater also provides valuable information concerning their respective distributions (Rogers and Murray 1997; Pierce et al. 2007).

As we will discover during the course of the book, it is often impossible or prohibitively expensive to clean up polluted groundwater effectively. This reality is why a better approach for ensuring groundwater quality is to map actual groundwater pollution and groundwater potentially vulnerable to pollution, and then to use this information as an integral part of land use planning and management (Zaporozec and Eaton 1996;Kaufman et al. 2011). Then, the information gathered through the process of geologic vulnerability assessment would allow decision makers to assess the current and future environmental risks associated with any particular site as long as it was contained within the area mapped.

4.3 METHODS

The evaluation of geologic vulnerability in an urban watershed using a subjective rating method requires a combination of geologic and hydrogeologic data, identification of the potential receptor sites, and political and economic information. The first and most crucial step is mapping the near-surface geology. Once the geologic map is created, the process of developing a geologic vulnerability map can be initiated.

Murray and Rogers (1999a; 1999b) and Kaufman et al. (2003; 2005) have developed a method for geologic vulnerability mapping using a modified DRASTIC model (Aller et al. 1987). This method contains a subjective numerical rating system and uses different weighting coefficients for various geologic and hydrogeologic parameters of concern and incorporates potential receptors and political data into the model. Listed below are those geologic and hydrogeologic factors, along with the political, ecological, and environmental factors, that are collected during environmental subsurface investigations at known or suspected sites of contamination (Rogers 1992; Rogers 1996; Murray and Rogers 1999a; Rogers 2002; Kaufman et al. 2005):

- Soil or sediment type, composition, color, texture, thickness, and soil moisture
- Stratigraphy of geologic units
- Horizontal and vertical extent of geologic units
- Variation within geologic units (heterogeneity and anisotropism)
- Type of primary and secondary features
- Presence, extent, and structure of unconformities
- Presence, depth, and relative abundance of groundwater
- Groundwater flow direction
- Areas of groundwater recharge
- Areas of groundwater discharge
- Groundwater–surface water interaction features
- Anthropogenic features including:
  - Storm sewer placement relative to groundwater elevations
  - Utility corridors
  - Building footings and other subsurface structures
  - Landfills
  - Artificial recharge basins
  - Surface water confinement features
  - Dams
  - Stormwater retention or detention basins
  - Large paved areas
  - Artificial surface drainage pattern alteration
  - Roads and road cuts
- Potential receptors or points of potential exposure:
  - Water supply wells
  - Irrigation wells
  - Surface water bodies such as lakes, streams, rivers, swamps, bogs, springs, groundwater seeps, etc.
  - Buildings and building foundations
  - Parks, schools, playgrounds, day care facilities, retirement communities, hospitals, long-term care facilities, etc.
- Potential sources of contamination:
  - State and federal lists of environmental contamination
  - Hazardous waste facilities
  - Abandoned dumps
  - Brownfield sites
  - Historical industrial manufacturing sites
  - Electrical generating facilities
  - Gasoline service stations
  - Dry cleaning facilities
  - Refineries
  - Other known or identified sources

Taken together, these factors provide a framework for the evaluation of the geologic vulnerability within any urban area. This framework is constructed through the development of a vulnerability matrix and scoring system, as shown in Table 4.2 (Kaufman et al. 2011).

A sequential description of the vulnerability matrix parameters is presented in Table 4.2.

- **Depth to groundwater.** The closer to the surface groundwater is encountered the higher the geologic vulnerability because contaminants have a shorter distance to migrate vertically before encountering saturated conditions.

  Depth to groundwater varies widely across the United States depending on geology, climate, and anthropogenic influence. For example, groundwater is routinely present at depths very near the surface along the east coast and southeast; from a few feet to a few tens of feet in the Midwest; to perhaps several hundred feet in the southwest. With respect to anthropogenic influence, the landscape in urban areas is characterized by extensive modifications. The natural hydrology is altered by impervious surface, the construction of stormwater retention basins, and the use of more permeable areas for artificial recharge or groundwater storage. In addition, leaks from sanitary sewers, storm sewers, potable water lines, and surface watering (especially golf courses) may profoundly impact the depth to groundwater and the direction of flow locally or over larger areas.

- **Composition, areal extent, and thickness of soil units in the unsaturated zone.** Composition of soil or sediment above the water table is important to evaluate because it helps to determine whether the soil or sediments above the water table will impede the vertical migration of contaminants through the soil or sediment column. For instance, clay soils are fine-grained and generally impede or slow migration, whereas sand and gravel deposits tend to facilitate contaminant migration. Interbedding of geologic units is also considered.

  Other significant factors include the primary and secondary porosity and the potential for unsaturated soils located above aquifers to be subject to anthropogenic disturbance. For instance, a clay layer that is 20-feet thick may appear to significantly impede the vertical migration of contaminants. However, if portions are excavated, or if the integrity of the clay has been compromised by vertical fractures, root fragments, or anthropogenic activities, these events increase the potential for contaminant migration through the overlying clay into groundwater (Murray et al. 1997).

  Scoring this value therefore involves a consideration of multiple factors. With the 10-point scoring system, sand and gravel typically receive ten points. Clay deposits of adequate integrity, thickness (generally more than 10- to 20-feet thick), and areal extent receive a score of three or less.

- **Composition, areal extent, and thickness of saturated zone.** Composition, areal extent, and thickness of aquifer materials are important factors because they influence the rate and direction of groundwater flow. Risk may be higher in a scenario where the saturated thickness is 75 feet and the material is composed of sand as opposed to a saturated thickness of 2 feet composed of silt. The areal extent and thickness of the aquifer must also be evaluated to establish whether multiple aquifers exist and are hydraulically connected. For instance, using the example above, if the 2-foot thick saturated silt layer was hydraulically connected to the 75-foot thick saturated sand layer, this would change the risk evaluation.

  In an urban environment, perched saturated zones often result from anthropogenic influences. Construction activities such as roads, building foundations, and utility corridors are backfilled with porous materials, and then become more porous than natural soils and sediments. Over time these porous backfilled materials become saturated with water. Therefore, investigative activities must be very detailed in nature to evaluate the significance of potential anthropogenic influences on the hydrology of the area being investigated and mapped.

TABLE 4.2
Geologic Vulnerability Matrix and Scoring System

Parameter Identification Number	Parameter Description	Rating Strength
1	**Depth to Groundwater**	
	Less than 10 feet below the ground surface	10
	10 to 30 feet	5
	Greater than 30 feet	1
2	**Composition, areal extent, and thickness of soil units in the unsaturated zone**	
	Thick and extensive sequence of sand and gravel	10
	Interbedded sands and clay deposits	5
	Thick and extensive sequence of clay	1
3	**Composition, areal extent, and thickness of saturated zone**	
	Thick and extensive sequence of sand and gravel	10
	Interbedded sands and clay deposits	5
	Thick and extensive sequence of clay	1
4	**Occurrence and relative abundance of groundwater**	
	25% or less likelihood before encountering an aquiclude	10
	25% to 74% likelihood	5
	Greater than 75% likelihood	1
5	**Area of groundwater recharge**	
	Significant area of recharge	10
	Moderate area of recharge	5
	Not a significant area of recharge	1
6	**Areas of groundwater discharge**	
	Significant area of recharge	10
	Moderate area of recharge	5
	Not a significant area of recharge	1
7	**Travel time and distance to point of potential exposure**	
	Less than 10 years	10
	10 to 25 years	5
	Greater than 25 years	1
8	**Source of potable water**	
	Current source of potable water	10
	Potential source of potable water	5
	Not a potential source of potable water	1

Source: Kaufman, M. M., Rogers, D. T. and Murray, K. S. 2011. Urban Watershed: Geology, Contamination, and Sustainable Development. CRC Press. Boca Raton, FL. 583 p.

Scoring this parameter is similar to scoring the composition of the unsaturated zone (parameter #2). In practice, however, the factors determining the score are the areal extent of the saturated zone and whether the aquifer is hydraulically connected to additional or larger aquifers. If a saturated zone is encountered and evaluated to be perched or discontinuous and therefore isolated, the scoring value of 1 may be assigned. However, if the saturated thickness is 75 feet and lies within a mappable geologic formation, the scoring value of 10 may be assigned.

- **Occurrence and relative abundance of groundwater.** This parameter is similar to the depth of groundwater (parameter #1) but focuses on whether groundwater is present at relatively shallow depths, and of a sufficient quantity to sustain a rate of withdrawal. A key aspect of this measure is whether any groundwater is encountered before an aquiclude. The abundance of groundwater relates to the transmissivity of the formation and is not intended to focus on whether groundwater is encountered.

With this parameter there is high scoring variability depending on the type of materials in the deposit. Consider an extensive deposit composed of saturated sand and gravel. This deposit would tend to have a much higher transmissivity value, and would likely receive a high vulnerability score of eight or more. When this deposit is compared to a similarly extensive and saturated silty clay deposit with a low relative transmissivity, the silty clay deposit would receive a score of five or less.

- **Area of groundwater recharge**. This parameter focuses on the source of groundwater. If the area being evaluated is a source for groundwater recharge, its risk will be higher than an area not considered a significant source of groundwater discharge.

  Special care should be taken in urban areas to evaluate the anthropogenic sources of groundwater recharge. Significant anthropogenic influences include storm water infiltration, sanitary sewer leakage, and water supply leakage. Other human impacts to consider are detention basins and wetland modification.

- **Areas of groundwater discharge**. This parameter requires an understanding of groundwater flow at different geographic scales; including specific sites, offsite areas of larger extent, and the entire watershed. The migration and final discharge locations for groundwater includes an understanding of the relationship between groundwater and surface water, and permits an evaluation of the potential for interconnecting aquifer systems and discharge to deeper aquifers. It is also essential to understand the geology of a region and the potential influence of unconformities associated with the depositional units within the watershed.

- **Travel time and distance to point of potential exposure**. In general, the longer the time and distance required for a contaminant to reach its potential point of exposure, the lower the risk. This is true for most contaminants that degrade naturally in the environment. Evaluating anthropogenic influence with this parameter is also important. For instance, if groundwater is used as a source of potable water and pumped from the ground using extraction wells, the residence time of potential contaminants in the aquifer may be greatly reduced.

- **Source of potable water**. If the aquifer encountered beneath a particular site is used as a source of potable water or is connected to a source of potable water, the risk increases. Evaluating this parameter must also include other potential uses besides the provision of drinking water. Further information may be necessary to identify those uses and accurately score this parameter.

4.4   MAP BUILDING EXAMPLE

Because of its diversity, a good example of building a geologic vulnerability map is provided by the Rouge River watershed located in southeastern Michigan (Figure 4.1).

The key features making the Rouge River watershed an ideal watershed to study include:

- It is an urban watershed
- It has varied geology and hydrogeology
- There are different stream patterns and densities throughout
- The population is approximately 1.5 million people
- The watershed has been studied extensively
- There is an abundance of geologic and hydrogeologic data
- It has a long history of industrial output

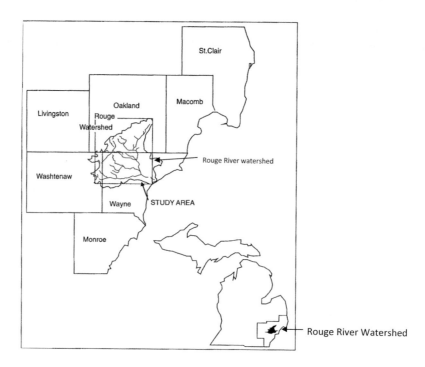

FIGURE 4.1   Rouge River watershed.

- It has significant environmental impacts
- It has similar geology to many other cities in the world

The Rouge River watershed has over 120 miles of streams, tributaries, lakes, and ponds and encompasses an area of approximately 450 square miles (Murray and Rogers 1999a). The fan-shaped watershed includes all or part of 47 municipalities in three counties. Over 66% of the watershed has been developed, and in the year 2000, 99% of the watershed's population lived within the U.S. Census-defined urbanized areas – making it the watershed with the highest population density in the eastern United States (Kaufman et al. 2003).

The near-surface geology of the watershed is dominated by glacial deposits, glacial lacustrine deposits, and the recent fluvial deposits from the Rouge River itself. The glacial deposits are generally greater than 200 feet in thickness, the glacial lacustrine deposits are rarely more than 30-feet thick, and the recent fluvial deposits are generally less than 10-feet thick (Farrand 1982; 1988; Rogers 1996; 1997a; 1997b). Distinctive geologic units identified and mapped within the watershed include:

- Four different surface moraine units that include the Fort Wayne Moraine, the Outer and Inner Defiance Moraine, and the Birmingham Moraine
- Glacial outwash deposits
- Several beach deposits composed of sand from historical glacial lakes
- Silt and clay deposits from historical glacial lakes
- Recent fluvial deposits from the Rouge River
- A ground moraine or lodgment till composed of clay underlies the entire watershed beneath the near-surface deposits described above
- A sand and gravel deposit is located beneath the ground moraine in the center portion of the watershed and represents a fluvial deposit from a large river that was present before the last glacial advance some 22,000 years before the present

- Weathered bedrock is encountered beneath the ground moraine at other locations
- Bedrock composed of shale, limestone, and sandstone of Paleozoic age is present beneath the unconsolidated units

The geologic map, stratigraphic column, and geologic cross-section for the watershed are included as Figures 4.2, 4.3, and 4.4 respectively. The geologic map, cross-section, and stratigraphic column were products of information from more than 3,000 subsurface investigations and an abundance of field work and historical literature from numerous sources (Rogers 1997b).

Anthropogenic effects within the Rouge River watershed have been numerous and significant. Surface water and groundwater within the watershed have been severely degraded. At one time,

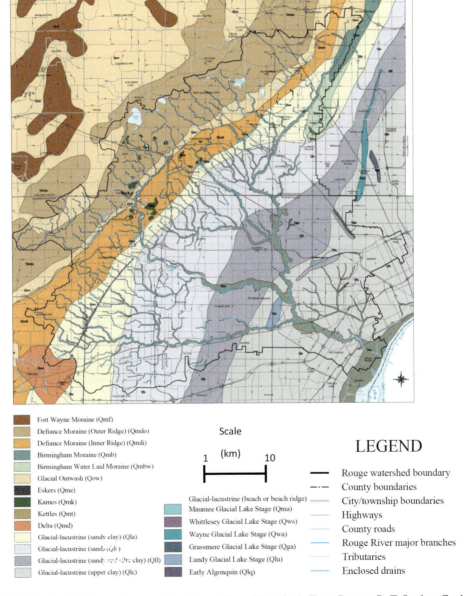

FIGURE 4.2   Surficial geologic map of the Rouge River Watershed. (From Rogers, D. T. Surface Geological Map of the Rouge River Watershed in Southeastern Michigan. Wayne County, MI. 1:62,500. 2 Sheets. 1997b.)

Era	Period	Epoch	Stage		Age (approximate years before present)	Symbol	Unit Name
Cenozoic	Quaternary	Holocene	Wisconsinan	Fluvial	<10,000		Qaa - Recent Rouge River Fluvial Deposits
		Pleistocene		Glacial lucustrine	<14,500		Qllu - Glacial Lacustrine Beach Deposit Lake Lundy
							Qlsg - Glacial Lacustrine Beach Deposit Lake Grassmere
							Qlswa - Glacial Lacustrine Beach Deposit Lake Wayne
							Qlc - Glacial Lacustrine Clay
							Qll - Glacial Lacustrine Silty Clay
							Qws - Glacial Lacustrine Beach Ridge Lake Whittlessey
							Qlsla - Glacial Lacustrine Sand, Lowest Lake Arkona
							Qlsha - Glacial Lacustrine Sand, Highest Lake Arkona
							Qv - Varves
							Qlahm - Glacial Lacustrine Sandy Clay Maumee Lake Deposit
							Qma - Glacial Lacustrine Beach Highest Lake Maumee
				Moraine	<14,500		Qmbw - Birmingham Waterlaid Moraine
							Qmb - Birmingham Moraine
				Delta	<14,500		Qmd - Delta
				Outwash	<14,500		Qowd - Outwash Defiance
				Esker	<14,500		Qme - Esker
				Moraine	<14,500		Qmdi - Inner Defiance Moraine
							Qmdo - Outer Defiance Moraine
				Glacial Outwash	<14,500		Qow - Outwash Fort Wayne
				Kame	<14,500		Qmk - Kame
				Kettle	<14,500		Qmt - Kettle
				Moraine	<14,500		Qmf - Fort Wayne Moraine
							Qmg - Ground Moraine / Lodgement Till
				Fluvial	22,000		Qfa - Fluvial (Interglacial) >22,000 ybp
				Moraine	22,000		Qmo - Pre-Late Wisconsinan Glacial Till Deposit >22,000 ybp
Paleozoic	Mississippian	Lower	Kinderhookian	Sunbury Shale	345,000,000		Ms - Sunbury Shale
				Bedford Shale	345,000,000		Mbd - Bedford Shale
	Devonian	Upper	Senecan	Antrim Shale	350,000,000		Da - Antrim Shale
			Erian	Traverse Group	355,000,000		Dt - Traverse Group
		Middle	Erian	Dundee Limestone	360,000,000		Dd - Dundee Limestone
			Ultsterian	Detroit River Dolomite	370,000,000		Ddr - Detroit River Formation

FIGURE 4.3 Stratigraphic column of the Rouge River watershed. (From Rogers, D.T. Surface Geological Map of the Rouge River Watershed in Southeastern Michigan. Wayne County, MI. 1:62,500. 2 Sheets. 1997b. Rogers 1997b.)

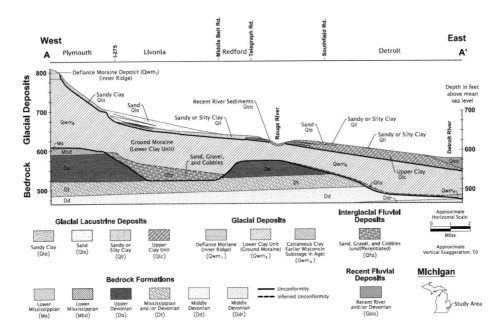

FIGURE 4.4   Geologic cross-section of the Rouge River watershed. (From Rogers, D. T. Surface Geological Map of the Rouge River Watershed in Southeastern Michigan. Wayne County, MI. 1:62,500. 2 Sheets. 1997b. Rogers 1997b.)

the Rouge River was ranked as one of the most toxic sites in Michigan (Michigan Department of Environmental Quality 2008). The watershed is also the focus of ongoing intense scientific study and restoration. It has been identified as an area of concern by the International Joint Commission (Hartig and Zarull 1991) and cited as a significant source of pollution to the lower Great Lakes (Murray and Bona 1993). Rouge River sediments have shown a significant presence of heavy metals and polychlorinated biphenyls (PCBs) (Michigan Department of Natural Resources 2008 ; Murray 1996; Murray et al. 1999).

Shallow groundwater within the Rouge River discharges to surface water and accounts for most of the base flow (Rogers and Murray 1997). Average annual precipitation amounts and the geology of the watershed determine this flow pattern. Climatically, the Rouge watershed is situated within the humid microthermal zone of the Midwestern United States, meaning the Rouge River is effluent and fed by groundwater entering as base flow. With respect to its geology, the lower clay unit (ground moraine or lodgment till) that completely underlies the watershed, very thick (ranging from 30- to more than 180-feet thick), has a very low hydraulic conductivity (less than 1×10^{-8} cm/s), and does not show signs of unconformities or features suggesting any significant secondary porosity (i.e., vertical fractures or root fragments). Therefore, the lower clay unit is a very effective aquiclude. This type of formation is not uncommon. Most other areas in North America glaciated during the Pleistocene have similar ground moraine or lodgment till deposits of similarly low hydraulic conductivity (Keller et al. 1989). This is a significant finding because any contamination that does not degrade and reaches groundwater will eventually migrate and discharge to the Rouge River. From there the contamination subsequently discharges into the lower Great Lakes (Rogers and Murray 1997;Rogers 1997a). Figure 4.5 shows this discharge of the watershed's surface water and groundwater.

Using this geologic, hydrogeologic, and anthropogenic-impact information, a vulnerability matrix was developed and is presented in Table 4.2 (modified from Murray and Rogers 1999a; Rogers 2002). The geologic units with the highest geological vulnerability (represented by the total score column) were the outwash unit, recent fluvial deposits associated with the Rouge River, and

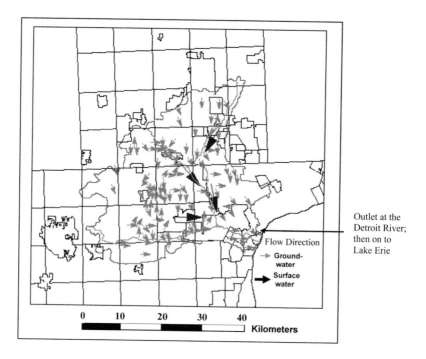

FIGURE 4.5 Surface water and groundwater flow patterns in the Rouge River watershed.

the sand unit of glacial lacustrine beach origin. The next group of units with moderate risk were the four surface moraine units. The geological units with the lowest risk were a sandy clay unit, silty clay unit, and an upper clay unit, all associated with the glacial lacustrine deposits.

The outwash, recent fluvial, and sand units were evaluated to be the most geologically vulnerable geologic units in the watershed because they are generally composed of coarse-grained sediments, have an abundance of groundwater, and account for the majority of the base flow of surface water to the Rouge River. The four moraine units were of moderate risk because they are composed of finer-grained deposits, have less abundance of groundwater, and do not serve to recharge groundwater or discharge to surface water as significantly as the outwash, recent fluvial, or sand units. The low vulnerability units are the sandy clay unit, sandy and silty clay unit, upper clay unit, and the lower clay unit. These geological units generally do not have an abundance of groundwater, are very fine-grained units, and generally impede the migration of groundwater. Using the surface geological map of the Rouge River watershed (Rogers 1997b) as a base map, and the information contributing to the construction of Table 4.3, Figure 4.6 was developed as the geologic vulnerability map for the Rouge River watershed (Rogers 1997c).

4.5 DEMONSTRATING THE SIGNIFICANCE OF VULNERABILITY MAPPING

As noted by Foster et al. (1994) and Loague et al. (1998), the significance of geologic vulnerability is not appreciated until it can be put into environmental, economic, or political perspective by actually cleaning up sites of environmental contamination. Therefore, to evaluate whether certain geologic units are vulnerable to contamination, a comparison between specific sites located in low geologic vulnerability areas to sites located in high geologic vulnerability areas must be conducted. If valid, the geologic vulnerability mapping should confirm that the sites situated above high vulnerability locations pose a greater risk of exposure than sites located above low vulnerability locations.

For this analysis, a site of low vulnerability located in a geological environment predominantly composed of clay sediments ("Site 1") is compared to a high vulnerability site located in a geological environment predominantly composed of sand ("Site 2"). Both sites are located in the Rouge River

TABLE 4.3

Rouge River Geologic Vulnerability Scoring

Geologic Unit	Parameters and Vulnerability Scoring								Total Score	Rank
	1	2	3	4	5	6	7	8		
Outwash unit	10	9	9	10	10	10	10	10	79	1
Recent fluvial	10	6	6	8	10	10	10	10	70	2
Main sand unit	10	9	9	10	10	10	6	1	65	3
Other sand units	10	8	8	10	10	10	6	1	63	4
Moraine unit 1	5	7	7	8	8	10	5	10	60	5
Moraine unit 2	5	6	6	7	7	10	5	10	56	6
Moraine unit 3	5	5	5	7	7	10	5	10	54	7
Moraine unit 4	5	4	4	6	6	10	5	10	50	8
Sandy clay unit	4	3	3	4	1	1	3	1	20	9
Sandy & silty clay unit	4	2	2	4	1	1	3	1	18	10
Upper clay unit	3	1	1	3	1	1	2	1	13	11
Lower clay unit (Ground moraine unit)	1	1	1	1	1	1	1	1	8	12

Source: Kaufman, M. M., Rogers, D. T. and Murray, K. S. 2011. Urban Watershed: Geology, Contamination, and Sustainable Development. CRC Press. Boca Raton, FL. 583 p.

watershed and are separated by only 7 miles. However, Site 1 is, a) significantly larger than Site 2 (approximately twice the size); b) had a much longer heavy industrial operational history (operated approximately 40 years longer); and c) had significantly more contamination and types of contaminants (nearly ten times the mass and three times as many contaminants) released into the environment.

Without considering the geology of each site, it would be logical to assume the environmental risks were higher at Site 1, and the associated cleanup costs would be higher and reflect its contamination history. We now determine if the vulnerability map predicts these outcomes.

4.5.1 SITE 1 – LOW VULNERABILITY SITE

Site 1 is a former heavy manufacturing facility located on approximately 16 acres of land that operated for approximately 70 years. A Phase I environmental site assessment was required by the lending institution and conducted due to a real estate transaction involving the property. This initial assessment identified five recognized environmental conditions (RECs). During the next investigational period (Phase II), several subsurface investigations were conducted at the facility and four main sources of contaminant release were identified that required remediation and included: (1) surface spills; (2) an aboveground storage tank; and (3) spills and leaks of hazardous liquids located in waste storage areas. Other sources or releases were identified during the course of evaluating the site but were not severe enough to warrant further action. The five RECs identified during the Phase I investigation included:

- Former chemical storage areas. Evidence of surface staining indicating some spillage of liquids was observed on bare ground near the two former storage areas. No staining was observed near the current storage area.
- Current storage area. The current storage area was located inside the main manufacturing building (northern building). The concrete flooring was heavily cracked, providing a potential pathway for spills and leaks to contaminate the ground beneath the building.
- A former aboveground storage tank that stored gasoline. A limited amount of surface staining was observed at the general location the tank had been located.

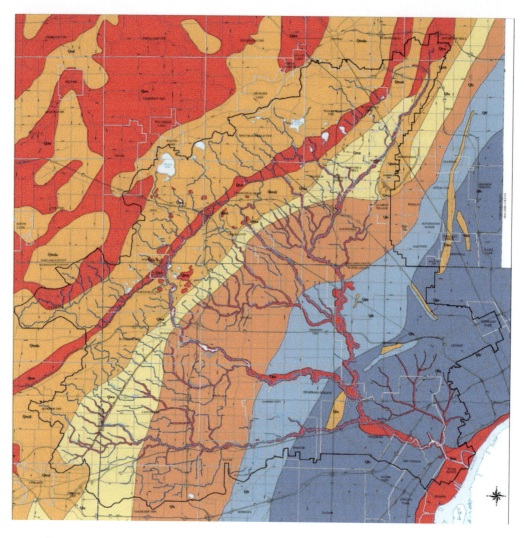

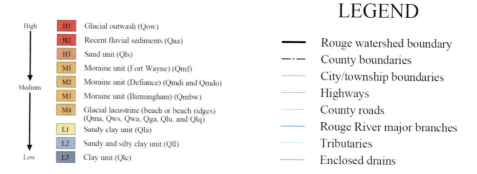

FIGURE 4.6 Geologic vulnerability map: Rouge River watershed. (From Rogers, D. T. Geologic Sensitivity Map of the Rouge River in Southeastern Michigan. Wayne County, Michigan. 1:62,500. 1 Sheet. 1997c.)

- Surface soil staining and stressed vegetation near a back door of the facility and close to an inside location where maintenance activities were conducted and dense nonaqueous phase liquids (DNAPL) solvents were used.
- Stained soil and stressed vegetation were observed at a location of bare ground where deliveries to the facilities were conducted, and where materials including liquids were offloaded from trucks.

The initial Phase II investigation involved drilling 15 soil borings in areas with the highest likelihood of detecting contamination. This initial subsurface investigation had the following objectives: 1) begin to characterize the subsurface geology of the site; 2) identify the contaminants that were suspected to have been released; and 3) evaluate the magnitude of contaminants present by collecting and analyzing "worst-case" samples from each area of suspected contamination. The results confirmed the presence of contamination at five of the six RECs identified during the Phase I environmental site assessment. Those compounds detected were at significant concentrations and required further evaluation. The current storage area was eliminated as an area of concern because contamination could not be confirmed through the drilling of three soil borings into the most visibly vulnerable areas of the concrete and the analysis of six soil samples.

Three additional investigation phases were conducted to define the nature and extent of the contamination. During the subsequent phases, a total of 40 additional soil borings were drilled, and six monitoring wells were installed to evaluate whether there was enough groundwater present for analysis and to establish the direction of groundwater flow. In many of the soil borings, more than two soil samples were analyzed to gather data on the vertical extent of the contamination. The maximum depth of the soil borings was 15 feet beneath the surface in impacted areas, yet the vertical extent of contamination did not exceed a depth of 5 feet. The monitoring wells indicated damp to moist soils existed at some locations within very thin layers of silt. These silty layers were just a few millimeters thick and were also observed in soil samples collected from some of the soil borings drilled during investigative activities. After the monitoring wells failed to detect any groundwater seepage, they were pulled from the ground and the boreholes sealed with a bentonite clay grout. Characterization of the geology at the site was accomplished by drilling a soil boring to a depth of 40 feet in a non-impacted area. In general, the site was underlain by a clay deposit from a Pleistocene age glacial lake that occupied the region more than 12,000 years before the present. Historical geological literature of the region indicates a ground moraine or lodgment till deposit extended to depths of approximately 180 feet beneath the ground at the site. Table 4.4 details the geology between the ground surface and 40 feet beneath the site.

Other pertinent technical and geological information concerning the site included the following:

- Storm sewers in the immediate vicinity did not intersect any of the contamination
- Surface water drainage was controlled by storm sewers
- No buried utilities intersected contaminated areas
- Potable water was supplied by the municipality and the source was more than 10 miles away
- The contamination did not extend beyond the property boundary

The types of contaminants detected at the facility included:

- Volatile organic compounds (VOCs) including:
  - DNAPLs commonly referred to as chlorinated solvents used to degrease and clean metal surfaces
  - Light nonaqueous phase liquids (LNAPLs) used as solvents, paint thinners, and cleaning products, and are common constituents in fuels such as gasoline
- Polynuclear aromatic hydrocarbons (PNAs or PAHs) commonly used as lubricants and motor oils and cutting fluids

TABLE 4.4

Description of Geology for Site 1

Geologic Unit	Depth (feet)	Color	Soil Class	Moisture Content	Description
Fill	0 to 2	Gray to brown	Fill	Dry	Fill material consisting of sand and silt and some construction debris including brick and wood fragments. Fill material ranges from 1- to 2-feet thick throughout the site and is not present beneath the buildings.
Clay with some silt	2 to 8	Light brown	CL	Dry to moist	Light brown clay with occasional very thin discontinuous silt layers indicating layered deposition. Silt layers range in thickness from less than a millimeter to not more than two millimeters.
Clay	8 to 40	Light olive-gray to blue-gray	CH	Damp	Blue to gray colored ground moraine clay. Upper portions very plastic. No visible signs of any silt or original depositional structures to indicate depositional layering or any sort. Very consistent in lithology and color with depth. No signs of larger grained materials such as pebbles. Groundwater was not observed at the contact between the upper lacustrine clay and the ground moraine clay.

- Polychlorinated biphenyls (PCBs) used in electrical equipment
- Heavy metals (arsenic, chromium, and lead) commonly used in paints and pigments, batteries, and metal plating

A list of specific chemical compounds, highest concentrations detected, and estimated contaminant mass remediated are listed in Table 4.5.

Soil excavation and disposal of the contaminated soils at a licensed landfill was the remedial method of choice for the contamination at this site. The overriding considerations for selecting this method were: the contaminated areas were less than 5 feet deep and were not located beneath any buildings, the activities could be conducted quickly without disturbing ongoing facility operations, and it represented the lowest cost alternative. Approximately 7,000 cubic yards – equivalent to 10,000 tons of soil – was excavated and transported to a local landfill for disposal. The total cost for investigation and remediation was approximately $400,000. This translates into a remediation cost per kilogram of the contaminant of nearly $2,000. After the remediation was verified by the regulatory authority through the collection and analysis of the soil samples taken from each area remediated, closure was granted and a "No Further Action Required" letter was issued for the site. The closure was deemed unrestricted, meaning the site had been remediated to comply with residential land use requirements. Eighteen months had elapsed since the Phase I environmental site assessment.

4.5.2 Site 2 – High Vulnerability Site

Site 2 is a former heavy manufacturing facility approximately 8 acres in size that operated for 30 years. The Phase I environmental site assessment identified four RECs:

- Two former chemical storage areas. Evidence of surface staining indicated some spillage of liquids at one storage area located on bare ground near an area storing waste paints. An additional waste storage area was located on stained asphalt pavement that was heavily cracked and broken.

- A former underground storage tank. This underground storage tank once stored gasoline and was identified as a REC because during its removal: 1) it was not physically inspected by a qualified professional; and 2) confirmatory soil samples were not collected and analyzed from the excavation pit to verify the tank did not leak.
- Surface staining. Surface staining and stressed vegetation were observed near the former location of a back door near the location where solvents were used inside the manufacturing building.

The multiple Phase II subsurface investigations conducted at the facility identified four main sources of contamination resulting in the release of contaminants and requiring remediation. These sources of contamination included four areas with a track record of prior spills. Other sources or releases were identified during site evaluation but were not severe enough to require further action.

The initial subsurface investigation conducted involved drilling 12 soil borings in areas with the highest likelihood of detecting contamination, and was performed with similar objectives to those at Site 1: characterize the geology; identify the contaminants; and analyze the worst-case samples.

TABLE 4.5
Contaminant Types, Concentration, Mass Remediated for Site 1

Contaminant	Maximum Concentration (ug/kg)	Estimated Contaminant Mass Remediated (kilograms)
Volatile Organic Compounds (VOCs)		80
DNAPL Compounds		
Tetrachloroethene (PCE)	60,100	
Trichloroethene (TCE)	45,000	
Cis-1,2-dichoroethene	20,000	
Trans-1,2-dichloroethene	3,000	
Methylene chloride	800	
LNAPL Compounds		
Ethyl benzene	10,000	
Xylenes	10,000	
Acetone	280	
Carbon disulfide	200	
Polynuclear Aromatic Hydrocarbons		115
Naphthalene	339,000	
Acenaphthalene	18,000	
Fluorene	22,000	
Phenanthrene	280,000	
Fluoranthene	156,000	
Pyrene	13,000	
Benzo(a)anthrcene	4,800	
Benzo(a)pyrene	3,200	
Dibenzo[a.h]anthracene	1,800	
Indeno[1,2,3-cd]pyrene	1,400	
Chrysene	11,000	
Polychlorinated Biphenyls (PCBs)	16,000	2
Heavy Metals		23
Arsenic	23,000	
Chromium	530,000	
Lead	930,000	

μg/kg = micrograms per kilogram

The results confirmed the presence of contamination at three of the four RECs identified in the Phase I environmental site assessment. Contaminant concentrations in near-surface soil were also detected at sufficient levels to require further investigation. The one location not pursued for further investigation was near the former underground storage tank. Four soil borings made at the location of this tank and the analysis of four soil samples taken from the soil beneath the tank did not confirm the presence of contamination above detectable concentrations.

Groundwater was encountered at a depth of 10 to 12 feet beneath the surface of the ground during the initial investigation. Temporary monitoring wells were installed at select locations to evaluate the possible presence of groundwater impacts and estimate the direction of groundwater flow. The analytical results suggested the presence of groundwater impacts likely originating from onsite sources. This finding was confirmed because there were levels of several contaminants exceeding applicable cleanup criteria, and the same contaminants were detected in near-surface unsaturated soil at the locations where the RECs were identified, but were not detected in soil or groundwater at upgradient locations.

Six additional investigation phases were conducted to define the nature and extent of the contamination. During these subsequent phases, a total of 132 additional soil borings were made, with many of the soil borings having multiple samples analyzed to help characterize the vertical extent of the contamination. A total of 80 monitoring wells were installed to define the nature and extent of impacts to groundwater.

The general subsurface geology of the site immediately beneath the surface consisted of a sand deposit originating from a Pleistocene age glacial lake that occupied the region more than 12,000 years before the present. Specific geology beneath the site consisted of sand from the surface to a depth of 26 to 32 feet. Beneath this glacial lacustrine beach sand deposit was a ground moraine or lodgment till deposit extending to a depth of at least 75 feet. Historical geological literature of the region indicated that the ground moraine or lodgment till deposit extended to depths of approximately 200 feet beneath the ground in the area. Table 4.6 describes the geology between the ground surface and 75 feet beneath the site.

During the investigation, multiple groundwater monitoring wells were installed at the same location but were screened at different depths within the saturated zone to evaluate the vertical distribution of contaminants within the aquifer. Several samples of the ground moraine deposit beneath the aquifer were also analyzed for the presence of contamination and for certain hydrologic parameters (such as hydraulic conductivity and grain size analysis) to evaluate whether the ground moraine

TABLE 4.6
Description of Geology for Site 2

Geologic Unit	Depth (feet)	Color	Soil Class	Moisture Content	Description
Sand	0 to 26	Medium orange	SW	Dry to 10 to 12 feet, then saturated	Medium- to coarse-grained sand with occasional pebbles. Evidence of bedding present. Thickness of bedding layers ranges between a few millimeters to three centimeters.
Clay with some silt	26 to 30	Light brown	SM	Saturated	Fine-grained sand. Immediately grades into blue clay. Sharp contact. Evidence of clay intraclasts in lower portion of sand indicating an erosional surface.
Clay	30 to 75	Light olive-gray to blue-gray	CH	Damp to dry	Blue to gray colored ground moraine clay. Upper portions very plastic. No visible signs of any silt or original depositional structures to indicate depositional layering or any sort. Very consistent in lithology and color with depth. No signs of larger grained materials such as pebbles.

deposit was an effective aquiclude preventing contaminant migration to deeper aquifers. In addition, three deep soil borings were drilled to a depth of 75 feet in non-impacted areas of the site to verify the horizontal distribution and thickness of the ground moraine deposit. Other technical and geologically related information relevant to the site analysis included the following:

- Storm sewers in the immediate vicinity did not intersect any of the contamination.
- Surface water drainage was controlled by storm sewers.
- No buried utilities intersected contaminated areas.
- Potable water was supplied by the municipality and the source was more than 15 miles away. However, some local residences used groundwater within the same aquifer for irrigation purposes.
- Contaminated groundwater extended beyond the property boundary approximately 1,200 feet.

The types of contaminants detected at the facility included:

- Volatile organic compounds (VOCs) including:
  - DNAPLs, commonly referred to as chlorinated solvents used to degrease and clean metal surfaces
  - Light nonaqueous phase liquids (LNAPLs) used as solvents, paint thinners, and cleaning products, and are common constituents in fuels such as gasoline
- Polychlorinated biphenyls (PCBs) used in electrical equipment

The specific chemical compounds with their highest concentrations detected and estimated contaminant mass remediated are listed in Table 4.7.

The highly permeable soil and shallow groundwater depth (i.e., the vulnerable geology) allowed VOCs to rapidly infiltrate and migrate to groundwater. In addition, the high groundwater seepage velocities resulted in a VOC plume extending one-third of a mile to a downgradient spring, and

---

TABLE 4.7

Contaminant Types, Concentration, Mass Remediated for Site 2 High Vulnerability Site

Contaminant	Maximum Concentration in Soil (ug/kg)	Maximum Concentration in Groundwater (ug/L)	Estimated Contaminant Mass Remediated (kilograms)
Volatile Organic Compounds (VOCs)			45
DNAPL Compounds			
Tetrachloroethene (PCE)	22,000	3,250	
Trichloroethene (TCE)	8,000	2,200	
Cis-1,2-dichoroethene	5,000	1,800	
Trans-1,2-dichloroethene	480	280	
1,1-dichloroethene	640	220	
1,1,1-trichloroethane	2,800	2,100	
Vinyl chloride	1,100	240	
LNAPL Compounds			
Ethyl benzene	32,000	130	
Xylenes	28,000		
Polychlorinated Biphenyls (PCBs)	7,000,000	Not Detected	**12**

μg/kg = micrograms per kilogram
ug/L = microgram per liter

ultimately discharged into a recreational lake and the Rouge River. The map in Figure 4.7 shows the extent of the VOCs in groundwater and the location of the downgradient spring.

PCBs were not detected in groundwater. Therefore, excavation and disposal of the PCB-contaminated soils at a licensed landfill was the remedial method of choice because the impacted soils were less than 3 feet in depth, were not located beneath any buildings, could be conducted quickly, did not disturb ongoing facility operations, and was the lowest cost alternative.

Excavation was also the remedial action of choice for soils highly impacted with VOCs. This method was chosen because there was not a substantial volume of impacted soil with VOCs, as the sandy soils at the site had a low capacity for retaining VOCs. As a result, the VOCs tended to migrate downward through the soil column and contaminate groundwater without absorbing to soil grains.

The remedial method chosen for groundwater was air sparging and soil vapor extraction since the contaminants in groundwater were VOC compounds and did not extend past the mid-portion of

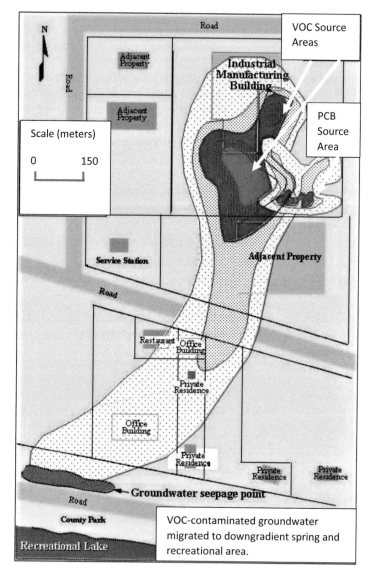

FIGURE 4.7   Contaminant migration at Site 2.

the saturated zone. Air sparging involved the injection of air beneath the impacted groundwater and then letting the air rise naturally through the saturated zone. As the air migrated upward through the saturated zone it volatilized the contaminants, and the vapors containing the VOCs were removed using a soil vapor extraction system in the vadose zone. The vapors were removed from the air by passing them through a granular activated carbon tank.

The VOCs contaminating groundwater at this site were DNAPL compounds having a specific gravity slightly greater than water. Therefore, when present at sufficient concentrations, DNAPL compounds may sink through the water column and contaminate lower portions of an aquifer. This sinking action did not occur at this site because its geology and hydrogeology – in effect a stratigraphic control – prevented the VOCs from migrating to the bottom of the aquifer. As listed and described in Table 6.5, the composition of the aquifer gradually became finer-grained with depth and the hydraulic conductivity decreased proportionately. This reduction of hydraulic conductivity, combined with low contaminant mass in groundwater, resulted in restricting contaminants to the upper portion of the aquifer. With the VOCs restricted to the upper portion of the aquifer, air sparging became the most practical remediation technique.

The time duration between the Phase I environmental site assessment and the receipt of the closure letter was approximately 14.5 years. The dollar costs for remediating this site stacked up this way: the PCB-contaminated soil, including investigation, was $1.1 million; $0.1 million for the VOC-contaminated soil; VOC-contaminated groundwater, including investigation, was $6.6 million. The total cost for investigation and remediation of this site was $7.8 million dollars, which translates into a cost of $60,000 per pound.

Remediation was verified by the regulatory authority through the collection and analysis of the soil samples from each area remediated. In addition, four iterations of groundwater samples over a period of 1 year were made until cleanup levels were achieved. Closure was then granted and a "No Further Action Required" letter was issued for the site. A risk assessment was also conducted for the site, and was designed to determine if the residual contamination at the site would pose an ongoing and unacceptable risk to human health and the environment. Locations included in this evaluation were the onsite areas with persistent soil contamination, and the onsite and offsite areas with contaminated surface water and groundwater.

The results of the risk assessment indicated that a deed restriction was appropriate for the site property and banned the use of groundwater for any purpose. A cap consisting of asphalt pavement was also required for portions of the property where some soil contamination remained in place, and if any soil were to become disturbed or exposed, another evaluation must be conducted to evaluate the need for further remedial actions. Land use for the property was restricted to industrial use.

4.5.3 Site Comparison Analysis

A profound difference between these two examples is seen in the cost per pound to remediate the contaminants. Despite the smaller acreage, shorter time of industrial operations, and low contaminant mass released to the environment, the cost to remediate a pound of the contaminant at the site of high vulnerability (Site 2) is 75 times greater than to remediate a pound of contaminant at the low vulnerability site (Site 1). This cost differential is solely due to the geology present at each site.

The high vulnerability site is located in a geologic area (a sand) that is environmentally vulnerable to contamination because: 1) the highly permeable soil allows contaminants to infiltrate readily and migrate to groundwater; and 2) contaminant plumes are transported at relatively high seepage velocities in the sand aquifer to potentially sensitive receptors that include potable water wells and surface water. The costs for the high vulnerability site would have been much higher had the contaminants migrated along the bottom portion of the aquifer; this is sometimes the case with the type of contaminants present (DNAPLs) since they are denser than water. Luckily, the decreased hydraulic conductivity within the lower portion of the aquifer prevented this from happening at the higher vulnerability site. In addition, the costs would have been significantly higher had there been

TABLE 4.8

Site Comparison Table

Parameter	Site 1 – Low Vulnerability	Site 2 – High Vulnerability
Predominant geology	Clay	Sand
Presence of shallow groundwater	No	Yes
Size of site	16 acres	8 acres
Length of operation	70 years	30 years
Land use	Heavy Industry	Heavy Industry
Types of contaminants	VOCs, PNAs, PCBs, and heavy metals	VOCs and PCBs
Number of contaminants remediated	27	9
Contaminant mass remediated	500 pounds	130 pounds
Cleanup criteria	Same as Site 2 for overlapping compounds	Same as Site 1 for overlapping compounds
Cleanup cost*	$400,000	$7,800,000
Remedial methods (soil)	Excavation	Excavation
Remedial methods (groundwater)	Remediation Not Required	Air Sparging and Soil Vapor Extraction
Other remedial control measures	None, unrestricted closure	Restricted closure included: • Deed restriction • Institutional controls • Industrial land use only
Timeframe	18 months	14.5 years
Cost per kilogram of contaminant	$2,000	$150,000
Vulnerability ranking Using Table 4.2	13	65

* Cleanup costs include costs for investigation and remediation

a completed human pathway represented by the ingestion of contaminated groundwater. Table 4.8 summarizes the major differences between these two sites of environmental contamination. Please note that the vulnerability map ranking accurately predicted the relative costs of remediation.

4.6 SUMMARY AND CONCLUSION

Through the construction of a vulnerability map, we have combined the knowledge gained from earlier chapters to explain the relationship between the natural environment and human influence in urban areas. The arrangement, thickness, and composition of the sediment layers beneath our feet have a profound influence on where cities are located, how buildings are constructed, where roads are built – and perhaps most important to the development and redevelopment of our urban centers – how contaminants behave and how they affect the environment and people. The two case studies presented in this chapter have highlighted this relationship – but they are a just a small subset of the thousands of examples of this important concept and its multi-faceted connections.

As demonstrated by comparing the two sites in this chapter and as we shall see in the next section of this book – once the environment has been contaminated at levels that pose a human or ecological risk – it is often very expensive to remediate, especially when groundwater is affected. Furthermore, it may be impossible to fully remediate some sites even with the most advanced technology. Therefore, minimizing wastes and preventing pollution have proven to be the most effective methods for reducing costs, and ultimately, preserving our environment. The two examples highlighted in this chapter are not uncommon. Tens of thousands of industrial and even commercial and residential sites in the United States have contaminated soil and groundwater to levels requiring one or more expensive remedial actions.

The realization that certain locations or areas within urban regions are especially vulnerable to contamination offers even greater promise for resolving future environmental issues. Geologic vulnerability analysis of urban regions produces essential information for evaluating the environmental and financial risks associated with development and redevelopment. By minimizing the impact of pollution once a release has occurred, certain geological features may play, if we so choose, a central role in the development and redevelopment of any urban area.

However, the story does not end here because geology alone is not responsible for dictating the environmental risks and the costs of remediating contamination. The physical chemistry of contaminants themselves also plays a central role. Until now we have concentrated on learning the geology of urban areas and how the geology of a particular region influences environmental risk. The next chapter introduces the next piece of the environmental risk puzzle – the contaminants themselves. As we will learn, the physical chemistry of specific contaminants is as important to geology when estimating the environmental risk.

REFERENCES

Albinet, M. and Margat, J. 1970. Cartographie de la Vulnerabilite a la Pollution des Nappes d'"neau Souterrians. *Bulletin Bur. Rech. Geol. Min. Sect 3 (Fr.)*. Vol. 2. pp. 13–22.

Aller, L., Bennett, T., Lehr, J.H., Petty, R.J., and Hackett, G. 1987. *DRASTIC: A Standardized System for Evaluating Ground Water Pollution Potential Using Hydrogeologic Settings*. United States Environmental Protection Agency USEPA-600/2-87-35. USEPA. Ada, OK.

Burn, S., Eiswirth, M., Correll, R., Cronin, A., DeSilva, D., Diaper, C., Dillon, P., Mohrlok, U., Morris, B., Rueedi, J., Wolf, L., Vizintin, G., and Vott, U. 2007. Urban Infrastructure and Its Impact on Groundwater Contamination. *In*: Howard, K.W.F. editor. *Urban Groundwater – Meeting the Challenge*. Taylor & Francis. London, England.

Eaton, T.T. and Zaporozec, A. 1997. Evaluation of Groundwater Vulnerability in an Urbanizing Area. *In*: J. Chilton et al. editors. *Groundwater in the Urban Environment*. Balkema Publishers. Rotterdam, The Netherlands.

Farrand, W.R. 1982. *Quaternary Geology of Southern (and Northern) Michigan*. Michigan Department of Natural Resources, Geological Survey Division. Lansing, MI.

Farrand, W.R. 1998. *The Glacial Lakes Around Michigan*. Bulletin #4. Michigan Department of Natural Resources. Lansing, MI.

Foster, S.S.D. and Hirata, R. 1988. *Groundwater Risk Assessment: A Methodology Using Available Data*. CEPIS. Lima, Peru.

Focazio, M.J., Reilly, T.E., Rupert, M.G., and Helsel, D.R. 2001. *Assessing Ground-Water Vulnerability to Contamination: Providing Scientifically Defensible Information for Decision Makers*. United States Geological Survey Circular 1224. Denver, CO.

Foster, S.S.D., Morris, B.L., and Lawrence, A.R. 1994. Effects of Urbanization on Groundwater Recharge. *In*: Wilkinson, W.B. editor. *Groundwater Problems in Urban Areas: ICE Conference Proceedings*. London, England.

Fresca, B. 2007. Urban – Enhanced Groundwater Recharge: Review and Case Study of Austin, Texas, USA. *In*: Howard, K.W.F. editor. *Urban Groundwater – Meeting the Challenge*. Taylor & Francis, London, England.

Hartig, J.H. and Zarull, M.A. 1991. Methods of Restoring Degraded Areas in the Great Lakes. *Reviews of Environmental Contamination and Toxicology*. Vol. 177. pp. 127–154.

Howard, K.W.F., Di Biase, S., Thompson, J., Maier, H., and Van Egmond, J. 2007. Stormwater Infiltration Technologies for Augmenting Groundwater Recharge in Urban Areas. *In*: Howard, K.W.F. editor. *Urban Groundwater – Meeting the Challenge*. Taylor & Francis. London, England.

Kaufman, M.M., Rogers, D.T., and Murray, K.S. 2003. Surface and Subsurface Geologic Risk Factors to Ground Water Affecting Brownfield Redevelopment Potential. *Journal of Environmental Quality*. Vol. 32. pp. 490–499.

Kaufman, M.M., Rogers, D.T., and Murray, K.S. 2005. An Empirical Model for Estimating Remediation Costs at Contaminated Sites. *Journal of Water, Air and Soil Pollution*. Vol. 167. pp. 365–386.

Kaufman, M.M., Rogers, D.T., and Murray, K.S. 2011. *Urban Watersheds: Geology, Contamination, and Sustainable Development*. CRC Press. Boca Raton, FL. 583p.

Keller, K.C., Van Der Kamp, G., and Cherry, J.A. 1989. A Multiscale Study of the Permeability of a Thick Clay Till. *Journal of Water Resources Research*. Vol. 25. No. 11. pp. 2299–2317.

Kibel, P.S. 1998. The Urban Nexus: Open Space, Brownfields, and Justice. *Boston College Environmental Affairs Law Review.* Vol. 25. No. 3. pp. 589–618.

Loague, K., Corwin, D.L., and Ellsworth, T.R. 1998. The Challenge of Predicting Nonpoint Source Pollution. *Environmental Science and Technology.* Vol. 26. pp. 127–154.

Michigan Department of Environmental Quality. 2008. *Michigan Sites of Environmental Contamination.* Lansing, MI.

Mohrlok, U., Cata, C., and Bucker-Gittel, M. 2007. Impact on Urban Groundwater by Wastewater Infiltration into Soils. *In*: Howard, K.W.F. editor. *Urban Groundwater – Meeting the Challenge.* Taylor & Francis. London, England.

Murray, J.E. and Bona, J.M. 1993. Rouge River National Wet Weather Demonstration Project, Wayne County. Detroit, MI.

Murray, K.S. 1996. Statistical Comparison of Heavy-Metal Concentrations in River Sediments. *Journal of Environmental Geology.* Vol. 27. pp. 54–58.

Murray, K., Bazzi, A., Carter, C., Ehlert, A., Harris, A., Kopec, M., Richardson, J., and Sokol, H. 1997. Distribution and Mobility of Lead in Soils at an Outdoor Shooting Range. *Journal of Soil Contamination.* Vol. 6. No. 1. pp. 79–93.

Murray, K.S., Lybeer, M., Cauvet, D., and Thomas, J.C. 1999. Particle Size and Chemical Control of Heavy Metals in Bed Sediment of the Rouge River, Southeastern Michigan. *Journal of Environmental Science and Technology.* Vol. 33. pp. 987–992.

Murray, K.S. and Rogers, D.T. 1999a. Groundwater Vulnerability, Brownfield Redevelopment and Land Use Planning. *Journal of Environmental Planning and Management.* Vol. 42. No. 6. pp. 801–810.

Murray, K.S. and Rogers, D.T. 1999b. Evaluation of Groundwater Vulnerability in an Urban Watershed. Proceedings of the 2nd International Congress on Water Resources and Environmental Research. Brisbane, Australia. pp. 877–883.

Pierce, S.A., Sharp, J.M., and Garcia-Fresca, B. 2007. Evaluating Groundwater Allocation Alternatives in an Urban Setting Using a Geographic Information System Data Model and Economic Valuation Technique. *In*: Howard, K.W.F. editor. *Urban Groundwater – Meeting the Challenge.* Taylor & Francis. London, England.

Robins, N., Adams, B., and Foster, S.S.D. 1994. Groundwater Vulnerability Mapping: The British Perspective. *Hydrogeologie.* Vol. 3. pp. 35–42.

Rogers, D.T. and Murray, K.S. 1997. Occurrence of Groundwater in Metropolitan Detroit, Michigan, USA. *In*: Chilton et al. editors. *Groundwater in the Urban Environment.* Volume 1. Balkema Publishers. The Netherlands. pp. 155–160.

Rogers, D.T. 1992. The Importance of Site Observation and Followup Environmental Site Assessments – A Case Study. Proceedings of the National Ground Water Association Phase I ESA Conference. Orlando, FL. pp. 218–227.

Rogers, D.T. 1996. *Environmental Geology of Metropolitan Detroit.* Clayton Environmental Consultants. Novi, MI.

Rogers, D.T. 1997a. The Influence of Groundwater in Surface Water in Michigan's Rouge River Watershed. Conjunctive Use of Water Resources: Aquifer Storage and Recovery. American Water Resources Association. Vol. 1. pp. 173–180.

Rogers, D.T. 1997b. *Surface Geological Map of the Rouge River Watershed in Southeastern Michigan.* Wayne County, MI. 1:62,500. 2 Sheets.

Rogers, D.T. 1997c. *Geologic Sensitivity Map of the Rouge River in Southeastern Michigan.* Wayne County, Michigan. 1:62,500. 1 Sheet.

Rogers, D.T. 2002. The Development and Significance of a Geologic Sensitivity Map of the Rouge River Watershed in Southeastern Michigan, USA. *In*: Bobrowsky, P.T. editor. *Geoenvironmental Mapping: Methods, Theory, and Practice.* A.A. Balkema Publishers. The Netherlands. pp. 295–319.

Rogers, D.T., Murray, K.S., and Kaufman, M.M. 2007. Assessment of Groundwater Contaminant Vulnerability in an Urban Watershed in Southeast Michigan, USA. *In*: Howard, K.W.F. editor. *Urban Groundwater – Meeting the Challenge.* Taylor & Francis. London, England.

Rogers, D.T. 2014. Scientists Call for a Renewed Emphasis on Urban Geologic Mapping. *American Geophysical Union. Earth and Space News. Eos.* Vol. 95. No. 47. pp. 431–432.

Rogers, D.T. 2016a. Next Generation of Urban Hydrogeologic Investigations. *Journal of the Italian Geological Society* Rome, Italy. Vol. 39. No. 1. pp. 349–353.

Rogers, D.T. 2016b. Scientific Advancements That Improve the Conceptual Site Model in Urban Hydrogeological Site Investigations. 35th International Geological Congress. Paper 3019. Cape Town, South Africa.

Rogers, D.T. 2018. Derivation of a Comprehensive Environmental Risk Model for Urban Groundwater Protection. *International Association of Hydrogeologists Congress.* Vol. 1. Daejeon, Korea.

Stiber, N.A., Small, M.J., and Fischbeck, P.S. 1995. The Relationship between Historic Industrial Site Use and Environmental Contamination. *Journal of the Air and Waste Management Association.* Vol. 48. No. 9. pp. 809–818.

United Nations. 2017. World Population Prospects. United Nations Department of Economic and Social Affairs. Population Division. https://esa.un.org/unpd/wpp/data. (accessed December 14, 2017).

Vuono, M. and Hallenbeck, R.P. 1995. Redeveloping Contaminated Properties. *Journal of Risk Management.* Vol. 42. pp. 58–69.

Wong, C.I., Sharp, J.M., Hauwert, N., Landrum, J., and White, K.M. 2012. Impact of Urban Development on Physical and Chemical Hydrogeology. *Journal of Elements.* Vol. 8. pp. 429–434.

Zaporozec, A. and Eaton, T.T. 1996. Groundwater Resource Inventory in Urbanized Areas. *In:* Howard, K. W. F. editor, *Hydrology and Hydrogeology of Urban and Urbanizing Areas.* American Institute of Hydrology (AIH) Annual Meeting Proceedings. St. Paul, MN.

5 Pollution Risk Factors

5.1 INTRODUCTION

What are the risks posed by pollutants to humans once they are released? Why do some cause more harm than others? The first step to answering this question involves combining three factors related to the physical chemistry of the contaminant: (1) toxicity or potency; (2) mobility; and (3) persistence. Next, the composite physical chemistry attributes are considered within the context of a region's geological vulnerability, which we discussed in the previous chapter. This framework provides a powerful tool for assessing the environmental risk of any location or region.

The risks posed by contaminants are not equal. Although it sounds like a paradox, an extremely toxic contaminant may not present as much risk as a moderately toxic contaminant. This outcome occurs if the more toxic contaminant does not migrate and degrades quickly, and the moderately toxic contaminant exhibits higher mobility and persistence that lasts for decades before degrading. The release location is also a factor in determining risk. If the extremely toxic contaminant is released at a certain location and under certain conditions it may inflict significant harm before it degrades. On the other hand, the moderately toxic contaminant may have much more opportunity to inflict harm because it lasts longer and is mobile.

Contaminants also behave differently in soil, water, and air. It is logical, therefore, to assess contaminant risk as a function of each environmental media. We begin the evaluation process by assessing the probability that a release will occur given certain land-use criteria. This is a critical step in the evaluation process, because not only must a contaminant be present for there to be a risk, it must also be released for there to be potential exposure. After examining the potential risks of a release, we will then examine the media releases target (air, water, or soil), and develop contaminant risk factors for groundwater, soil, and air. When these factors are combined with surface risk and geologic vulnerability, environmental risk can be estimated.

This is where the predictive power lies; knowledge of the physical attributes of the chemicals used combined with the geology of a region and the probability of a release occurring. Together, these pieces form the basis of a scientifically grounded environmental assessment process that can lead to greatly enhancing the sustainability of any urban region.

5.2 SURFACE RISK FACTOR

Surface risk, as employed here, is the probability of any given site contaminating the environment given the best available data from public sources. Surface risk evaluations have traditionally used spatially generalized categories of land use to represent various level of risk, such as industrial, commercial, residential, and recreational (Barringer et al. 1990; Eckhardt and Stackelberg 1995; Secunda et al. 1998).

The use of general land-use categories for vulnerability assessments is problematic because of their inadequate **spatial resolution**; defined as the smallest identifiable element in a sequence (Tobler 1988). In urban and urbanizing areas, mixed land uses within small areas such as city blocks are common, so the variable risks may be obscured by generalizations when the capture zones for water supply wells, termed **wellhead protection zones**, are delineated. For example, the 10-year capture zone is the subsurface and surface areas from where water (and any contamination it carries) will reach the well over a time period of 10 years. Figure 5.1 shows an example capture zone of a water supply well in an urban area (Wisconsin Department of Natural Resources 1999). Figure 5.2 is an example of a recharge area in cross section (USGS 1998).

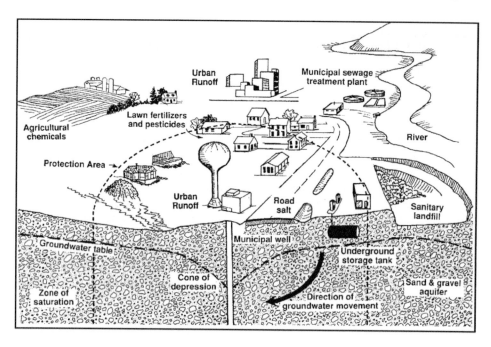

FIGURE 5.1   Example capture zone. (From WDNR. Wellhead Protection. WDNR Publication PUB-DG-039 99REV. Madison, WI. 1999.)

To demonstrate this problem of inadequate spatial resolution, we can consider a water well in an area designated for residential land use, as depicted in Figure 5.3 (Kaufman et al. 2003). Within this zone of low risk is a single and small industrial establishment engaged in metal plating. Metal plating activities exhibit a high incidence rate of soil and groundwater contamination, but this specific risk is masked by the generalization of the area within the capture zone as a lower-risk residential category. This is depicted in Figure 5.3a.

In Figure 5.3b, two adjacent but different zones create edge effects. At the edge, the residential zone becomes exposed to the higher risks associated with industrial land, but the sources and

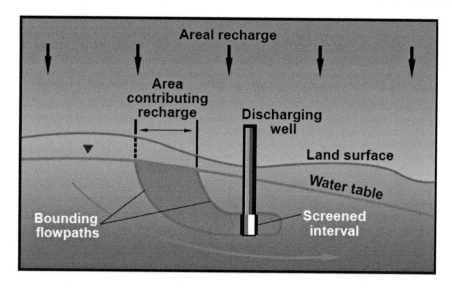

FIGURE 5.2   Example recharge area. (From USGS. Estimating Areas Contributing to Recharge of Wells. USGS Circular 1174. Denver, CO. 1998.)

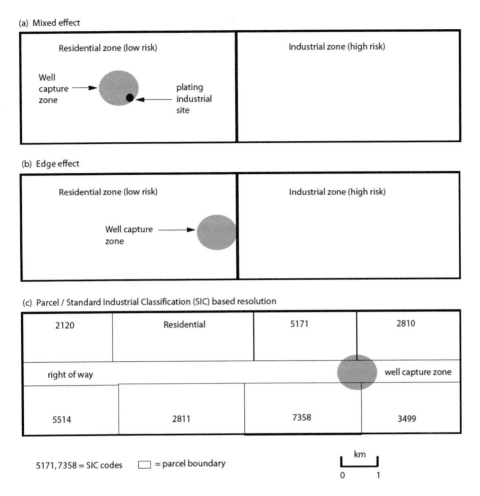

FIGURE 5.3 Effects of different spatial resolutions across multiple land-use types on surface risk. (From Kaufman, M.M. Rogers, D.T. and Murray, K.S. Urban Watersheds: Geology, Contamination, and Sustainability. CRC Press. Boca Raton, FL. 2011.)

amounts of the actual risk from the industrial land near the edge are unknown. In Figure 5.3c, the use of the Standard Industrial Classification (SIC) code or equivalent for each establishment within a general land-use category permits a greatly improved parcel-level spatial resolution of the relative risks of contamination.

The SIC code is a four-digit code defined as (United States Office of Management and Budget 1997)[1]:

- The first two digits identify a major group, such as agriculture, retail trade, and manufacturing
- The third digit denotes industry groups within each major category, such as agricultural crop production
- The fourth digit identifies a specific industry code, such as metal plating

A normalized measure of risk between different establishment types is achieved through the use of contamination incident rates (Kaufman 1997). Incidence rates are obtained by:

1. Assigning a SIC code to each source of contamination appearing on a known list of contaminated sites. Lists of known contaminated sites are available through private companies or are publicly available either through local municipalities or state or federal environmental agencies, such as the Michigan Department of Environmental Quality (MDEQ 2008).

2. Obtaining the total number of establishments for each SIC code within the study area (United States Bureau of the Census 2019).
3. Dividing the number of contaminated sites with a specific SIC code by the total number of establishments with the same SIC code in the study area.

To scale the scores equivalently to the other risk factors discussed in the following sections, these rates are multiplied by ten and converted to scores between 0 and 10. These scores are then summed for a circular area encompassing each water well within the study area. Figure 5.4 shows an example surface risk calculation for a brownfield site situated above an area where the subsurface geology is composed of sand.

The computed risk includes only those lighter-shaded establishments contained within the geological unit composed of sand, because the bounding geological units (sandy clay, sandy and silty clay) have much lower hydraulic conductivities yielding much lower contaminant migration potentials. The legend box in Figure 5.4 shows the SIC and risk scores for three of the many establishments within the radius (Kaufman et al. 2003). For example, SIC 2822 represents an establishment producing synthetic rubber with a risk score of 4.00. This particular risk score was calculated by dividing the two SIC 2822 establishments known to be sites of contamination by the five SIC 2822 establishments in the region and then multiplying the result by ten. The risk scores for the other establishment types shown in Figure 5.4 are computed similarly. The dots in the legend box indicate there are many more scores to add within the radius to obtain the total surface risk score indicated at the bottom.

The circle shown in Figure 5.4 represents the 10-year capture zone of a pumping well located in the center of the circle. Circular areas may be effective in modeling wellhead protection areas under small regional hydraulic gradients and low groundwater velocities (Barringer et al. 1990; Camp and Outlaw 1998). In this example, the capture zone was calculated by using existing data from hydrogeologic investigations conducted near the study site, under the assumptions of a pumping rate

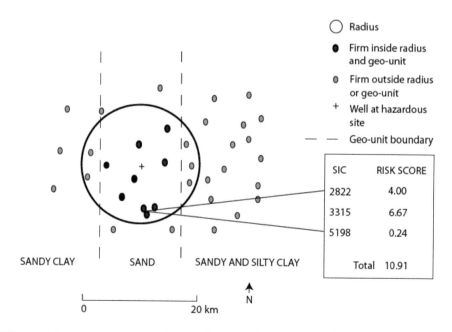

FIGURE 5.4 Calculation of surface risk using the risk values computed for specific establishment types within the capture radius around a contaminated brownfield site. (From Kaufman, M.M. Rogers, D.T. and Murray, K.S. Urban Watersheds: Geology, Contamination, and Sustainability. CRC Press. Boca Raton, FL. 583p. 2011.)

that induces a hydraulic gradient of 0.1 m/m in all directions toward the pumped well, a relatively level water table, and predominately horizontal groundwater flow. It is also possible to derive the capture zone if the necessary information on gradient, hydraulic conductivity, and effective porosity are obtained at the site. After an appropriate radius is established, an automated procedure for the surface risk can be performed. This can be achieved by: (1) geocoding a file containing the street addresses of public, commercial, and industrial establishments (labeled "Firms" in Figure 5.4) each coded by their SIC designation; and (2) using the capabilities of a geographic information system (GIS) to sum the risk scores for each firm within the specified radius on a digital map image (Kaufman 2000).

Table 5.1 lists the 10-year capture zones and mean surface risk values for four of the most common soil types encountered in the Rouge River watershed in southeastern Michigan. This watershed is a good example because it is heavily urbanized and includes a broad spectrum of geological units including clay, silty clay, sandy silty clay, and sand (Kaufman et al. 2003).

The mean surface risk values are highest in the sand for three reasons: (1) there is a high density of sites with a propensity for releases of contaminants to occur; (2) the larger capture zone for sand translates into potentially more sites of contamination than the other geological units; and (3) the subsurface geology (i.e., sand) does not significantly impede the migration of contaminants.

Industrial location patterns also help to explain other mean surface risk values in this watershed. For example, the Silty Clay Unit has a mean surface risk value of 33.7 and the Sandy Silty Clay Unit has a surface risk value of 0.42, yet the Silty Clay Unit has a 10-year capture zone smaller than the Sandy Silty Clay Unit. This discrepancy is due to the significantly greater number of potential contaminant sources located within the Silty Clay Unit.

The next section examines individual contaminants and their migration potential in groundwater within different subsurface geological environments.

5.3 GROUNDWATER POLLUTION RISK FACTOR

The risk posed to groundwater by contaminants themselves is often overlooked or underrepresented (Kaufman et al. 2005; Rogers et al. 2007a). Considering specific types of contamination through the use of vulnerability models is important because each contaminant has unique physical-chemical properties significantly influencing its behavior when released into the environment.

Contaminant fate and transport evaluations require interdisciplinary analyses involving chemical, geologic, hydrologic, and biological factors (USEPA 1989;1992). Correspondingly, the development of groundwater vulnerability models for specific contaminants requires an interdisciplinary process (Rogers et al. 2007b). The critical physical/chemical attributes influencing contaminant risk are associated with mobility and persistence and include the following factors (USEPA 1989; USEPA 1996; Wiedemeier et al. 1999, USGS 2006; Rogers et al. 2007a):

TABLE 5.1
Surface Risk Values for the Rouge Watershed

Geologic Unit	10-Yr Capture Zone (km)	Mean Surface Risk Value
Clay	0.03	0.04
Silty clay	0.06	33.70
Sandy silty clay	1.07	0.42
Sand	597.00	343.64

Source: Kaufman, M.M. Rogers, D.T., and Murray, K.S. Urban Watersheds: Geology, Contamination, and Sustainability. CRC Press. Boca Raton, FL. 583p. 2011.

1. Solubility
2. Vapor pressure
3. Density
4. Chemical stability
5. Persistence
6. Adsorption potential

As noted previously, toxicity is an important factor when examining risk. Mobility and persistence are also critically important because these two factors dictate a chemical's ability to migrate from its point of release in the environment to a distant point where human exposure may occur, such as a drinking-water supply or a surface water body. The environmental risk posed by specific contaminants to contaminant groundwater, termed **Contaminant Risk Factor for Groundwater** (CRF$_{GW}$), can be developed as a function of these three factors (Kaufman et al. 2005; Rogers et al. 2007a). Contaminants released into the environment only present a risk to humans if there is a completed exposure pathway. In general terms, the CRF$_{GW}$ is expressed in Equation 5.1 (Kaufman et al. 2005; Rogers et al. 2007a).

$$CRF_{GW} = Toxicity \times mobility \times persistence \tag{5.1}$$

Toxicity values are obtained from the USEPA Integrated Risk Information System (IRIS)(2019a). This database is updated weekly, and often more frequently. The toxicity values selected should be the most conservative for each exposure pathway – ingestion, dermal adsorption, and inhalation. Using the most conservative value is appropriate since exposure to contaminated groundwater can occur in each of these pathways. For instance, dermal adsorption can occur during washing, ingestion can occur through drinking, and inhalation can occur during showering.

Mobility is derived from Henry's Law constant and the retardation factor shown in Equation 5.2 (Kaufman et al. 2005; Rogers et al. 2007a).

$$M = (H)(R) \tag{5.2}$$

where:

M = mobility
H = Henry's Law constant
R = retardation factor

Henry's Law constant (H) (atm. mol^{-1} m^{-3}) is a measure of the tendency for substances to volatilize, and is very useful in assessing the mobility of specific contaminants because solubility affects the volatilization of contaminants into the atmosphere (Sander 1999). It is related to vapor pressure (VP) (atm.), molecular weight (MW) (g/mol), and solubility in water (W$_s$) (g/l) and is expressed as Equation 5.3 (Kaufman et al. 2005; Rogers et al. 2007a).

$$H = (VP)(MW)(W_s) \tag{5.3}$$

where:

VP = vapor pressure
MW = molecular weight
W$_s$ = water solubility

Henry's Law constants can be obtained from several sources including USEPA (1996), Sander (1999), Wiedemeier (1999), Suthersan and Payne (2005), and Payne et al. (2008).

The retardation factor is represented by Equation 5.4.

$$R = 1 + \frac{(\rho b)(Kd)}{\eta} \qquad (5.4)$$

where:

R = retardation factor
ρb = bulk density of aquifer matrix (g/cm^3)
Kd= distribution coefficient (mL·g^{-1})
η = effective porosity (calculated as a percent value)

The distribution coefficient is calculated using Equation 5.5.

$$Kd = (Foc)(Koc) \qquad (5.5)$$

where:

Kd = distribution coefficient
Foc = organic carbon partition coefficient (kg/kg)
Koc = fraction of total organic carbon in soil (l/kg)

Values for the organic carbon partition coefficient can be obtained from numerous sources, including USEPA (1996), Wiedemeier (1999), USEPA (2002a), Suthersan and Payne (2005), MDEQ (2019), and USEPA (2019b). To obtain the best representation, values for the fraction of organic carbon should be collected in the field. If field collection is not possible, standard values and ranges can be obtained from USEPA (1996), Wiedemeier (1999), USEPA (2002a), and Suthersan and Payne (2005).

The retardation factor represents the ratio between the rate of groundwater movement and the rate of contaminant movement. When the retardation value equals 1, the rate of groundwater movement equals the rate of contaminant movement and no retardation is expected. A retardation value >1 indicates groundwater movement is greater than contaminant movement; so increasing values indicate greater contaminant retardation (USEPA 1989, 2002b).

Finally, the CRF$_{GW}$ is calculated in Equation 5.6 by multiplying the inverse of the chemical compound's toxicity (T), by the inverse of its mobility (M) and its persistence (P).

$$CRF_{GW} = \frac{1}{T} \times \frac{1}{M} \times (P) \qquad (5.6)$$

where:

CRF$_{GW}$ = Contaminant Risk Factor for Groundwater, and:
T = toxicity
M = mobility
P = persistence

The inverse of the toxicity value must be used because of the integer values assigned for toxicity decrease with increasing carcinogenicity (USEPA 2019a). The inverse of the mobility values must also be used because of the calculated values of retardation decrease with increasing mobility.

Here is an example calculation of the CRF$_{GW}$ for a chemical XYZ in a geologic unit composed of sand:

$$CRF_{GW} \text{ for chemical XYZ} = \frac{1}{(T)} \times \frac{1}{(M)} \times (P) \qquad (5.7)$$

Step 1: Obtain toxicity value:
    The toxicity of XYZ chemical was obtained from the literature and has a value of 0.04.

Step 2: Determine the mobility value.

We need Henry's Law constant and the retardation factor. Let's calculate the retardation factor first. We start with the distribution coefficient (Kd) using Equation 5.5:

$$Kd = (Foc)(Koc)$$

Foc was obtained through analysis of several soil samples in the study area and was found to be 0.0003 kg/kg. Koc for chemical XYZ was obtained from the literature and has a value of 58.9 L/kg (USEPA 2002a). We can now calculate the distribution coefficient:

$$Kd = (Foc)(Koc) = (0.0003)(58.9) = 0.017$$

The remaining values necessary to calculate the retardation value are the bulk density of the aquifer material and the effective porosity. The bulk density of the aquifer material was obtained through the collection and analysis of soil samples and was 1.7 g/cm^3. The effective porosity was estimated from literature values to be 25% or 0.25.

All the necessary information has been obtained, and the retardation value of 1.11 is calculated using Equation 5.4:

$$XYZ \text{ chemical retardation} = R = 1 + \frac{(\rho b)(Kd)}{\eta}$$

$$XYZ \text{ chemical retardation} = R = 1 + \frac{(1.7)(0.017)}{0.25} = 1.11$$

Now we can complete the mobility calculation by plugging in Henry's Law constant (H) obtained from the literature into Equation 5.2. The Henry's Law constant for XYZ chemical is 0.228.

$$M = (H)(R)$$

$$M = 0.228 \times 1.11 = 0.253$$

Step 3: Determine the persistence value. The persistence value was obtained from the literature and is 0.2 years.

Step 4: With toxicity, mobility, and persistence values obtained, the CRF$_{GW}$ is calculated using Equation 5.6.

$$CRF_{GW} \text{ for chemical XYZ} = \frac{1}{(T)} \times \frac{1}{(M)} \times (P)$$

$$CRF_{GW} \text{ for chemical XYZ} = \frac{1}{(0.04)} \times \frac{1}{(0.253)} \times (0.2) = 19.76$$

$$CRF_{GW} \text{ for chemical XYZ} = 19.76$$

Tables 5.2, 5.3, and 5.4 display the groundwater contaminant risk factors (CRF$_{GW}$) for common VOC LNAPL, VOC DNAPL, and PAH compounds, respectively (Kaufman et al. 2005; Rogers et al. 2007a). Table 5.5 summarizes those values and includes other selected compounds. The values span the four most common types of soils encountered in urban areas of the United States, including

TABLE 5.2
CRF$_{GW}$ for Common VOC LNAPLs

LNAPL Compound	Soil Type	CRF$_{GW}$
Benzene	Clay	3.64
	Silty Clay	14.50
	Sand Silty Clay	15.40
	Sand	19.40
Toluene	Clay	0.63
	Silty Clay	4.10
	Sand Silty Clay	4.53
	Sand	7.53
Ethyl benzene	Clay	0.27
	Silty Clay	2.10
	Sand Silty Clay	2.40
	Sand	4.80
Xylenes	Clay	1.10
	Silty Clay	7.80
	Sand Silty Clay	8.10
	Sand	13.90
Mean LNAPL CRF$_{GW}$	Clay	1.41
	Silty Clay	7.12
	Sand Silty Clay	7.60
	Sand	11.40

Source: Kaufman, M.M. Rogers, D.T., and Murray, K.S. Urban Watersheds: Geology, Contamination, and Sustainability. CRC Press. Boca Raton, FL. 583p. 2011.

clay, silty clay, sandy silty clay, and sand. As such, their relative magnitudes can be used as a starting point for risk assessments in other urbanized watersheds with similar geologic units.

In Table 5.2, the CRF$_{GW}$ values for benzene are greater than other VOCs within the LNAPL group. This is because benzene exhibits the highest combined values of toxicity, mobility, and persistence in groundwater compared to the other LNAPL compounds examined.

Table 5.3 lists the CRF$_{GW}$ for common VOCs from the DNAPL group (Kaufman et al. 2005; Rogers et al. 2007a).

Examination of the CRF$_{GW}$ for DNAPL compounds indicates they are much greater than those of the LNAPL compounds listed in Table 5.2. The comparatively higher CRF$_{GW}$ of DNAPLs stems not from their relative toxicity, but from their higher persistence and mobility in groundwater compared to the LNAPL compounds (Rogers et al. 2007a).

Table 5.4 lists the CRF$_{GW}$ for common PAHs.

The CRF$_{GW}$ for PAH compounds is significantly lower than the CRF$_{GW}$ for DNAPL and LNAPL VOCs. This difference occurs because PAHs strongly sorb to soil particles and are much less soluble in water compared to the VOC compounds examined (Rogers et al. 2007a). Table 5.5 contains a summary of CRF$_{GW}$ for the LNAPL and DNAPL VOCs, PAHs, and also includes the CRF$_{GW}$ for total PCBs, the pesticide chlordane, and the heavy metals chromium VI, lead, mercury, and arsenic. The contaminant with the highest CRF$_{GW}$ is chromium VI, followed by DNAPL VOCs, LNAPL VOCs, mercury, and lead, with their CRF$_{GW}$ ranging from 100 to 10,000 times less than chromium VI. Contaminants with the lowest CRF$_{GW}$ include chlordane, PAHs, and the PCBs having a CRF$_{GW}$ more than a billion times lower than chromium VI.

TABLE 5.3

CRF$_{GW}$ for Common VOC DNAPLs

LNAPL Compound	Soil Type	CRF$_{GW}$
Tetrachloroethene (PCE)	Clay	98.0
	Silty Clay	601.0
	Sand Silty Clay	657.0
	Sand	1,048.0
Trichloroethene (TCE)	Clay	148.0
	Silty Clay	933.0
	Sand Silty Clay	1,018.0
	Sand	1,647.0
Cis-1,2-Dichloroethene	Clay	228.0
	Silty Clay	702.0
	Sand Silty Clay	730.0
	Sand	851.0
Trans-1,2-Dichloroethene	Clay	131.0
	Silty Clay	495.0
	Sand Silty Clay	520.0
	Sand	647.0
Vinyl chloride	Clay	860.0
	Silty Clay	1,911.0
	Sand Silty Clay	1,962.0
	Sand	2,132.0
1,1,1-Trichloroethane (1,1,1-TCA)	Clay	2.1
	Silty Clay	11.0
	Sand Silty Clay	12.0
	Sand	17.3
Mean DNAPL CRF$_{GW}$ for degradation sequence from PCE to vinyl chloride[a]	Clay	333.0
	Silty Clay	773.0
	Sand Silty Clay	1,091.0
	Sand	1,274.0

[a] Represents cumulative risk.

Source: Kaufman, M.M. Rogers, D.T., and Murray, K.S. Urban Watersheds: Geology, Contamination, and Sustainability. CRC Press. Boca Raton, FL. 583p. 2011.

The distributions of CRF$_{GW}$ are shown in Figure 5.5.

Examination of the CRF$_{GW}$ values in Figure 5.5 reveals a grouping of contaminants into three distinct levels. Chlordane, PCBs, and PAHs form the group with low CRF$_{GW}$; lead, mercury, and LNAPLs comprise the moderate range; and Chromium VI and DNAPL compounds appear on the top of the figure with the highest CRF$_{GW}$. Chlordane, PCBs, and PAHs are very toxic and persistent contaminants in the environment, but each strongly sorbs to soil and are not very soluble in water. As a result, they have very low CRF$_{GW}$ compared to the other contaminants listed.

The middle grouping has moderate CRF$_{GW}$ resulting from a combination of factors unique to each contaminant. LNAPLs have moderate mobility and the ability to degrade in the environment. Mercury is not very mobile, but may become transformed in the environment to methyl mercury; a change allowing it to be adsorbed by organisms that increases its environmental risk. Lead has low solubility in water but is very persistent. Arsenic has a much higher CRF$_{GW}$ compared to lead because it is more soluble and toxic.

TABLE 5.4

CRF$_{GW}$ for Select PAHs

PAH Compound	Soil Type	GWCRF
Naphthalene	Clay	0.0004
	Silty Clay	0.003
	Sand Silty Clay	0.004
	Sand	0.01
Chrysene	Clay	0.000002
	Silty Clay	0.00002
	Sand Silty Clay	0.00002
	Sand	0.0009
Benzo[b]fluoranthrene	Clay	<0.00001
	Silty Clay	<0.00001
	Sand Silty Clay	<0.00001
	Sand	<0.00001
Benzo[k]fluoranthrene	Clay	<0.00001
	Silty Clay	<0.00001
	Sand Silty Clay	<0.00001
	Sand	<0.00001
Phenanthrene	Clay	0.00002
	Silty Clay	0.0002
	Sand Silty Clay	0.0002
	Sand	0.0007
Benzo[g,h,i] perylene	Clay	<0.00001
	Silty Clay	<0.00001
	Sand Silty Clay	<0.00001
	Sand	<0.00001
Benzo(a)pyrene	Clay	<0.00001
	Silty Clay	<0.00001
	Sand Silty Clay	<0.00001
	Sand	<0.00001
Mean PAH CRF$_{GW}$	Clay	0.0001
	Silty Clay	0.0008
	Sand Silty Clay	0.001
	Sand	0.002

Source: Kaufman, M.M. Rogers, D.T., and Murray, K.S. Urban Watersheds: Geology, Contamination, and Sustainability. CRC Press. Boca Raton, FL. 583p. 2011.

The group consisting of chromium VI and DNAPL compounds has the highest CRF$_{GW}$ values. Some of these values may be more than a million times greater than the contaminants in the lowest group. This magnitude of difference occurs because chromium VI and DNAPL compounds have relatively high toxicity, mobility, and persistence in the environment.

Another characteristic of these data is the correspondence between the increase in the risk factor for each contaminant and the increase in mean grain size from clay to sand. Larger mean grain sizes increase permeability and raise the mobility factor, especially if a specific contaminant's physical chemistry has a high relative solubility and low sorptive potential. If this relationship is correct, then contaminants with a high CRF$_{GW}$ should be detected in groundwater at greater distances from their source than contaminants with low CRF$_{GW}$. Specifically, sites contaminated with chromium VI and

TABLE 5.5

Summary of CRF$_{GW}$ for LNAPL and DNAPL VOCs, PAHs, and Other Select Compounds

Compound	Soil Type	CRF$_{GW}$
Mean LNAPL CRF$_{GW}$	Clay	1.41
	Silty Clay	7.12
	Sand Silty Clay	7.60
	Sand	11.40
Mean DNAPL CRF$_{GW}$ for degradation sequence from PCE to vinyl chloride	Clay	333.00
	Silty Clay	773.00
	Sand Silty Clay	1,091.00
	Sand	1,274.00
Mean PAH CRF$_{GW}$	Clay	0.0001
	Silty Clay	0.0008
	Sand Silty Clay	0.001
	Sand	0.002
PCBs	Clay	0.00002
	Silty Clay	0.00026
	Sand Silty Clay	0.0003
	Sand	0.0009
Pesticide chlordane	Clay	0.0035
	Silty Clay	0.035
	Sand Silty Clay	0.041
	Sand	0.046
1,2,3-Trichloropropane (TCP)	Clay	545.17
	Silty Clay	616.70
	Sand Silty Clay	890.40
	Sand	1,230.00
Methyl tert butyl ether (MTBE)	Clay	114.57
	Silty Clay	339.20
	Sand Silty Clay	465.70
	Sand	920.56
Chromium VI	Clay	948.00
	Silty Clay	2,080.00
	Sand Silty Clay	2,116.00
	Sand	2,300.00
Lead	Clay	0.03
	Silty Clay	0.30
	Sand Silty Clay	0.34
	Sand	1.10
Mercury	Clay	1.57
	Silty Clay	2.52
	Sand Silty Clay	2.77
	Sand	3.10
Arsenic	Clay	9.88
	Silty Clay	11.26
	Sand Silty Clay	21.12
	Sand	50.25

Source: Kaufman, M.M. Rogers, D.T., and Murray, K.S. Urban Watersheds: Geology, Contamination, and Sustainability. CRC Press. Boca Raton, FL. 583p. 2011.

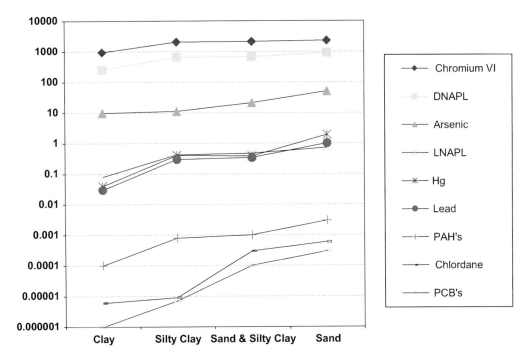

FIGURE 5.5 Distribution of CRF$_{GW}$. (From Kaufman, M.M. Rogers, D.T. and Murray, K.S. Urban Watersheds: Geology, Contamination, and Sustainability. CRC Press. Boca Raton, FL. 583p. 2011.)

VOC DNAPLs should exhibit greater contamination extents than sites contaminated by the other chemical compounds evaluated.

To test this hypothesis, 83 contaminated sites within the Rouge River watershed and 127 additional sites located outside the Rouge River watershed were evaluated. The Rouge sites were located on different types of geological units, and the external sites were located in areas of varied subsurface geology and contaminant type (Kaufman et al. 2005). Of the 127 sites located outside the Rouge River watershed, 117 were distributed among 32 different states, with the remaining ten sites located in other countries including Canada, Italy, England, France, Belgium, South Africa, and Australia. All 210 sites were located in urban areas with varying geology composed of unconsolidated sediments originating from lacustrine, fluvial, or glacial processes.

The critical information gathered from each of the 210 sites is described in Table 5.6 (Kaufman et al. 2011). A summary of the data is listed in Table 5.7 (Kaufman et al. 2011).

A scan down the rightmost column (average extent from source) of Table 5.7 strongly validates the hypothesis: sites contaminated with chromium VI and VOC DNAPLs do exhibit greater contamination extents than sites contaminated by the other chemical compounds. The top four average extents are: 2,200 m (eight world sites, chromium in sand); 1012 m (eight Rouge watershed sites, DNAPL in sand); 975 m (15 world sites, DNAPL in sand); and 625 m (four Rouge watershed sites, DNAPL in moraine). The next ranked average extent is almost twice as small (328 m); found in eight LNAPL sand sites within the Rouge watershed. Strong empirical evidence indicates a positive association between a chemical's CRF$_{GW}$ and its likelihood to migrate. Since contamination affects other media such as soil and air, we must also develop contaminant risk factors for those media.

5.4 SOIL POLLUTION RISK FACTOR

In most cases, contaminants released to the ground surface migrate downward through the upper soil layers. Over time, they may or may not contaminate groundwater. Therefore, evaluating the

TABLE 5.6

Critical Information Obtained from Sites of Environmental Contamination

Category	Description
Contaminant	General chemical category (e.g., DNAPL)
Type of facility	Primary activity at the site (dry cleaning, foundry, etc.)
Geology	Composition, stratigraphy, and other information on subsurface units
Remedial technology for groundwater	Contaminant abatement method (e.g., air sparging, pump and treat, in situ chemical or biological treatment, natural attenuation, etc.)
Remedial technology for soil	Contaminant abatement method (e.g., excavation, capping, soil vapor extraction, institutional controls, etc.)
Mass (kg)	Total mass of contamination at site
Cost	Total cost of investigation and remediation from start to finish
Media remediated	Soil, water, or other (e.g., building decontamination, demolition, etc.)
Extent (meters)	Measured extent of contamination in each media affected
Cost/kg	Cost of investigation and remediation per kilogram of contaminant
Years of operation	Number of years the facility had been in operation
Geologic vulnerability	Geologic vulnerability rating (see Table 4.2)
Surface risk factor	Average surface risk (see Section 5.2 and Table 5.1)
CRF_{GW}	CRF_{GW} calculated for each contaminant (see Section 5.3, Equation 5.6)
Soil cost	Amount of the total remediation cost attributed to soil
Soil cost/kg	Cost per kilogram to remediate soil
Groundwater cost	Amount of the total remediation cost attributed to groundwater
Groundwater cost/kg	Cost per kilogram to remediate groundwater
Groundwater mass (kg)	Mass of groundwater remediated at the site

Source: Kaufman, M.M., Rogers, D.T., and Murray, K.S. 2011. Urban Watershed: Geology, Contamination, and Sustainable Development. CRC Press. Boca Raton, FL. 583 p.

potential for a chemical to contaminate the soil should also be conducted whenever there is a potential for it to contaminate groundwater. Development of the soil contaminant risk factors (CRF_{SOIL}) should be a high priority in urban watersheds because it provides an additional piece for characterizing the total risks posed by contaminants in the environment. The migration potential of a contaminant in soil is dependent upon the same physical and chemical attributes as those found in groundwater: solubility, vapor pressure, density, chemical stability, persistence, and adsorption potential. There is also biological interaction between the contaminant and the soil environment to which the chemical is released (Schnoor 1996), and as with the CRF_{GW}, is accounted for within the persistence factor. Given the similarities between the two processes, developing the CRF_{SOIL} requires rather simple modifications to the CRF_{GW} equation. The CRF_{SOIL} is calculated by multiplying the inverse of a chemical's toxicity (T), by its mobility (M) and persistence (P) (Equation 5.8). The change in the equation is reflected in the mobility factor, where it is not necessary to multiply by the inverse because the calculated values increase with increasing retardation for soil.

$$CRF_{SOIL} \text{ for chemical XYZ} = \frac{1}{(T)} \times (M) \times (P) \qquad (5.8)$$

where:

CRF_{SOIL} = Soil Contaminant Risk Factor;
T = Toxicity;
M = Mobility;
and P = Persistence.

TABLE 5.7

Data Summary

Rouge River Watershed Data

Contaminant of Concern	Number of Sites	Soil Type	Number of Sites	Geologic Setting	Average Cost ($/Kg)	Extent from Source (Meters)
DNAPL	23	Moraine	4	Glacial	145,000	625
		SSC	3	GL	3,260	42
		SC	2	GL	1,366	50
		Sand	8	GL	116,400	1,012
		Clay	6	GL	721	30
LNAPL	27	Moraine	5	Glacial	6,411	270
		SSC	4	GL	669	43.5
		SC	3	GL	518	45.6
		Sand	8	GL	2,627	328
		Clay	7	GL	319	38
PAHs`	22	Moraine	0	Glacial	–	–
		SSC	2	GL	203	32
		SC	7	GL	841	27.4
		Sand	5	GL	444	30
		Clay	8	GL	964	16.3
Lead	11	Moraine	0	Glacial	–	–
		SSC	2	GL	538	42
		SC	3	GL	230	25
		Sand	2	GL	442	27
		Clay	4	GL	68.7	52.5

Worldwide Data

Contaminant of Concern	Number of Sites	Soil Type	Number of Sites	Geologic Setting	Average Cost ($/Kg)	Average Extent from Source (Meters)
DNAPL	27	Clay	9	Fl, GL,L	474	40
		SC	3	Fl, GL	224	95
		Sand	15	Fl, GL, L	98,269	975
LNAPL	27	Clay	10	Fl, GL, L	197.8	35
		SC	4	Fl, GL	416	23
		SSC	3	GL	700	25
		Sand	10	Fl, GL, L	1,255	175
PAHs	26	Clay	10	Fl, GL, L	366	30.5
		SC	2	GL	125	22
		SSC	3	Fl	340	20
		Sand	11	Fl, GL	213	25
Lead	13	Clay	5	Fl, GL, L	190	125
		SC	4	Fl, GL, L	930	81
		Sand	4	Fl	500	100
Chromium	19	Clay	11	Fl, GL, L	474	40
		Sand	8	Fl, GL, L	81,713	2,200
Mercury	3	Clay	1	GL	1,000	10
		SC	2	Fl	3,000	75

(Continued)

TABLE 5.7 (CONTINUED)
Data Summary

				Worldwide Data		
Contaminant of Concern	Number of Sites	Soil Type	Number of Sites	Geologic Setting	Average Cost ($/Kg)	Average Extent from Source (Meters)
Chlordane	2	Clay	1	GL	600	10
		Sand	1	GL	830	15
PCBs	8	SC	3	Fl	1,200	5
		Sand	5	Fl, GL, L	2,053.9	13.75
Arsenic	2	Clay	1	Fl	780	30
		Sand	1	GL	960	40

SC = Sandy clay; SSC = Sandy and silty clay; GL = Glacial lacustrine; Fl = Fluvial; L = Lacustrine

Source: Kaufman, M.M., Rogers, D.T., and Murray, K.S. 2011. Urban Watershed: Geology, Contamination, and Sustainable Development. CRC Press. Boca Raton, FL. 583 p.

An example calculation of a CRF_{SOIL} for a chemical XYZ in a geologic unit composed of sand:

Step 1: Obtain toxicity value:

The toxicity of XYZ chemical was obtained from the literature and has a value of 0.04.

Step 2: Determine the mobility value.

Calculate the distribution coefficient (Kd) using Equation 5.5:

$$Kd = (Foc)(Koc)$$

Foc was determined through analysis of several soil samples in the study area and was found to be 0.0003 kg/kg. Koc for chemical XYZ was obtained from the literature and has a value of 58.9 L/kg (USEPA 2002a). Therefore, we now have enough information to calculate the distribution coefficient as follows:

$$Kd = (Foc)(Koc) = (0.0003)(58.9) = 0.017$$

The remaining values necessary to calculate the retardation value are the bulk density of the aquifer material and the effective porosity. The bulk density of the aquifer material was obtained through the collection and analysis of soil samples and was 1.7 g/cm³. The effective porosity was obtained from literature values and was 25% or 0.25.

All the necessary information has been obtained, and the retardation value of 1.11 is calculated using Equation 5.4:

$$\text{XYZ chemical retardation} = R = 1 + \frac{(\rho b)(Kd)}{\eta}$$

$$\text{XYZ chemical retardation} = R = 1 + \frac{(1.7)(0.017)}{0.25} = 1.11$$

Plug in Henry's Law constant (H) obtained from the literature into Equation 5.2; the Henry's Law constant for XYZ chemical is 0.228.

$$M = (H)(R)$$

$$M = 0.228 \times 1.11 = 0.253$$

Step 3: Determine the persistence value. The persistence value was obtained from the literature and is 0.2 years.

Step 4: With toxicity, mobility, and persistence values obtained, the CRF_{GW} is calculated using Equation 5.8.

$$CRF_{SOIL} \text{ for chemical XYZ} = \frac{1}{(T)} \times (M) \times (P)$$

$$CRF_{SOIL} \text{ for chemical XYZ} = \frac{1}{(0.04)} \times (0.253) \times (0.2) = 1.265$$

$$CRF_{SOIL} \text{ for chemical XYZ} = 1,265$$

Contaminant risk factors in soil for common LNAPL VOC compounds in the Rouge River watershed are listed in Table 5.8 (Rogers et al. 2007b; Kaufman et al. 2011). As with the CRF_{GW}, the values span the four most common types of soils encountered in urban areas of the United States, so their relative magnitudes can be used as a starting point for risk assessments in other urbanized watersheds with similar geologic units.

TABLE 5.8
CRF_{SOIL} Values for Common LNAPL VOCs

LNAPL Compound	Soil Type	CRF_{SOIL}
Benzene	Clay	10.00
	Silty Clay	2.50
	Sand Silty Clay	2.30
	Sand	1.90
Toluene	Clay	12.76
	Silty Clay	1.97
	Sand Silty Clay	1.79
	Sand	1.07
Ethyl benzene	Clay	74.80
	Silty Clay	9.60
	Sand Silty Clay	8.40
	Sand	4.20
Xylenes	Clay	23.13
	Silty Clay	3.43
	Sand Silty Clay	3.10
	Sand	1.80
Mean LNAPL CRF_{SOIL}	Clay	30.17
	Silty Clay	4.37
	Sand Silty Clay	3.89
	Sand	2.24

Source: Kaufman, M.M. Rogers, D.T., and Murray, K.S. Urban Watersheds: Geology, Contamination, and Sustainability. CRC Press. Boca Raton, FL. 583p. 2011.

As shown in Table 5.8, the values for LNAPL compounds are greatest for soil composed of clay, and reflect the tendency of LNAPLs to sorb more strongly to finer-grained soils. Table 5.9 lists the CRF_{SOIL} for common DNAPL VOCs (Kaufman et al. 2009).

The CRF_{SOIL} for DNAPLs are much greater those than of LNAPL compounds listed in Table 5.8. Not only are the DNAPL compounds more toxic, they are also much more persistent in the environment than the LNAPL compounds (Kaufman et al. 2009). For instance, the LNAPL compound benzene – a very toxic chemical – has a half-life of 0.2 years, whereas the DNAPL compound tetrachloroethene – also very toxic – has a half-life of 18 years because of its sequential degradation to vinyl chloride. If released into the environment at the same time, the DNAPL tetrachloroethene would last approximately 90 times longer than the LNAPL benzene. Table 5.10 lists the CRF_{SOIL} for common PAHs (Kaufman et al. 2005; Rogers et al. 2007a).

The CRF_{SOIL} values for PAH compounds are significantly higher than the CRF_{SOIL} for LNAPL VOCs, because PAHs strongly sorb to soil particles and are much less soluble in water (Kaufman et al. 2009). Table 5.11 summarizes the CRF_{SOIL} for the LNAPL and DNAPL VOCs, PAHs, and also

TABLE 5.9
CRF_{SOIL} Values for Common DNAPL VOCs

LNAPL Compound	Soil Type	CRF_{SOIL}
Tetrachloroethene (PCE)	Clay	3,955.0
	Silty Clay	647.0
	Sand Silty Clay	591.0
	Sand	371.0
Trichloroethene (TCE)	Clay	255.0
	Silty Clay	40.0
	Sand Silty Clay	37.0
	Sand	23.0
Cis-1,2-Dichloroethene	Clay	3,698.0
	Silty Clay	1,205.0
	Sand Silty Clay	1,159.0
	Sand	993.0
Trans-1,2-Dichloroethene	Clay	491.0
	Silty Clay	111.0
	Sand Silty Clay	99.0
	Sand	87.0
Vinyl chloride	Clay	5,727.0
	Silty Clay	2,575.0
	Sand Silty Clay	2,508.0
	Sand	2,308.0
1,1,1-Trichloroethane (1,1,1-TCA)	Clay	13.7
	Silty Clay	2.5
	Sand Silty Clay	2.3
	Sand	1.6
Mean DNAPL CRF_{SOIL} for degradation sequence from PCE to vinyl chloride[a] (see Figure 2.4)	Clay	2,825.0
	Silty Clay	915.6
	Sand Silty Clay	878.8
	Sand	756.4

[a] Represents cumulative risk

Source: Kaufman, M.M. Rogers, D.T., and Murray, K.S. Urban Watersheds: Geology, Contamination, and Sustainability. CRC Press. Boca Raton, FL. 583p. 2011.

TABLE 5.10

CRF$_{SOIL}$ Values for Select PAHs

PAH Compound	Soil Type	CRF$_{SOIL}$
Naphthalene	Clay	11.8
	Silty Clay	1.24
	Sand Silty Clay	1.06
	Sand	0.37
Chrysene	Clay	2,407.00
	Silty Clay	240.00
	Sand Silty Clay	204.00
	Sand	61.00
Benzo[b]fluoranthrene	Clay	146.00
	Silty Clay	14.60
	Sand Silty Clay	12.40
	Sand	3.70
Benzo[k]fluoranthrene	Clay	126.00
	Silty Clay	12.60
	Sand Silty Clay	10.70
	Sand	3.21
Phenanthrene	Clay	70.00
	Silty Clay	7.00
	Sand Silty Clay	5.90
	Sand	1.80
Benzo[g,h,i] perylene	Clay	1,080.00
	Silty Clay	108.00
	Sand Silty Clay	91.00
	Sand	27.50
Benzo(a)pyrene	Clay	797.30
	Silty Clay	79.70
	Sand Silty Clay	67.50
	Sand	20.20
Mean PAH CRF$_{SOIL}$	Clay	662.58
	Silty Clay	66.16
	Sand Silty Clay	55.23
	Sand	16.82

Source: Kaufman, M.M. Rogers, D.T., and Murray, K.S. Urban Watersheds: Geology, Contamination, and Sustainability. CRC Press. Boca Raton, FL. 583p. 2011.

lists the CRF$_{SOIL}$ for total PCBs, the pesticide chlordane, and the heavy metals chromium VI, lead, mercury, and arsenic (Rogers et al. 2007b; Kaufman et al. 2009).

Contaminants with the highest CRF$_{SOIL}$ are mercury, chlordane, PCBs, arsenic, and PAHs, respectively. These contaminants strongly sorb to soil particles and tend not to be very soluble in water. DNAPL compounds also have rather elevated CRF$_{SOIL}$, especially in soil composed of clay because they also strongly sorb to fine-grained soils. LNAPL VOCs and chromium VI had the lowest CRF$_{SOIL}$. LNAPL VOCs have rather short half-lives in soil and chromium VI does not strongly sorb to soil grains and is rather soluble. When the data in Table 5.11 are graphed, it becomes clear the highest CRF$_{SOIL}$ in each group occurs in soils composed of clay. Clay soils have the lowest permeability and have more surface area available for contaminants with high sorptive potential.

TABLE 5.11

Summary of CRF$_{SOIL}$ Values for the LNAPL and DNAPL VOCs, PAHs, and CRF$_{SOIL}$ for Other Selected Compounds

Compound	Soil Type	CRF$_{SOIL}$
Mean LNAPL CRF$_{SOIL}$	Clay	30.17
	Silty Clay	4.37
	Sand Silty Clay	3.89
	Sand	2.24
Mean DNAPL CRF$_{SOIL}$ for degradation sequence from PCE to vinyl chloride	Clay	3,408.00
	Silty Clay	1,116.00
	Sand Silty Clay	1,073.00
	Sand	965.00
Mean PAH CRF$_{SOIL}$	Clay	662.58
	Silty Clay	66.16
	Sand Silty Clay	55.23
	Sand	16.82
PCBs	Clay	17,992.00
	Silty Clay	2,800.00
	Sand Silty Clay	2,524.00
	Sand	1,457.60
Pesticide chlordane	Clay	28,850.00
	Silty Clay	2,663.00
	Sand Silty Clay	2,215.00
	Sand	1,876.00
Chromium VI	Clay	10.20
	Silty Clay	4.00
	Sand Silty Clay	3.50
	Sand	3.00
Lead	Clay	234.00
	Silty Clay	11.70
	Sand Silty Clay	9.94
	Sand	3.03
Mercury	Clay	41,800.00
	Silty Clay	19,140.00
	Sand Silty Clay	17,600.00
	Sand	15,600.00
Arsenic	Clay	7,800.00
	Silty Clay	2,950.00
	Sand Silty Clay	2,850.00
	Sand	2,400.00

Source: Kaufman, M.M. Rogers, D.T., and Murray, K.S. Urban Watersheds: Geology, Contamination, and Sustainability. CRC Press. Boca Raton, FL. 583p. 2011.

Therefore, contaminants with low water solubility and high relative sorptive potentials will have high CRF$_{SOIL}$ values in clay-rich soils (Figure 5.6).

Figure 5.6 indicates mercury has the highest CRF$_{SOIL}$ and chromium VI has the lowest. Chromium VI's low CRF$_{SOIL}$ results from its high relative water solubility and low sorptive potential. On the other hand, mercury has a high sorptive potential and low water solubility, and checks in with a

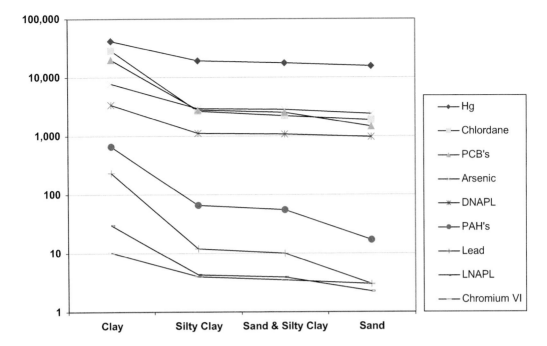

FIGURE 5.6 CRF_{SOIL} distribution by contaminants type. (From Kaufman, M.M. Rogers, D.T. and Murray, K.S. Urban Watersheds: Geology, Contamination, and Sustainability. CRC Press. Boca Raton, FL. 583p. 2011.)

high CRF_{SOIL}. It should also be noted that for every contaminant, the CRF_{SOIL} value is the highest in clay, a consequence of this soil texture acting to impede the migration of contaminants. Table 5.12 compares the CRF_{GW} with the CRF_{SOIL}.

From Table 5.12 it is clear the contaminants exhibiting high soil risk factors (e.g., chlordane, mercury, and PCBs) have a low groundwater contaminant risk factor. This behavior is primarily attributed to their low mobility. Computing the ratio of the data in columns 3 and 4 yields the likelihood of a particular contaminant type contaminating soil and/or groundwater. This computation is shown in column 5; and the values range from 1.0×10^{-9} to 766 – a difference in magnitude of more than a billion.

A ratio value of 1.0 in column 5 indicates a particular chemical compound has an equal probability of contaminating soil and groundwater; values less than 1.0 indicate a higher likelihood of contaminating soil only, and values greater than 1.0 mean the compound is more likely to contaminate groundwater. The high mean (arithmetic average) CRF_{GW} /CRF_{SOIL} ratios of chromium VI and DNAPL clearly suggests these compounds have the highest likelihood of contaminating groundwater; whereas the low mean ratios of PCBs, mercury, and chlordane positions them as posing the greatest contamination threat to soil. This conclusion is supported by the average extents of contamination shown in column 6, which were measured using analytical data from hundreds of sites (Kaufman et al. 2005; Rogers et al. 2007b; Kaufman et al. 2009). The measured extents of contamination listed in the right-most column represent contaminants at concentrations exceeding the applicable regulatory action level from an identified anthropogenic source. As predicted by the computed CRF_{GW} /CRF_{SOIL} ratios, the average extents of contamination were the largest for chromium VI and DNAPL compounds.

CRF_{SOIL} values were significantly higher for each contaminant group where the soil was composed of clay, but lowest for each contaminant group in sandy soils. This observation is the result of several physical/chemical factors, including:

- Density
- Grain size
- Composition

TABLE 5.12

Comparison of CRF$_{SOIL}$ to CRF$_{GW}$

Contaminant Type	Geologic Unit	Groundwater Contaminant Risk Factor[a] (CRF$_{GW}$)	Soil Contaminant Soil Risk Factor (CRF$_{SOIL}$)	Groundwater/Soil Contaminant Risk Factor Ratio CRF$_{GW}$/CRF$_{SOIL}$	Average Extent of Contamination from Source (meters)
DNAPL[b]	Clay	333	3,408	0.09	30
	SC[c]	773	1,116	0.69	42
	SSC[d]	1,091	1,073	1.01	50
	Sand	1,419	965	1.47	1,012
	Moraine	1,274	982	1.30	625
LNAPL[b]	Clay	0.08	30.17	0.002	38
	SC	0.42	4.37	0.09	43.5
	SSC	0.47	3.89	0.12	45.6
	Sand	0.72	2.24	0.32	328
	Moraine	0.65	2.52	0.25	270
PAHs[b]	Clay	0.0001	576.44	1.7×10^{-7}	25
	SC	0.0008	57.55	3.7×10^{-6}	25
	SSC	0.001	48.79	2.0×10^{-5}	30
	Sand	0.002	12.49	1.6×10^{-4}	40
	Moraine	0.002	17.47	1.1×10^{-4}	40
Lead	Clay	0.03	234	1.2×10^{-4}	25
	SC	0.3	11.7	0.02	25
	SSC	0.34	9.94	0.03	25
	Sand	1.1	3.03	0.36	30
	Moraine	1.0	3.56	0.28	30
Mercury	Clay	1.57	41,800	3.70×10^{-5}	30
	SC	2.52	19,140	1.31×10^{-4}	20
	SSC	2.77	17,600	1.57×10^{-4}	20
	Sand	3.10	15,600	1.98×10^{-4}	No Data
	Moraine	2.85	16,328	1.74×10^{-4}	No Data
Chromium VI	Clay	948	10.2	93	40
	SC	2,080	4	520	No Data
	SSC	2,116	3.5	604	No Data
	Sand	2,300	3.0	766	2,200
	Moraine	2,190	3.2	684	No Data
Arsenic	Clay	9.88	7,800	1.26×10^{-3}	30
	SC	11.26	2,950	3.81×10^{-3}	No Data
	SSC	21.12	2,850	7.41×10^{-3}	No Data
	Sand	50.25	2,400	2.09×10^{-2}	40
	Moraine	43.65	2,615	1.66×10^{-3}	No Data
Chlordane	Clay	0.0035	28,850	1.21×10^{-7}	25
	SC	0.035	2,663	1.31×10^{-5}	No Data
	SSC	0.041	2,215	1.85×10^{-5}	No Data
	Sand	0.046	1,876	2.45×10^{-5}	No Data
	Moraine	0.043	1,950	2.21×10^{-5}	25

(*Continued*)

TABLE 5.12 (CONTINUED)
Comparison of CRF$_{SOIL}$ to CRF$_{GW}$

Contaminant Type	Geologic Unit	Groundwater Contaminant Risk Factor[a] (CRF$_{GW}$)	Soil Contaminant Soil Risk Factor (CRF$_{SOIL}$)	Groundwater/Soil Contaminant Risk Factor Ratio CRF$_{GW}$/ CRF$_{SOIL}$	Average Extent of Contamination from Source (meters)
Total PCBs	Clay	0.000026	17,992	1.0×10^{-9}	25
	SC	0.00026	2,800	9.3×10^{-8}	25
	SSC	0.0003	2,524	1.2×10^{-7}	25
	Sand	0.0009	1,457	6.2×10^{-7}	25
	Moraine	0.0008	1,548	5.2×10^{-7}	25

[a] Values obtained from Kaufman et al. 2005
[b] Contaminant risk factors listed represent an average value of specific compounds listed in Tables 5.5 and 5.11
[c] SC = Sandy clay
[d] SSC = Sandy and silty clay
Source: Kaufman, M.M. Rogers, D.T., and Murray, K.S. Urban Watersheds: Geology, Contamination, and Sustainability. CRC Press. Boca Raton, FL. 583p. 2011.

- Permeability
- pH
- Redox

These factors work to increase the retention capacity of soils composed of clay and decrease the capacity in sandy soils. For example, clay soils generally have a higher contaminant retention capacity and a much lower permeability than sandy soils. Thus, the CRF$_{SOIL}$ shows a decrease in risk as the mean grain size of soil increases.

5.5 AIR POLLUTION RISK FACTOR

Atmospheric contaminants vary dramatically and originate from anthropogenic and natural sources. Anthropogenic sources are most prevalent in urban areas, and include stationary emitters such as manufacturing facilities, power generating plants, dry cleaners, and mobile emitters such as automobiles, trucks, buses, and airplanes (USEPA 2018). Natural sources include particles from volcanic eruptions, biological decay, forest fires, and pollen. Volatile organic compounds (VOCs) are also emitted by plants and trees. As an example of atmospheric contaminant variety, consider the VOC isoprene emitted by some tree species (e.g., oaks) (Sharkey et al. 2008). Isoprene is carcinogenic to humans if inhaled as a concentrated vapor within a closed space (USEPA 2002b), but concentrations in the atmosphere do not reach levels of concern.

Contaminants are released into the atmosphere as a gas, particulate matter, or as part of a water droplet through sorption or solution (USEPA 2018). Primary contaminants include heavy metals, volatile and semi-volatile compounds, and particulate matter. Secondary contaminants are formed in the atmosphere through photochemical or chemical reactions (USEPA 2018). Many of these contaminants travel between the air, soil, and surface water, and change states. For example, contaminants initially released to the soil or water may volatilize and become airborne. Or in some cases, contaminants airborne in the vapor phase may become deposited on the soil and sorb onto a soil particle only to become airborne again by wind action.

Given the composition and behavior of air pollutants, developing a model for evaluating the contaminant risks to air must address contaminants in the vapor phase and contaminants attached to particulate matter. Once airborne, a combination of these seven factors control whether there is an adverse effect from an air contaminant (McKone and Enoch 2002; USEPA 2018):

- Toxicity
- Mobility
- Persistence
- Volume released
- Time period of the release
- Distance to a specific receptor being evaluated
- Wind speed and direction

The last four factors in this list are environmental factors dependent upon site-specific criteria, so they are not included in the development of the air contaminant risk factor (CRF_{AIR}). These environmental factors are applicable when evaluating the impacts of an actual release, as done with the definition of the nature and extent of a contaminant plume within soil and groundwater. Three familiar factors remain (toxicity, mobility, and persistence). In addition, there are similarities between the physical and chemical attributes controlling the migration potential of a contaminant in air, groundwater, and soil including: solubility, vapor pressure, density, chemical stability, persistence, and adsorption potential (Kaufman et al. 2005; USGS 2006; Rogers et al. 2007a). On the basis of these similarities, the environmental risk posed by specific contaminants to air, termed the air contaminant risk factor, (CRF_{AIR}) can be computed by modifying the equations of the CRF_{GW} and CRF_{SOIL} (Kaufman et al. 2005; Rogers et al. 2007). The CRF_{AIR} is expressed in general terms as Equation 5.9 (Kaufman et al. 2011; Rogers et al. 2012; Rogers 2018).

$$CRF_{AIR} = \text{Toxicity} \times \left[\left(\text{Mobility}_{gas} \times \text{Persistence}_{gas} \right) + \left(\text{Mobility}_{particulate} \times \text{Persistence}_{particulate} \right) \right] \tag{5.9}$$

Toxicity values are obtained from the USEPA Integrated Risk Information System (IRIS) (2019a). The values selected should be the most conservative for each exposure pathway – ingestion, dermal adsorption, and inhalation. Using the most conservative value is appropriate since exposure to contaminated air can occur in each of these pathways. For instance, dermal adsorption of soil with sorbed contaminants can occur during any outdoor activity, ingestion can occur through hand to mouth contact and from swallowing, and inhalation can occur through inhalation of contaminants in the gas or particulate matter, especially if the particulate matter is very fine.

Mobility for air is represented separately for contaminants in the gas phase and for solid particulate matter, and is combined with persistence. This separation is necessary because the half-life of many contaminants is significantly different depending on whether the contaminant is in the gas phase or solid particulate matter. For instance, xylene has a half-life of 6 days in the vapor phase, but its half-life is 1 year when present in the particulate phase. To evaluate mobility in the gas phase, Henry's Law constant alone is sufficient because a retardation factor is not necessary. The mobility of atmospheric contaminants in the gas phase can be expressed as Equation 5.10.

$$M = (H) \tag{5.10}$$

where:
M = mobility
H = Henry's Law constant

As with the CRF_{GW} and CRF_{SOIL}, Henry's Law constants are derived from Equation 5.3, or from the literature including USEPA (1996), Sander (1999), Wiedemeier (1999), Suthersan and Payne (2005), and Payne et al. (2008). The mobility of a contaminant sorbed to particulate matter is represented by Equation 5.11.

$$M_{(particulate)} = \frac{1}{SpG} \times Koc \qquad (5.11)$$

where:

$M_{(particulate)}$ = mobility of a contaminant sorbed to particulate matter
Spg = Specific gravity
Koc = partitioning coefficient

Specific gravity is an important determinant of mobility because as the specific gravity of a contaminant increases, its buoyancy in the atmosphere decreases. This relationship explains why the inverse of the specific gravity is used. The partitioning coefficient is a measure of a contaminant's tendency to sorb to particulate matter. This factor is important because the higher the partitioning coefficient, the higher the likelihood a contaminant will attach to particulate matter. For example, a contaminant with a low specific gravity and high sorptive potential has the potential to be very mobile in the atmosphere compared to one with a high specific gravity and low sorptive potential. Values for the partitioning coefficient can be obtained from numerous sources, including USEPA (1996), Wiedemeier (1999), USEPA (2002a), and Suthersan and Payne (2005). Persistence of a contaminant in the air is generally expressed in the gas phase and particulate phase (McKone and Enoch 2002; USEPA 2018). Persistence times for most organic compounds are much less in the gas phase than when they are attached to particulate matter. Finally, the CRF_{AIR} is calculated by multiplying the inverse of the chemical compound's toxicity (T), by mobility (M) and persistence (P) in the gas and particulate phases (Equation 5.12). Multiplying by the inverse of the mobility value is not necessary.

$$CRF_{AIR} = \frac{1}{(T)} \times \left[\left(M_{gas} \times P_{gas} \right) + \left(M_{particulate} \times P_{particulate} \right) \right] \qquad (5.12)$$

where:

CRF_{AIR} = Soil Contaminant Risk Factor
T = Toxicity
M_{gas} = Mobility of a contaminant in the gas phase
P_{gas} = Persistence of a contaminant in the gas phase
$M_{particulate}$ = Mobility of a contaminant sorbed to particulate matter
$P_{particulate}$ = Persistence of a contaminant sorbed to particulate matter

The inverse of the toxicity value must be used because the integers assigned for toxicity values decrease with increasing carcinogenicity (USEPA 2019a). The presence of mobility and persistence factors for gas and particulates accounts for the significant differences in the half-lives depending on whether the contaminant is present as a gas, or if it is sorbed onto particulate matter (USEPA 2018). Using Equation 5.12, an example calculation of the CRF_{AIR} for a chemical XYZ is written as:

$$CRF_{AIR} \text{ for chemical XYZ} = \frac{1}{(T)} \times \left[\left(M_{gas} \times P_{gas} \right) + \left(M_{particulate} \times P_{particulate} \right) \right]$$

Step 1: Obtain toxicity value:
The toxicity of XYZ chemical was obtained from the literature and has a value of 0.04.
Step 2: Determine the mobility and persistence values for the gas phase.

Mobility in the gas phase $\left(M_{gas} \right)$ = Henry's Law constant $\left(H \right)$ = 0.228

Persistence in the gas phase $\left(P_{gas}\right)$ = literature value of 4 days

$$M_{gas}\left(0.228\right)\times P_{gas}\left(4\right)=0.228\times 4=0.912$$

Step 3: Determine the mobility and persistence values for the particulate phase using Equation 5.11:

$$\text{Mobility in the pariculate phase}\left(M_{particulate}\right)=\frac{1}{SpG}\times Koc$$

SpG for Chemical XYZ = 0.88

Koc for Chemical XYZ = 1.92

Therefore:

$$\text{Mobility in the pariculate phase}\left(M_{particulate}\right)=\frac{1}{0.88}\times 1.92=2.18$$

Persistence in the particulate phase $\left(P_{gas}\right)$ = literature value of 1.0 years

$$M_{particulate}\left(2.18\right)\times P_{particulate}\left(0.2\right)=2.18\times 1.0=2.18$$

Step 4: Determine the CRF_{air} using Equation 5.12:

$$CRF_{AIR}\text{ for chemical XYZ}=\frac{1}{(T)}\times\left[\left(M_{gas}\times P_{gas}\right)+\left(M_{particulate}\times P_{particulate}\right)\right]$$

$$CRF_{AIR}\text{ for chemical XYZ}=\frac{1}{(0.04)}\times\left[\left(0.228\times 4\right)+\left(2.18\times 1.0\right)\right]$$

$$CRF_{AIR}\text{ for chemical XYZ}=\frac{1}{(0.04)}\times\left[\left(0.91\right)+\left(2.18\right)\right]$$

$$CRF_{AIR}\text{ for chemical XYZ}=77.25$$

This sample calculation for CRF_{AIR} demonstrates how the risk posed by the chemical in the particulate phase is much greater than the risk posed by the chemical in the gas phase (2.18 vs. 0.912). This occurs in most cases because the contaminant is much more persistent as a particulate than in the gas phase, and as a result, there is an increased probability of exposure with a particulate form of a chemical. This trend is common with most contaminants, especially organics. Table 5.13 displays the CRF_{AIR} values for LNAPLs, DNAPLs, PAHs, selected metals, PCBs, and the pesticide chlordane.

As shown in column 2, lead had the lowest contaminant risk factor for air compared to any of the contaminants evaluated. The reason is not because lead is less toxic, but because lead is not volatile and has the second-highest specific gravity of any compound evaluated. The specific gravity of lead is 11.34; mercury is the only compound evaluated with a higher specific gravity (13.6). Lead is routinely released into the environment from air emission sources (USEPA 2018), but most airborne lead is rapidly deposited on the ground surface.

The particulate to gas ratios shown in column 5 are significant because the contaminants shown here are commonly found in US urban areas, where particulate matter tends to concentrate. With the exception of benzene and mercury, the risk of exposure to contaminants in the atmosphere is highest from those contaminants sorbed to particulate matter. This is dramatically demonstrated by PAHs and PCBs – compounds having a mean particulate to gas ratio of 1,000 to 1. Also displaying impressively large particulate to gas ratios are chlordane (500:1) and DNAPL compounds (100:1). If an equal weighting were used between the gas phase and particulates, the differential would be much greater. This is because the values of persistence in the gas phase used to calculate the CRF_{AIR} values are in days and the values of persistence in the particulate phase are in years. The LNAPL VOCs provide a good example of how the presence of particulate matter can transform a short-lived group of chemicals into a significant atmospheric risk. This group of compounds includes benzene, toluene, ethyl benzene, and xylenes and is commonly referred to as BTEX. They are common components of gasoline and highly volatile, toxic, and flammable, but degrade rapidly in the gaseous phase due to photolysis. Their persistence increases significantly when sorbed onto a particle, and this is reflected by their moderate values for CRF_{AIR} in column 2. Without this sorption, these values would be very low.

5.6 DISCUSSION AND IMPLICATIONS

Table 5.14 compares the contaminant risk factors for groundwater, soil, and air. DNAPLs and chromium VI have the highest potential to contaminate groundwater; mercury, PCBs, and chlordane have the highest potential to contaminate and remain in soil or sediment, and several compounds including LNAPLs and DNAPLs, PCBs, PAHs, chlordane, arsenic, chromium VI, and mercury have the potential to contaminate the air, especially when attached to particulate matter.

An examination of the DNAPL compounds listed in Tables 5.3, 5.9, and 5.13 reveals that the compound vinyl chloride exhibits the highest groundwater, soil, and air contaminant risk factors compared to the other DNAPL compounds evaluated. This dubious distinction occurs because vinyl chloride has these physical/chemical attributes: 1) it is the most toxic DNAPL compound evaluated (USEPA 2019a; 2019c) and is also the most mobile; 2) it is the most soluble DNAPL compound in water compared to the other DNAPL compounds evaluated; and 3) it has the highest vapor pressure so it evaporates readily (ATSDR 2006b).

Chromium VI is also a compound of special note. It has the highest groundwater and air risk factors because of its high toxicity and relatively high solubility in water. Because of its high solubility, we would expect chromium VI to have the lowest soil risk factor; and it does as a result of its low sorptive potential. Examining the contaminant risk factors for many compounds helps explain their fate and transport; specifically, where contaminants are found after being released into the

environment (their sinks), and why they have time to migrate or degrade after being released. For instance, mercury and PCBs tend to accumulate in river and lake sediments and soil, but are rarely detected in water. When they are detected in water it is at typically low concentrations. This fate and transport are predicted by the contaminant risk factor scores for mercury and PCBs – they score very high for soil, and very low for groundwater (Figure 5.7).

Two more examples of the predictive power of the risk scores are seen with the DNAPL compounds and chromium VI in groundwater. As shown in Table 5.7, groundwater contaminant plumes for DNAPL compounds and chromium VI are far longer than any of the other compounds evaluated. When the CRF_{GW} and the geologic vulnerability assessment described in Chapter 6 are combined, it results in a synergistic effect. This outcome is likely when chromium VI or DNAPLs are released (often continuously over a long duration) into a vulnerable geologic environment (e.g., soils composed of sand with shallow groundwater used as a source of potable water). In these cases, significant adverse environmental and human health effects occurred (Rogers et al. 2006).

Another view of the data is provided by the pie charts shown in Figure 5.8. For each type of contaminant risk factor, certain contaminants favor certain environmental media.

Table 5.15 provides an interpretation of the data presented by the pie charts in Figure 5.8, and can be used as a summary reference for CRFs by major contaminant groups and compounds of selected interest.

TABLE 5.13

Summary of CRF$_{air}$ Values and Exposure Risk Potential

Chemical Compound	Contaminant Risk Factors for Air (CRF$_{air}$)	Percent of Potential Exposure Risk from Gas Phase	Percent of Potential Exposure Risk from Particulate Phase	Approximate Ratio of Exposure Risk Particulate:gas
LNAPL compounds				
Benzene	36.05	55.69	44.31	4:5
Ethyl benzene	18.18	11.00	89.00	9:1
Toluene	7.75	13.22	86.88	8.5:1.5
Xylenes	18.95	2.07	97.93	97:3
Mean LNAPL	20.23	81.98	318.02	
		20.49%	79.51%	4:1
DNAPL compounds				
PCE	91.09	1.63	98.37	98:2
TCE	9.37	1.05	98.94	99:1
111-TCA	7.71	0.08	99.92	99:1
Cis-12-DCE	34.10	0.18	99.82	99:1
Trans -12-DCE	18.90	0.10	99.90	99:1
Vinyl chloride	243.97	8.30	91.70	92:8
Mean DNAPL	67.52	11.34	588.66	
		1.89%	98.11%	98:2
PAHs				
Naphthalene	30.44	0.003	99.997	1,000:1
Chrysene	162.09	0.0003	99.9997	1,000:1
Benzo (b) fluoranthene	167.19	0.0004	99.9996	1,000:1
Benzo (k) fluoranthene	97.74	0.01	99.99	1,000:1
Phenanthrene	40.40	0.0009	99.9991	1,000:1
Benzo (g,h,i) perylene	184.13	0.0002	99.9998	1,000:1
Benzo (a) pyrene	94.59	0.000008	99.999992	100,000:1
Mean PAHs	124.82	0.0148	699.985	
		0.003%	99.997%	1,000:1
Metals				
Arsenic	44.50	3.6	96.4	24:1
Chromium VI	210.25	44.59	55.41	5:4
Mercury	166.33	83.36	16.64	1:5.6
Lead	3.45	12.68	87.32	7:1
Mean metals	106.13	144.23	255.77	
		36.06%	63.94%	16:9
Mean PCBs	37.02	0.009	99.99	1,000:1
Chlordane	173.02	0.017	99.983	500:1

Source: Kaufman, M.M. Rogers, D.T., and Murray, K.S. Urban Watersheds: Geology, Contamination, and Sustainability. CRC Press. Boca Raton, FL. 583p. 2011.

TABLE 5.14

Contaminant Risk Factors for Groundwater, Soil, and Air for Each Contaminant Group

Contaminant Type	Groundwater CRF (CRF_{GW})	Soil CRF (CRF_{SOIL})	Air CRF (CRF_{AIR})
DNAPL	978.00	1,508.00	67.52
LNAPL	0.47	10.16	20.23
PAH	0.001	142.00	124.82
Total PCBs	0.0004	5,264.00	37.00
Chlordane	0.0001	7,501.00	173.00
Arsenic	27.24	3,723.00	44.00
Chromium VI	1,926.00	5.17	210.00
Lead	0.55	52.44	3.45
Mercury	2.56	22,013.00	166.00

Source: Kaufman, M.M. Rogers, D.T., and Murray, K.S. Urban Watersheds: Geology, Contamination, and Sustainability. CRC Press. Boca Raton, FL. 583p. 2011.

Additional support for the validity of the risk scores would be their ability to explain a high amount of the variation in remediation costs, and to also explain a high amount of the variation in remedial costs across a wide range of geological environments. When the air, groundwater, soil, and surface risk factors are combined, a total risk characterization of a chemical's environmental risks is achieved. Does more risk result in higher remediation cost? To answer this question, a stepwise regression was used to search for the variables that could explain the most variance in remediation

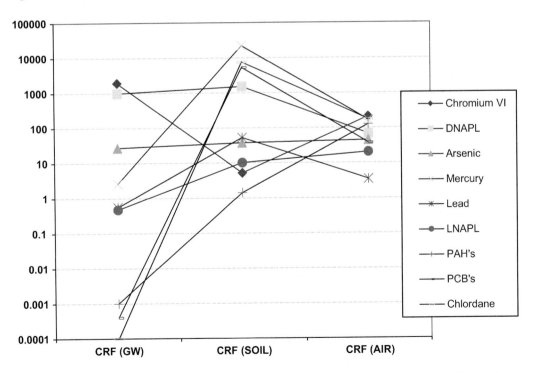

FIGURE 5.7 Groundwater, soil, and air CRF distribution. (From Kaufman, M.M. Rogers, D.T. and Murray, K.S. Urban Watersheds: Geology, Contamination, and Sustainability. CRC Press. Boca Raton, FL. 583p. 2011.)

costs at 79 sites in an urbanized watershed within a variety of geological environments. Using as input the variables listed in Table 5.6, early iterations of the stepwise procedure on the Rouge River watershed data set retained extent of contamination and the total risk scores to explain remediation cost. Examination of the distribution of the dependent and independent variables indicated the data were positively skewed. After a logarithmic transformation, and the multiplication of the total risk scores by plume length, the River Rouge data set indicates a strong correlation between the remediation cost and the total risk given by Equation 5.13 and represented by Figure 5.9 (Kaufman et al. 2005).

$$\text{Lg cost} = 5.107 + 0.4949 \text{ lg risk weight} \tag{5.13}$$

When Equation 5.8 was tested against 127 sites located worldwide in various geological environments, there was a very strong correlation between the estimated values derived from Equation 5.13 for remediation cost and the actual cost values ($r = 90$; $F = 334.1$, $p < 0.0001$). The results are shown

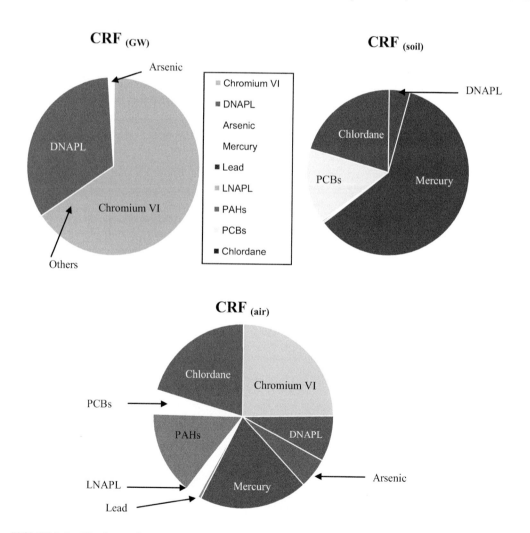

FIGURE 5.8 Pie charts of groundwater, soil, and air CRFs. (From Kaufman, M.M. Rogers, D.T. and Murray, K.S. Urban Watersheds: Geology, Contamination, and Sustainability. CRC Press. Boca Raton, FL. 583p. 2011.)

TABLE 5.15
Relative CRF Risk Rankings of Contaminants by Media

Compound	CRF_{GW}	CRF_{SOIL}	CRF_{AIR}
Chromium VI	Extremely high	Very low	Very high
DNAPL	High	High	Moderate
Arsenic	Low	Very low	Moderate
LNAPL	Very low	Very low	Moderate
Mercury	Very low	Extremely high	Very high
Lead	Very low	Very low	Low
PAHs	Extremely low	Very low	Very high
Chlordane	Extremely low	Very high	Very high
PCBs	Extremely low	Very high	Moderate

Source: Kaufman, M.M. Rogers, D.T., and Murray, K.S. Urban Watersheds: Geology, Contamination, and Sustainability. CRC Press. Boca Raton, FL. 583p. 2011.

in Figure 5.10. It is reasonable to conclude that total risk scores do help explain much of the variation in remediation costs.

For all the sites tested, those contaminated with chromium VI and DNAPL compounds located within a geologic setting of sand exhibited the highest extent of contamination and cost of remediation per kilogram of contaminant recovered.

There still is much work needed on risk scores. Contaminant risk factors of other contaminants such as ozone, sulfur dioxide, greenhouse gases, viruses, bacteria, acids and bases, and fertilizers have not yet been calculated. The main reasons for this are: 1) there is a lack of data required to calculate a CRF; 2) some uncertainties exist in assessing risk for certain contaminants such as bacteria and carbon dioxide; and 3) the jury is still out on whether to include any weighting factors in calculating CRFs for these compounds.

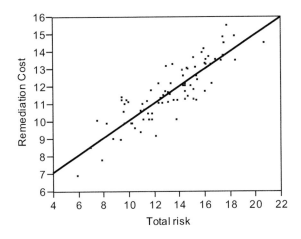

FIGURE 5.9 Regression analysis of remedial cost and CRF_{GW}. (From Kaufman, M.M. Rogers, D.T. and Murray, K.S. Urban Watersheds: Geology, Contamination, and Sustainability. CRC Press. Boca Raton, FL. 583p. 2011.)

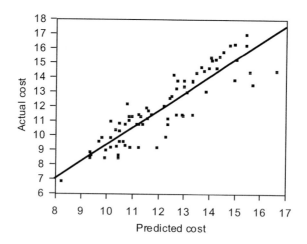

FIGURE 5.10   Actual vs. predicted remedial costs using the CRF$_{GW}$. (From Kaufman, M.M. Rogers, D.T. and Murray, K.S. Urban Watersheds: Geology, Contamination, and Sustainability. CRC Press. Boca Raton, FL. 583p. 2011.)

5.7   SUMMARY AND CONCLUSION

CRFs for air, groundwater, and soil provide a powerful tool for assessing risks posed by contaminants anywhere in the environment. They are derived by combining the three physical/chemical attributes of toxicity, mobility, and persistence. CRF analysis has demonstrated why some contaminants prefer certain locations or sinks. For instance, CRF analysis has shown why PCBs and mercury prefer soil or sediments instead of being present as dissolved constituents in groundwater.

Combining geologic vulnerability analysis with CRF analysis can reveal and explain potential synergistic effects. Examples of synergy occur when DNAPLs and chromium VI are released in a vulnerable geologic setting, such as one with a sandy soil, shallow groundwater, and current use as a potable source of water. Under these conditions, the potential for adverse human health and ecological impact greatly increases, and is unfortunately often realized. With CRF and geological vulnerability analysis now in hand, these situations can be avoided in the future with proper urban planning.

CRF analysis employing the risks posed by surface releases, added to the risk factors for groundwater, soil, and air, also has predictive power for evaluating and assessing future cleanup costs for sites of contamination. This ability has major implications for sustaining urban areas, which we explore in later chapters.

NOTE

1. The 4-digit SIC codes were replaced by the six-digit North American Industry Classification System (NAICS) in 1997. The NAICS was developed to allow for a higher level of comparability in business statistics among Mexico, Canada, and the US. Some governmental agencies still use the SIC code, and the authors prefer them because the descriptions are more complete than those provided by the NAICS.

REFERENCES

Agency for Toxic Substances and Disease Registry (ATSDR). 2006b. *Vinyl Chloride: CAS Registry Number 75-01-4*. ATSDR. Atlanta, GA.

Barringer, T.H., Dunn, D., Battaglin, W.A., and Vowinkel, E.F. 1990. Problems and Methods Involved in Land Use to Groundwater Quality. *Water Resources Bulletin*. Vol. 26. pp. 1–9.

Camp, C.V. and Outlaw, J.E. 1998. Stochastic Approach to Delineating Wellhead Protection Areas. *Journal of Water Resources, Planning and Management*. Vol. 124. pp. 199–208.

Eckhardt, D.A.V. and Stackelberg, P.E. 1995. Relation of Groundwater Quality to Land Use on Long Island, New York. *Journal of Ground Water*. Vol. 33. pp. 1019–1033.

Kaufman, M.M. 1997. Spatial Assessment of Risk to Groundwater from Surface Activities. *Proceedings of the Applied Geography Conference*. Vol. 20. pp. 135–142.

Kaufman, M.M. 2000. Automated Procedure for Groundwater Risk Assessment from Surface Sources. *Proceedings of the Applied Geography Conference*. Vol. 23. pp. 277–284.

Kaufman, M.M., Murray, K.S., and Rogers, D.T. 2003. Surface and Subsurface Geologic Risk Factors to Ground Water Affecting Brownfield Redevelopment. *Journal of Environmental Quality*. Vol. 32. pp. 490–499.

Kaufman, M.M., Rogers, D.T., and Murray, K.S. 2005. An Empirical Model for Estimating Remediation Costs at Contaminated Sites. *Journal of Water, Air, and Soil Pollution*. Vol. 167. pp. 365–386.

Kaufman, M.M., Rogers, D.T., and Murray, K.S. 2009. Using Soil and Contaminant Properties to Assess the Potential for Groundwater Contamination to the Lower Great Lakes, USA. *Journal of Environmental Geology*. Vol. 56. pp. 1009–1021.

Kaufman, M.M., Rogers, D.T., and Murray, K.S. 2011. *Urban Watersheds: Geology, Contamination, and Sustainability*. CRC Press. Boca Raton, FL. 583p.

McKone, T.E. and Enoch, K.G. 2002. *CalTox*TM, *A Multimedia Total Exposure Model Spreadsheet Users Guide*. Lawrence Berkeley National Laboratory. LBNL 47399. University of California. Berkeley, CA.

Michigan Department of Environmental Quality (MDEQ). 2008. *State List of Contaminated Sites*. MDEQ. Lansing, MI.

Michigan Department of Environmental Quality. 2019. *Chemical Update Worksheets*. Remediation and Redevelopment Division. Lansing, MI. https://www.michigan.gov/documents/deq/deq-rrd-chem-Ben zeneDatasheet_527763_7.pdf. (accessed March 31, 2019).

Payne, F.C., Quinnan, J.A., and Potter, S.T. 2008. *Remediation Hydraulics*. CRC Press. Boca Raton, FL.

Rogers, D.T., Kaufman, M.M., and Murray, K.S. 2006. Improving Environmental Risk Management through Historical Impact Assessments. *Journal of Air and Waste Management*. Vol. 56. pp. 816–823.

Rogers, D.T., Murray, K.S., and Kaufman, M.M. 2007a. Assessment of Groundwater Contaminant Vulnerability in an Urban Watershed in Southeast Michigan, USA. *In*: Howard, K.W.F. editor. *Urban Groundwater – Meeting the Challenge*. Taylor & Francis. London, England.

Rogers, D.T., Murray, K.S., and Kaufman, M.M. 2007b. *Improving the Assessment of the Potential for Groundwater Contamination Using an Analytical Risk Factor Model*. International Union of Geodesy and Geophysics World Congress. Perugia, Italy.

Rogers, D.T., Murray, K.S., and Kaufman, M.M. 2012. Environmental Risk Analysis through Integration of Geologic Vulnerability and Air, Water, and Soil Contaminant Risk Factor Derivation. International Geological Congress. p. 3003. Brisbane, Australia. p. 193–195.

Rogers, D.T. 2018. Derivation of a Comprehensive Environmental Risk Model for Urban Groundwater Protection. International Association of Hydrogeologists Congress. Vol. 1. Daejeon, Korea. p. 55.

Sander, R. 1999. Compilation of Henry's Law Constants for Inorganic and Organic Species of Potential Importance in Environmental Chemistry (Version 3). http://www.henrys-law.org. (accessed December 20, 2018).

Schnoor, J.L. 1996. *Environmental Modeling: Fate and Transport of Pollutants in Water and Soil*. John Wiley & Sons. New York.

Secunda, S., Collin, M.L., Melloul, M.L., and Abraham, J. 1998. Groundwater Vulnerability Assessment Using a Composite Model Combining DRASTIC with Extensive Agricultural Land Use in Israel's Sharon Region. *Journal of Environmental Management*. Vol. 54. pp. 39–57.

Sharkey, T.D, Wiberley, A.E., and Donohue, A.R. 2008. Isoprene Emission from Plants: Why and How. *Annals of Botany*. Vol. 101. pp. 5–18.

Suthersan, S.S. and Payne, F.C. 2005. *In Situ Remediation Engineering*. CRC Press. Boca Raton, FL.

Tobler, W. 1988. Resolution, Resampling and All That. *In*: Mounsey, H. and Tomlinson, R. Editors. *Building Database for Global Science*. Taylor & Francis. London, UK. pp. 129–137.

United States Census Bureau. 2019. Standard Industrial Classification (SIC) Code Description. https://www.census.gov/programs-surveys/cbp/technical-documentation/reference/naics-descriptions/sic-code-descriptions.html. (accessed April 20, 2019).

United States Environmental Protection Agency (USEPA). 1989. *Transport and Fate of Contaminants in the Subsurface*. USEPA. EPA/625/4-89/019. Washington, DC.

USEPA. 1992. *Dense Nonaqueous Phase Liquids*. USEPA. Office of Research and Development. Washington, DC.

USEPA. 1996. *Bioscreen. Natural Attenuation Decision Support System, Version 1.4.* USEPA. Office of Research and Development. Washington, DC.

USEPA. 2002a. *Supplemental Guidance for Developing Soil Screening Levels for Superfund Sites.* USEPA. Office of Solid Waste and Emergency Response. OSWER 9355.4-24. Washington, DC.

USEPA. 2002b. *Health Assessment of 1,3-butadiene.* National Center for Environmental Assessment, USEPA/600/P-98/001F. Washington, DC.

USEPA. 2018. National Air Quality: Status and Trends of Key Air Pollutants. https://www.epa.gov/air-trends. (accessed April 20, 2019).

USEPA. 2019a. Integrate Risk Information System (IRIS). http://www.epa.gov/ncea/iris/intro.htm. (accessed March 31, 2019).

USEPA. 2019b. *EPA Online Tools for Site Assessment Calculation. Ecosystems Research Division.* USEPA. http://www.epa.gov/athens/learn2model/part-two/onsite/retard.html. (accessed March 31, 2019).

USEPA. 2019c. EPA's Report on the Environment (ROE). https://www.epa.gov/report-environment. (accessed March 24, 2019).

United States Geological Survey (USGS). 1998. *Estimating Areas Contributing to Recharge of Wells.* USGS Circular 1174. Denver, CO.

USGS. 2006. *Volatile Organic Compounds in Nation's Ground Water and Drinking-Water Supply Wells.* USGS Circular 1292. Washington, DC.

Wiedemeier, T.H., Rifai, H.S., Newell, C.J., and Wilson, T.J. 1999. *Natural Attenuation of Fuels and Chlorinated Solvents in the Subsurface.* John Wiley & Sons. New York.

Wisconsin Department of Natural Resources (WDNR). 1999. *Wellhead Protection.* WDNR Publication PUB-DG-039 99REV. Madison, WI.

Part Two

Environmental Regulations

6 Environmental Regulations and Pollution of the United States

6.1 INTRODUCTION

Many believe that environmental regulations in the United States are the most comprehensive and protective on Earth. In many respects, this is true but there are significant exceptions and gaps and we will examine many in this chapter.

The purpose of environmental regulations are to protect human health and the environment. However, protecting human health and the environment is difficult if the regulations are not comprehensive and all-encompassing. To assist in explaining this point, it is best to demonstrate the need for comprehensive environmental regulations through example scenarios. For example, if a person were placed in a situation where there are ten detrimental environmental circumstances that may cause harm to that person and the person is only protected through regulations against nine of the items, that person is still harmed. Some may interpret that the regulations were ineffective in this example because the person was still harmed. This is a common frustration of applying environmental regulations; there always seems to be something no one thought of that leads to an unacceptable exposure. This is one of the reasons why environmental regulations are complex; they have to be in order to be truly protective.

Another challenge is that environmental regulations of the United States only apply to the United States. What if the person in the example above who was protected from nine of the ten potential hazards but the one hazard that the person was not protected originated in a different country. This is more difficult to address because our laws only apply to our country. This scenario is central to one of our themes in this book, which is *pollution does not respect boundaries*. This is another reason why environmental regulations are so complex and sometimes ineffective.

The third challenge from our example above would be that the person was harmed not because the regulations were inadequate, but because something new was introduced into the situation that was not present when the regulations took effect which modified the situation enough to cause harm. This is why environmental regulations always seem to need to be updated and improved. New information and facts or new circumstances arise that must be accounted for and addressed. This is yet another example of why environmental regulations are so complex and become more complex with time; they need to be comprehensive.

Lastly, environmental regulations are in large part based on single incidents that cause large scale harm to human health and the environment. Therefore, they are reactive rather than proactive, which tends to narrow the focus of regulations and does not fully address other circumstances that cause similar harm.

At the end of this chapter we shall discuss the state of environmental regulations in the United States. We shall see that some regulations in certain subject areas, such as solid waste, for example, have perhaps matured to a point where further improvement may not be necessary at this time. We shall also discover that there are areas where additional regulations are necessary or are totally lacking and are in need of more immediate attention.

Factors that influence the effectiveness of environmental regulations vary from country to country and within each country as well. The United States is no different. Factors that influence the degree to which a country has effective environmental regulations include (Bates and Ciment 2013; United Nations 2017a):

- Political will
- Environmental tragedies, harm to the environment, and lessons learned
- Cost

- Geography and climate
- Enforcement
- Incentives
- Lack of overriding social factors such as war, political structure, greed, poverty, hunger, lack of infrastructure, educational awareness, and corruption

Along with our example situations, each of the previously listed factors either influences the effectiveness of environmental regulations in a positive way or undermine environmental regulations. At the end of Chapter 7 and after the discussion of the environmental regulations and pollution challenges throughout the world, we will evaluate and rank the effectiveness of environmental regulations of each country evaluated, including the United States.

While reading this chapter, keep in mind the three challenges mentioned at the beginning of this section and the three basic environmental fundamentals introduced in Chapter 1, which are referred to numerous times throughout this book:

- Protect the air
- Protect the water
- Protect the land

This chapter is organized by history and subject. The first section is a background and introduction into environmental regulations. The second section is a brief history of the environmental movement in the United States. The final and largest section is a detailed examination of the most important environmental regulations of the United States. We then review and summarize what we have learned. An important section at the end of this chapter is dedicated to examining the need for improvements to existing regulations and where there are needs for new environmental regulations altogether in the United States and beyond. We then move on to the next chapter where we will apply what we have learned in this chapter and explore the environmental regulations of other countries and pollution of the Earth.

6.2   GEOGRAPHY AND CLIMATE

As we learned in Chapter 2, geography and climate influence the types and behavior of pollution throughout the world. Therefore, it's only fitting that we include an overview of the geography and climate of each country that we evaluate. This will provide some context as to the unique attributes each country has when the connection between pollution and specific geographic and climatic influences within each country is realized.

The United States encompasses an area of 9,826,675 square kilometers with 93% being land and 7% being rivers, streams, and lakes (see Figure 6.1). The climate of the United States is highly variable ranging from subtropical to tundra (see Figure 6.2). Land use in the United States consists of the following (United States Department of Agriculture 2011):

- 2,748,717 square kilometers or 30% forest
- 2,473,845 square kilometers or 27% pasture and rangeland
- 1,649,230 square kilometers or 18% crop land
- 1,282,734 square kilometers or 14% parks and wildlife areas
- 824,615 square kilometers or 8.5% swamps and tundra
- 274,871 square kilometers or 2.5% urban

A significant portion of the 30% of land in the United States made up of forest is under the jurisdiction of the National Forest Service or the Department of the Interior. This forested land is routinely harvested for lumber and other pulp and paper products (United States Forest Service 2018). The United States' population is estimated at 329 million (United States Census Bureau 2019). The

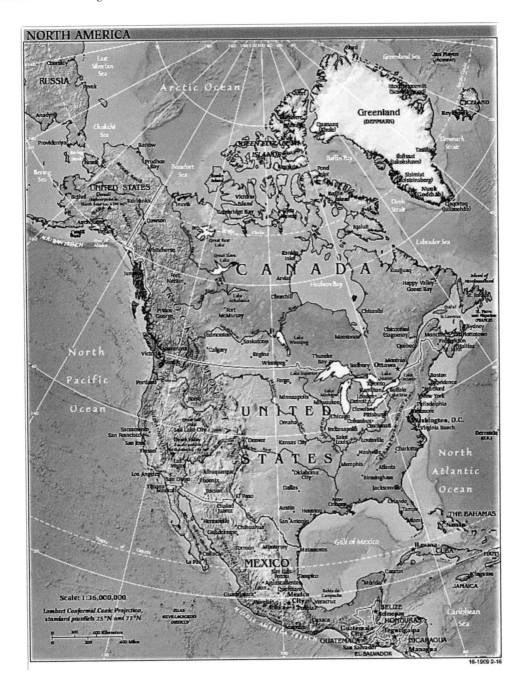

FIGURE 6.1 Map of the United States. (From United States Central Intelligence Agency. World Fact Book. Map of North America. 2019. www.cia.gov/library/publications/the-world-factbook/attachments/images/large/north_america-physical.jpg?1547145652.) (Accessed February 16, 2019.)

economy of the United States is the third largest in the world, just behind China and the European Union. The estimated gross domestic product of the United States is $19.3 trillion (United States Department of Commerce 2018). Per capita income in the United States is approximately $60,000 and ranks 19th in the world. The United States' economic drivers are natural resources (e.g., mining, petroleum, timber), agriculture, manufacturing, well-developed infrastructure, and high productivity (United States Department of Commerce 2018).

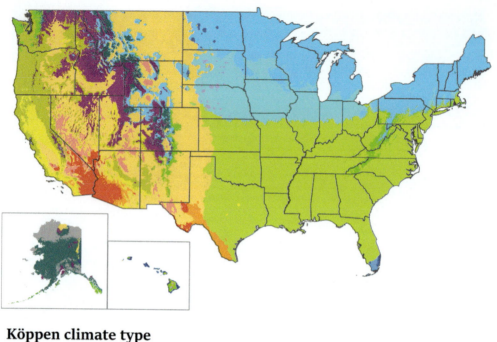

Köppen climate type

EF (Ice-cap)	Dsb (Warm-summer mediterranean continental)	Csa (Hot-summer mediterranean)
ET (Tundra)	Dsa (Hot-summer mediterranean continental)	BSk (Cold semi-arid)
Dfc (Subarctic)	Cfc (Subpolar oceanic)	BSh (Hot semi-arid)
Dfb (Warm-summer humid continental)	Cfb (Oceanic)	BWk (Cold desert)
Dfa (Hot-summer humid continental)	Cfa (Humid subtropical)	BWh (Hot desert)
Dwc (Subarctic)	Cwb (Subtropical highland)	Aw (Savanna)
Dwb (Warm-summer humid continental)	Cwa (Humid subtropical)	Am (Monsoon)
Dwa (Hot-summer humid continental)	Csc (Cold-summer mediterranean)	Af (Rainforest)
Dsc (Dry-summer subarctic)	Csb (Warm-summer mediterranean)	

FIGURE 6.2 Climate Map of the United States. (From National Oceanic and Atmospheric Administration (NOAA). Koppen-Geiger Climate Changes. 2019. https://sos.noaa.gov/datasets/koppen-geiger-climate-changes-1901-2100/.) (Accessed February 16, 2019.)

6.3 OVERVIEW OF ENVIRONMENTAL REGULATIONS OF THE UNITED STATES

To begin our journey through environmental regulations, we must first consider the meaning of the term **environment**. According to the United Nations (2019a), the environment is a broad term that encompasses all living and nonliving things on Earth or some regions thereof. This definition identifies the natural world existing within the atmosphere, hydrosphere, lithosphere, and biosphere, and also includes the built or developed world (defined as the **anthrosphere**).

Second, we must define the difference between a law and a regulation. A **law** is a written statute passed by an appropriate governing body. A **regulation**, which is often used interchangeably with a law, is a set of standards and rules adopted by administrative agencies that govern how laws will be enforced. In our case, that primarily means the United States Environmental Protection Agency (USEPA). Regulations often have the same force as such laws, since, without them, regulatory agencies wouldn't be able to enforce laws. Therefore, the USEPA is called a regulatory agency because the United States Congress authorized the USEPA to write regulations that explain the technical, operational, and legal details necessary to implement laws passed by the United States Congress (USEPA 2019a).

TABLE 6.1

Select List of Countries with Established Environmental Regulations Based on the United States Model

AFRICA	EUROPE	Malta
Egypt	Austria	Netherlands
Kenya	Belgium	Norway
South Africa	Bulgaria	Poland
ASIA	Cyprus	Portugal
China	Czech Republic	Romania
India	Denmark	Russia
Japan	Estonia	Slovakia
Kyrgyzstan	Finland	Slovenia
Pakistan	France	Spain
NORTH AMERICA	Germany	Sweden
Canada	Greece	Switzerland
Mexico	Iceland	Turkey
United States	Ireland	United Kingdom
SOUTH AMERICA	Italy	**OCEANIA**
Bolivia	Latvia	Australia
Brazil	Lithuania	New Zealand
Chile	Luxembourg	Indonesia

Source: United Nations. The United Nations Environment Programme. 2019a. www.unenvironment.org/environment-you. (accessed March 3, 2019.)

Although laws focused on protecting human health or the environment can be traced back more than 2,000 years, it was the United States in the early 1970s that set the example when it enacted the most significant and comprehensive environmental regulations and laws in the world (Hahn 1994). Based on the example set by the United States, over 50 countries have established laws that protect human health and the environment on the same platform as USEPA. Table 6.1 is a short list of the countries that have well-established laws that protect human health and the environment.

Environmental laws and regulations in the United States have methodically become more numerous and complex to the point that even many scientists and other professionals need assistance at interpreting, understanding, and applying many of these environmental laws and regulations. The current cost of protecting human health and the environment in the United States is generally considered to be 2% of the gross national product (USEPA 2018a).

The growth of the number and complexity of environmental laws and regulations can be more easily understood by simply applying the principle of the scientific term called entropy. Entropy is a principle that states that as time progresses forward there is more disorder which increases complexity. This can be applied to our human civilization as well. Our civilization has become more complex with time as a result of our technological advances, population increase, and other factors. This in turn has put pressure on the natural world and our response has been the passage of environmental laws and regulations as an attempt to establish order in a disordered and out-of-equilibrium environment that humans caused.

The basic overriding principle of environmental law of the United States is to protect human health and the environment. This principle is broken down into categories that track some of the laws and how humans impact the environment and other living species with which we share our planet (Lazarus 2004). These principles include:

- Environmental Impact Assessment
- Water Quality
- Contaminant Investigation
- Risk Assessment
- Chemical Transport
- Mineral Resources
- Wildlife
- Soil
- Hunting
- Equity
- Public Transparency
- Prevention

- Air Quality
- Waste Management
- Contaminant Cleanup
- Chemical Safety
- Water Resources
- Forest Resources
- Plants
- Fish
- Sustainable Development
- Transboundary Responsibility
- Public Participation
- Polluters Pay

6.4   THE BIRTH OF THE ENVIRONMENTAL MOVEMENT AND REGULATIONS IN THE UNITED STATES

The term **environmental regulation** refers to a set of laws, sometimes referred to as an amalgam of principles, treaties, laws, regulations, directives, ordinances, policies, at the local, state, or federal level that are enacted to protect the environment in some shape or form (USEPA 2019a). Environmental regulations or laws can be traced to the Roman Empire when the Roman Senate passed legislation to protect the city's supply of water which addressed concern over fresh water and sewage disposal (Journal of Roman Studies 2019).

In the 14th century, England prohibited the burning of coal in London and the disposal of waste into waterways, specifically the Thames River (Smithsonian 2019). In December 1682, in what would become the United States, William Penn ordered that one acre of forest be preserved for every five acres cleared for settlement (Pennsylvania Historical Association 2019). In addition, Benjamin Franklin widely wrote about the adverse effects of lead and dumping of waste (Environmental Education Associates 2019). However, it was not until 1899 that the United States government passed what is commonly referred to as the first environmental law in the United States, Rivers and Harbors Act, which was intended to clear the harbors of garbage so that ships would not be impeded in transporting goods (USEPA 2019a).

It was not until the Industrial Revolution starting around 1760 that environmental degradation of the air, water, and land became a significant issue. The Industrial Revolution marked a time of rapid transition from making products by hand to making products with machines. The stationary steam engine invented by James Watt led to widespread use of steam power that used coal. New chemical manufacturing methods, iron production processes, the development of machine tools, and the rise of the factory system all came about in a relatively short time, which sparked the rise of the Industrial Age and resulted in the discharge of large amounts of waste emitted into the air, into waterways, and discarded onto the land. Figure 6.3 shows air emissions in the late 1880s in Pennsylvania (United States Research Archives 2017).

As recently as the first half of the 20th century, environmental awareness and legislation were generally lacking not only in the United States but world-wide. There were a few international agreements that primarily focused on boundary waters, navigation, and fishing rights along shared waterways between countries. However, they ignored pollution and ecological issues. In fact, the most convenient and least costly method of waste disposal at the time was "up the stack or down the river" (Haynes 1954). One of the first international agreements was between 12 European countries for the protection of valuable birds called the Convention for the Protection of Valuable Birds Useful to Agriculture in 1902.

It was not until the publishing of Rachel Carson's book, *Silent Spring* in 1962, which described the environmental effects of polychlorinated biphenyls (PCBs) on birds, that catapulted the environmental movement in the United States into the public eye.

The 1960s and 1970s are generally described as the age of environmentalism in the United States. **Environmentalism** is defined as a value system that seeks to redefine humankind's relationship with

FIGURE 6.3 Air emissions from industrial plants in Pennsylvania in the 1880s. (United States Research Archives, 2017. www.UnitedStates.research.archives.gov.) (Accessed February 24, 2017.)

nature (Tarlock 2001). This is when the United States realized that environmental degradation had significantly affected the air and water quality of many urban areas, especially the air in Los Angeles and the water in Lake Erie, which is one of the Great Lakes. Starting on Earth Day in 1971, the Ad Council and Keep America Beautiful campaign aired public awareness commercials on television depicting a Native American shedding a tear when looking out over the polluted landscape of the United States. The commercial stated "people start pollution, people can stop pollution" (see Figure 6.4).

Two significant sites of environmental contamination were discovered about this time and were Times Beach in Missouri and Love Canal near Buffalo, New York. These two sites would become

FIGURE 6.4 Pollution: it's a crying shame. (From United States Advisory Council. 2017. (www.adcouncil.org/ Our-Campaigns/The-Classics/pollution-keep-america-beautiful-iron-eyes-cody. (accessed February 27, 2017.)

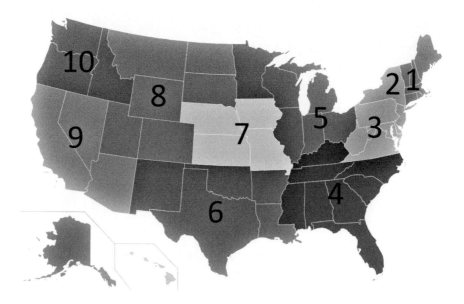

FIGURE 6.5 USEPA regions. (From United States Environmental Protection Agency. Summary of Environmental Regulations of the United States. 2017b. https://epa.gov/summary-environmental-laws.) (Accessed February 24, 2017.)

the first Superfund Sites in the United States. Currently, there are over 1,500 Superfund Sites (see Section 6.6.11) (USEPA 2019a).

6.5 FORMATION OF NATIONAL POLICY AND THE ENVIRONMENTAL PROTECTION AGENCY

On December 2, 1970, then-President Richard Nixon signed an executive order that established the United States Environmental Protection Agency (USEPA) and a cavalcade of environmental legislation ensued. The creation of the USEPA originated from the passage of the United States National Environmental Policy Act (NEPA) of 1969. NEPA required the federal government to give environmental issues priority when planning any type of large scale projects and also provided for the establishment of councils and agencies to monitor and protect the environment. NEPA was developed to establish consistency in the application of environmental laws and standards throughout the United States (Eccleston 2008; USEPA 2017a).

Prior to its passage and the creation of the USEPA, some states had pollution statutes, but many did not. For example, Ohio, in particular, had an active and highly regarded environmental program but surrounding states did not. The result created significant confusion because what was considered permissible at a location in one state that resulted in the release of pollutants to the environment was not permissible in another state and often they shared a common border.

NEPA developed standards for conducting environmental impact statements (EIS) (Eccleston 2008). An EIS describes investigation and remediation methods necessary in order to mitigate or minimize any and all potentially detrimental effects of any new development (see Section 6.6.3). Another method of evaluation is environmental auditing, modeled similar to that of a financial audit and which examines processes and outcomes of environmental impacts.

Environmental laws, along with all other federal laws, are organized by subject into what is termed United States Code. The United States Code (U.S.C.) contains all the current enacted statutory language. The official United States Code is maintained by the Office of the Law Revision Counsel in the United States House of Representatives (USEPA 2017a).

TABLE 6.2
USEPA Regions

Region	States
Region 1	Maine, New Hampshire, Vermont, Rhode Island, Massachusetts, Connecticut
Region 2	New York, New Jersey
Region 3	Pennsylvania, West Virginia, Virginia, Maryland, Delaware, District of Columbia
Region 4	Tennessee, Kentucky, Georgia, North Carolina, South Carolina, Florida, Alabama, Mississippi
Region 5	Illinois, Wisconsin, Michigan, Ohio, Indiana, Minnesota
Region 6	Texas, Oklahoma, New Mexico, Arkansas, Louisiana
Region 7	Kansas, Nebraska, Iowa, Missouri
Region 8	North Dakota, South Dakota, Wyoming, Colorado, Utah, Montana
Region 9	California, Arizona, Nevada, Hawaii
Region 10	Washington, Oregon, Idaho, Alaska

Source: United States Environmental Protection Agency. Summary of Environmental Regulations of the United States. 2017b. https://epa.gov/summary-environmental-laws. (accessed February 24, 2017.)

USEPA is divided into ten regions (see Figure 6.5). Table 6.2 shows each region and the states within each region.

6.6 ENVIRONMENTAL LAWS OF THE UNITED STATES

Laws that are commonly attributed to protecting human health and the environment include (USEPA 2017b):

- Rivers and Harbors Act of 1899
- Antiquities Act of 1906
- Atomic Energy Act of 1946
- Atomic Energy Act of 1954
- Brownfield Revitalization Act of 2002
- Clean Air Act (CAA) of 1970
- Clean Water Act (CWA) of 1972
- Coastal Zone Management Act of 1972
- Comprehensive Environmental Response, Compensation, and Liability Act (CERCLA) of 1980
- Emergency Planning and Community Right-to-Know Act of 1986
- Endangered Species Act of 1973
- Energy Policy Act of 1992
- Energy Policy Act of 2005
- Federal Power Act of 1935
- Federal Feed, Drug, and Cosmetic Act (FFDCA) of 1938
- Federal Insecticide, Fungicide, and Rodenticide Act (FIFRA) of 1947
- Fish and Wildlife Coordination Act of 1968
- Food Quality Protection Act of 1996
- Fisheries Conservation and Management Act of 1976
- Global Climate Protection Act of 1987
- Hazardous Materials Transportation Act (HMTA) of 1975
- Insecticide Act of 1910

- Lacey Act of 1900
- Marine Protection Act of 1972
- Marine Mammal Protection Act of 2015
- Migratory Bird Treaty Act of 1916
- Mineral Leasing Act of 1920
- National Environmental Policy Act of 1969
- National Forest Management Act of 1976
- National Historic Preservation Act of 1966
- National Parks Act of 1980
- Noise Control Act of 1974
- Nuclear Waste Policy Act of 1982
- Ocean Dumping Act of 1988
- Occupational Safety and Health Act (OSHA) of 1970
- Oil Spill Prevention Act of 1990
- Pollution Prevention Act (PPA) of 1990
- Refuse Act of 1899
- Resource, Conservation, and Recovery Act (RCRA) of 1976
- Rivers and Harbors Act of 1899
- Safe Drinking Water Act (SDWA) of 1974
- Superfund Amendments and Reauthorization Act (SARA) of 1986
- Surface Mining Control and Reclamation Act of 1977
- Toxic Substance Control Act (TSCA) of 1976
- Wild and Scenic Rivers Act of 1968

The number of environmental laws and regulations is a very long list and encompasses hundreds of thousands of pages of rules. We shall discuss each act listed above separately in the sections below and will go into substantial detail for those that focus on addressing pollution.

6.6.1 Rivers and Harbors Act of 1899

The Rivers and Harbors Act of 1899 is generally considered the oldest federal environmental law in the United States. The passage of the Rivers and Harbor Act made it a misdemeanor to excavate, fill, or alter the course, condition, or capacity of any port, harbor, channel, or other types of waterways without a permit. Included in the act is a provision that building any structure in the channel or along the banks, such as a bridge, that changes the course, conditions, location, or capacity of the waterway requires a permit. The act also made it a misdemeanor to discharge refuse of any kind into any navigable waters or tributaries of the United States without a permit. This specific section became known as the Refuse Act (USEPA 2017b).

The definition of **navigable waters** has come to be defined as those waters that are under the influence of tidal fluctuations and are used, or have been used in the past or potentially used in the future, to transport interstate or foreign commerce. This includes coastal and inland water, lakes, rivers, or streams that are navigable or are connected to navigable waters and territorial seas (USEPA 2017b). As you can surmise, the definition of a navigable waterway is rather broad and essentially can be simply interpreted as to encompass any flowing water within the United States, even a human-made ditch adjacent to railroad tracks. The Rivers and Harbors Act is enforced by the Army Corps of Engineers, and some jurisdiction also falls under the responsibility of the Coast Guard (USEPA 2017b). The Refuse Act of 1899 was initially focused on controlling debris that obstructed navigation in the waterways of the United States. In 1970, President Richard Nixon issued an executive order creating a new permit program under the Refuse Act. The focus of the program was on pollution from industrial sources and was later rolled into the Clean Water Act, which is described later in this chapter (USEPA 2017b).

6.6.2 Insecticide Act, Feed, Drug, and Cosmetic Act, and Federal Insecticide, Fungicide, and Rodenticide Act

Discussion of the Insecticide Act of 1910, the Feed, Drug, and Cosmetic Act of 1938, and the Federal Insecticide, Fungicide and Rodenticide Act of 1972 are discussed together in this section because they are each closely related and evolved together in development of protecting human health from chemical ingredients in pesticides, fungicides, rodenticides, and cosmetics.

6.6.2.1 The Insecticide Act of 1910

The Insecticide Act of 1910 was established to protect the sale of pesticides from misbranding and established a set of minimal standards for pesticide products sold within the United States (USEPA 2017c). Specifically, the chemicals referred to as Paris green, lead arsenate, and various fungicides were targeted. This act was passed in response to concerns from the United States Department of Agriculture (USDA) and many farm groups about the sale of fraudulent or substandard pesticide products listed, which was common at the time. This act represented the first set of regulations that required that listed ingredients must be within certain percentages contained within the product. Pesticide containers were required to be marked with a serial number that was specific to the manufacturer. The labels indicated that the product was guaranteed under the act and were often embossed with a brand and identity of the product. The purpose of the Act was to stop what was termed "home-brew" operations (Conn et al. 1984).

6.6.2.2 The Food, Drug, and Cosmetic Act

The Food, Drug, and Cosmetic Act of 1938 gave authority to the Food and Drug Administration (FDA) to oversee the safety of food, drugs, and cosmetics. The act was passed as a result of the deaths of over 100 patients due to prescribed medication that turned out to be poison. The act was amended in 2002 under Section 408, which authorized the FDA to set tolerance limits, or maximum residue limits, of pesticides on food (USEPA 2017c). To determine what a tolerance or residue limit should be requires USEPA to evaluate the toxicity of the pesticide itself and its breakdown products (USEPA 2017c). The Office of Pesticide Programs (OPP) regulates the use of all pesticides in the United States and has established maximum levels for pesticide amounts in food with the intended purpose of safeguarding the nation's food supply (USEPA 2017c).

The projects and programs managed by OPP include (USEPA 2017c):

- Federal Insecticide, Fungicide, and Rodenticide Act (FIFRA)
- Pesticide Registration and Extension Act
- Key parts of the Food Quality Protection Act
- Key parts of the Food, Drug, and Cosmetic Act
- Key parts of the Endangered Species Act
- Assessing pesticide risks
- Bed bugs
- Biotechnology under FIFRA
- Endangered Species Protection Program
- Ingredients used in pesticides
- Insect repellents
- Integrated pest management in schools
- Mosquito control
- Pesticide labels
- Pesticide registration
- Pesticide registration review
- Pesticide tolerances

- Pesticide Environmental Stewardship Program
- Protecting pets from fleas and ticks
- Protecting pollinators
- Reducing pesticide drift
- Worker safety protection of pesticides

6.6.2.3   The Insecticide, Fungicide, and Rodenticide Act of 1947

Little change occurred in the regulations of pesticides until after World War II when the use of synthetic organic pesticides in agriculture became widespread in the United States. In response to the large increase in the use of pesticides, the United States passed the Federal Insecticide, Fungicide, and Rodenticide Act (FIFRA) of 1947. This act amended and consolidated the previous regulations for the distribution, sale, and use of pesticides (USEPA 2017b).

FIFRA required manufacturers to demonstrate that when using the specific product according to its labeled specifications and applications that it would not generally cause unreasonable adverse effects on the environment beyond that which it was targeted to control. As defined under FIFRA, the term **pesticide** is generally defined as any substance or mixture of substances intended for preventing, destroying, repelling, or mitigating any pest, and any substance or mixture of substances intended for use as a plant regulator, defoliant, or desiccant. The term **pest** is defined as any living organism (plant or animal) occurring where they are not wanted or causing damage to crops, humans, animals, or plants. Specifically, FIFRA requires that pesticide products have a certain label that provides information to the user that included (Conn et al. 1984):

- Manufacturer's name
- Manufacturer's address
- List of ingredients
- Directions for proper use
- Warning statements to protect users and the public
- Potential effects on non-targeted species or plants and animals

FIFRA was significantly amended in 1972 by USEPA and was retitled the Federal Environmental Pesticide Control Act (FEPCA). The most significant revision to FIFRA made by FEPCA was the requirement that manufacturers of pesticides must demonstrate that any product would not cause unreasonable adverse effects on the environment (USEPA 2017c).

FEPCA defines the term "unreasonable adverse effects on the environment" to mean (USEPA 2017c):

- Any unreasonable risk to humans or the environment, taking into account the economic, social, and environmental costs and benefits of the use of any pesticide
- A human dietary risk from residues that result from a use of a pesticide in or on any food inconsistent with the standard under Section 408 of the Federal Food, Drug, and Cosmetic Act of 1938

The significant revisions to FIFRA in 1972 were the direct result of a book by Rachel Carson titled *Silent Spring* (1962), which documented the harmful effects of Dichlorodiphenyl-trichloroethane (DDT). DDT is a synthetic organic compound first developed in 1874. The insecticide properties of DDT were not discovered until 1939. DDT was commonly used during World War II for prevention of malaria, typhus, body lice, and bubonic plague. Reportedly, due to its use, cases of malaria dropped from 400,000 in 1946 to virtually none in 1950 (USEPA 2017c). Once it was discovered that DDT worked effectively as an insecticide, its use and applications increased significantly. From 1950 until 1970, more than 40,000 tons were used each year world-wide. In addition, it is estimated that 1.8 million tons or more have been produced globally since 1940

(USEPA 2019a). DDT is still in use in South America, Africa, and Asia as an insecticide to prevent malaria.

Trillions of pounds of petrochemicals are applied to the environment every year in the United States (Vallianatos 2014). The American Academy of Pediatrics has concluded that current regulations are not enough, especially with respect to the very young, very old, and minorities because it's these populations that are either more sensitive or are more heavily exposed (American Academy of Pediatrics 2011).

6.6.3 NATIONAL ENVIRONMENTAL PROTECTION ACT

The National Environmental Protection Act (NEPA) of 1969 is considered the "environmental Magna Carta" in that it established a pathway for the development of the national environmental policy and the emergence of the United States Environmental Protection Agency. NEPA grew out of the increased awareness and concern for the environment from human events that caused harm to the environment that included (USEPA 2017b):

- Rachel Carson's chronicled impact of DDT on the environment, especially birds, in her book published in 1962
- An oil spill off the coast of Santa Barbara in 1969
- Freeway revolts in the 1960s which consisted of a series of protests in response to the destruction of many communities and ecosystems during the construction of the interstate highway system

The purpose of NEPA was to establish three goals, which were (USEPA 2017b):

1. The nation's environmental policies and goals
2. Provisions for the federal government to enforce such policies
3. The Council of Environmental Quality in the executive branch of the president

The declaration of the national environmental policy contained in Section 2, Title I of the NEPA states (USEPA 2017b):

> The Congress, recognizing the profound impact of man's activity on the interrelations of all components of the natural environment, particularly the profound influences on population growth, high-density urbanization, industrial expansion, resource exploitation, and new expanding technological advances and recognizing further the critical importance of restoring and maintaining environmental quality to the overall welfare and development of man, declares that is the continuing policy of the Federal Government in cooperation with State and local governments, and other concerned public and private organizations, to use all practicable means and measures, including financial and technical assistance, in a manner calculated to foster and promote the general welfare, to create and maintain conditions under which man and nature can exist in productive harmony, and fulfill the social, economic, and other requirements of present and future generations of Americans.

Of special note in the paragraph above is the phrase "..., particularly the profound influences on population growth...". This will be discussed in greater detail in subsequent chapters when we discuss the influence of continued population growth on sustainability. Specifically, the question to be discussed is how humans can maintain sustainability with the environment with an ever-increasing population and when will the stresses caused by human population growth on the quality of air, water, land, and other species become too great. Just since NEPA was written, our population in the United States has risen from 202 million (in 1969) to over 323 million (in 2019) or an increase of more than 37% (United States Census Bureau 2019). During this same period of time, the world population has increased from 3.6 billion to more than 7.2 billion or 100% (United Nations 2017b).

In essence, the purpose of NEPA was to ensure that environmental factors are weighted equally compared to other factors in the decision-making process undertaken by federal agencies and to establish a national environmental policy. NEPA also promotes the Council of Environmental Quality (CEQ) to advise the president in the preparation of an annual report on the progress of federal agencies in implementing NEPA. A significant environmental benefit as a result of NEPA is that it required federal agencies to prepare an **environmental impact statement** (EIS) to accompany reports and recommendations for congressional funding. An EIS is a systematic scientific evaluation of the relevant environmental effects of a federal project or action, such as building a dam or an interstate highway (USEPA 2017b).

6.6.4 Occupational Safety and Health Act

The Occupational Safety and Health Act was passed in 1970 and was designed to protect people from injury in the workplace. Authority for enforcing workplace health and safety fell to the Occupational Safety and Health Administration (OSHA), which is within the United States Department of Labor (2017). However, the first law addressing workplace safety was a state law in Massachusetts passed in 1877. This law focused on elevator safety and establishing fire exits in factories (United States Department of Labor 2017).

OSHA first focused its efforts on safety hazards in the workplace in five different industries that included (United States OSHA 2017):

1. Marine cargo handling
2. Roofing and sheet metal work
3. Meat and meat products
4. Transportation equipment (primarily mobile homes)
5. Lumber and wood products

Along with establishing safety rules in the workplace, OSHA also addressed health hazards in the workplace that included (United States OSHA 2017):

- Asbestos
- Lead
- Silica
- Carbon monoxide
- Cotton dust

The Occupational Safety and Health Act grants OSHA the authority to issue workplace health and safety regulations. These regulations include (United States OSHA 2017):

- Limits on chemical exposure
- Employee access to hazard information
- Requirements to prevent falls
- Preventing hazards from operating dangerous equipment
- Preventing trench cave-ins
- Preventing exposure to infectious diseases
- Ensuring the safety of workers entering confined spaces
- Machine guarding
- Providing respirators and other safety equipment
- Job training

For chemical or substance health risks, OSHA established what is termed a permissible exposure limit or PEL. A PEL is a legal limit for exposure of an employee to a chemical substance or physical agent such as noise. A PEL is usually expressed as a time-weighted average (TWA). However, there is also an additional exposure measurement termed short-term exposure limit (STEL). The TWA is usually a measure over a time period of 8 hours, which is a typical work shift. Units of measure are usually expressed as parts per million (ppm), milligrams per kilogram (mg/kg), or milligrams per cubic meter (mg/m^3) (United States OSHA 2017).

OSHA also requires that employers must do the following (United States OSHA 2017):

- Find and correct safety and health problems
- Eliminate or reduce hazards by changing working conditions
- Use of personal protective clothing when elimination is not feasible
- Use of less harmful chemicals
- Enclosing processes that emit harmful chemicals
- Use of proper ventilation
- Inform workers of chemical hazards
- Provide appropriate training
- Record keeping
- Provide personal protective equipment
- Provide hearing exams
- Post OSHA citation
- Conduct appropriate testing such as air sampling
- Post illness and injury summary data annually
- Notify OSHA within 8 hours of a workplace fatality
- Notify OSHA within 24 hours of all work-related inpatient hospitalizations, all amputations, and all losses of an eye
- Prominently display the official OSHA Job Safety and Health "It's the Law Poster" that describes the rights of employees and responsibilities of employers under the Occupational Safety and Health Act
- Not retaliate or discriminate against workers for using their rights under the law, including their right to report a work-related injury or illness

Note that OSHA requires employers to evaluate whether the use of less harmful chemicals is possible in an effort to add protection to employee safety. This item is significant, not only as an OSHA requirement but as an important protective action for the environment and will be described in greater detail later in this chapter, as well as in subsequent chapters.

6.6.5 Clean Air Act

At the time of its inception, the Clean Air Act (CAA) was by far the most complex and lengthy piece of environmental legislation of its kind in the world. The CAA has grown enormously since 1970 into a complex behemoth that comprises tens of thousands of pages of statutory requirements, regulations, and rules (Cornell Law 2017). The CAA and its amendments, including a significant amendment in 1990, cover numerous topics that have generated intense political controversy. The CAA had its beginnings in the Air Quality Act of 1967, which initially focused on regulation of ambient air quality to protect public health and welfare. The purpose of the initial law was to (USEPA 2007a; 2007b):

- Protect and enhance the quality of the nation's resources so as to promote the public health and welfare and the productivity capacity of its population
- Initiate and accelerate a national research and development program to achieve the prevention and control of air pollution

- Provide technical and financial assistance to state and local governments in connection with the development and execution of their air pollution and prevention and control programs
- Encourage and assist the development and operation of regional air pollution control programs

The center piece of this legislation was the development of air quality "criteria,", which at the time was headed up by the Department of Health, Education, and Welfare since USEPA had not yet been formed and would not be for a few more years. The goal of the "criteria" was to accurately reflect the latest in scientific knowledge on the health and welfare effects of individual pollutants, such as sulfur dioxide (SOx), nitrogen oxides (NOx), and particulate matter (PM). These criteria would eventually evolve into what will be later described as "criteria pollutants," which are discussed in greater detail later in this chapter (USEPA 2017b).

These early experiences with attempting to regulate air pollution revealed many scientific uncertainties and technical difficulties that challenged early regulatory agencies. The criteria development process proved long and cumbersome and states had difficulty translating the information made available into source-specific standards. Among other things, techniques for relating air quality concentrations to source-specific emissions (e.g., atmospheric dispersion models) were not well developed or were not developed at all. In addition, it became apparent that since air quality is influenced by regional as well as local factors, more sophisticated regulatory and control techniques and technologies were required. Hence, the CAA of 1970 was developed and passed and the United States established the Environmental Protection Agency to administer the regulations with the purpose of protecting human health and the environment (USEPA 2017b).

The CAA attempts to resolve how areas in the United States that have been unable to attain the goal of clean air should be brought into compliance. The act also describes the penalties that should be applied to communities that do not achieve these goals. It addresses how much more we as a nation should spend on controlling emissions from automobiles, and in what ways the fuels that we burn in our automobiles should be improved. It attempts the almost impossible task to allocate among different regions of the United States the multi-billion dollar burden of reducing emissions that emit or contribute to hazardous air pollutants, acid rain, smog, ozone, ozone layer depletion, and lastly to climate change, all the while with an increasing population. Simply put, the CAA has provided standards that govern the day-to-day operation of virtually every industry, car, truck, train, plane, refrigerator, furnace, air conditioner, light bulb, and power plant in the United States (USEPA 2017b).

The CAA essentially required that every industrial facility that had air emissions must obtain a permit to operate. It required that industrial facilities be designed and operated to prevent or to minimize releases of hazardous air pollutants. Those facilities that own or operate larger sources of emissions must not modify those sources or add new sources without first obtaining preconstruction approvals. In addition, industries that offered electricity for sale must purchase or otherwise obtain pollution allowances in order to operate.

The CAA is a comprehensive law that regulates air emissions from stationary and mobile sources. Among other things, the law authorizes the EPA to establish national ambient air quality standards (NAAQS) to protect public health and public welfare and to regulate emissions of what are considered criteria pollutants and hazardous air pollutants (HAPs) (USEPA 2017d). Other provisions in the CAA that we will examine include ozone-depleting substances, smog-forming chemicals, acid rain, and climate change.

6.6.5.1 Stationary Sources

A **stationary source or point source** is defined as a fixed pollution source such as a coal-fired electrical power generation plant or a refinery. Just in the State of California alone, there are over 13,000 permitted stationary sources (California Environmental Protection Agency 2017).

6.6.5.1.1 NAAQS Standards

The CAA identifies two types of NAAQS standards and includes **primary standards** which are provided for public health protection, including the protection of sensitive populations such as asthmatics, children, and the elderly. **Secondary standards** which provide for public welfare protection include protection against decreased visibility and damage to animals, crops, vegetation, and buildings (USEPA 2017e). An additional subdivision is **primary and secondary pollutants**. A primary pollutant is an air pollutant emitted directly from a source. A secondary pollutant is a pollutant that is not directly emitted from a source but forms when other pollutants (primary pollutants) react with the atmosphere, such as acid rain, ozone, or what is commonly referred to as smog (USEPA 2017e).

Criteria pollutants or criteria air pollutants (CAPs) were the first set of pollutants recognized by USEPA as requiring standards established on a national level. USEPA uses the term criteria pollutants because it regulates them by developing human health-based or environmentally-based criteria (science-based guidelines) for establishing permissible levels. The set of limits based on human health comprise primary standards. Another set of limits intended to prevent environmental and property damage is called secondary standards. A geographic area with air quality lower than the established standards is termed an "attainment area." Conversely, a geographic area with air quality exceeding one or more of the standards is termed "non-attainment area" (USEPA 2017f).

The six criteria pollutants include (USEPA 2017e):

1. Ozone (ground level)
2. Particulate matter (PM)
3. Lead
4. Carbon monoxide
5. Sulfur oxides (SOx)
6. Nitrogen oxides (NOx)

USEPA has established standards for these six principal pollutants under NAAQS. Periodically, the standards are reviewed and are revised when appropriate or necessary. The current standards are presented in Table 6.3. Units of measure for the standards presented in Table 6.3 are parts per million (ppm) by volume, parts per billion (ppb) by volume, and micrograms per cubic meter of air (ug/m^3) (USEPA 2017e). Particle pollution, also known as particulate matter (PM), includes very fine dust soot, smoke, and droplets that are formed from chemical reactions and are produced when fuels such as coal, wood, or oil are burned (Brownwell and Zeugin 1991).

With the CAA amendments of 1990, PM was divided into two parts: PM10 and PM2.5. PM10 refers to a particle size greater than 10 micrometers in diameter, and PM2.5 refers to a particle size less than 2.5 micrometers in diameter. The logic behind this distinction is that these smaller particles are more likely to harm our health (USEPA 2017e). One of the goals of the act was to set and achieve NAAQS in every state by 1975 in order to address the public health and welfare risks posed by certain widespread pollutants. The setting of these pollutant standards was coupled with directing the states to develop state implementation plans (SIPs), applicable to appropriate industrial sources in such state, in order to achieve these standards (USEPA 2017f).

The act was amended in 1977 and in 1990 primarily to set new goals (dates) for achieving attainment of NAAQS since many areas of the country had failed to meet the deadlines. Figure 6.6 shows counties within the United States that are in non-attainment for one or more hazardous air pollutants as of February 2017 (USEPA 2017f). In addition to the CAPs listed above, the 1990 amendments to the CAA added a list of 188 compounds on the USEPA termed hazardous air pollutants (HAPs), which are addressed in Section 112 of the amended act. The list can be partially viewed in Table 6.4 and the whole list is available at www.epa.gov/haps (USEPA 2017g). The list of HAPs is important for air permits as we shall see when we discuss Title V permits.

Prior to 1990, the CAA established a risk-based program under which only a few standards were developed. The 1990 amendments to the CAA revised Section 112 to first require issuance of

TABLE 6.3
NAAQS Table of Select Criteria Air Pollutants

Pollutant		Primary/Secondary	Time Period	Level	Form
Carbon Monoxide (CO)		Primary	8 hours	9 ppm	Not to be exceeded more than once
			1 hour	35 ppm	per year
Nitrogen Dioxide (NO2)		Primary	Rolling 3-month average	0.15 ug/m3	98th percentile of 1-hour daily maximum concentrations averaged over 3 years
			1 hour	100 ppb	
		Primary and Secondary	1 year	53 ppb	Annual mean
Ozone		Primary and Secondary	8 hours	0.070 ppm	Annual fourth-highest daily maximum 8-hour concentration, averaged over 3 years
Particle Pollution (PM)	PM 2.5	Primary	1 year	12.0 ug/m3	Annual mean averaged over 3 years
		Secondary	1 year	15.0 ug/m3	Annual mean averaged over 3 years
		Primary and Secondary	24 hours	35 ug/m3	98th percentile averaged over 3 years
	PM 10	Primary and Secondary	24 hours	150 ug/m3	Not to be exceeded more than once per year on average over 3 years

Source: United States Environmental Protection Agency (USEPA). NAAQS Table. 2017e. www.epa.gov/criteria-air-pollu tants/naaqs-table. (accessed March 3, 2017.)

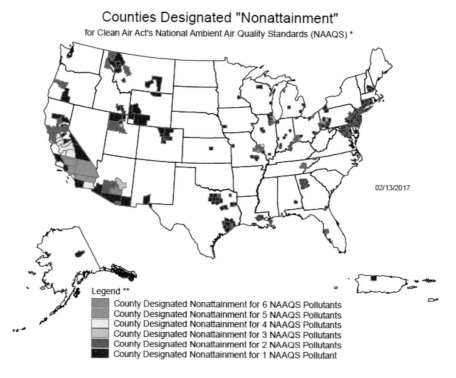

FIGURE 6.6 US counties that are in non-attainment with one or more hazardous air pollutants as of February 2017. (From United States Environmental Protections Agency (USEPA). Non-attainment areas of the United States. 2017f. www.jpg.gov/nonattainmentmap.) (Accessed March 2, 2017.)

TABLE 6.4

Select Hazardous Air Pollutants (HAPs)

CAS Number	Compound	CAS Number	Compound	CAS Number	Compound
75070	Acetaldehyde	79447	Dimethyl carbamoyl chloride	598920	N-Nitrosomorpholine
60355	Acetamide	68122	Dimethyl formamide	56382	Parathion
75058	Acetonitrile	57147	1,1-Dimethyl hydrazine	82688	Pentachloronitrobenzene
98862	Acetophenone	131113	Dimethyl phthalate	87865	Pentachlorophenol
53963	2-Acetylaminofluorene	77781	Dimethyl sulfate	108952	Phenol
107028	Acrolein	534521	4,6-Dinitro-o-cresol	106503	p-Phenylenediamine
79061	Acrylamide	51285	2,4-Dinitrophenol	75445	Phosgene
79107	Acrylic acid	121142	2,4-Dinitrotoluene	7803512	Phosphine
107131	Acrylonitrile	123911	1,4-Dioxane	7723140	Phosphorus
107051	Allyl chloride	122667	1,2-Diphenylhydrazine	85449	Phthalic anhydride
92671	4-Aminobiphenyl	106898	Epichlorohydrin	1336363	Polychlorinated biphenyls
62533	Aniline	106887	1,2-Epoxybutane	1120714	1,3-Propane sultone
90040	o-Anisidine	140885	Ethyl acrylate	57578	Beta-Propiolactone
1332214	Asbestos	100414	Ethyl benzene	123386	Propionaldehyde
71432	Benzene	51796	Ethyl Carbamate	114261	Propoxur
92875	Benzidine	75003	Ethyl chloride	78875	Propylene dichloride
98077	Benzotrichloride	106934	Ethylene dibromide	75569	Propylene oxide
100447	Benzyl chloride	107062	Ethylene dichloride	75558	1,2-Propylenimine
92524	Biphenyl	107211	Ethylene glycol	91226	Quinoline
117817	Bis(2-ethylhexyl) phthalate	151564	Ethylene imine	106514	Quinone
542881	Bis(chloromethyl) ether	75218	Ethylene oxide	100425	Styrene
75252	Bromoform	96457	Ethylene thiourea	96093	Styrene oxide
106990	1,3-Butadiene	75343	Ethylidene dichloride	1746016	2,3,7,8-Tetrachlorodibenzo-p-dioxin
156627	Calcium cyanamide	50000	Formaldehyde	79345	1,1,2,2-Tetrachloroethane
105602	Caprolactam	76448	Heptachlor	127184	Tetrachloroethylene
133062	Captan	118741	Hexachlorobenzene	7550450	Titanium tetrachloride
63252	Carbaryl	87683	Hexachlorobutadiene	108883	Toluene
75150	Carbon disulfide	77474	Hexachlorocyclopentadiene	95807	2,4-Toluene diamine
56235	Carbon tetrachloride	67721	Hexachloroethane	584849	2,4-Toluene diisocyanate
463581	Carbonyl sulfide	822060	Hexamethylene-1,6-diisocyanate	95534	o-Toluidine
120809	Catechol	680319	Hexamethylphosphoramide	8001352	Toxaphene
133904	Chloramben	110543	Hexane	120821	1,2,4-Trichlorobenzene
57749	Chlordane	302012	Hydrazine	79005	1,1,2-Trichlrorethane
7782505	Chlorine	7647010	Hydrochloric acid	79016	Trichloroethylene
79118	Chloroacetatic acid	7664393	Hydrogen fluoride	95954	2,4,5-Trichlorophenol
532274	2-Chloroacetophenone	7783064	Hydrogen sulfide	88062	2,4,6-Trichlorophenol
108907	Chlorobenzene	123319	Hydroquinone	121448	Triethylamine
510156	Chlorobenzilate	78591	Isophorone	65014	Vinyl chloride

Source: United States Environmental Protection Agency (USEPA). Hazardous Air Pollutants (HAPs). 2017g. www.epa.gov/haps. (accessed February 20, 2019.)

technology-based standards for major sources and certain area sources. The term **major source** is defined as a stationary source or group of stationary sources that emit or have the potential to emit 10 tons per year or more of a hazardous air pollutant or 25 tons per year or more of a combination of hazardous air pollutants. An **area source** is defined as any stationary source that is not a major source (USEPA 2017h). The 1990 amendments to the CAA also expanded the hazardous air pollution program and shifted its focus from a risk-based regulatory provision to one that is primarily based on control technology regulation (i.e., installing a baghouse or other air pollutant capture system). For major sources, Section 112 of the CAA requires that EPA establish emission standards that require the maximum degree of reduction in emission of hazardous air pollutants. These emission standards are commonly referred to as **maximum achievable control technology** or **MACT** standards (USEPA 2017h).

Eight years after the technology-based MACT standards are issued for a source category, EPA is required to review those standards to determine whether any residual risk exists for that source category, and if necessary, revise the standards to address the risk. The 1990 amendments to the CAA define MACT standards for both existing and new sources. MACT must be equal to at least the level of control achieved by the best performing 12% of regulated sources within the same category. Where a category has less than 30 sources, MACT is defined as at least the level of control activity achieved by the best five sources. New source standards may be more stringent than existing source standards, and existing sources are subject to statutory schedules for retrofitting with upgraded technology, as necessary (USEPA 2017h).

Non-major sources subject to National Emission Standards for Hazardous Air Pollutants (NESHAPS) are required to install and operate air pollution control equipment subject to MACT (USEPA 2017h). These types of operations include (USEPA 2017h):

- Hazardous waste combustors
- Portland cement manufacturers
- Mercury cell chlor-alkali plants
- Secondary lead smelters
- Carbon black production
- Chemical manufacturing: chromium compounds
- Primary copper smelting
- Secondary copper smelting
- Nonferrous metal area sources: zinc, cadmium, and beryllium
- Glass manufacturing
- Electric arc furnace (EAF) steelmaking facilities
- Gold mine ore processing and production

In addition to MACT standards, there is also an additional set of standards called Generally Available Control Technology or GACT standards, which are less stringent standards and are discretionary for area sources by the appropriate regulatory authority (USEPA 2017d).

6.6.5.1.2 Major Source

A major stationary source is a facility that emits, or has the potential to emit (PTE), any criteria pollutant or hazardous air pollutant (HAP) at levels equal to or greater than specific emission thresholds defined by USEPA. PTE is defined for functional purposes as the maximum capacity of a facility to emit any air pollutant under its physical and operational design. Table 6.5 presents the Title V PTE emission threshold levels established by USEPA (USEPA 2017h). If a facility has the PTE above the thresholds, the facility will require what is termed a Title V Air Permit (USEPA 2017h).

Title V of the 1990 amendments of the CAA created the operating permit program that is implemented by the states. The final rule for Title V permits was promulgated in 40 Code of Federal Regulation (CFR) Part 70 (57 Federal Register 32250) on July 21, 1992. Any stationary source of a listed HAP (see Table 6.4) that is a major source must obtain a Title V permit. The purpose of the Title V permitting process is to improve compliance by clarifying what facilities or air pollutant sources must do to control air pollution (USEPA 2017h).

Title V permits are comprehensive in nature and include all requirements for controlling air emissions at a stationary source that is required to have a Title V permit. This amended requirement streamlined the process requiring just one permit. Previous to the amendments in 1990, an air permit was required for each source of air pollution, which at some stationary sources meant that perhaps 50 or more permits were necessary, each being different and with unique compliance monitoring and reporting obligations. The Title V permit process brought all sources of air pollutants from a particular facility under one permit, which simplified compliance and regulatory enforcement (USEPA 2017h). Title V permits generally include the following terms (USEPA 2017h):

- Permit duration (usually 5 years)
- List of all emissions sources covered under the permit
- Emission source diagram showing the location of each source
- Emission limits for each source
- List air pollution control equipment for each source
- Acceptable operating ranges for each source
- Monitoring requirements for each source
- Record keeping requirements
- Reporting requirements
- Provision stating that non-compliance constitutes a violation
- Authorization for modifying, revoking, reopening, reissuing, or terminating a permit
- Identification of terms that are federally enforceable
- Compliance certification
- Authorization for emergencies to constitute an affirmative defense to enforcement action

Other related items that may be listed in a Title V permit may include (USEPA 2017h):

- Provisions for alternative operating scenarios
- Provisions for emission caps
- Provisions for a permit shield

In addition to major sources, other types of sources are required to obtain a Title V permit as well. These include solid waste incineration units of the following (USEPA 2017h):

- Municipal waste combustors
- Hospital/medical/infectious waste incinerators
- Commercial and industrial solid waste incinerators
- Other solid waste incinerators
- Sewage sludge incinerators

Once a facility has been built and has obtained a Title V permit and a modification to operations is considered, the facility must conduct an evaluation to assess if the physical or operational change considered will result in a significant net increase in emissions of a regulated pollutant. That consideration must include all aspects of operations potentially affected by such modification. If the modification results in increased emissions for pollutants listed in Table 6.5, the facility must apply for a permit called a Prevention of Significant Deterioration (PSD) permit. Table 6.6 presents the PSD thresholds. If the modification results in increased emissions but are below the thresholds listed in Table 6.6, the facility must still apply for what is termed a permit to construct (USEPA 2017h).

TABLE 6.5
Title V PTE Emission Threshold Levels

Pollutant	Level (tons per year [tpy])
Volatile organic compounds (VOCs)	10
Nitrogen oxides (NOx)	10
Sulfur oxides (SOx)	100
Carbon monoxide (CO)	50
Single hazardous air pollutant	10
Combination of hazardous air pollutants	25
Particulate matter less than 10 microns (PM10)	70

Source: United States Environmental Protection Agency (USEPA). Title V Operating Permits. 2017h. www.epa.gov.title-v-operating-permits. (accessed March 3, 2017.)

These thresholds may be locally lower if the facility is located in an area of non-attainment. Non-attainment areas also vary depending on the degree to which they are in non-attainment. USEPA classifies non-attainment as moderate, significant, severe, and extreme. Non-attainment pollutants include (USEPA 2017h):

- Carbon monoxide (CO)
- Nitrogen oxides (NOx)

TABLE 6.6
PSD Permit Thresholds

Pollutant	Threshold (tons per year [tpy])
Carbon monoxide (CO)	100
Nitrogen oxides (NOx)	40
Sulfur dioxide (SOx)	40
Total particulate matter (PM)	25
Particulate matter (PM10)	10
Ozone (VOCs)	40
Lead	0.6
Asbestos	0.007
Beryllium	0.0004
Mercury	0.1
Vinyl chloride	1
Fluorides	3
Sulfuric acid mist	7
Hydrogen sulfide	10
Total reduced sulfur (including H2S)	10
Reduced sulfur compounds (including H2S)	10

Source: United States Environmental Protection Agency (USEPA). Title V Operating Permits. www.epa.gov.title-v-operating-permits. 2017h. (accessed March 3, 2017.)

- Sulfur dioxide (Sox)
- Lead
- Particulate matter (PM)
- Ozone

Therefore, contacting the local air quality management district should be the first step in evaluating whether a Title V permit is required because emission limits that trigger the need for a Title V permit vary from location to location (USEPA 2017h).

6.6.5.1.3 Area Source

As required by the 1990 amendments of the CAA, the USEPA is required to regulate emissions of HAPs and criteria pollutants from a published list of industrial source categories. The USEPA has developed a list of source categories that must meet the control technology requirements for these toxic air pollutants that are known or suspected to cause cancer or other serious health problems, commonly known as national emission standards hazardous air pollutants (NESHAPs) (USEPA 2017i).

Under the 1990 amendments of the CAA, if a facility was not defined as a major source of HAPs or criteria pollutants, meaning that a facility did not emit more than 10 tons of a single HAP or 25 tons of one or more HAPs, and were below the thresholds for criteria pollutants but the facility still emitted one or more criteria pollutants or HAP, the facility may be subject to area source rules and would be required to obtain an air permit, but not a Title V permit (USEPA 2017h).

The list of area source categories include the following 70 types of operations or sites (USEPA 2017j):

- Acrylic and modacrylic fibers production
- Agricultural chemicals and pesticides manufacturing
- Aluminum foundries
- Asphalt processing and asphalt roofing manufacturing
- Autobody refinishing
- Brick and structural clay
- Carbon black production
- Chemical manufacturing: chromium compounds
- Chemical preparations
- Chromic acid anodizing
- Clay ceramics
- Commercial sterilization facilities
- Copper foundries
- Cyclic crude and intermediate production
- Decorative chromium electroplating
- Dry cleaning facilities
- Fabricated metal products, electrical and electronic products-finishing operations
- Fabricated metal products not elsewhere classified
- Fabricated metal products (boiler shops)
- Fabricated metal products, structural metal manufacturing
- Fabricated metal products, heating equipment, except electric
- Fabricated metal products, industrial machinery, and equipment-finishing operation
- Fabricated metal products, iron and steel forging
- Fabricated metal products, primary metal products manufacturing
- Fabricated metal products, valves, and pipe fitting
- Ferroalloys production: ferromanganese and silicomanganese
- Flexible polyurethane foam production
- Flexible polyurethane foam fabrication
- Gasoline distribution Stage I

- Halogenated solvent cleaners
- Hard chromium electroplating
- Hazardous waste incineration
- Hospital sterilizers
- Industrial boilers
- Industrial inorganic chemical manufacturing
- Industrial organic chemical manufacturing
- Inorganic pigments manufacturing
- Institutional/commercial boilers
- Iron foundries
- Lead acid battery manufacturing
- Medical waste incinerators
- Mercury cell chlor-alkali plants
- Miscellaneous coatings
- Miscellaneous organic chemical manufacturing
- Municipal landfills
- Municipal waste combustors
- Nonferrous foundries
- Oil and natural gas production
- Other solid waste incineration
- Paint stripping
- Paints and allied products manufacturing
- Pharmaceutical production
- Plastic materials and resins manufacturing
- Plating and polishing
- Polyvinyl chloride and copolymers production
- Portland cement manufacturing
- Prepared feeds manufacturing
- Pressed and blown glass and glassware manufacturing
- Primary copper smelting
- Primary nonferrous metals-zinc, cadmium, and beryllium
- Publicly owned treatment works
- Secondary copper smelting
- Secondary lead smelting
- Secondary nonferrous metals
- Sewage sludge incineration
- Stainless and non-stainless steel manufacturing: electric arc furnaces (EAF)
- Stationary internal combustion engines
- Steel foundries
- Synthetic rubber manufacturing
- Wood preserving

6.6.5.1.4 Permit Exemptions

There are exemptions to air permitting requirements. Some sources of air pollution are exempt because the air emissions from certain sources are low or below what is termed "de-minimis" amounts. Other sources are specifically listed as exempt sources. Finally, some sources are eligible for coverage under a "permit by rule" (PBR), which means that the facility would be required to submit a notification form instead of a more complex permit application. Contacting the local air pollution regulatory authority is recommended since each air pollution control district in each state can be different, especially if it has recently been designated as non-attainment (Ohio Environmental Protection Agency 2019). Given the rather extensive operating conditions and circumstances of operating with or without a permit, conducting an inventory of all air emission sources and then contacting the local air permitting authority is recommended.

6.6.5.2 Mobile Sources

Up to this point, the discussion has focused on stationary sources of air pollution. Let's now turn to the other portion of the CAA and discuss mobile sources. A **mobile source** is defined as all on-road vehicles such as automobiles and trucks and off-road vehicles such as trains, ships, aircraft, and farm equipment (USEPA 2016a). Today, mobile sources are responsible for (USEPA 2016a):

- More than 50% of smog-forming VOC emissions
- More than 50% of nitrogen oxides (NOx)
- More than 50% of the other toxic air emissions in the United States

The total vehicle miles traveled in the United States increased by 178% between 1978 and 2005 and continue to increase at a rate of 2–3% per year. Currently, there are approximately 210 million cars and light trucks in the United States. During the past 35 years, Americans are driving more vans, trucks, and sport utility vehicles (SUVs), which typically pollute the air three to five times more than a small car (USEPA 2016a).

The CAA takes a comprehensive approach to reducing pollution from these types of mobile sources by requiring manufacturers to build cleaner operating engines, refiners to produce cleaner-burning fuels, and certain areas with air pollution problems to adopt and run passenger vehicle inspection and maintenance programs. In addition, operating cleaner buses that use propane instead of gasoline, hybrid buses, and electric buses further reduce harmful emissions, especially in heavily populated cities and urban areas where exposure to harmful air pollutants is at its greatest (USEPA 2007a; 2017d; 2017f). Last, USEPA has issued a series of regulations affecting diesel trucks and non-road vehicles such as lawn mowers, garden equipment, recreational vehicles, and boats so that as people buy new vehicles and equipment, overall emissions per vehicle will be reduced. Of course, overall emissions depend on a combination of number of mobile sources and amount of emitted pollutants per source.

The challenge is that as our population increases, so does the number of mobile sources which in large part offsets any efficiencies gained through improved technology. However, since 1970, new cars purchased today are over 90% more efficient. These improvements are from improved engine efficiency, changes in refueling operations, changes in the composition of fuels, and improved catalytic converters. Improvements must continue because the number of cars in the world exceeded 1 billion in 2010 and is expected to reach 2.5 billion by 2040 (USEPA 2016a).

One of USEPA's earliest accomplishments after the CAA was passed in 1970 was the elimination of leaded gasoline. In the mid-1970s, USEPA began its lead phase-out effort by proposing to limit the amount of lead that could be used in gasoline. By the summer of 1974, leaded gasoline was widely unavailable in the United States. This effort was followed by even stronger restrictions in the 1980s and by 1996, leaded gasoline was finally banned as a result of the Clean Air Act (USEPA 2017d).

However, the news hasn't always been positive. The CAA revisions in 1990 required the use of oxygenated gasoline in certain areas of the United States that were prone to excessive smog and carbon monoxide formation as a result of the incomplete combustion of fuel in automobiles (USEPA 2019b). The chemical of choice to add to gasoline was methyl tertiary butyl ether (MTBE). MTBE is a volatile organic compound (VOC) that is formed by a chemical reaction between methanol and isobutylene and was used to raise the oxygen level in gasoline so that it burned at a higher temperature and resulted in more complete combustion and less pollution emitted by the automobile. MTBE is a chemical that is toxic, mobile, and persistent with a high CRF value. Shortly after MTBE was used as an additive to gasoline, it started showing up in groundwater, some of which was used as a source of drinking water. As a result of groundwater impacts, Congress mandated the elimination of MTBE as a gasoline additive in 2005 but the damage was done. Numerous groundwater aquifers in the United States were contaminated with MTBE and can no longer be used as a source of drinking water and are being cleaned up at a cost of hundreds of millions of dollars and will continue for perhaps decades (USEPA 2006).

6.6.5.3 Acid Rain Reduction

The CAA Amendments of 1990 established what is called an allowance market system known today as the Acid Rain Program. The Acid Rain Program is a market-based initiative created by the USEPA with the purpose of reducing overall atmospheric levels of the principal acid rain-causing chemicals, which are sulfur dioxide (SOx) and nitrogen oxide (NOx). The program is an implementation of emission trading that primarily targets coal-burning power plants, allowing them to buy and sell emission credits called "allowances" according to individual plant needs and costs (USEPA 2008).

Phase I of the program required that significant reductions be achieved by January 1, 1995, largely requiring 110 electric power generating plants to reduce sulfur dioxide emission rates by 2.5 pounds per million British thermal unit (BTU). Each of the electric power generating plants were identified in the statute by name, address, and quantity of emissions allowances in tons of allowable sulfur dioxide emissions per year. New generating plants built after 1978 were required to limit sulfur dioxide to an emission rate of approximately 0.6 pounds per million BTUs (USEPA 2008). As an incentive for reducing emissions, for each ton of sulfur reduced below the applicable emissions limit, owners of the generating unit received an emissions credit that they could use at another unit or sell. This legitimized a market for sulfur dioxide emission credits and was administered by the Chicago Board of Trade. Power generating units that installed flue-gas desulfurization equipment (e.g., scrubbers) or other qualified equipment, which reduced sulfur emission by 90%, qualified for an extension of 2 years, provided that they already owned allowances to cover their total actual emissions for each year of the extension period (USEPA 2008; 2015a).

Phase II of the program required that all-fossil-fired units over 75 million megawatts in size were required to limit emissions of sulfur dioxide to 1.2 pounds per million BTUs generated by January 1, 2000. Thereafter, they were required to obtain an emissions credit for each ton of sulfur emitted, and were subject to a mandatory fine of $2,000 for each ton emitted in excess of the credits held. USEPA distributes credits equivalent to nearly 9 million tons each year (USEPA 2015a).

To reduce nitrogen oxide (NOx) emissions, many electric power generating plants installed low NOx burner retrofits and reduced NOx emissions by approximately 50%. This technology was readily available, so installation of NOx control was considerably less expensive and easy to install and comply with the required CAA amendments. Since the program was initiated, SOx emissions have been reduced by 40% and acid rain levels have dropped by 65% when compared to 1976 levels. USEPA has estimated that the cost of continued compliance ranges from 1 to 2 billion US dollars per year (USEPA 2015a).

6.6.5.4 Greenhouse Gases and Climate Change

In 2009, USEPA concluded that under Section 202(a) of the CAA, greenhouse gases threaten both the public health and the public welfare and that greenhouse gas emissions from motor vehicles have contributed to that threat (USEPA 2011). This action by USEPA had two distinct findings, which are (USEPA 2017k):

1. The **endangerment finding** in which the administrator concluded that the mix of atmospheric concentrations of six key, well-mixed greenhouse gases threatens both the public health and the public welfare of current and future generations. These six greenhouse gases are considered by USEPA to be the compounds that constitute the air pollutants that threaten the public health and welfare. The compounds include:
   a. Carbon dioxide (CO_2)
   b. Methane (CH_4)

    c. Nitrous oxide (N2O)
    d. Hydrofluorocarbons (HFCs)
    e. Perfluorocarbons (PFCs)
    f. Sulfur hexafluoride (SF6)

2. The **cause or contribute finding** in which the administrator concluded that the combined greenhouse gas emissions from new motor vehicles and motor vehicle engines contribute to the atmospheric concentrations of the six listed greenhouse gases and hence to the threat of climate change.

The USEPA issued these endangerment findings in response to a 2007 United States Supreme Court case of Massachusetts v. EPA, when the court ruled that greenhouse gases are air pollutants according to the Clean Air Act (USEPA 2017k).

In 2010 and in response to the endangerment finding, USEPA required industry to track facility emissions and report GHG emissions of the facilities if the sources of GHG emissions were greater than 25,000 metric tons of carbon dioxide equivalent to the newly formed USEPA Greenhouse Gas Reporting Program under 40 CFR Part 98 (USEPA 2013).

Greenhouse gases originate from both stationary sources and mobile sources. An example of a stationary source of greenhouse gas emissions would be a coal-fired power plant. An example of a mobile source would be an automobile. Greenhouse gas emissions from mobile sources are now greater than any other source passing electrical generation as of December 2018 (USEPA 2018a). USEPA and the National Highway Traffic Safety Administration (NHTSA) have established economy standards for light-duty vehicles, commercial trucks and buses, aircraft, and federal fleets that reduce the generation of greenhouse gas emissions over time (USEPA 2018b).

Individual opportunities to lower air pollution include (USEPA 2018b):

- Conserve energy – turn off appliances and lights when you leave a room
- Recycle paper, plastic, glass bottles, cardboard, and aluminum cans
- Keep woodstoves and fireplaces well maintained. Consider replacing old wood-burning stoves with an EPA-certified model
- Plant deciduous trees in locations around your home to provide shade in the summer and light into your home during the winter
- Purchase green electricity, if possible, from your utility company
- Wash clothes with warm and cold water instead of hot water
- Lower the thermostat in your home by even just a couple degrees
- Lower the thermostat on your hot water heater
- Connect your outside lights to a timer
- Use solar lighting when and where possible
- Use low-VOC or water-based paints, stains, finishes, and paint strippers
- Test your home for radon
- Do not smoke in your home
- Purchase "Energy Star" products
- Choose vehicles to purchase that are low-polluting
- Choose consumer products that have less packaging and that are recyclable
- Shop with a reusable canvas bag instead of using paper or plastic bags
- Purchase rechargeable batteries
- Keep vehicle tires properly inflated
- Fill gas tanks in the evening during the summer months
- Avoid spilling gasoline or topping off the tank while refueling
- Attempt to limit or avoid drive-thru lines
- Use public transportation
- Walk

- Ride a bike
- Get regular engine tune-ups and regular maintenance checks
- Use an energy-conserving motor oil
- Request flexible work hours
- Request to work from home, if possible
- Report vehicles with excess exhaust
- Join a carpool
- Check daily air quality forecasts
- Remove indoor asthma triggers
- Avoid outdoor asthma triggers
- Minimize sun exposure

6.6.5.5   Upper Atmospheric Ozone

Ozone in the stratosphere, a layer of the atmosphere 10 to 30 miles above the Earth, serves as a shield, protecting life on the surface of the Earth from ultraviolet radiation. The 1990 amendments to the CAA required USEPA to establish a program for phasing out production and use of ozone-destroying compounds termed ozone-depleting substances (ODS), which were used as aerosol propellants in consumer products such as hairsprays and deodorants, and as coolants in refrigerators and air conditioners (USEPA 2019a).

Specific ODS compounds include the following (USEPA 2019a):

- Chlorofluorocarbons (CFCs)
- Hydrochlorofluorocarbons (HFCs)
- Halons
- Methyl bromide
- Carbon tetrachloride
- Methyl chloroform

Scientists have been monitoring ozone levels in the upper atmosphere since the 1970s and in 1987, 190 countries, including the United States and other industrialized nations, signed the Montreal Protocol which called for an elimination of chemicals that destroy the ozone layer.

In 1996, production of ODS in the United States ceased for many of the compounds capable of the most harm, such as CFCs, halons, and methyl chloroform. Scientists estimate that it will take another 60 years for the upper atmosphere to recover. The largest observed ozone hole attributed to ODS was estimated to be 28.3 million square kilometers in September 1998 and has been shrinking ever since (NASA 2019).

In addition to phasing out ODS compounds, the CAA includes other steps to protect the ozone layer. The CAA encourages development of ozone-friendly substitutes, and because of this, many products have been reformulated. For instance, aerosol propellants and refrigerators no longer use ODS compounds. However, there are a few examples where alternatives for ODS compounds have not been developed. These include some applications in the medical field and an effective substitute for methyl bromide, which is a pesticide that is used by farmers in the United States (USEPA 2019a).

6.6.5.6   Summary of the Clean Air Act

Between 1970 and 1990, USEPA established regulations for only six pollutants and these were termed "Criteria Pollutant." During this time period, USEPA regulated only one chemical compound at a time with limited success at actually reducing pollution. The 1990 CAA amendments took a completely different approach to addressing air pollution. The 1990 Amendments required USEPA to regulate a list of chemical compounds labeled toxic air pollutants, which numbered approximately 188 different chemicals. In addition, the amendments required USEPA to take steps to reduce pollution by requiring air pollution sources to install controls or change production processes. These requirements, which were a different direction for regulations of the day, ultimately made excellent sense to regulate by category rather than by compound since many individual sources of air pollution most often emitted a mixture of chemical compounds rather than just one. Developing controls and process changes for industrial source categories resulted in major reductions in releases of multiple air pollutants to the air at a time (USEPA 2007b).

USEPA has published regulations covering a wide range of industrial categories, including chemical plants, incinerators, dry cleaners, and manufacturers of wood furniture, just to name a few. Harmful air toxins from large industrial sources have been reduced by nearly 70%. This number is even more significant when considering that industry has grown significantly. These regulations apply to mostly "major" sources of air pollution but also to smaller sources called "area" sources. In some cases, USEPA does prescribe a specific control technology, but sets a performance level based on a technology or other practices already used by the better-controlled and lower-emitting sources in an industry (USEPA 2019a).

The CAA has greatly influenced mobile sources of air pollution, especially automobiles which, according to USEPA has achieved a 90% reduction in air pollutant since the CAA was enacted in 1970. However, as demonstrated by the example of MTBE, efforts to reduce air pollution have had detrimental and harmful collateral effects in causing groundwater pollution.

The CAA has also reduced ODS emissions which has reduced the destruction of the ozone layer. However, the USEPA was not alone in this endeavor. Nearly 200 other countries also participated in efforts to eliminate ODS use and restore the ozone layer (2019a). We shall explore in later chapters the future role of global efforts in addressing pollution issues on Earth.

One last item that the CAA has introduced into the United States environmental arsenal is the single operating permit for air, which is termed the Title V permit. Environmental regulations are very complex and streamlining the process under one permit has made it much easier for industry and the regulatory authority as well. We shall again explore in later chapters opportunities to further improve environmental regulations to increase the efficiency of implementing and understanding environmental regulations and ultimately better protecting human health and the environment.

6.6.6 CLEAN WATER ACT

In the late 1960s there were several indications that surface waters of the United States were heavily impacted and polluted through human activities. Some of these were (USEPA 2018c):

- Pollution in the Chesapeake Bay killed fish that resulted in an estimated loss to the fishing industry of $3 million annually in 1968.
- Bacteria levels in the Hudson River were 170 times greater than what was considered the safe limit as measured in 1969.
- An estimated 26 million fish were killed in Lake Thonotosassa, Florida, due to wastewater discharges from four food processing plants in 1969.
- A floating oil slick on the Cuyahoga River, just southeast of Cleveland, Ohio, caught fire causing significant damage to two railroad trestles. The cause of the fire was never determined but investigations concluded that discharges of volatile organic compounds to the river with low flashpoints provided enough fuel that a passing train, for instance, could have provided an ignition source to start the fire.
- The Department of Health, Education and Welfare's Bureau of Water Hygiene reported that 30% of drinking water samples had detected chemicals exceeding recommended public health service limits.
- A study by the FDA in 1971 concluded that over 70% of samples of swordfish had mercury at concentrations that were considered unfit for human consumption.

By the time the Clean Water Act (CWA) was passed in 1972, the USEPA estimated that nearly two-thirds of the United States' lakes, rivers, and coastal waters had become unsafe for fishing or swimming due to untreated sewerage being dumped into open water. To address these concerns, USEPA has divided the CWA into four subgroups which include (USEPA 2018c):

1. Point source discharges to surface water
2. Wetland protection
3. Stormwater runoff
4. Pollution prevention

We will describe each separately in the following sections.

6.6.6.1   Point Source Surface Water Discharge

The CWA authorized the USEPA to regulate what are called **point sources** that discharge pollutants in the waters of the United States through what was termed the National Pollutant Discharge Elimination System (NPDES) permit program. The term "point sources" are defined as those locations where wastewater is generated that originated from a variety of locations and activities such as municipal and industrial operations, including treated wastewater, process water, cooling water, and stormwater runoff from drainage systems. A total of 46 of the 50 states actually implement the program under authorization from Congress and oversight of the USEPA. USEPA directly implements the program in Idaho, New Mexico, Maine, and New Hampshire as well as Native American lands (USEPA 2018c).

Under the CWA, the USEPA has implemented pollution control programs such as establishing wastewater treatment standards for certain industries and water quality standards for contaminants in the surface waters of the United States. The CWA made it unlawful to discharge any pollutant from a point source into navigable waters, unless a permit was obtained through the NPDES program. Individual homes that are connected to a municipal system that has a septic system, or do not discharge to surface water, do not require a permit under the program (USEPA 2018d).

USEPA has compiled water quality criteria for human health, aquatic life, and organoleptic effects, such as taste and odor for approximately 150 pollutants (USEPA 2017k). USEPA developed these criteria to provide guidance to the states and Native American tribes to use to establish water quality standards and provide a basis for controlling discharges or releases of pollutants to the environment. Table 6.7 presents select compounds on the current list while the entire list can be viewed at www.epa.gov/wqc/national-recommended-water-quality-criteria-human-health-criteria-table (USEPA 2019c). USEPA has established concentrations of each listed pollutant for "water + organism" and "organism" only. The concentrations recommended by USEPA are those levels that are not expected to cause adverse health effect to humans (USEPA 2019c). A water quality standard defines the water quality goals of a water body, or portion thereof, by designating the use or uses to be made of the water body and by setting the minimum criteria that protect the designated uses that include (USEPA 2010):

* Propagation of fish, shellfish, and wildlife
* Recreation in or on the water
* Consideration as a drinking water source
* Agricultural
* Industrial
* Navigation
* Other uses

There are two basic types of NPDES permits: individual and general permits. These permit types share the same components but are used under different circumstances and involve different permit issuance processes (USEPA 2010).

An individual permit is a permit that is specifically tailored to an individual facility. Most industrial businesses or locations have an individual permit. After submitting the required information

TABLE 6.7

Recommended USEPA Water Quality Standards for Human Health (USEPA 2019c)

Compound	CAS Number	Water+Organism (ug/L)	Organism Only (ug/L)	Compound	CAS Number	Water+Organism	Organism Only
Acenaphthene	83329	70	90	Methoxyoclor	72435	0.02	0.02
Acrolein	107028	3	400	Methyl Bromide	74839	100	10,000
Acrylonitrile	107131	0.061	7	Methylene Chloride	75092	20	1,000
Aldrin	309002	7.7E-7	7.7E-7	Nickel	7440020	610	4,600
Alpha-Hexachlorocyclo-hexane	319846	0.00036	0.00039	Nitrates	1479755	10,000	Under Review
Alpha-Endosulfan	959988	20	30	Nitrobenzene	98953	10	600
Anthracene	120127	300	400	Nitrosamines	None	0.0008	1.24
Antimony	7440360	5.6	640	Nitroso-dibutylamine	924163	0.0063	0.22
Arsenic	7440382	0.018	0.14	Nitroso-diethylamine	55185	0.0008	1.24
Asbestos	1332214	7E+7 fibers/L	Under Review	Nitroso-pyrrolidine	930552	0.016	34
Barium	7440393	1,000	Under Review	N-Nitroso-dimethylamine	62759	0.00069	3.0
Benzene	71432	0.58	16	N-Nitrosodi-n-Propylamine	621647	0.005	0.51
Benzidine	92875	0.00014	0.11	N-Nitroso-diphenylamine	86306	3.3	6.0
Benzo(a)anthracene	56553	0.0012	0.0013	Pentachloro-benzene	608935	0.1	0.1
Benzo(a)pyrene	50328	0.00012	0.00013	Pentachloro-phenol	87865	0.03	0.03
Benzo(b) fluoranthene	205992	0.0012	0.0013	pH	None	5-9	Under Review
Benzo(k) fluoranthene	207089	0.012	0.013	Phenol	108952	4,000	300,000
Beryllium	744047	4	4	PCBs	See Arochlor	6.4e-5	6.4e-5
Beta- Hexachlorocyclo-hexane	319857	0.008	0.14	Pyrene	129000	20	30
Beta-Endosulfan	3321365	20	40	Selenium	7782492	170	4,200
Bis(2-Chloro-1-Methylethyl) Ether	108601	200	4,000	Dissolved Solids	None	250,000	Under Review
Bis(2-Chloroethyl)Ether	111444	0.03	2.2	Salinity	None	25,000	Under Review
Bis(2-Ethylhexyl) Phthalate	117817	0.32	0.37	Tetrachloroethene	127184	10	29
Bis(Chloroethyl) Ether	542881	0.00015	0.017	Thallium	7440280	0.24	0.47
Bromoform	75252	7	120	Toluene	108883	57	520
Butylbenzyl Phthalate	85678	0.10	0.10	Toxaphene	8001352	0.00070	0.00071

(Continued)

TABLE 6.7 (CONTINUED)
Recommended USEPA Water Quality Standards for Human Health (USEPA 2019c)

Compound	CAS Number	Water + Organism (ug/L)	Organism Only (ug/L)	Compound	CAS Number	Water + Organism	Organism Only
Cadmium	7440439	5	5	Trichloroethylene	79016	0.6	7
Carbon tetrachloride	56235	0.4	5	Vinyl Chloride	75014	0.022	1.6
Chlordane	57749	0.00031	0.00032	Zinc	7440666	7,400	26,000
Chlorobenzene	108907	100	800	1,1,1-Trichloroethane	71556	10,000	200,000

ug/l = microgram per liter

Source: United States Environmental Protection Agency (USEPA). National Recommended Water Quality Criteria-Human Health Criteria Table. 2019c. www.epa.gov/wqc/national-recommended-water-quality-criteria-human-health-criteria-table. (accessed February 17, 2019.)

to the regulatory authority, the agency develops a permit for that facility on the basis of information from the permit application and other sources, which may include (USEPA 2010):

- Previous permit requirements
- Discharge monitoring reports
- Technology and water quality standards
- Total maximum daily loads
- Ambient water quality data
- Other source materials and reports

The permit authority then issues the permit to the facility for a defined period of not more than 5 years, after which the permitted facility must reapply usually no later than within 6 months prior to the expiration date of the permit or other designated time frame defined within the permit (USEPA 2010).

A general permit covers multiple facilities in a specific category of discharges or of sludge use or disposal practices. General permits are typically for sewer districts, urbanized areas, cities, or towns. The regulation also allows a general permit to cover any other appropriate division or combination of such boundaries. For instance, USEPA has issued general permits to cover multiple states, territories, and Native American tribes where USEPA is the permitting authority (USEPA 2010).

The major components of a NPDES permit include the following (USEPA 2010):

- Cover page. Contains the name and location of the permittee, a statement authorizing the discharge, and a listing of the specific locations for which the discharge is authorized.
- Effluent limitations. The primary mechanism for controlling discharges of pollutants to receiving water are the effluent limitations. Effluent limitations typically include a list of each pollutant in the effluent and an acceptable concentration for each pollutant. In addition, the permit typically will include daily and or monthly maximums of loading to the receiving waters.
- Monitoring and reporting requirements. Monitoring and reporting requirements are used to characterize:
  - Waste streams
  - Receiving waters
  - Wastewater treatment efficiency
  - Compliance with permit conditions
- Special conditions. Special conditions are developed to supplement numeric effluent limitations. Examples may include:
  - Additional monitoring activities
  - Additional or special studies or sampling
  - Best management practices (BMPs)
  - Compliance schedules
- Standard conditions. Standard conditions are those conditions that apply to all NPDES permits and delineate the legal, administrative, and procedural requirements of the NPDES permit.

6.6.6.2 Stormwater

Stormwater runoff occurs when precipitation from rain or snowmelt flows over the ground. Impervious surfaces like driveways, sidewalks, parking lots, streets, and roofs of houses and buildings prevent some or all stormwater from naturally soaking into the ground. These discharges may contain pollutants of various types and concentration depending upon the chemistry of the rain water, potential contaminants, or surfaces the rain water comes into contact with, and for how long the contact exists. Therefore, USEPA requires that municipalities, industry, and construction sites obtain a permit that outlines what controls are necessary to prevent or minimize the potential for pollutants to adversely affectsurface water quality (California Environmental Protection Agency 2001; USEPA 2009).

The NPDES Stormwater Program, existing as a separate set of regulations, was put into place in the Clean Water Act amendments of 1987 and 1990, regulates discharges from municipal storm sewer systems, construction activities, industrial activities, and those designated by USEPA due to water quality impact. The stormwater regulation went into effect in two phases (USEPA 2009).

Phase I required permits for the following five categories (USEPA 2010):

1. Discharges permitted before February 4, 1987
2. Discharges associated with industrial activity
3. Discharges from large municipal separate storm sewer systems (MS4s) (systems serving a population of 250,000 or more)
4. Discharges from medium MS4s (system serving a population of greater than 100,000, but less than 250,000)
5. Discharges evaluated by the permitting authority to be significant sources of pollution or which contribute to a violation of a water quality standard

Phase II required smaller jurisdictions of less than a population of 100,000 to come into compliance by 2008, and to implement USEPA's six minimum control measures that included (USEPA 2010):

1. Public information and education
2. Public involvement and participation
3. Illicit discharge detection
4. Construction site stormwater runoff control
5. Post-construction stormwater management
6. Pollution prevention/housekeeping

Phase II also required numerous small MS4s, construction sites of one to five acres, and industrial facilities owned or operated by small MS4s which were previously exempted in Phase I to also obtain a stormwater permit (USEPA 2010).

The CWA and related regulations define the specific industrial and municipal stormwater sources that must apply and obtain an NPDES permit. The CWA also recognizes that other sources, such as commercial properties, may need to be regulated on a case by case or category by category basis based on additional information or localized conditions (USEPA 2010).

The authority to regulate other sources based on the localized adverse impact of stormwater on water quality through the NPDES permits is commonly referred to as "Residual Designation" authority. In 2008, USEPA issued records of decision requiring additional NPDES stormwater permits in specific areas within the Charles River watershed in Massachusetts and the Long Creek watershed in Maine. In July 2013, the Conservation Law Foundation, Natural Resources Defense Council, and American Rivers filed a petition seeking USEPA to render a decision that commercial, industrial, and institutional sites contribute to violations of water quality standards and therefore required NPDES permits (USEPA 2009).

A stormwater permit is termed a Stormwater Pollution Prevention Plan (SWPPP). An outline of a typical SWPPP is outlined below (Wisconsin Department of Natural Resources 2017):

1. Overview
   a. General facility information including name, location, property size, operations, physical characteristics, and site map
   b. Statement of Introduction. This states the citation and reference to applicable stormwater regulations that apply to the specific facility

   c. Statement of Objectives. States the primary goal of the stormwater permit which is to improve the quality of surface waters by reducing the amount of pollutants potentially contained in stormwater runoff.

2. Storm Water Pollution Prevention Team. This section identifies specific individuals at the facility who will be responsible for developing, implementing, maintaining, and revising the SWPPP.

3. Potential Sources of Pollution. This section includes a comprehensive assessment of all potential sources of stormwater sources at the facility and includes the following:

   a. Site map. A site is a critical item included in any SWPPP. The map or maps typical include:

     i. Facility property boundaries

     ii. A depiction of the storm drain collection and disposal system, including all known surface and subsurface conveyances that are clearly labeled

     iii. Any secondary containment structures

     iv. The location of all outfalls, which should also be numbered for reference purposes, that discharge channelized flow to surface water, groundwater, or wetlands

     v. The drainage area boundary for each storm water outfall

     vi. The surface area in acres draining to each outfall, including the percentage that is impervious defined as those areas that are paved, roofed, or highly compacted soil, and the percentage that is pervious defined as those areas that are grass and woods

     vii. Any structural stormwater controls

     viii. Name and location of receiving waters

     ix. Location of activities and materials that have the potential to contaminate stormwater shall also be identified. These locations typically refer to any outside storage of materials, drums, and other containers (even if empty), machines, waste storage locations, maintenance shops, used oil storage, loading and unloading locations, onsite vehicle storage and parking (e.g., forklifts).

   b. Summary of Sampling Data. Not every facility will have sampling data, especially if applying for a first time permit and are not renewing a permit. If renewing a permit, historical data is required to be provided. Sampling data will vary from site to site since operations and chemical use may vary greatly depending on operations. However, typical historical data that may be available may include the following:

     i. Heavy metals

     ii. Total solids

     iii. Total dissolved solids

     iv. Oil and grease

     v. Total petroleum hydrocarbons

     vi. Volatile organic compounds

     vii. Semi-volatile organic compounds

     viii. Polychlorinated biphenyls

     ix. pH

     x. Temperature

     xi. Turbidity

     xii. Other compounds

   c. Inventory of Potential Sources of Contamination. The following have been identified as potential sources of stormwater contamination:

     i. Outdoor manufacturing areas

     ii. Rooftops contaminated from industrial activity or emission deposition from a pollution control device

     iii. Areas of significant soil erosion, discoloration, or stressed vegetation

     iv. Industrial plant storage yards

        v. Onsite access roads or rail lines and spurs
      vi. Material handling locations (storage, loading, unloading, transportation, or conveyance of any raw material, finished product, intermediate product, byproduct, solid or liquid water)
     vii. Shipping and receiving locations
   viii. Manufacturing buildings
     ix. Residual treatment, storage, and disposal locations
      x. Storage areas (including tank farms) for raw materials, finished or intermediate materials
     xi. Refuse sites
    xii. Disposal locations
   xiii. Areas containing residual pollutants from past industrial activity, including previous areas of leaks or spills
    xiv. Vehicle maintenance and cleaning areas
     xv. Any other area having the capability or possibility of contaminating storm water runoff

4. Other Plans Incorporated by Reference. These may include:
   a. Preparedness, Prevention, and Contingency Plan
   b. Spill Control and Countermeasures Plan
   c. National Pollution Discharge Elimination System Permit
   d. Toxic Organic Management Plan
   e. Occupational Safety and Health Administration Emergency Action Plan
   f. Preventative Maintenance Plan
5. Best Management Practices. These items describe storm water management controls labeled best management practices that would be implemented to reduce the amount of pollutants in storm water discharged from the facility and would typically include the following:
   a. Source Elimination (e.g., removing the risk by relocation indoors, thereby eliminating the chance that rain water can come into contact with the source
   b. Source Area Controls for those areas that can't be eliminated including:
      i. Erosion control measures
      ii. Improved housekeeping
      iii. Preventative maintenance
      iv. Quarterly visual comprehensive inspections and documentation
      v. Spill prevention and response procedures
      vi. Employee training
      vii. Bulk storage management improvements
   c. Residual pollutants (e.g., capping with an imperious layer)
   d. Stormwater treatment may be necessary and may include:
      i. Preventative measures (e.g., redirecting stormwater)
      ii. Diversions
      iii. Containment
      iv. Other controls
   e. Facility monitoring includes:
      i. Visual inspections
      ii. Sampling and analysis
6. Record Keeping. All reports and records pertaining to the permit coverage shall be retained 5 years from the date of the permit. The records are to be maintained onsite and shall be made available upon request of the permit authority.
7. Certification of the SWPPP. To be made by the plan preparer and an authorized representative of the facility in charge.

6.6.6.3   Wetlands

Placement of dredged or fill material into wetlands, lakes, streams, rivers, estuaries, and certain other waters are also regulated under the CWA in Section 404. The goal of Section 404 is to avoid and minimize losses of wetland and other waters and to compensate for unavoidable loss through mitigation and restoration. Section 404 is jointly implemented by USEPA and the Army Corps of Engineers (United States Army Corps of Engineers 2017).

In addition, property development may also be subject to wetlands regulations if wetlands are present. **Wetlands** are defined by the United States Army Corps of Engineers (USACE) and the USEPA as those areas that are inundated or saturated by surface or groundwater as a frequency and duration sufficient to support, and under normal circumstances, do support a prevalence of vegetation typically adapted for life in saturated soil conditions. In order for an area to be classified as a wetland, a wetland evaluation must typically be conducted by a wetland professional to determine the presence of hydrophytic vegetation hydric soils, and if wetland hydrology indicators are present, the area is typically considered a wetland (United States Army Corps of Engineers 2017). If a proposed development has the potential to destroy or otherwise influence or impact any portion of a wetland, a permit will be required from either the USACE or perhaps even the applicable state agency where the property is located (United States Army Corps of Engineers 2017).

6.6.6.4   Pollution Prevention

As part of the pollution prevention/housekeeping requirements of the CWA, USEPA requires facilities that store oil or other petroleum products prepare a Spill Prevention Control and Countermeasures (SPCC) Plan (USEPA 2012). The purpose of a SPCC plan is to prevent the release of oil or petroleum product to a navigable waterway of the United States. It is not the purpose of a SPCC Plan that describes actions to be conducted to clean up a spill once it occurs. This would be a separate plan commonly termed a Spill Response Plan (SRP) (USEPA 2018c).

SPCC requirements are classified into three tiers that include small facilities, mid-size facilities, and large facilities. What determines which class a specific facility would be included is based on storage capacity of petroleum products. Regardless of the classification, all SPCC plans must be prepared in accordance with the oil pollution guidelines in the Federal Code of Regulations (CFR) 40 CFR Part 112 (USEPA 2012).

Tier I requirements include the following (USEPA 2012):

- Above ground petroleum product storage capacity of 10,000 gallons or less.
- Storage is calculated by adding up all above-ground storage containers of petroleum product at a capacity of 55 gallons or more. This includes transformers if they contain oil and are not the dry type and contain more than 55 gallons capacity.
- No single discharge of petroleum to navigable waters or adjoining shorelines each exceeding 42 gallons within any 12 month period, 3 years before the certification date.
- No two discharges of petroleum to navigable water or adjoining shorelines each exceeding 42 gallons.
- Having little or no piping and very few oil transfer stations.

If the above criteria are met, an owner or operator may complete and self-certify a plan template instead of a full professional engineer certified plan.

Tier II requirements include the following (USEPA 2012):

- Above ground petroleum product storage capacity of 10,000 gallons or less
- Storage is calculated by adding up all above-ground storage containers of petroleum product at a capacity of 55 gallons or more. This includes transformers if they contain oil and are not the dry type and contain more than 55 gallons capacity.

- No individual above ground petroleum storage container is greater than 5,000 gallons
- Meets all other Tier I conditions

If the above Tier II criteria are met, an owner or operator may self-certify a spill plan in accordance with requirements outlined in 40 CFR, Part 112.7, in lieu of a professional engineer certified plan.
    Tier III requirements include the following (USEPA 2012):

- Above ground petroleum storage capacity greater than 10,000 gallons
- Meets all other Tier I and II conditions

A licensed professional engineer must prepare and certify the plan for Tier III sites.
    There are 12 basic requirements of any SPCC plan, which are (USEPA 2012):

1. Description of any spills in the past 12 months, including:
    a. Corrective action measures undertaken
    b. Plans to prevent reoccurrence
2. Layout of facility, including diagrams marking locations and contents of:
    a. All oil and petroleum storage containers and capacities, including transformers, if applicable
    b. All underground or buried tanks
    c. All transfer stations, if any
    d. All connecting piping
3. Predictions of the direction, rate of flow, and total quantity of oil or petroleum that could be discharged
4. A complete explanation of spill containment and any diversionary structures or equipment including:
    a. Dikes
    b. Berms
    c. Retaining walls
    d. Curbing
    e. Culverts, gutters, or other drainage systems
    f. Spill diversion/retention ponds
    g. Double-walled tanks with interstitial monitors
    h. Absorbent materials
    i. Tools (i.e., shovels and other implements)
    j. Personal protective clothing (i.e., boots, shields, gloves, etc.)
    k. Location and contents of spill kits
5. Facility maintenance of containment area drainage that includes:
    a. Storm water in berms and dikes
    b. Dike drainage practices
    c. Management of areas that are not contained
6. Bulk storage practices that include:
    a. Verification that tank material and construction are compatible with material stored
    b. Secondary containment means such as double-walled tanks or physical containment with a capacity equal to the largest capacity container plus an additional 10%
    c. Procedure to ensure that drainage of secondary containment areas do not release oil or petroleum product
7. Transfer protocol method that includes:
    a. Methods to limit corrosion of buried piping, if any
    b. Methods to inspect and maintain above-ground valves and piping
    c. Procedures, barriers, and methods that warn vehicles to avoid damaging above-ground piping and storage areas

8. Tank truck loading and unloading practices that include:
   a. Documentation that loading and unloading procedures meet Department of Transportation (DOT) requirements
   b. Loading and unloading area containment capacity is calculated and sufficient
   c. Containment methods for unloading and loading areas are satisfactory
   d. Methods to prevent vehicle departure before transfer lines are disconnected
9. Inspection and documentation methods that include:
   a. Method to ensure plan is being implemented properly
   b. Retain records for at least 3 years
10. Site security that includes:
    a. Restricting access to oil handling and storage areas
    b. Methods to secure tank valves, pumps, and loading and unloading connections when in standby status
11. SPCC training programs that include:
    a. Operation and maintenance of equipment
    b. Overview of applicable environmental regulations and requirements
    c. Designation of an SPCC Plan coordinator
    d. Training schedule
    e. Personal training records
12. A "Certification of Applicability of Substantial Harm Criteria" form. If the potential exists that a release could cause substantial harm then a facility must prepare a site-specific response plan to the applicable regulatory authority.

6.6.6.5   Summary of the Clean Water Act

The Clean Water Act established the basic structure for regulating discharges of pollutants into the water of the United States. It also established water quality standards for surface waters. The Clean Water Act made it unlawful to discharge any pollutant from a point source into navigable waters, unless a permit was obtained. USEPA has estimated that cleaning up and preventing polluted discharges to surface water, mainly through construction of publicly owned treatment works (POTWs), has cost over $56 billion since the Clean Water Act was passed in 1972 (USEPA 2018d). There is no doubt that the Clean Water Act has been one of many great success stories of environmental regulations in the United States thus far. However, much work remains especially from nonpoint sources of water pollution (USEPA 2018d).

Nonpoint source water pollution occurs when atmospheric deposition, rainfall, or snowmelt flows over land or through the ground, dissolves or carries pollutants, and deposits them into rivers, lakes, streams, and coastal waters or introduces them into groundwater (USEPA 2018e). Regulating nonpoint source (NPS) of pollution is much more difficult to regulate because the pollution is generally, by the nature of its definition, diffuse and widespread throughout the environment. According to USEPA (2018d), NPS pollution includes:

- Fertilizers including phosphorus and nitrogen
- Herbicides
- Insecticides
- Oil
- Grease
- Salt from roads and irrigation
- Bacteria from livestock and septic tanks

One of the best examples that demonstrates the need for NPS regulations under the Clean Water Act in the Gulf of Mexico hypoxic zone. It is commonly referred to as the Gulf of Mexico Dead Zone because the oxygen levels within the zone are too low to support marine life. Sometimes it is even referred to as Red Tide. The Dead Zone was first documented as early as the 1970s and originally occurred every few years (Carlisle 2000). Now it occurs every year and in some years is so huge that it is roughly equivalent in size to the state of Georgia, which is nearly 100,000 square kilometers, and has also appeared as far away as the Florida coast (NOAA 2018). NOAA now releases routine bulletins to inform boaters and beach visitors of the presence of what they term Gulf of Mexico harmful algal bloom (NOAA 2018). The main cause has been traced to NPS of pollution, mainly rural and agricultural land in the Mississippi River Basin and other drainage basins that discharge to the Gulf of Mexico that carry dissolved fertilizers. Once in the Gulf of Mexico, the fertilizers provide nutrients favorable for algae growth and deplete the oxygen supply in the water, resulting in massive (in the millions or more) marine life deaths (Malakoff 1998; Carlisle 2000).

6.6.7   SAFE DRINKING WATER ACT

The Safe Drinking Water Act (SDWA) of 1974 was established to protect the quality of drinking water in the United States. This law focuses on all waters actually or potentially designed for use as drinking water from above ground or below ground sources, which includes groundwater (USEPA 2017l).

The SDWA authorizes the USEPA to establish minimum standards to protect tap water and requires all owners or operators of public water systems to comply with these standards, termed Primary Drinking Water Standards. **National Primary Drinking Water Standards** (NPDWS) are defined as those drinking water standards that are based on health-related criteria established by USEPA. The 1996 amendments to the SDWA required USEPA to consider a detailed risk assessment, cost assessment, and best available peer-reviewed science when revising the standards (USEPA 2017l). USEPA and state governments, which can be approved by USEPA to implement these rules on behalf of USEPA, also encourage compliance with an additional set of standards termed secondary standards. Secondary standards are commonly referred to as nuisance-related compounds (USEPA 2017l). Currently, there are over 155,000 public water supplies in the United States that are subject to the SDWA. Private wells are not covered under the act. The SDWA also does not apply to bottled water. Bottled water is regulated by the FDA under the Federal Food, Drug, and Cosmetic Act (USEPA 2017l).

NPDWS regulations include both mandatory levels, termed maximum contaminant levels or MCLs, which are enforceable and non-enforceable health goals termed maximum contaminant level goals or MCLGs for each contaminant listed. MCLs have additional significance because they can be used under the Superfund Law, which we will discuss later in the chapter, as "Applicable or Relevant and Appropriate Requirements" in setting cleanup requirements at contaminated sites on the National Priority List (NPL). NPDWS are organized into six groups of compounds, which include (USEPA 2017l):

1. Microorganisms
2. Disinfectants
3. Disinfection byproducts
4. Inorganic chemicals
5. Organic chemicals
6. Radionuclides

Table 6.8 lists the current MCLs or MCLGs for the chemicals regulated under the primary drinking water standards. Table 6.9 lists the current MCLs or MCLGs for secondary standards. The complete lists of each are at www.epa.gov/dwstandardsregulations (USEPA 2019c).

Although state health agencies and public water systems often decide to monitor and treat their water supplies for secondary contaminants, it is not a requirement under federal regulations. Corrosion control is perhaps the single most cost-effective method a water distribution system can use to treat iron, copper, zinc, and sometimes lead (USEPA 2017l).

TABLE 6.8
MCLGs for Select Parameters

Contaminant	MCLG	Sources of Contaminant
Cryptosporidium	Zero	Human and animal fecal waste
Legionella	Zero	Present naturally in water
Total coliform (*E. Coli*)	Zero	*E. Coli* only originate from human and animal fecal waste
Total trihalomethanes	0.080 Mg/L	Byproduct of disinfection
Chlorine (as Cl2)	4.0 mg/L	Byproduct of disinfection
Arsenic	0.10 mg/L	Naturally occurring, farming, and industry
Asbestos	7E+7 fibers/L	Naturally occurring, decay of water mains
Barium	2 mg/L	Naturally occurring and drilling waste
Cadmium	0.005 mg/L	Naturally occurring, water pipe corrosion
Chromium (total)	0.1 mg/L	Naturally occurring, industrial discharges
Copper	1.3 mg/L	Naturally occurring, household plumbing
Cyanide	0.2 mg/L	Fertilizer plants, industrial discharges
Lead	0.015 mg/L	Naturally occurring, water pipe corrosion
Mercury	0.002 mg/l	Naturally occurring
Nitrate (N)	10 mg/L	Naturally occurring. fertilizer
Selenium	0.05 mg/L	Naturally occurring
Acrylamide	Zero	Sewerage treatment chemical
Alachlor	Zero	Herbicide, agricultural runoff
Atrazine	0.003 mg/L	Herbicide, agricultural runoff
Benzene	0.005 mg/L	Gasoline, industrial discharges
Benzo(a)pyrene (PAHs)	0.0002	Oil and industrial discharges
Carbon tetrachloride	0.005	Industrial discharges
Chlordane	0.002	Banned termiticide
Chlorobenzene	0.1 mg/L	Industrial discharges
2,4-D	0.07	Herbicide, farming runoff
Dalapon	0.2 mg/L	Herbicide, farming runoff
1,2-Dibromo-3-chloropropane	Zero	Agricultural runoff
o-Dichlorobenzene	0.6 mg/L	Industrial discharges
1,2-Dichloroethane	0.005 mg/L	Industrial discharges
Cis-1,2-Dichloroethylene	0.07 mg/L	Industrial discharges
Trans-1,2-Dichloroethylene	0.1 mg/L	Industrial discharges
Dichloromethane	0.005 mg/L	Industrial discharges
1,2-Dichloropropane	0.005 mg/L	Industrial discharges
Di(2-ethylhexyl) phthalate	0.006 mg/L	Industrial discharges
Dioxin	3.0E-8 mg/L	Waste incineration and other combustion
Diquat	0.02 mg/L	Herbicide, agricultural runoff
Endrin	0.002 mg/L	Insecticide, agricultural runoff
Trichloroethylene	0.005 mg/L	Industrial discharges
Ethylbenzene	0.7 mg/L	Gasoline and industrial discharges
Tetrachloroethylene	0.005 mg/L	Industrial discharges

mg/L = milligram per liter

Source: United States Environmental Protection Agency (USEPA). 2019d. Drinking Water Contaminants – Standards and Regulations. 2019d. www.epa.gov/dwstandardsregulations. (accessed February 17, 2019.)

TABLE 6.9

Secondary Drinking Water Standards

Contaminant	Secondary MCL	Effect
Aluminum	0.05 to 0.2 mg/L	Colored water
Chloride	250 mg/L	Salty taste
Color	15 color units	Visible tint
Copper	1.0 mg/L	Metallic taste, blue-green staining
Corrosivity	Non-corrosive	Metallic taste, corroded pipes/fixtures, staining
Fluoride	2.0 mg/L	Tooth discoloration
Foaming agents	0.5 mg/L	Frothy, cloudy, bitter taste, odor
Iron	0.3 mg/L	Rusty color, sediment, metallic taste, reddish or orange staining
Manganese	0.05 mg/L	Black to brown color, staining, bitter metallic taste
Odor	3 TON (threshold odor number)	Rotten egg odor, musty or chemical odor
pH	6.5–8.5	Low pH: bitter taste, corrosion
		High pH: slippery feel, soda taste, deposits
Silver	0.1 mg/L	Skin discoloration, graying of the white portion of the eye
Sulfate	250 mg/L	Salty taste
Total dissolved solids (TDS)	500 mg/L	Hardness, deposits, colored water, staining, salty taste
Zinc	5 mg/L	Metallic taste

mg/L = milligram per liter

Source: United States Environmental Protection Agency. 2017m. Drinking Water Standards. 2017m. www.epa.gov/safe-drinkingwater (accessed March 15, 2017.)

The benefits include the following (USEPA 2017l):

- Reduction of contaminants at the point of consumption or exposure
- Cost savings due to extending the useful life of water mains and service lines
- Energy savings from transporting water more easily through smoother, less corroded distribution systems
- Reduced water losses through leaks and broken water mains

Corrosion control is most often a function of monitoring and controlling the pH of the water supply. A pH that is less than 7 may lead to an increase of corrosion and lead to leach of metals from the pipes and water mains. This leads to an increase in concentration of some heavy metals in the water that include copper, iron, zinc, and lead. Lead may increase in concentration in older water systems since lead was more widely used as solder and pipe connections historically. This resulted in the enactment of the Lead and Copper Rule by USEPA in 1991 in an effort to reduce the amount of lead and copper in drinking water from plumbing and also lowered the maximum contaminant level (MCL) for lead in drinking water from 50 ug/l to 15 ug/l (USEPA 1992).

Corrosion control is not used for treatment to remove heavy metals from the water supply but has a secondary benefit of preventing the leaching of heavy metals from the water distribution system. Therefore, corrosion control should be viewed as not necessarily increasing the concentration of heavy metals not already present from the source (USEPA 2017l).

Applying USEPA regulations for safe drinking water is complex due to numerous factors that can influence water quality as evidenced by the situation that occurred in Flint, Michigan when the water supply was switched to the Flint River as its source and ultimately resulted in higher lead concentrations in the drinking water supply (USEPA 2016b).

6.6.8 HAZARDOUS MATERIALS TRANSPORTATION ACT

In the 1970s, numerous landfills throughout the United States began to refuse acceptance of wastes considered hazardous because of liability concerns from what would later become known as Superfund liability, which greatly increased the disposal costs of wastes at those waste sites that did accept hazardous wastes for disposal. This led to increased illegal dumping of wastes in vacant lots, along highways, and in many isolated and wooded areas and farms. At the same time, there were many incidents and accidents involving hazardous materials during transportation causing property damage and endangering the public (USEPA 2017n). During this time, the United States Department of Transportation estimated that 75% of hazardous waste shipments within the United States violated the regulations in effect at the time because of the following (USEPA 2017n):

- Inconsistencies in the regulations between individual states
- Lack of inspectors to enforce those regulations in effect at the time
- Lack of proper training of those inspectors that were present at the time
- Poor coordination between those federal and applicable state agencies responsible for regulating transportation routes and methods including the United States Coast Guard, Federal Aviation Authority, Federal Highway Administration, and the Federal Railroad Administration

As a result of the issues highlighted above, Congress passed the Hazardous Materials Transportation Act (HMTA) in 1975. The HMTA is the principal law governing the transportation of hazardous materials. Its purpose is to protect against the risks to life, property, and the environment that are inherent in the transportation of hazardous material within the United States and beyond. The responsibility for enforcing the act has been delegated to the United States Department of Transportation, not the USEPA (2017n).

USDOT estimates that currently there are 500,000 shipments of hazardous materials in the United States every day and more than 90% of these shipments are transported by truck. According to USDOT, approximately 50% of those materials that are shipped daily contain corrosive or flammable petroleum products, while the remaining shipments represent any of the 2,700 other chemicals considered hazardous by USDOT. The Act was passed to improve the uniformity of existing regulations for transporting hazardous materials and to prevent spills and illegal dumping or disposal thus endangering the public and the environment, which was a problem that was sometimes made worse by fragmented regulations (USEPA 2017n).

The definition of hazardous materials includes those materials designated by the Secretary of the United States Department of Transportation as posing an unreasonable threat to the public and the environment. The term "Hazardous Material" includes all of the following (USEPA 2017n):

- Hazardous substance
- Hazardous waste
- Marine pollutants
- Elevated temperature material
- Materials identified in Part 172.101 of the act
- Materials meeting the definitions contained in Part 173 of the act

The act has four main elements that include (USEPA 2017n):

1. Procedures and policies
2. Material designations and labeling
3. Packaging requirements
4. Operational rules

HMTA regulations apply when an individual or company satisfies any one or more of the following (USEPA 2017n):

- When hazardous materials are transported in commerce
- When a company or individual designs, manufactures, fabricates, inspects, marks, maintains, reconditions, repairs or tests a package, container, or packaging component that is represented, marked, certified, or sold as qualified for use in transporting hazardous material in commerce
- When a company or individual prepares or accepts hazardous material for transportation in commerce
- When a company or individual is responsible for the safety of transporting hazardous material in commerce
- When a company or individual certifies compliance with any requirement under the act
- When a company or individual misrepresents whether such person or company is engaged in any activity under the above-listed requirements

Essentially, even though every person involved in the preparation of the transportation of hazardous materials shares some responsibility to ensure its safety, the primary burden of liability falls on the shipper of the material. Carriers are only responsible to ensure that required information accompanying hazardous materials packaging is immediately available to personnel who would respond to an incident or conduct a hazardous materials investigation. The act is implemented through several agencies and is based on the type of transportation and type of hazardous material being shipped (USEPA 2017n).

Procedures of proper handling and preparation for handling hazardous materials, as well as finding instructional information about implementing the act correctly, includes (USEPA 2017n):

- Training requirements. The act describes what an employer who ships hazardous material must do to train employees on the safe loading, unloading, handling, storing, and transporting of hazardous material and emergency preparedness for responding to an accident or incident involving the transportation of hazardous material.
- Beginning and completing training. A hazardous material employer shall begin training of employees no later than 6 months after employment begins.
- Certification of training. After training is complete, each employer shall certify, with documentation, that the employees have received training and have been tested on an appropriate area of responsibility.

A record of current training is required to be maintained by the employer (USEPA 2017n):

- The hazmat employee's name
- The most recent training completion date of the hazmat employee's training
- A description, copy, or the location of the training materials used to meet the requirements
- The name and address of the person who provided the training
- Certification that the hazmat employee has been trained and tested

Each package, freight container, and transport vehicle carrying a hazardous material must have markings that are (USEPA 2017n):

- Durable
- In English
- Printed or affixed on the surface of the shipping package, or on a label, tag, or sign on the package

- Displayed on a background of sharply contrasting color
- Not obscured by other labels or attachments
- Located away from any other marking that could reduce its effectiveness

Each non-bulk package, container, or small tank must be labeled with a label code corresponding to the hazard class of the hazardous material being transported and must also follow design and placement requirements. Each bulk packaging, freight container, unit load device, transport vehicle or rail car containing any quantity of a hazardous material must be placarded corresponding to the hazard class of the hazardous material being transported and must follow design and placement requirements (USEPA 2017n).

There are nine hazardous classes that include (see Figure 6.7) (USDOT 2013):

1. Class 1: Explosives
2. Class 2: Gases
3. Class 3: Flammable liquid and combustible liquid
4. Class 4: Flammable solid, spontaneously combustible, and dangerous when wet
5. Class 5: Oxidizer and organic peroxide
6. Class 6: Poison (toxic) and poison inhalation hazard
7. Class 7: Radioactive
8. Class 8: Corrosive
9. Class 9: Miscellaneous

The contents of any package containing a hazardous material and the material of construction of the package itself must be resistant to significant chemical or galvanic reactions that can potentially compromise the integrity of the package. In addition, hazardous materials may not be mixed together with other materials regardless if they are hazardous or not if they potentially can create a reaction causing the following (USDOT 2017):

- Combustion or generating heat
- Flammable gas
- Poisonous gas

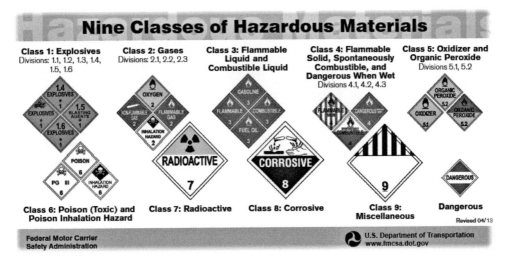

FIGURE 6.7 USDOT hazardous classes. (From United States Department of Transportation. Federal Motor Carrier Safety Administration (FMCSA) Placard Requirements. 2013. www.fmcsa.dot.gov.) (Accessed March 22, 2017.)

- Asphyxiant gas
- Formation of unstable or corrosive materials

It is the responsibility of the shipper of a hazardous material to determine that the compatibility between the hazardous material and the packaging is sufficient for safe transportation (USDOT 2017).

Regulations providing for immediate emergency response information in an incident, as well as requirements for the development and implementation of security plans, must be adhered to by any person who offers transportation in commerce or transports in commerce hazardous materials. Placarding on vehicles is contained in 49 CFR Subpart F Part 172 and is required on all four sides of a vehicle carrying a hazardous material. An example placard is presented in Figure 6.8. Placards are composed of four squares that form a diamond-shaped placard (USDOT 2017).

The upper portion of a placard contains information concerning the fire hazard of the material on a scale of 0 (the material will not burn) to 4 (the material has a flashpoint of 73 degrees F° or less and is typically colored red). The left portion of the placard contains information concerning health hazard information on a scale of 0 (normal materials) to 4 (the material is considered deadly). The right portion of the placard contains information of the potential reactivity of the hazardous material on a scale of 0 (the material is considered stable) to 4 (the material is not stable and may detonate under normal conditions). The final portion of the placard is at the bottom and contains information about the specific hazard of the material (USDOT 2017).

Examples of specific hazards that may present information such as whether the material includes (USDOT 2017):

- Acid
- Alkali
- Corrosive
- Oxidizer
- Use no water or water reactive

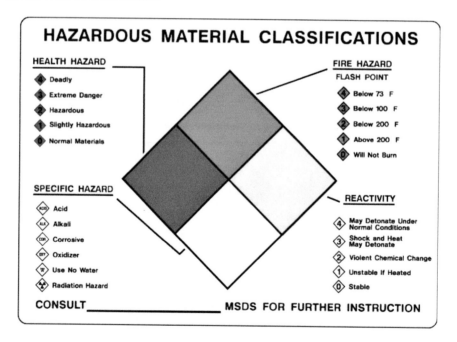

FIGURE 6.8 USDOT hazardous material classifications and example placard. (From United States Department of Transportation. Summary of the Federal Hazardous Material Transportation Act. Office of Hazardous Materials Safety. 2017. www.USDOT.gov./HMTA.) (Accessed March 18, 2017.)

- Radiation hazard
- Explosive
- Flammable, etc.

Immediate notification of a hazardous material incident by the carrier is required at the earliest practical moment for incidents that occur during the course of transportation (including loading, unloading, and temporary storage) in which as a direct result of the hazardous materials any one of the following occurs (USDOT 2017):

- A person is killed
- A person receives an injury requiring admittance to a hospital
- The general public is evacuated for 1 hour or more
- A major transportation artery or facility is closed or shut down for 1 hour or more
- Fire, breakage, spillage, or suspected contamination occurs involving an infectious substance other than a diagnostic specimen or regulated medical waste
- A release of a marine pollutant occurs in a quantity exceeding 450L (119 gallons) for a liquid or 400 kg (882 pounds) for a solid
- A situation exists of such a nature (e.g., a continuing danger of life exists at the scene of the incident) that, in the judgment of the person in possession of the hazardous material, it should be reported to the National Response Center even though it does not meet the other criteria listed above

Each notice shall be given via telephone to the National Response Center at 800-424-8802. Incidents involving etiologic agents should also be made to the Centers for Disease Control (CDC) at 800-232-1024. A written report is also required to be submitted to USDOT and others as directed using DOT Form F 5800.1 for all incidents involving the transportation of hazardous materials unless specifically directed by USDOT (USDOT 2017).

6.6.9 Toxic Substance Control Act

The Toxic Substance Control Act (TSCA) authorizes USEPA to screen existing and new chemicals used in manufacturing and commerce to identify dangerous products or uses that should be subject to federal control. TSCA was first passed in 1976 and has been amended in 1986, 1988, 1990 (twice), 1992, 2007, 2008, 2010, and 2016. Both naturally occurring and synthetic chemicals are subject to TSCA, with the exception of chemicals already regulated under other federal laws that include food, drugs, cosmetics, firearms, ammunition, pesticides, herbicides, and tobacco (Schierow 2013).

The purpose of TSCA was to protect the public from unreasonable risk of injury to health or the environment by regulating the manufacture and sale of chemicals. TSCA does not address wastes produced as byproducts of manufacturing. Instead, TSCA attempted to exert direct government control over which types of chemicals could and could not be used in actual use and production. For example, the use of chlorofluorocarbons in manufacturing is now strictly prohibited in all manufacturing processes in the United States, even if no chlorofluorocarbons are released into the atmosphere. The types of chemicals regulated under TSCA are in two broad groups: new and existing. New chemicals are defined as any chemical substance which is not included in the chemical substance list compiled and published under TSCA, Section 8(b). This list included all of the chemical substances manufactured or imported into the United States prior to 1979, which were grandfathered into TSCA and were considered safe and included over 62,000 chemicals (USEPA 2017o).

When first passed, TSCA limited the manufacture, processing, commercial distribution, use, and disposal of chemical substances including polychlorinated biphenyls (PCBs), asbestos, radon, and lead-based paint. At first USEPA could only limit the commercial use of new chemicals, and even then only immediately after the chemicals were introduced in commerce. Once a chemical substance established itself in the commercial world, USEPA could only restrict new uses of the substance. Thus, USEPA's influence was limited under the 1976 version of TSCA. This created much criticism of TSCA from environmental groups through the years that focused on the "after the fact focus" of TSCA in that it failed to protect individuals before substances were available in products. In addition, TSCA was difficult to implement because of the number of substances, totaling approximately 62,000 that were grandfathered and remained on the market. Last, fully assessing the risks posed by a new chemical was expensive and time consuming (Markell 2014).

To put things into perspective, as of 2014 the list of chemicals in use in the United States is greater than 84,000 and the number of chemicals that USEPA has what they consider to be fully evaluated numbers only 250 (USEPA 2014). Even though TSCA gives the authority to USEPA to test existing chemicals, USEPA has had difficulty in obtaining the data needed from industry to determine their risks and the cost to conduct the tests themselves has been evaluated as too costly (Markell 2014).

USEPA has been successful in restricting only nine chemicals in its history of over 40 years since it was promulgated in 1976 and none since 1984. Those substances currently restricted include (USEPA 2017o):

- Polychlorinated biphenyls (PCBs)
- Chlorofluorocarbons (CFCs)
- Dioxin
- Asbestos
- Hexavalent chromium
- Four nitrite compounds that are either:
  - Mixed mono and diamides of an organic acid
  - Triethanolanime salt of a substituted organic acid
  - Triethanolanime salt of tricarboxylic acid
  - Tricarboxylic acid

Even though many critics seem to believe that TSCA has failed in many respects, it still represents a potentially powerful tool for pollution prevention and can and should play a vital role in our society (Markell 2014).

In June 2016, TSCA was amended significantly and gave USEPA the authority and responsibility to proactively evaluate the risks of all chemical substances, which will be conducted in a multi-step process that is outlined as the following (National Law Review 2016):

- Developing a screening process to identify high-priority chemical substances
- Designating chemical substances as high or low priority
- Evaluating the risks of high priority chemicals
- Determining whether any chemical presents unreasonable risk
- Issuing rules to restrict the use of substances that present an unreasonable risk

The 2016 amendments define high priority substances as those chemicals that may present an unreasonable risk of injury to health or the environment because of a potential hazard and a potential route of exposure including an unreasonable risk to a potentially exposed or susceptible subpopulation that includes but is not limited to any of the following (National Law Review 2016):

- Infants
- Workers
- Children

- Elderly
- Pregnant women

Accordingly, USEPA has stated that its intent is to accord preference to those substances with persistence and bioaccumulative tendencies and known human carcinogens with high acute and chronic toxicity (USEPA 2017l). In the near term, USEPA has identified ten chemicals as high priority and are currently conducting risk evaluations. They include the following (USEPA 2017o):

• 1,4-Dioxane	• 1-Bromopropane
• Asbestos	• Carbon tetrachloride
• Cyclic Aliphatic Bromide Cluster	• Methylene chloride
• N-methylpyrolidone	• Pigment Wiolet 29
• Tetrachloroethylene	• Trichloroethylene

Analysis of the list above include four of substances that belong to the same chemical group called chlorinated or halogenated volatile organic compounds (CVOCs or HVOCs respectively), and are also known as chlorinated solvents. Those include carbon tetrachloride, methylene chloride, tetrachloroethylene, and trichloroethylene. In addition, 1,4-Dioxane is associated with the breakdown of tetrachloroethylene and trichlorethylene. Therefore, of the ten chemicals USEPA has initially placed in its high priority list, half belong or are associated with the same group of compounds, CVOCs or chlorinated solvents. We shall discuss this in much greater detail in the upcoming chapters.

6.6.10 RESOURCE CONSERVATION AND RECOVERY ACT

The Resource Conservation and Recovery Act (RCRA) was enacted in 1976 and is the principal law in the United States addressing land-based disposal of solid and hazardous waste (USEPA 2017p). It was enacted to address problems that the United States faced from its growing volume of municipal and industrial waste.

In very general terms, RCRA defines a waste as anything that is discarded (USEPA 2017p). RCRA recognizes that a solid waste may not be a solid. Under RCRA, many solid wastes are liquid, semi-solid, or contained gaseous material (USEPA 2017p). RCRA set goals for:

- Protecting human health and the environment from the hazards of waste disposal
- Energy conservation and natural resources
- Reducing the amount of waste generated through source reduction and recycling
- Ensuring the management of waste in an environmentally sound manner

RCRA also established standards for the treatment, storage, and disposal of hazardous waste in the United States. In the United States, RCRA is responsible for (USEPA 2017p):

- Managing approximately 2.5 billion tons of solid, industrial, and hazardous waste
- Overseeing 6,600 facilities with over 20,000 process units in the full permitting universe
- Working to address more than 3,700 existing contaminated facilities in need or cleanup
- Reviewing a possible 2,000 additional facilities that may need cleanup in the future
- Providing nearly $100 million in grant funding to assist states in implementing authorized waste programs
- Providing incentives and opportunities to reduce or avoid greenhouse gas emissions through material and land management practices

It is important to note and highlight some of RCRA's accomplishments nation-wide to understand how RCRA is currently viewed and what the future may hold. The major accomplishments of RCRA include (USEPA 2017p):

- Developing a comprehensive nation-wide system to manage waste from its point of generation to its final resting place, which is commonly referred to as "cradle to grave"
- Establishing the framework for states to implement effective municipal solid waste and non-hazardous waste management programs
- Preventing contamination from adversely impacting communities and becoming future Superfund sites
- Restoring 18 million acres of contaminated lands, nearly equal to the size of South Carolina
- Creating partnership and award programs to encourage companies to modify manufacturing practices to generate less waste and reuse materials safely
- Enhancing perceptions of wastes as valuable commodities that can be part of new products through USEPA's sustainable materials management efforts
- Creating and enhancing the nation's recycling infrastructure and increasing municipal solid waste recycling/composting rate from less than 7% to nearly 35%

RCRA divides those directly affected by the regulation into three categories that include (USEPA 2017p):

1. Generators. Generators must determine if they are creating any wastes that would be classified as hazardous. If so, they must obtain a tracking number and manifest, ensure proper storage and labeling of the waste, and keep records.
2. Transporters. Transporters must ensure that they comply with USEPA and USDOT requirements for transportation of hazardous materials (HAZMAT) and must ensure proper packaging, labeling, reporting, and record keeping.
3. Treatment, Storage, and Disposal Facilities (TSD). Treatment, Storage, and Disposal Facilities have by far the most complex and stringent requirements. They must obtain permits, which require inspections and monitoring, as well as comply with the manifest system.

There are three main components of RCRA called subtitles. The first describes and defines a non-hazardous waste, the second defines and describes a hazardous waste, and the third addresses petroleum underground storage tanks. In addition to the three subtitles, RCRA also includes methods for analytical testing of solid wastes under RCRA, which is called Test Methods for Evaluating Solid Waste (SW-846). This is significant because what the SW-846 provision has created is a consistent set of methods for fully characterizing any solid waste (USEPA 2017q).

The three main components of RCRA and SW-846 are described in greater detail in the following sections.

6.6.10.1   Subtitle D – Non-Hazardous Waste

Regulations under Subtitle D addresses non-hazardous waste, which essentially banned open dumping of waste and set minimum federal criteria for the operation of municipal waste and industrial waste landfills, including (USEPA 2017p):

- Design criteria
- Location restrictions
- Financial assurance
- Corrective action or cleanup
- Closure requirements

States play a lead role in implementing these regulations and may set more stringent requirements but cannot set less stringent requirements. States may not set less stringent requirements. USEPA approves state programs and if a state does not have an approved program, then the waste facilities must demonstrate that they meet the requirements.

RCRA defines a solid waste in broad terms and includes solids, sludges, liquids, semisolids, or contained gaseous material. In defining waste, USEPA focuses on the actual waste, not materials that were still part of the manufacturing process. USEPA intended to exclude recycling, but also wanted to prevent fraudulent recycling efforts. Thus, USEPA uses the following five-factor test to determine the definition of a waste (USEPA 2017p):

1. Whether the material is typically discarded on an industry-wide basis
2. Whether the material replaces raw material when it is recycled and the degree to which its composition is similar to that of the raw material
3. The relation of the recovery practice to the principal activity of the facility
4. If the material is handled prior to reclamation in a secure manner that minimizes loss and prevents releases to the environment
5. Other factors, such as the length of time the material is accumulated

Non-hazardous solid wastes include certain hazardous wastes which are exempted from the Subtitle C regulations, such as hazardous wastes from households and from conditionally exempt small quantity generators, which we will define later. Oil and gas exploration and production wastes, such as drilling, cutting, produced water, and drilling fluids are categorized as "special wastes" and are also exempt from Subtitle C. Subtitle D also includes garbage (e.g., food containers, coffee grounds), non-recycled household appliances, residue from incinerated automobile tires, refuse such as metal scrap, construction materials, and sludge from industrial and municipal wastewater facilities and drinking water treatment plants (USEPA 2017p).

6.6.10.2   Subtitle C – Hazardous Waste

Regulations under Subtitle C address hazardous waste and were enacted to ensure that hazardous waste is managed safely from the moment it is generated to its final disposal. This is termed "cradle-to-grave." Subtitle C regulations set criteria for hazardous waste generators, transporters, and treatment, storage and disposal facilities. This also includes permitting requirements, enforcement, and corrective action or cleanup. An important definition central to RCRA and its enforcement is what defines a hazardous waste. From a universal perspective, a hazardous waste is defined as a waste that could pose a threat to human or public health or the environment if it were to be released. From this basic definition, the potential list of hazardous wastes is large and diverse and indeed this is a true statement. Therefore, USEPA also has the following definition of hazardous waste (USEPA 2017p; 2019e):

> A solid waste, or combination of solid waste, which because of its quantity, concentration, or physical, chemical, or infectious characteristics may (a) cause, or significantly contribute to, an increase in mortality or an increase in serious irreversible, or incapacitating reversible, illness; or (b) pose a substantial present or potential hazard to human health or the environment when improperly treated, stored, transported, or disposed of, or otherwise managed.

This definition is broad as well. However, it does provide a general indication of which wastes USEPA intended to regulate as hazardous, but it obviously does not provide the clear scientific distinctions necessary for waste generators to determine whether their wastes pose a significant threat to warrant regulation or not in most cases. Therefore, USEPA developed more specific criteria for defining hazardous waste and decided on two definitions: a statutory definition and a regulatory definition. The statutory definition is presented above and serves as a general guideline in what would become the regulatory definition, which we discuss in much greater detail below. Hazardous wastes can be solids, liquids, and contained gases. They can be byproducts of manufacturing processes, discarded used materials, or discarded unused commercial products, such as residual cleaning fluids (solvents), cleaning products (bleach or ammonia), paints, pigments, pesticides or a multitude of other items (USEPA 2017p). RCRA divides the regulatory definition of a hazardous waste into three major categories (USEPA 2017p):

1. Characteristic waste
2. Listed waste
3. Universal waste

There are a couple other categories of hazardous wastes or wastes that are sometimes managed as hazardous and they include used oil and wastes that are considered hazardous because of what is called the "contained-in rule" or the "derived-from policy." In addition, there are what is called "Land Disposal Restrictions" (LDRs), which must be met for hazardous wastes. We will discuss each in the sections below along with manifesting hazardous wastes in its own section.

6.6.10.2.1 Characteristic Hazardous Waste

Under RCRA (USEPA 2017p), a characteristic waste are materials that are known or tested to exhibit one or more of the following four hazardous traits:

1. Ignitibility
2. Reactivity
3. Corrosivity
4. Toxicity

Ignitibility are wastes that can create fires under certain conditions, undergo spontaneous combustions, or have a flashpoint of less than 60°C (140°F). Examples include used oil and solvents. Test methods to determine whether a waste is ignitable and therefore a hazardous waste include the (1) Pensky-Martens closed-cup method, (2) the Staflash closed-cup method, and (3) ignitibility of solids, which are each described in SW-846 Sections 1010, 1020, and 1030, respectively (USEPA 2005a; 2017p).

Corrosivity are materials, including solids, that are acids or bases, or that produce acidic and alkaline solutions. Aqueous wastes with a pH of less than or equal to 2.0 or greater than or equal to 12.5 are corrosive and therefore hazardous wastes. A liquid may also be corrosive if it is able to corrode metal containers, such as storage tanks, drums, or barrels. Spent battery acid is an example. Test methods to determine whether a waste exhibits the characteristic of corrosivity are pH electronic measurement and corrosivity towards steel in SW-846 Methods 9040 and 1110, respectively (USEPA 2005a; 2017p).

Reactivity are wastes that are unstable under normal conditions. They can cause explosions or release toxic fumes, gases, vapors when heated, compressed, or mixed with water. Examples include lithium-sulfur batteries and unused explosives. Wastes are evaluated for reactivity using criteria set forth in the hazardous waste regulations in SW-846 (USEPA 2017r).

Toxicity are wastes that are harmful or fatal when ingested or absorbed (e.g., wastes containing mercury, lead, DDT, PCBs, etc.). When toxic wastes are disposed, the toxic constituents may leach from the waste and impact groundwater. The characteristic of toxicity contains eight subsections described below. A waste is a toxic hazardous waste if it is identified as being toxic by any of the eight subsections that include (USEPA 2017q):

1. TCLP. Toxic as defined through application of a laboratory test procedure called Toxicity Characteristic Leaching Procedure or TCLP test. The analytical method is described in USEPA SW-846 Method 1311. The TCLP identifies wastes (as hazardous) that may leach

hazardous concentration of toxic substances into the environment. The result of the TCLP test is compared to the regulatory level (RL) (USEPA 2017o). Table 6.10 lists some common chemical compounds with their RLs.

2. Total and WET. Toxic is defined through application of laboratory test procedures called the "total digestion" and the "Waste Extract Test," commonly referred to as the WET. The results of each of these laboratory tests are compared to their soluble threshold limit concentrations (STLCs) (USEPA 2017o).

3. Acute Oral Toxicity. Toxic because the waste either is an acutely toxic substance or contains an acutely toxic substance, if ingested. A waste identified as being toxic would have an acute oral LD50 less than 2,500 mg/kg or simply calculated LD50. We will define and explain LD50 when we discuss health risk assessments in a later chapter.

4. Acute dermal toxicity. Toxic because the waste either is an acutely toxic substance or contains an acutely toxic substance, if inhaled. A waste is identified as being toxic would have an dermal LC50 less than 4,300 mg/kg or simply calculated LC50. Again, we will define and explain LC50 when we discuss health risk assessments in a later chapter.

5. Acute inhalation toxicity. Toxic because the waste either is an acutely toxic substance or contains an acutely toxic substance, if inhaled. A waste is identified as being toxic if it has a dermal LC50 less than 10,000 mg/kg. Again, we will define and explain LC50 when we discuss health risk assessments in a later chapter. USEPA test method 3810 in

TABLE 6.10

Select Maximum Concentrations of Contaminants for Toxicity Characteristic

Contaminant	Hazardous Waste Code	Regulatory Level (RL) (mg/l)	Contaminant	Hazardous Waste Code	Regulatory Level (RL) (mg/l)
Arsenic (As)	D004	5.0	Barium (Ba)	D005	100.0
Benzene	D018	0.5	Cadmium (Cd)	D006	1.0
Carbon tetrachloride	D019	0.5	Chlordane	D020	0.03
Chlorobenzene	D021	100.0	Chloroform	D022	6.0
Chromium (Cr)	D007	5.0	o-Cresol	D024	200.0
m-Cresol	D024	200.0	p-Cresol	D025	200.0
Cresol (total)	D026	200.0	2,4-D	D016	10.0
1,4-Dichlorobenzene	D027	7.5	1,2-Dichloroethane	D028	0.5
1,-Dichloroethylene	D09	0.7	2,4-Dichlorotoluene	D030	0.13
Endrin	D012	0.02	Heptachlor	D031	0.008
Hexachlorobenzene	D032	0.13	Hexachlorobutadiene	D033	0.5
Hexachloroethane	Do34	3.0	Lead (Pb)	D008	5.0
Lindane	D013	0.4	Mercury (Hg)	D009	0.2
Methoxychlor	D014	10.0	Methyl ethyl ketone	D035	200.0
Nitrobenzene	D036	2.0	Pentachlorophenol	D037	100.0
Pyridine	D038	5.0	Selenium (Se)	D010	1.0
Silver (Ag)	D011	5.0	Tetrachloroethylene	D039	0.7
Toxaphene	D015	0.5	Trichloroethylene	D040	0.5
2,4,5-Trichlorophenol	D041	400.0	2,4,6-Trichlrophenol	D42	2.0
2,4,5-TP (Silvex)	D017	1.0	Vinyl chloride	D043	0.2

mg/l – milligram per liter

Source: United States Environmental Protection Agency (USEPA). Resource Conservation and Recovery Act Listed and Characteristic Hazardous Wastes. Office of Solid Waste. Washington, D.C. 2017s. www.epa.gov/hazarodus-wastes. (accessed March 26, 2017.)

SW-846 termed headspace is the recommended analytical test to evaluate inhalation toxicity.

6. Acute aquatic toxicity. Toxic because the waste is toxic to fish. A waste is aquatically toxic if it produces an LC50 less than 500 mg/L when tested using the static acute bioassay procedures for hazardous waste samples (USEPA 2017o).

7. Carcinogenicity. Toxic because it contains one or more carcinogenic substances. A waste is identified as being toxic if it contains any of the specified carcinogens at a concentration of greater than or equal to 0.001% by weight.

8. Experience or testing. A waste may be toxic, and therefore a hazardous waste, even if it is not identified as toxic by any of the seven criteria listed above. At the present time, only wastes containing ethylene glycol (e.g., spent antifreeze) have been identified as toxic in this subsection.

Let's look at a few examples of waste characterization.

Example 1. An auto repair shop uses mineral spirits as a parts washer solvent. The solvent does not contain any halogenated or listed solvents and its flashpoint is greater than 140°F. When the solvent becomes dirty, it is distilled. The distilled solvent is placed back into use. The residual solids left over from the distillation process are the waste that must be characterized. The waste was transferred to an appropriate container and labeled as "Parts Washer Waste, Analysis Pending." A representative sample of the waste is collected and sent to a certified laboratory under chain-of-custody for waste characterization. The laboratory conducted the required tests and a portion of the results are as follows:

- Lead   0.8 mg/L
- Cadmium   0.5 mg/L
- Chromium   8.0 mg/L
- Selenium   0.5 mg/L
- Silver   0.1 mg/L

Examination of the results presented above show that lead, cadmium, selenium, and silver do not exceed regulatory levels. However, chromium does exceed the regulatory level. Therefore, the waste would be classified as hazardous for chromium D007.

Example 2. Auto body shop. The exhaust filters in the paint spray booth became saturated with overspray from the painting operation. Since the body shop uses many different types of paint and primers, it's difficult to determine whether the filters are hazardous or not without laboratory testing. A representative filter is removed and sampled. The remaining filters are placed into a container and are labeled "Waste Filters Pending Analysis." The sample is sent to a certified laboratory under chain of custody for analysis. The laboratory conducted the required tests and a portion of the results are as follows:

- Lead   9.1 mg/L
- Chromium   0.4 mg/L
- Barium   0.8 mg/L
- Methyl ethyl ketone   10 mg/L

Only lead exceeded the regulatory limit and the filters were characterized as a hazardous waste. The container label was then immediately changed to a hazardous waste label.

Example 3. General manufacturing facility. The facility receives large steel components which the facility re-manufactures. The process requires that the components be dismantled, and the surfaces are prepared for re-finishing. The metal components are placed into a sand blasting machine and are cleaned under high pressure blasting sand. After a few weeks of use, the blasting media (sand) is no longer effective and must be replaced. A representative sample of the sand is sampled and the used sand from the operation is placed in a metal 55-gallon drum and labeled "Waste Sand from Sand Blasting – Analysis Pending." The sample collected for waste characterization is sent to a certified laboratory under chain of custody for analysis. The laboratory conducted the required tests and a portion of the results are as follows:

- Arsenic   0.5 mg/L
- Cadmium   3.5 mg/L
- Chromium   25 mg/L
- Lead   40 mg/L

The results of the analysis indicated that the sand waste is indeed a hazardous waste for cadmium, chromium, and lead. The drum was immediately labeled with an appropriate hazardous waste label for D006, D007, and D008 hazardous waste.

6.6.10.2.2   Listed Hazardous Waste

Under RCRA (USEPA 2017m), a hazardous waste can also be what is called a "Listed Waste." A listed waste is a waste that appears on one of four RCRA hazardous waste lists described as (USEPA 2017s):

1. **F-Listed Wastes**. F-listed wastes identify wastes from many common manufacturing and industrial processes, such as solvents that have been used for cleaning or degreasing. Since the processes producing these wastes occur in many different industry sectors, the F-listed wastes are known as wastes from non-specific sources. Non-specific means that the waste does not originate from one specific industry or one specific industrial or manufacturing process.
2. **K-Listed Wastes**. K-listed wastes include certain wastes from specific industries, such as petroleum refining or pesticide manufacturing. Also, certain sludges and wastewaters from treatment and production in these specific industries are examples of source-specific wastes.
3. **P-Listed Wastes**. P-listed wastes include specific commercial chemical products that have not been used, but that will be (or have been) discarded. Industrial chemicals, pesticides, and pharmaceuticals are examples of commercial chemical products that appear on the P-list and become hazardous waste when discarded.
4. **M-Listed Wastes**. M-listed wastes include certain wastes known to contain mercury, such as fluorescent lamps, mercury switches, and the products that house mercury switches, and mercury-containing novelties. These types of wastes may also be included as types of universal wastes, which we will discuss later in this chapter. Therefore, clarification from the applicable regulatory authority may be required.

Now that a waste has been characterized as hazardous it must be disposed of in a timely fashion. Typically, waste containers such as a 55-gallon drum that are full and have been characterized as hazardous must be affixed with an appropriate label that identifies it as hazardous. Figure 6.9 is an example of an appropriate label. For some wastes, it may take a few weeks or even longer before enough waste is accumulated to fill a container. The maximum accumulation time allowed ranges between 90 and 270 and is dependent upon (USEPA 2017s):

- How much waste is accumulated in any month
- The generator status of the facility (i.e., small quantity generator, large quantity generator, etc.)
- Whether the waste is acutely hazardous or extremely hazardous

FIGURE 6.9 Hazardous waste label. (From United States Environmental Protection Agency (USEPA). Resource Conservation and Recovery Act Listed and Characteristic Hazardous Wastes. Office of Solid Waste. Washington, D.C. 2017s. www.epa.gov/hazarodus-wastes.) (Accessed March 26, 2017.)

Typically, the "clock" begins on the first date on which any amount of hazardous waste begins to accumulate during that month (USEPA 2017s).

A universal waste is a category of waste materials designated as hazardous waste but containing materials that are very common. Universal wastes are widely produced by households and businesses alike and typically include the following (USEPA 2017s):

- Televisions
- Computers
- Cell phones
- Other electronic devices such as VCRs, portable DVD players, etc.
- Fluorescent tubes and bulbs, high-intensity discharge lamps, sodium vapor lamps, and electric lamps that contain added mercury, as well as other lamps that exhibit a characteristic of a hazardous waste (e.g., typically lead)
- Pesticides
- Mercury-containing equipment including thermostats, switches, mercury thermometers, pressure and vacuum gauges, dilators and weighted tubing, mercury gas flow regulators, dental amalgams, counterweights, dampers, mercury-added novelties such as jewelry, ornaments, and footwear. Of special note of clarification is that mercury-containing equipment may also be considered a listed M-waste. In these cases, clarification from the applicable regulatory authority may be required.
- Cathode ray tubes (CRTs), which are typically the glass picture tubes removed from devices such as televisions and computer monitors
- Non-empty aerosol cans

6.6.10.2.3   Used Oil

Some states, such as California, regulate used oil as a hazardous waste even if they do not exhibit any characteristic of a hazardous waste. The term **used oil** is a defined term meaning any oil that has been refined from crude oil, or any synthetic oil that has been used and, as a result of use, is contaminated with physical or chemical impurities. Other materials that contain or are contaminated with used oil are also commonly considered a hazardous waste as well and are subject to used oil regulations under RCRA 40 CFR Part 279.22 (USEPA 2017t).

In general, used oil must be kept in approved containers with secondary containment that does not leak and must be properly labeled as "used oil" (USEPA 2017t).

6.6.10.2.4   Contained-In Policy and Derived-From Rule

Materials can be hazardous even if they are not specifically listed or don't exhibit any characteristic of a hazardous waste. This is often confusing so let's try and make sense of a couple of hazardous waste concepts to understand if a waste is classified as a hazardous waste in a step-by-step approach. For example, let's discuss what is considered "the contained-in rule" (USEPA 2017p).

The contained-in rule or contained-in policy under RCRA empower hazardous waste with the power to transform solids, otherwise not hazardous, into a hazardous waste. USEPA separates the contained-in rule from the mixture or derived-from rule as separate and distinct. This power to render a solid waste into a hazardous waste stems from USEPA's general policy that "once a hazardous waste always a hazardous waste." Once a hazardous waste is hazardous, it is presumed to be forever hazardous regardless of changes in its form or its combination with other substances. This rule provides that any mixture of a listed hazardous waste and a solid waste is itself, in totality, a hazardous waste regardless of the concentration of the hazardous constituents in the waste even if it no longer exhibits any toxicity (USEPA 2017p).

The derived-in rule differs slightly in concept from the contained-in rule but the outcome is the same. To best explain the derived-from rule is through an example. Let's say ash is created through the incineration of a listed hazardous waste. Since the ash is created from a listed hazardous waste, the ash itself is considered a listed hazardous waste. Thus, the ash produced by burning the listed hazardous waste bears the same hazardous waste code and regulatory status as the original listed waste, regardless of the ash's actual physical and chemical properties (USEPA 2017p).

USEPA's basic position on termination of a material's status as a listed hazardous waste is simple. Once a listed hazardous waste has been classified as a hazardous waste it generally remains a hazardous waste forever. There are two explicit bases for termination of this status. First, an unlisted waste that is hazardous only because of a characteristic ceases to be a hazardous waste if it no longer exhibits a hazardous waste characteristic. Second, a listed waste which is a waste containing a listed waste, or a waste derived from a listed waste, ceases to be a hazardous waste only if it has been "delisted" from classification. Although a characteristic waste can lose its status as a hazardous waste without action by USEPA, in most cases a listed waste will remain hazardous until USEPA affirmatively grants a petition to reclassify the material as non-hazardous (USEPA 2017p). Some states, such as California, have a process by which a listed hazardous waste that impacts a media such as soil can be delisted if it is demonstrated that the listed waste is present in insignificant concentrations based on a risk-based evaluation (California Department of Toxic Substance and Control [DTSC] 2017).

By adopting the mixture and derived-from rules, USEPA established incentives for generators to keep waste streams separate and to limit the amount of hazardous waste generated since the cost for treatment and disposal are much higher for a hazardous waste compared to a non-hazardous waste (2017p).

6.6.10.2.5   Land Disposal Restrictions

The 1984 amendments to RCRA, called the Hazardous and Solid Waste Amendments (HSWA), required USEPA to establish treatment standards termed **land disposal restrictions** (LDRs) for all listed and characteristic hazardous wastes destined for land disposal. For wastes that are restricted, the amendments to RCRA required USEPA to set concentration levels or methods of treatment, both of which are called treatment standards, that substantially diminish the toxicity of the wastes or reduce the likelihood that hazardous constituents contained within the wastes will migrate from the disposal site and potentially impact human health or the environment in an adverse way (USEPA 1991a). The major reason for established LDRs was for the prevention of groundwater impacts, usually through the leaching of contaminants from hazardous wastes from landfills and potentially impacting potable water supplies.

LDR provisions focus on eliminating or greatly reducing the probability of impacting groundwater through proper treatment of hazardous or toxic constituents in hazardous waste before land disposal. The processes may include (USEPA 1991a):

- Incineration, which destroys organic hazardous compounds
- Stabilization or immobilization, which binds toxic metals to the mass itself making it less likely to migrate. An example would be mixing the toxic material with concrete, mortar, a bentonite clay, or other material to render the waste immobile.

There are three main sections of the LDR program and they include (USEPA 1991a):

1. Disposal prohibition. This requires waste-specific treatment standards be met before disposal of the waste on land. A facility can meet such standards by either:
   a. Treating the hazardous chemicals in the waste to meet appropriate levels. Many such methods are approved by USEPA for use, excluding dilution.
   b. Treating hazardous waste with a USEPA approved technology. Once the waste is appropriately treated, it can then be land disposed.
2. Dilution prohibition. Ensures that wastes are treated properly and that diluting the waste to meet LDRs is not permitted. This is because of the assumption that dilution does not reduce the overall toxicity of a hazardous waste.
3. Store prohibition. Hazardous waste cannot be stored indefinitely.

From the moment a hazardous waste is generated, it is subject to LDRs. If a hazardous waste generator produces more than 220 pounds (or 2.2 pounds of acute hazardous waste) of hazardous waste in a calendar month, they must identify the type and nature of the waste, and also determine the course of applicable treatment of the waste before land disposal. Generators that produce less than these amounts and are conditionally exempt small quantity generators (CESQG) of hazardous waste are exempt from LDR requirements, as are waste pesticides and container residues which are disposed of by farmers on their own land, and newly listed wastes which USEPA has not yet established treatment standards (USEPA 2017s; Ohio EPA 2014).

6.6.10.2.6 Analytical Methods

Much of what defines a hazardous waste is dependent upon what chemical compounds are contained within the waste. Therefore, for consistency, USEPA developed a manual describing acceptable analytical sampling methods and quality control for assuring that wastes were properly characterized. The document USEPA published was entitled Test Methods for Evaluating Solid Waste SW-846 or simply SW-846. It was issued in 1980 and has changed through time as technology is advanced in analytical procedures. Currently the SW-846 manual is over 3,500 pages (USEPA 2017r). SW-846 has become one of the most important documents for not only assessing waste but also in conducting investigations at potentially hazardous waste sites and in conducting environmental investigations in general.

6.6.10.2.7 Hazardous Waste Manifests

RCRA requires that generators of hazardous waste and owners or operators of hazardous waste treatment, storage, or disposal facilities use what is called the Uniform Hazardous Waste Manifest (USEPA Form 8700-22), and if necessary the continuation sheet (EPA Form 8700-22A) for both interstate and intrastate transportation. General instructions for completing a manifest are provided at www.epa.gov/hwgenerators/uniform-hazardous-waste-manifest-instructions-sample-form-and-continuation-sheet.

The return receipt of the manifest from the disposal facility must be returned to the generator within 45 days from when the waste was signed and shipped from the generator facility (USEPA 2017p; 2019f).

6.6.10.3 Subtitle I – Petroleum Underground Storage Tanks

The operation of underground storage tanks (USTs) became subject to RCRA with the enactment of the Hazardous and Solid Waste Amendments of 1984, which amended RCRA by inserting Subtitle I. It was estimated at the time that there were approximately 2 million USTs in the United States. Currently, the estimate is 560,000. The UST regulations under Subtitle I cover tanks storing petroleum or a listed hazardous waste, and define the types of tanks permitted. The UST requirements under Subtitle I set standards for (USEPA 2017p):

- Groundwater monitoring
- Double-walled tanks
- Release detection, prevention, and correction
- Spill control
- Overfill control

Up until the middle portion of the 1980s, USTs were largely unregulated, except for those that contained hazardous wastes. It was not until the catastrophic failure of USTs and large-scale groundwater contamination occurred in Provincetown, Massachusetts; Northglenn, Colorado; and Dover-Walpole, Massachusetts did Congress recognize the need for UST regulation. Subtitle I imposed new obligations and potential liabilities on a large and diverse number of entities and businesses including (USEPA 2017p):

• Service stations	• Petroleum bulk tanks
• Petroleum bulk plants	• Municipal warehouses
• Shopping centers	• School bus yards
• Farms	• Factories
• Automobile dealerships	• Delivery services

The UST regulations apply to owners and operators of regulated UST systems. An **operator** is defined as any person in control of, or having responsibility for, the daily operation of a UST system. The definition includes current operators only, and implies that one must actively manage a tank system to be considered an operator (USEPA 2017p).

An **owner** is defined as any person who owns a UST system, unless that system was taken out of service prior to November 8, 1984. For UST systems no longer in use at that date, the owner is the person who owned the system immediately prior to is discontinuation. A person is defined broadly to include an individual, trust, firm, joint-stock company, federal agency, corporation, state, municipality, commission, state political subdivision, any interstate body, or the government of the United States (USEPA 2017p).

The UST regulations impose joint and several liability on UST owners and operators and do not place primary responsibility on one over the other (USEPA 2017p).

When USTs are currently being operated, it is not difficult to identify at least one person responsible for compliance with the UST regulations. By definition, there is no operator of an abandoned or out-of-service UST system. If a UST was in use after November 8, 1984, and the UST remains in the ground, the current property owner may fall within the statutory definition as an owner. If the UST was abandoned before November 8, 1984, by a previous property owner, it may be impossible to locate records identifying the owner of the UST immediately before discontinuation of use because UST registration was not historically required and UST closures were not reported because that was not required (USEPA 2017p).

In amending Subtitle I of RCRA, USEPA required every owner of a UST system to notify the designated state or local agency of the existence of the UST system, age, size, type, locations, product stored, and uses for each UST. Initial notification forms were due by May 8, 1986. In addition to the reporting requirement for operating USTs, owners of USTs taken out of service after January 1974 but still in the ground were required to notify a designated state agency of the existence of the UST by May 8, 1986, as well. This notice also required a UST owner to specify, to the extent known, the date that the UST was taken out of service, the age of the UST when it was taken out of service, size, type, product stored, location, and the type and quantity of substances that remained stored in the UST (USEPA 2017p).

USTs brought into service after December 22, 1988, must meet design and construction requirements. RCRA specified that owners or operators notify the designated state agency within 30 days of the existence of the new UST. The notification form for new USTs requires the owners or operators to certify compliance to the following requirements (USEPA 2017p):

- That the UST system was properly installed by a state-certified UST contractor
- That cathodic protection was installed for steel tanks and associated piping
- That financial responsibility requirement was met
- That release detection was installed and meets regulatory standards
- That spill and overfill prevention measures were installed and meet regulatory standards

6.6.10.4 Summary of the Resource Conservation and Recovery Act

RCRA, along with the CAA, SWA, and the SDWA, form the basic fundamentals of environmental protection in the United States with the exception of cleaning up legacy sites, which we will address next. RCRA was created essentially to deal with huge increases in the amount of wastes that were being created throughout the United States as a direct result of our industrialization and population increase. The success of RCRA can't be understated. RCRA has changed our behavior in many ways and has involved everyone in participating in taking care of our environment. Almost all of us now recycle much of our discarded materials that were once considered garbage. These materials are now transformed into new products that we use and includes used tires, wood and pulp products, plastic, glass, and especially scrap metal, which is now considered a commodity and has grown into an industry of itself.

Looking back, there are some challenges that RCRA has failed to address proactively, and one of the most significant is municipal or household wastes. We have discussed at length what industry faces in characterizing and disposing of wastes properly in order to protect human health and the environment, but have not addressed with the same vigor our everyday household garbage, which is generally exempted from RCRA but contains many of the same components and chemicals as many hazardous wastes but is not regulated.

Figure 6.10 shows some common household products that typically become hazardous wastes if they were generated by industry as opposed to households. This to many scientists does not make sense at all when a chemical which is known to be harmful to human health and the environment and is a hazardous waste but is not fully regulated as such. The Earth does not care how or by whom the material is released into the environment, it's toxic to the Earth just the same.

Next we will discuss the Comprehensive Environmental Response Compensation and Liability Act, which focuses on addressing cleaning up contaminated sites.

6.6.11 Comprehensive Environmental Response, Compensation, and Liability Act

The Comprehensive Environmental Response, Compensation, and Liability Act (CERCLA) of 1980 is commonly referred to as Superfund or Polluters Pay Law. CERCLA authorized USEPA to recover natural resource damages caused by hazardous substances released into the environment. Created out of CERCLA was the Toxic Substance and Disease Registry (ATSDR) (USEPA 2019g).

FIGURE 6.10 Common household chemicals. (From Kaufman, M.M., Rogers, D.T. and Murray, K.S. 2011. Urban Watersheds: Geology, Contamination, and Sustainability. CRC Press. New York. New York. 583 pages.)

6.6.11.1 Potential Responsible Parties

USEPA may identify responsible parties for contamination from releases of hazardous substances to the environment (polluters) and either compel them to investigate and cleanup sites, or it may undertake the investigation and cleanup themselves using the Superfund (a trust fund) and recover all response costs and even more in some instances from polluters by referring to the Department of Justice for cost recovery proceedings. As of 2016, there were 1,328 sites listed, 391 sites delisted, and 55 new sites proposed for a total of 1,774 Superfund sites. Figure 6.11 is a map showing the distribution of CERCLA sites in the United States (USEPA 2019g).

FIGURE 6.11 CERCLA or Superfund sites in the United States. (From United States Environmental Protection Agency (USEPA). CERCLA Sites in the United States. 2017u. www.epa.gov/CERCLA-sites.) (Accessed March 26, 2017.)

Approximately 70% of CERCLA investigation and cleanup activities have been paid for by responsible parties (PRPs). The exceptions occur when the responsible party cannot be found or is unable to pay for the cleanup. The majority of the funding came from a tax on the petroleum and chemical industries, but since 2001, most of the funding now comes from taxpayers (USEPA 2019g).

USEPA's enforcement program under CERCLA is based on the "**polluters pay principle**," which stipulates that the party responsible for the pollution pays for cleaning up the pollution. Over the past 35 years, USEPA has recovered over US$35 billion from potentially responsible parties for cleanup at CERCLA sites. This principle is one of many highlights of USEPA environmental protection that has been copied world-wide (USEPA 2019g).

6.6.11.2   CERCLA History

CERCLA was enacted in 1980 in response to threats from sites of environmental contamination, including three sites in particular: Times Beach in Missouri, Love Canal in New York, and the Valley of Drums in Kentucky. As part of CERCLA, USEPA established a hazardous ranking system (HRS) to prioritize each Superfund site related to its threat to human health and the environment. In 1982, USEPA published its first list of Superfund sites and titled it the national priority list (NPL).

Other prominent sites that have shaped not only Superfund, but have had significant contributions to environmental science include:

- Toms Rivers (also referred to as Ciba-Geigy) and the Reich Farm Superfund site in New Jersey
- Wells G&H Superfund site in Woburn, Massachusetts
- PG&E Hinckley site in California

A brief discussion of each of the six sites mentioned above are covered in the following sections below.

6.6.11.2.1   Times Beach

Times Beach is now a ghost town located approximately 17 miles southwest of St. Louis, Missouri. Times Beach was a town of more than 2,000 people before they were all evacuated in 1983 because of dioxin contamination. During the 1960s, a pharmaceutical and chemical company was located in Times Beach and produced some thick oily wastes that contained dioxin. Approximately 18,500 gallons of the dioxin-containing oily waste ended up with a waste oil contractor who mixed the material with other sources of waste oil and sprayed the material at three local horse arenas and 23 miles of roads in Times Beach as a dust suppressant (USEPA 2017v).

The Centers for Disease Control (CDC) initially became involved in August 1971 when local citizens reported the deaths of numerous horses that used one or more of the horse arenas sprayed with the dust suppressant. The CDC discovered an old tank containing approximately 4,300 gallons of oily waste that contained dioxin at a concentration of 340 ppm. In 1974, the CDC collected soil samples from the area and detected trichlorophenol, PCBs, and dioxin. USEPA did not become heavily involved until 1979 when they discovered 90 drums of oily waste at the site with dioxin concentrations as high as 2,000 ppm. In May and June 1982, USEPA collected soil samples from the area and detected dioxin as high as 0.3 ppm in soil. The CDC recommended cleanup level for dioxin at the time was as low as 0.001 ppm. On December 23, 1982, the CDC publicly recommended that Times Beach not be inhabited. Subsequently, 800 residential properties and 30 businesses were abandoned. In 1997, the cleanup was completed at a cost of nearly $200 million (Sun 1983; USEPA 2017v).

6.6.11.2.2   Love Canal

Love Canal is a neighborhood within Niagara Falls, New York. The site that became the epicenter of contamination was a 29-acre parcel that was used as a landfill. Originally planned as a canal, the site was abandoned at the turn of the 20th century and was used as a dump by the city of Niagara Falls. In 1942, the Niagara Power and Development Company granted permission to Hooker Chemical to dump wastes in the canal. In 1948, the city of Niagara ended disposal of refuse into the canal and the Hooker Chemical Company became the sole user and owner of the site. Hooker Chemical ceased dumping in 1952 and capped the dump with clay. During its lifespan of 5 years, an estimated total of 21,800 tons of chemicals mostly consisting of sludges, caustics, alkalines, chlorinated hydrocarbons, and volatile organic hydrocarbons were disposed in the landfill, many in 55-gallon drums (Beck 1979; USEPA 2017w).

By the early 1950s, the city of Niagara Falls was in the midst of a population increase and purchased the site from Hooker Chemical to build a school. In 1955, 99th Street School opened with 400 children enrolled. The area surrounding the dump continued to develop in the 1950s with a total of 800 private residences built in the surrounding area. Soon after development, residences began to complain of odors and black substances in their yards and public playgrounds (USEPA 2017w).

In 1977, the New York State Department of Health and Environmental Conservation conducted a large scientific study of the air, groundwater, and soil in the area and detected a number of organic compounds in the basements of several residences adjacent to the former dump. Compounds detected included benzene, chlorinated hydrocarbons, PCBs, and dioxin (Beck 1979; USEPA 2017s). By 1978, Love Canal became a national media event and on August 7, 1978, then President Jimmy Carter announced that Love Canal was a federal health emergency. In total 950 families were evacuated. This was the first time that emergency funds were used for a situation that was not a natural disaster. When CERCLA was passed, Love Canal was placed first on the list of Superfund sites. USEPA finished the cleanup of Love Canal in 2004 at an estimated cost of $400 million (USEPA 2004, 2017w).

6.6.11.2.3 Valley of Drums

The Valley of Drums Site is a 23-acre site in northern Kentucky near Louisville. The site became a collection site for wastes dating back into the 1960s. It initially caught the attention of local government officials when some drums that were onsite caught fire and took firefighters more than a week to extinguish. However, no action was taken at the time since no laws existed at the time to take any further action. In 1978, a Kentucky Department of Natural Resources and Environmental Protection (KDNREP) investigation of the property estimated that over 100,000 drums of waste were located onsite and that an estimated 27,000 drums were buried and the remaining containers were discharged into onsite pits and trenches (USEPA 2017x). Figure 6.12 shows some of the drums discovered.

FIGURE 6.12 Photograph of Valley of Drums Superfund site, circa 1979. (From United States Environmental Protection Agency (USEPA). Superfund Site Profile: Valley of Drums, Kentucky. 2017x. www.epa.gov/valley-of-drums-profile.) (Accessed March 29, 2017.)

The KDNREP investigation revealed that many of the drums onsite were deteriorated and their contents spilled onto the ground and were washed into a nearby creek. Analysis of collected samples detected the presence of several heavy metals, PCBs, and 140 other chemical substances. USEPA initiated an emergency cleanup in 1979 but quickly realized that the magnitude of the project was beyond what they were prepared for and that was used by members of Congress as an additional reason why CERCLA was needed (USEPA 2017x).

The Valley of Drums site was placed on the Superfund List in 1983 when a large-scale cleanup of the site began and ended in 1990. In 1996, the site was removed from the Superfund List. However, more buried drums were discovered during an inspection of the site in 2008 outside the cleanup area. Currently, USEPA is monitoring the site and has authorized an additional $400,000 in funds to remove and properly dispose of more drums discovered buried drums (USEPA 2017x).

6.6.11.2.4   Wells G&H Superfund Site in Woburn, Massachusetts

Groundwater extraction Wells G&H were developed by the City of Woburn, Massachusetts in 1964 and 1967, respectively. The wells were capable of pumping 2 million gallons of water per day and were initially intended to supplement an existing water system that supplied water to the residents of Woburn. In 1979, the Massachusetts Department of Environmental Protection (MDEP) conducted analytical testing of water from the two wells and discovered the presence of several chlorinated volatile organic compounds in the samples that included (USEPA 1989a):

- 1,1,1-trichloroethane (TCA)
- Trans-1,2-dichloroethene (trans DCE)
- Tetrachloroethene (PCE)
- Trichloroethene (TCE)

Concentrations ranged from 1 to 400 ppb. MDEP immediately shut the wells down. In 1981, USEPA conducted a hydrogeologic investigation of a 10 square mile area surrounding the wells to evaluate the source of the contamination and to define the extent of groundwater contamination. Five facilities were identified as sources of contamination and the site was placed on the Superfund List in December 1982 (USEPA 1989a). Remedial response actions were first initiated by USEPA in 1983 and are still being conducted more than 35 years later. USEPA has been maintaining close community relations since 1986. Community involvement has been significant and was even the subject of a movie titled "A Civil Action."

6.6.11.2.5   Ciba-Geigy and Reich Farm Site in Toms River, New Jersey

Ciba-Geigy Superfund site is located in Toms River, New Jersey. The site is approximately 1,400 acres in size, of which 320 acres were developed and were formerly used to manufacture synthetic organic pigments. From 1952 to 1866, the facility discharged treated effluent into Toms River, which then discharged directly into the Atlantic Ocean. Beginning in 1966, the facility discharged effluent directly into the Atlantic Ocean through a pipeline. In addition, the facility disposed of solid wastes onsite at several locations (USEPA 1989b).

During the 1970s and early 1980s, the facility conducted several investigations at the direction of the New Jersey Department of Environmental Protection (NJDEP). USEPA eventually listed the site as a Superfund site in 1982. In 1984, NJDEP discovered that the facility was illegally disposing of drums containing hazardous waste onsite, which led to a criminal investigation. As a result of the criminal investigation, several company officials were charged with illegal disposal of chemical waste, filing false documents, making false and misleading statements, and illegal disposal of hazardous waste (1989b).

In 1985, USEPA began conducting a remedial investigation of the site to fully define the extent of contamination in soil and groundwater, both onsite and offsite. The investigation was completed in 1988 (USEPA 1989b). Remedial activities included excavation of drums, contaminated soil, and groundwater treatment that continues to this day more than 30 years later.

The Reich Farm Superfund Site in Toms River, New Jersey, is in close proximity to the Ciba-Geigy Superfund Site and was designated a Superfund site in 1983. According to USEPA (1988), a 3-acre portion of the farm was leased for temporary storage of used 55-gallon drums in 1971. However, it was subsequently discovered that wastes from the drums may have been dumped into trenches and that this activity contaminated groundwater (USEPA 1988).

Compounds detected in groundwater included (USEPA 1988):

- Styrene-Acrylonitrile Trimer
- Tetrachloroethene (PCE)
- Trichloroethene (TCE)
- 1,1,1-trichloroethane (TCA)
- Chlorobenzene
- 1,2-dichloroethene (DCE)
- Other volatile organic compounds

Contaminants listed above and many others not listed contaminated groundwater, which residents of Toms River used as a source of drinking water. This led to a significant toxicological risk evaluation that is still ongoing (USEPA 2017y).

6.6.11.2.6   PG&E Hinkley Site in California

The PG&E site was located 2 miles from the town of Hinkley, California, and 12 miles west of Barstow, California. Between 1952 and 1966, PG&E used hexavalent chromium as corrosion protection in its cooling tower water. The wastewater from this process was discharged to unlined ponds onsite which subsequently contaminated groundwater. The resulting groundwater contaminant plume of hexavalent chromium migrated from the site and was estimated to be 8 miles long and 2 miles wide and impacted drinking water in Hinkley (California Regional Water Quality Control Board 2018). In 1993, citizens of Hinkley filed suit against PG&E and settled in 1996 for $333 million (San Francisco Chronicle 1996). This site was also the subject of a movie titled *Erin Brockovich* (2000). Remediation at the site is ongoing.

Although this site is not actually a federal Superfund site, it does follow a similar profile to the five other sites briefly described above. Similarities include a significant toxicological risk study, large groundwater contaminant plume, and a high remedial cost.

6.6.11.2.7   Summary of CERCLA History

The six sites highlighted above have been often referred to as the sites that influenced the formation and passage of CERCLA or have demonstrated a lack of knowledge of how contaminants behave in the subsurface geological environment. In addition, four of the six sites involve agriculture or agricultural land.

These examples have provided scientific information that have highlighted:

- The need for groundwater protection zones
- The need, development, and procedures for conducting toxicological risk evaluations
- The development of groundwater modeling
- The difficulty in remediating contaminated groundwater
- The enormous cost of remediating contaminated groundwater
- The toxicity, mobility, and persistence of chlorinated VOCs
- The toxicity, mobility, and persistence of hexavalent chromium

6.6.11.3   Site Scoring

USEPA developed the hazard ranking system (HRS) to score potential sites of environmental contamination. A score of 28.5 was usually high enough to be placed on the NPL or Superfund List.

The HRS is a numerical process whereby it assigns integers to factors that relate to risk based on site conditions. The factors are grouped into three categories (USEPA 2019g):

1. Likelihood that a site has released or has the potential to release hazardous substances into the environment
2. Characteristics of the waste, including toxicity, quantity, etc.
3. Potential human exposure pathways

There are four pathways for the potential human exposure category that include (USEPA 2019g):

1. Surface water migration (i.e., drinking water sources, sensitive habitat such as a wetland, human food chain, etc.)
2. Soil exposure (i.e., dermal contact, dust, etc.)
3. Groundwater migration (i.e., drinking water)
4. Air migration

After scores are calculated for one or all the pathways, they are combined using a root-mean-square equation to determine overall site score. It should be noted that some sites score high for just one pathway if the potential exposure is severe enough and the toxicity of the contaminant or contaminants of concern are also high.

The identification of a site on the NPL is intended primarily to guide USEPA in the following (USEPA 2019g):

- Determining which sites warrant further investigation to assess the nature and extent of the risks to human health and the environment
- Identifying what CERCLA-financed remedial actions may be appropriate
- Notifying the public of sites which USEPA believes warrant further investigation
- Notifying PRPs that USEPA may initiate a CERCLA-financed remedial action

A potential responsible party (PRP) is a possible polluter who may eventually be liable under CERCLA for contamination or misuse of a particular property or resource. Four classes of PRPs may be liable for contamination at a Superfund site and they include (USEPA 2019g):

1. The current owner or operator of the site
2. The owner or operator of a site at the time that disposal of a hazardous substance, pollutant, or contaminant occurred
3. A person who arranged for the disposal of a hazardous substance, pollutant, or contaminant at a site
4. A person who transported a hazardous pollutant or contaminant to a site, who also has selected that site for the disposal of the hazardous substances, pollutants, or contaminants

Inclusion of a site on the Superfund List does not itself require PRPs to initiate action to clean up a site, nor does it assign liability to any one entity or person. The Superfund List serves primarily for informational purposes, notifying the government and the public of those sites or releases that appear to warrant a remedial action. Under Superfund, the key difference in the authority to address hazardous substances and pollutants or contaminants is that the cleanup of pollutants or contaminants, which are not hazardous substances, cannot be compelled by unilateral administrative order (USEPA 2019g).

An important provision of CERCLA is that if a PRP spends money to clean up a CERCLA site, that party may sue other PRPs in a contribution action. CERCLA liability has generally been judicially established as joint and several among PRPs for cleanup costs in that each PRP is hypothetically responsible for all costs, but CERCLA liability is allocable among PRPs in contribution based on comparative fault. An "orphan share" is the share of costs at a Superfund site that is attributed to a PRP that is unidentifiable or insolvent. USEPA as a matter of long-standing policy attempts to treat all PRPs equitable and fairly (USEPA 2019g).

6.6.11.4 Remediation Criteria

Most states have passed their own versions of CERCLA so that sites that do not qualify to become a Superfund site but have had a release of hazardous substances above cleanup levels get addressed. Cleanup levels established at the state level vary from state to state depending on various factors. However, much is based on USEPA exposure scenarios and toxicological data. States have established cleanup levels for different types of land use, typically residential and non-residential (industrial or commercial) and by media or exposure pathway including:

- Groundwater or drinking water
- Soil
- Groundwater – surface water interface
- Inhalation to indoor air
- Inhalation to outdoor or ambient air

Table 6.11 lists the generic residential cleanup levels for select chemical compounds for the state of Michigan (MDEQ 2016). The exposure pathways listed in the table are residential direct contact, commercial-industrial direct contact, and construction worker direct contact. The complete list can be viewed at www.michigan.gov/documents/deq/deq-rrd-chem-CleanupCriteriaTSD_527410_7.pdf (MDEQ 2016). The values listed under the groundwater-surface water interface in Table 6.11 are values within a zone where groundwater becomes surface water before mixing with surface water or exposure to the atmosphere. The values listed for air are presented in micrograms per kilogram in soil immediately beneath a building for indoor air and immediately beneath the groundwater surface if an obstruction is not present (MDEQ 2016).

The values listed for indoor and ambient air should be considered screening values since they are not actual air samples collected from a typical breathing zone. In Michigan, groundwater is defined as any water encountered beneath the surface of the ground regardless of quantity or if it occurs in a naturally occurring formation.

As a comparison, Table 6.12 presents soil cleanup criteria for New Jersey (New Jersey Department of Environmental Protection [NJDEP] 2008; 2015; 2017), Illinois (Illinois Environmental Protection Agency [IEPA] 2017; California Department of Toxic Substances [DTSC] Control Chemical Look-Up Table [California DTSC 2013; California Environmental Protection Agency 2005]; USEPA Regional Soil Screening Levels [RSLs] [USEPA 2019h]).

Criteria presented in Table 6.12 generally correspond to a carcinogenic risk of 1 in 100,000 and a hazard quotient of 1 for residential properties. The values presented for New Jersey are not for protection of groundwater. Soil remediation values for New Jersey protective of groundwater are calculated by NJDEP on a site-by-site basis. The lessons from examining Tables 6.11 and 6.12 are the following:

- Cleanup or remediation criteria vary substantially by media which include soil, groundwater, surface water, indoor air, and outdoor air
- Cleanup or remediation criteria vary substantially based on land use which typically include residential, commercial, industrial, construction worker, or recreational
- Cleanup or remediation criteria vary between each state and with USEPA

6.6.11.5 Summary of Comprehensive Environmental Response, Compensation, and Liability Act

CERCLA was originally designed as a mechanism to fund cleanup of "orphaned" sites in the United States using the "Polluters Pay Principle." However, CERCLA has meant much more than

TABLE 6.11

Generic Residential Cleanup Levels for Select Compounds

Contaminant	Cas No.	Soil (ug/kg)	Ground-Water (ug/l)	Surface Water Interface (ug/l)	Indoor Air (ug/kg)	Ambient Air (ug/kg)
Volatile Organic Compounds						
Benzene	71432	100	5	4,000	1,600	13,000
Bromobenzene	108861	550	18	NA	3.10 E+5	4.50 E+5
Bromochloromethane	74755	1,600	80	ID	1,200	9,100
Bromodichloromethane	75274	1,600	80	ID	1,200	9,100
Bromoform	75252	1,600	80	ID	1.50 E+5	9.0 E+5
Bromomethane	74839	200	10	35	860	11,000
n-Butylbenzene	104518	2.60 E+5	80	44,000	5.4 E+7	2.90 E+7
sec-Butylbenzene	135988	1,600	80	ID	1.0 E+5	1.0 E+5
tert-Butylbenzene	98066	78,000	80	ID	3.1 E+8	9.70 E+9
Carbon tetrachloride	56235	100	5	900	190	3,500
Chlorobenzene	108907	2,000	100	25	1.3 E+5	7.70 E+5
Chloroethane	75003	8,600	430	1,100	2.9 E+6	3.0 E+7
Chloroform	67663	1,600	80	350	7,200	45,000
Chloromethane	74873	5,200	260	ID	2,300	40,000
2-Chlorotoluene	95498	3,300	150	ID	2.70 E+5	1.20 E+6
4-Chlorotoluene	106434	900	150	360	4.3 E+5	9.60 E+6
1,2-Dibromo-3-chloropropane	96128	100	.2	ID	260	260
1,2-Dibromoethane	106934	20	80	ID	1.1 E+5	2.60 E+6
Dibromomethane	74953	1,600	80	ID	7,000	7,000
1,2-Dichlorobenzene	95501	14,000	600	13	1.1 E+7	3.90 E+7
1,3-Dichlorobenzene	541731	170	6.6	28	26,000	79,000
1,4-Dichlorobenzene	106467	1,700	75	17	19,000	77,000
Dichlorodifluoromethane	75718	95,000	80	ID	5.3 E+7	5.50 E+8
1,1-Dichloroethane	75353	18,000	880	740	2.30 E+5	2.10 E+6
1,2-Dichloroethane	10762	100	5	360	2,100	6,200
1,1-Dichloroethene	75354	140	7	130	62	1,100
cis-1,2-Dichloroethene	156592	1,400	70	620	22,000	1.80 E+5

(Continued)

TABLE 6.11 (CONTINUED)
Generic Residential Cleanup Levels for Select Compounds

Compound	CAS				
trans-1,2-Dichloroethene	156592	2,000	100	1,500	23,000	2.80 E+5
1,2-Dichloroporopane	78875	100	5	230	4,000	25,000
1,4-Dioxane	123911	1,700	85	2,800	NLV	NLV
Ethylbenzene	100414	1,500	74	18	87,000	7.20 E+5
Hexachlorobutadiene	87673	26,000	50	ID	1.30 E+5	1.30 E+5
Isopropylbenzene	98828	91,000	800	28	4.0 E+5	1.70 E+6
Methylene chloride	75092	100	5	1,500	45,000	2.10 E+5
Methyl-tert butyl ether (MTBE)	163404	800	40	7,100	9.9 E+6	2.50 E+7
1,1,1,2-Tetrachloroethane	630206	1,500	8.5	78	36,000	54,000
1,1,2,2-Tetrachloroethane	79345	170	8.5	35	4,300	10,000
Tetrachloroethene (PCE)	127184	100	5	60	11,000	1.70 E+5
Toluene	108883	16,000	790	270	3.3 E+5	2.80 E+6
1,2,4-Trichlorobenzene	120821	4,200	70	99	9.6 E+6	2.80 E+7
1,2,3,-Trichlorobenzene	87616	4,200	70	99	9.6 E+6	2.80 E+7
1,1,1-Trichloroethane (TCA)	75556	4,000	220	89	2.50 E+5	3.80 E+6
1,1,2-Trichloroethane	79005	100	5.0	330	4,600	17,000

ug/kg = microgram per kilogram
ug/L = microgram per liter
NA = Not available
ID = Insufficient data
NLV = Not likely to volatilize

Source: Michigan Department of Environmental Quality. Cleanup Criteria and Screening Levels Development and Application. 2016. www.michigan.gov/documents/deq/deq-rrd-chem-CleanupCriteriaTSD_527410_7.pdf. (accessed February 17, 2019.)

TABLE 6.12
Soil Cleanup Levels for New Jersey, Illinois, New York, California, and USEPA

Contaminant	CAS ID No	New Jersey[1] (mg/kg)	Illinois[2] (mg/kg)	California[3] (mg/kg)	USEPA[4] (mg/kg)
Acetone	67641	1,000	25	20	2.9
Benzene	71432	3	0.03	5	0.0023
Bromobenzene	108861	NL	NL	NL	0.042
Bromochloromethane	74755	11	NL	NL	0.021
Bromodichloromethane	75274	NL	0.6	NL	0.00036
Bromoform	75252	86	0.8	NL	0.00087
Bromomethane	74839	79	NL	NL	0.0019
Methyl ethyl ketone	789833	1,000	17	NL	1.2
tert-Butylbenzene	98066	NL	NL	NL	1.6
Carbon tetrachloride	56235	2	0.07	NL	0.00018
Chlorobenzene	108907	37	1	NL	0.053
Chloroethane	75003	NL	NL	NL	0.081
Chloroform	67663	19	0.6	NL	0.000061
Chloromethane	74873	520	NL	NL	0.049
2-Chlorotoluene	95498	NL	NL	NL	0.23
4-Chlorotoluene	106434	NL	NL	NL	0.24
1,2-Dibromo-3-chloropropane	96128	NL	NL	NL	0.081
1,2-Dibromoethane	106934	NL	NL	NL	0.0000021
Dibromomethane	74953	NL	NL	NL	0.0021
1,2-Dichlorobenzene	95501	5,100	17	NL	0.03
1,3-Dichlorobenzene	541731	5,100	NL	NL	0.00082
1,4-Dichlorobenzene	106467	570	2	NL	0.00046
Dichlorodifluoromethane	75718	NL	NL	NL	0.03
1,1-Dichloroethane	75353	8	0.06	NL	0.00078
1,2-Dichloroethane	10762	6	0.02	NL	0.000048
1,1-Dichloroethene	75354	8	0.06	5	0.00078
cis-1,2-Dichloroethene	156592	4	0.4	5	0.011
trans-1,2-Dichloroethene	156592	4	0.7	NL	0.11
1,2-Dichloroporopane	78875	10	0.03	NL	0.00015
2,2-Dichloropropane	594207	NL	NL	NL	0.013
1,3-Dichloropropane	142289	NL	2NL	NL	0.13
1,4-Dioxane	123911	NL	NL	10	0.000094
Ethylbenzene	100414	1,000	13	5	0.0017
Hexachlorobutadiene	87673	1	NL	5	0.00027
Isopropylbenzene	98828	NL	NL	NL	0.084
Methylene chloride	75092	49	0.02	10	0.0029
Methyl-tert butyl ether (MTBE)	1634044	NL	0.32	NL	0.0032
1,1,1,2-Tetrachloroethane	630206	170	NL	NL	0.00022
1,1,2,2-Tetrachloroethane	79345	34	NL	NL	0.00003
Tetrachloroethene (PCE)	127184	4	0.06	5	0.0051
Toluene	108883	1,000	12	5	0.76
1,2,4-Trichlorobenzene	120821	68	5	NL	0.0034
1,2,3,-Trichlorobenzene	87616	NL	NL	NL	0.021
1,1,1-Trichloroethane (TCA)	75556	210	2	NL	2.8

(Continued)

TABLE 6.12   (CONTINUED)
Soil Cleanup Levels for New Jersey, Illinois, New York, California, and USEPA

Contaminant	CAS ID No	New Jersey[1] (mg/kg)	Illinois[2] (mg/kg)	California[3] (mg/kg)	USEPA[4] (mg/kg)
Trichloroethene (TCE)	75694	23	0.06	5	0.00018
Trichlorofluoromethane	96184	NL	NL	NL	3.3

mg/kg = milligram per kilogram

NL = Not listed

Source: [1] New Jersey Department of Environmental Protection (NJDEP). Introduction to Site-Specific Impact to Ground Water Soil Remediation Standards Guidance Document. Trenton, New Jersey. 10 pages. 2008. www.nj.gov/dep/srp/guidance/rs/igw. (accessed May 5, 2017.); New Jersey Department of Environmental Protection (NJDEP). Site Remediation Program: Remediation Standards. 2015. www.nj.gov/dep/srp/guidance. (accessed May 4, 2017.); and New Jersey Department of Environmental Protection (NJDEP). 2017. Soil Cleanup Criteria. 2017. www.nj.gov.srp/guidance/scc. (accessed May 5, 2017.)

[2] Illinois Environmental Protection Agency (IEPA). 2017. Tiered Approach to Corrective Action Objectives (TACO). Springfield, Illinois. 284 pages.

[3] California Department of Toxic Substances and Control (DTSC) 2013. Chemical Look-Up Table. Technical Memorandum. Sacramento, California. 5 pages; and California Environmental Protection Agency (CalEPA). 2005. Use of Human Health Screening Levels (CHHSLs) in Evaluation of Contaminated Properties. Sacramento, California. 65 pages.

[4] United States Environmental Protection Agency (USEPA). Regional Screening Levels. 2019h. www.epa.gov/RSLs. (accessed February 20, 2019.)

just a funding mechanism. CERCLA also was a catalyst in developing investigative techniques and protocols, risk assessment evaluations, remediation technologies, and remediation target values for hundreds of chemical compounds in different media. Over its more than 35 year history, USEPA has recovered over $35 billion to fund cleanup at thousands of CERCLA sites in the United States (USEPA 2019i). In addition, the "Polluters Pay Principle" is used world-wide as the preferred method for protecting human health and the environment at difficult and complex sites of contamination.

6.6.12   SUPERFUND AMENDMENTS AND REAUTHORIZATION ACT (SARA) AND EMERGENCY PLANNING AND COMMUNITY RIGHT-TO-KNOW ACT

Superfund Amendments and Reauthorization Act (SARA) of 1986 revised CERCLA largely in response to a 1984 tragedy in Bhopal, India, where thousands of residents were killed or injured by exposure to a chemical, methyl isocyanate (MIC), gas that leaked from a Union Carbide plant (USEPA 2017z). In response to continuing community concerns regarding hazardous materials and the Bhopal, India, tragedy, Title III of SARA was developed and is known as the Emergency Planning and Community Right-to-Know Act (EPCRA).

During the early morning hours of December 3, 1984, a Union Carbide plant in a village just south of Bhopal, India, released approximately 40 tons of MIC into the air. MIC is used in manufacturing pesticides and is considered a lethal chemical. The gas quickly and silently diffused near the ground surface and in the end killed thousands of people and injured tens of thousands of residents (Eckerman 2005). Title III of SARA, commonly called EPCRA, was a direct result of the tragedy in India.

There are many examples of incidents in the United States that involve hazardous substances and the purpose of EPCRA was to prevent or minimize the potential for an incident like what occurred in India from ever happening in the United States (USEPA 2017z). The purpose of EPCRA includes:

- To encourage and support emergency planning for responding to chemical accidents
- To provide local governments and the public with information about chemical hazards in their communities

To facilitate cooperation between industry, interested citizens, environmental and other public-interest groups and organizations, and the government at all levels, EPCRA establishes an ongoing forum at the local level called the Local Emergency Planning Committee (LEPC). LEPCs are governed by the State Emergency Response Commission (SERC) in each state. Under EPCRA, SERCs and LEPCs are responsible for (USEPA 2017aa):

- Preparing emergency plans to protect the public from chemical accidents
- Establishing procedures for war and, if necessary, evacuate the public in case of an emergency
- Providing the public citizens and local governments with information about hazardous chemicals and accidental releases of chemical in their communities
- Assisting in the preparation of public reports on annual release to toxic chemicals into the air, water, and soil

EPRCRA does not place limits on which chemicals can be stored, used, released, disposed, or transferred at a facility. It only requires a facility to document, notify, and report information (USEPA 2017aa). Emergency planning under EPCRA is found in Sections 301–303 and are to ensure that state and local communities are prepared to respond to a potential chemical accident. As a first step, each state had to establish a SERC. In turn the SERC designated local emergency planning districts. For each district, the SERC appoints supervisors and coordinates the activities of the LEPCs. The LEPC in turn must develop an emergency response plan for its district and review it annually. The membership of the LEPC includes representatives from public and private organizations as well as a representative from every facility subject to EPCRA emergency planning requirements, which we will discuss later in this section (USEPA 2017aa).

The plan developed by the LEPC must have the following information (USEPA 2017aa):

- Identify affected facilities and transportation routes
- Describe emergency notification and response procedures
- Designate community and facility emergency coordinators
- Describe methods to determine the occurrence and extent of a release
- Identify available response equipment and personnel
- Outline evacuation plans
- Describe training and practice programs and schedules
- Contain methods and schedules for exercising the plan

USEPA has established the requirements of when a facility is subject to EPCRA emergency planning reporting. USEPA has published a list of what it terms "extremely hazardous substances (EHS)." For each EHS, the list includes its name, the Chemical Abstract Service number of the substance, and a number that is called the threshold planning quantity (TPQ) and is expressed in pounds. If a facility has within its boundaries an amount of an EHS equal to or greater than the TPQ, the facility is subject to EPCRA and must notify both the SERC and the LEPC. The facility must also appoint an emergency response coordinator to work with the LEPC in developing and implementing the local emergency plan at the facility (USEPA 2017aa).

Emergency release notification requirements are described in Section 304 of EPCRA and defines if a facility is subject to release reporting requirements even if it is not subject to the provisions described in Sections 301–303. Section 304 applies to a facility which stores, produces, or uses a **hazardous chemical**, defined as any chemical which is a physical hazard or health hazard, and releases a reportable quantity (RQ) of a substance contained in either the list of extremely hazardous substances or the list of CERCLA hazardous substances (USEPA 2017aa).

Under EPCRA, the RQ is the determining factor in whether a release must be reported. This is a number that is expressed in pounds that is assigned to each chemical in USEPA's list of extremely hazardous substances or the CERCLA list of hazardous substances. If the amount of a chemical released to the environment exceeds its corresponding RQ, the facility must immediately report the release to the appropriate LEPC and SERC and provide written follow-up as soon as practicable. In addition to immediate notification, the facility is required to provide a written follow-up report that includes an update to the original notification, any additional information in the response action conducted, known, anticipated, or health risk incurred, and any information on medical care needed by any exposed person or population (USEPA 2017aa).

Community Right-to-Know Reporting Requirements are described in Section 311 and 312. The purpose of these requirements is to increase community awareness of chemical hazards and to assist in emergency planning, when and if ever needed. Section 311 and 312 apply to any facility that is required by the OSHA under its Hazard Communication Standard to prepare or have available a Safety Data Sheet (SDS) for a hazardous chemical or that has onsite, for any one day in a calendar year, an amount of a hazardous chemical equal to or greater than the following threshold limits established by USEPA (USEPA 2017aa):

- 10,000 pounds (4,500 kg) for hazardous chemicals
- Lessor of 500 pounds (230 kg) or the threshold planning quantity (YPQ) for extremely hazardous substances

If a facility is subject to reporting under these sections, it must submit information to the SERC and LEPC and the local fire department with jurisdiction over the facility under two categories that include SDS reporting and inventory reporting. SDS reporting requirements specifically provide information to the local community about mixtures and chemicals present at a facility and their associated hazards. For all substances whose onsite quantities exceed the threshold values, the facility must submit the following (USEPA 2017aa):

- A copy of the SDS for each above-threshold chemical onsite or a list of the chemicals grouped into categories
- Submit any change within 3 months, an SDS or list of additional chemicals which meet the reporting criteria

Inventory reporting is designed to provide information on the amounts, location, and storage condition of hazardous chemicals and mixtures containing hazardous chemicals present at any facility. The inventory has two forms. The first is called Tier I. The Tier I form contains aggregate information for applicable hazard categories and must be submitted yearly by March 1. The Tier II form contains more detailed information, including specific names of each chemical. This form is submitted upon request of the agencies authorized to receive the Tier I form. It can also be submitted yearly in lieu of the Tier I form (USEPA 2017aa). Toxic Chemical Release Inventory Reporting is described in Section 313 of EPCRA. Under this section, the USEPA is required to establish a Toxics Release Inventory (TRI), which is an inventory of routine toxic chemical emission from certain facilities. The original requirements under this section have been expanded by the passage of the Pollution Prevention Act of 1990 and will be described further in a later section. The TRI must now include information on source reduction, recycling, and treatment. To obtain this data, EPCRA requires each required facility to submit a Toxic Chemical Release Inventory Form (Form R) to the USEPA and designated state officials each year on July 1 (USEPA 2017aa).

A facility must file a Form R if the facility (USEPA 2017aa):

- Has more than 10 full-time employees
- Is in a specified Standard Industrial Classification Code
- Manufactures more than 25,000 lb/year of a listed toxic chemical
- Processes more than 25,000 lb/year of a listed toxic chemical
- Otherwise uses more than 10,000 lb/year of a listed toxic chemical
- Manufactures, processes or otherwise uses a listed persistent bio-accumulative toxic (PBT) chemical above the respective PBT's reporting threshold. PBT reporting thresholds can vary anywhere from 0.1 grams for dioxin compounds to 100 pounds (45 kg) for lead.

USEPA has updated some aspects of the Form R rules in 1999, 2001, and in 2006, which related to adding chemicals or adjusting thresholds for chemical on either the list of extremely hazardous substances or the list of CERCLA hazardous substances. Therefore, companies should review the lists routinely for changes and modifications (USEPA 2017aa).

Table 6.13 is the partial list of extremely hazardous substances defined in Section 302 of EPCRA. The full list can be viewed at www.epa.gov/epcra/final-rule-extremely-hazardous-substance-list-and-threshold-planning-quantities-emergency (USEPA 2019h).

6.6.13 Pollution Prevention Act

The Pollution Prevention Act (PPA) of 1990 created a national policy to prevent or reduce pollution at the source whenever possible. As discussed in the previous section, it also expanded the Toxics Release Inventory under SARA and EPCRA. The PPA focused on reducing and preventing pollution and waste, not just for industry but also the government itself and the general public, and encouraged and emphasized recycling efforts (USEPA 2017ab).

Up to this point in the rapid enactments of environmental laws beginning just 20 years before the PPA, the focus was on compliance. Now, with the PPA, there is a shift in focus from compliance to prevention and reducing amounts of waste generated. However, as we discussed earlier in this chapter, it was in 1971 when public service announcements first were aired on television drawing attention to waste disposal and recycling. Why did it take so long? The answer is of course, like many other aspects of human interaction with nature, complex and has much to do with how and to what extent humans impact the natural environment, which we are still learning about. The PPA is also considered the first real legislative effort in the United States to take up the subject of sustainability. Pollution prevention as we shall see in subsequent chapters is addressed as a subset of sustainability.

Emphasis was placed on preventing pollution from being produced, in other words, preventing pollution at the source was preferred through technological innovation and advancements in manufacturing. If it could not be prevented, then the focus should be on recycling in an environmentally safe manner. The PPA goes on to also state that disposal of waste or other types of releases into the environment should be conducted only as a last resort (USEPA 2017ab).

6.6.14 Brownfield Revitalization Act

The Brownfield Revitalization Act of 2006 amended CERCLA to provide funding to assess and clean up brownfields, clarified CERCLA liability protections, and provided funds to enhance state and Native American Tribal response programs. USEPA defines a **brownfield** as a property, the expansion, redevelopment, or reuse of which may be complicated by the presence or potential presence of a hazardous substance, pollutant, or contaminant (USEPA 2019j; 2019k). USEPA estimates that there are more than 450,000 brownfield sites in the United States.

TABLE 6.13

Partial List of Extremely Hazardous Substances

Compound	Compound	Compound
Acetone cyanohydrin	Acetone thiosemicarbazide	Acrolein
Acrylamide	Acrylonitrile	Acryloyl chloride
Adiponitrile	Aldicarb	Aldrin
Allyl alcohol	Allylamine	Aluminum phosphide
Aminopterin	Amiton	Amiton oxalate
Ammonia	Amphetamine	Aniline
Aniline, 2,4,6-trimethyl	Antimony pentafluoride	Antimycin A
ANTU (Alpha-Naphthlthiourea)	Arsenic pentoxide	Arsenous oxide
Arsenous trichloride	Arsine	Azidoazide azide
Azinphos-ethyl	Azinphos-methyl	Benzal chloride
Benzenamine, 3-(trifluoromethyl)-	Benzenearsonic acid	Benzimidazole, 4,5-dichloro-2-(trifluoromethyl)-
Benzothrochloride	Benzyl chloride	Benzyl cyanide
Bicyclo(2.2.1) heptane-2-carbonitrile	Bis(chloromethyl) ketone	Bitoscanate
Boron trichloride	Boron trifluoride	Boron trifluoride w/methyl ether
Bromadiolone	Bromine	Cadmium oxide
Cadmium stearate	Calcium arsenate	Camphechlor
Cantharidin	Carbachol chloride	Carbamic acid
Carbofuran	Carbon disulfide	Carbophenothion
Chlordane	Chlorfenvinfos	Chlorine
Chlorine trifluoride	Chlormethos	Chlormequat chloride
Chloracetic acid	2-chlorothanol	Chloroethyl chloroformate
Chloroform	Chloromethyl methyl ether	Chlorophacinone
Chloroxuron	Chlorthiophos	Chromin chloride
Cobalt carbonyl	Cobalt, (2,2-(1,2-ethanediylbis)	Colchicine
Coumaphos	Cresol, -o	Crimidine
Crotonaldehyde	Crotonaldehyde, (E)	Cyanide
Cyanogen bromide	Cyanigen iodide	Cyanophos
Cyanuric fluoride	Cycloheximide	Cyclohexylamine
Decaborane	Demeton	Demeton-S-methyl
Dialifor	Diborane	Dichloroethyl ether
Dichloromethylphenylsilane	Dichlorvos	Dicrotophos
Diepoxybutane	Diethyl chlorophosphate	Digitoxin
Diglycidyl ether	Digoxin	Dimefox
Dimethoate	Dimethyl mercury	Dimethyl phosphorochloridothioate
Dimethyl-p-phenylenediamine	Dimethyl-p-phenylebediamine	Dimethylcadmium
Dimethyldichlorosilane	Dimethylhydrazine	Dimetilan
Dinitrocresol	Dinoseb	Dineterb
Dioxathion	Diphacinone	Disulfoton
Dithiazanine iodide	Dithobiuret	Endosulfan
Endoothion	Endrin	Epichlorohydrin
EPN	Ergocalciferol	Ergotamine tartrate
Ethanesulfonyl chloride, 2-chloro-	Ethanol, 1,2-dichloro-. Acetate	Ethion
Athoprophos	Ethylbis(2-chloroethyl)amine	Ethylene oxide
Ethylemediamine	Ethyleneimine	Ethylthiocyanate
Fenamiohos	Fenitrothion	Fensulfothion

(Continued)

TABLE 6.13 (CONTINUED)
Partial List of Extremely Hazardous Substances

Compound	Compound	Compound
Fluenetil	Fluorine	Fluoroacetamide
Fluoroacetic acid	Floroacetyl chloride	Fluoroantimonic acid
Fluorouracil	Fonofos	Formaldehyde

Source: United States Environmental Protection Agency, USEPA. 2015b. List of Lists. EPA 55-B-15-001. Office of Solid Waste and Emergency Response. Washington, DC. 125p.

The Brownfield Revitalization Act was enacted in large part so that these properties, many of which are impacted with hazardous substances, can be investigated, cleaned up, and be redeveloped (USEPA 2019j; 2019k).

However, CERCLA had to be revised because under CERCLA, United States courts have ruled that a buyer, lessor, or lender may be held responsible for remediation of hazardous substance residues, even if a prior owner caused the contamination. Therefore, CERCLA was amended to include what is termed "All Appropriate Inquiry," which outlined procedures for the performance of a Phase I Environmental Site Assessment (ESA). By conducting a Phase I ESA, a new property owner may create a safe harbor or protection from liability, known as the "Innocent Landowner Defense" for new purchasers or lenders for transactions involving properties. In the United States, Phase I ESAs have become the standard type of environmental investigation employed when initially investigating a property (USEPA 2019j).

Standards for conducting Phase I ESAs were originally published by the American Society for Testing Materials (ASTM) in 1993 and were revised in 1997, 2000, and 2005 (ASTM 2005). On November 1, 2006, the USEPA published federal standards for conducting Phase I ESAs, termed "Standards and Practices for All Appropriate Inquiries" (AAI), by amending the Comprehensive Environmental Response, Compensation and Liability Act (CERCLA) of 1980, commonly known as the Brownfield Revitalization Act. As stated above, the Phase I Environmental Site Assessment (ESA) is typically the first environmental investigation conducted at a specific property (Kaufman et al. 2011).

A Phase I ESA is conducted with the objective of qualitatively evaluating the environmental condition and potential environmental risk of a property or site. A **property** is defined here as a parcel of land with a specific and unique legal description. A **site** is defined here as a parcel of land including more than one property or easement and typically refers to an area of contamination potentially affecting more than one property. As the first environmental investigation, the Phase I ESA is often regarded as the most important activity because all subsequent decisions concerning the property are, in part, based on the results of the Phase I ESA (Rogers 1992). Therefore, great care, scrutiny, scientific inquiry, and objectivity should be exercised while conducting the Phase I ESA. According to USEPA (2005b) requirements, an environmental professional, such as a geologist or environmental scientist, must conduct the Phase I ESA. General requirements for conducting a Phase I ESA include:

- Extensive review of current and historical written records, operations, and reports
- Extensive site inspection
- Interviews with knowledgeable and key onsite personnel
- Assessment of potential environmental risks from offsite properties
- Data gap or data failure analysis

A Phase I ESA is typically a non-invasive assessment, performed without sampling or analysis. In some instances, limited sampling may be conducted on a case-by-case basis if in the professional judgment of the person conducting the assessment – or due to other requests or mitigating factors – sampling is justified. In most cases, collecting and analyzing samples is usually deferred to the Phase II investigation, but if it does occur, the sampling conducted during a Phase I ESA may include the following:

• Sediment	• Drinking water
• Surface water	• Groundwater
• Waste material	• Soil
• Lead-based paint	• Suspect asbestos-containing material
• Mold	• Radon

The environmental professional conducting the Phase I ESA must perform extensive research and review all available written records. These researching activities apply not only to the property in question, but also to the surrounding properties within a radius of up to 1 mile of the investigated property or site. The research/review process typically involves (Kaufman et al. 2011):

• Title search and environmental liens	• Historical operations documents
• Historical chemical ordering documents	• Historical photographs
• Historical aerial photographs	• Engineering reports
• Engineering diagrams	• Historical environmental reports
• Previous environmental incident reports	• Safety data sheets (SDSs)
• Environmental compliance documents	• Hazardous substance inventories
• Fire insurance maps	• Soil Conservation Service maps
• USGS topographic maps	• USGS investigations reports
• USGS geologic maps	• Environmental agency maps
• Native American Tribal records	• Local governmental records
• Building permits	• Construction diagrams and blueprints

The site inspection consists of a walk-through of the property or site. Items to evaluate and document during the site inspection include (modified from Rogers 1992):

• Interviews with key personnel	• Stressed vegetation
• Areas absent of vegetation	• Stained soil or pavement
• Aboveground storage tanks (ASTs)	• Underground storage tanks (USTs)
• Chemical storage areas	• Back doors
• The "back 40" (the rear of the facility)	• Signage
• Storage sheds	• Refuse storage and containers
• Special labeling	• Evidence of fill or mounding
• Recent excavations or land disturbance	• Depressions in the land surface
• General topography	• Wetlands
• Mold	• Animal scat
• Insects	• Pits and trenches
• Sumps	• Floor and roof drains
• Broken concrete	• Areas not inspected or inaccessible
• Weather conditions	• Recent precipitation events
• Nearest water body	• Evidence of wells or borings
• Potential asbestos-containing materials	• Potential septic tanks
• Utilities	• Onsite and offsite dumping or fill
• Offsite inspection	• Soil type(s)
• Potential contaminant migration pathways	• Potential ecological and human receptor pathways

Once the environmental professional has completed the data collection and site inspection portion of the Phase I ESA, an evaluation of whether there is evidence of an existing release, a past release, or a material threat of a release of any hazardous substance or petroleum is conducted. If a product has been released and made its way into structures on the property or into the ground, groundwater, or surface water of the property, then this situation is termed a Recognized environmental condition (REC). **Recognized environmental condition (REC)** is defined as the presence or likely presence of any hazardous substance or petroleum product on a property or site under conditions that may materially affect or threaten the environmental condition of the property or site, human health, or the environment.

If a REC is discovered, further investigation will likely be recommended to evaluate its potential significance. Many sites have more than one REC, and many of these RECs may require further evaluation. An example of a REC with significant amounts of oil-stained soil, stressed, and dead vegetation is presented in Figure 6.13. RECs are intended to exclude *de minimis* conditions, generally not presenting a threat to human health or the environment and typically would not be the subject of an enforcement action if brought to the attention of appropriate governmental agencies. However, there may be impacts encountered whose perceived severity falls between a *de minimus* condition and a REC. In these situations, the term **environmental concern** is applied. Items listed as environmental concerns become RECs if left unattended or lead to a release. Figure 6.14 is a photograph of a paved parking space with a residual amount of what appears to be a petroleum product discharged from an automobile. In the opinion of the environmental professional who conducted a Phase I ESA at this property, this condition was not evaluated to be a REC but was characterized as *de minimis*.

FIGURE 6.13   Recognized environmental condition. (Photograph by Daniel T. Rogers.)

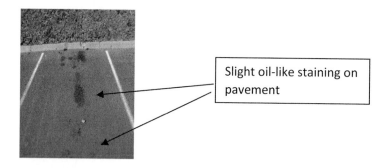

FIGURE 6.14   Example of a de minimis release. (Photograph by Daniel T. Rogers.)

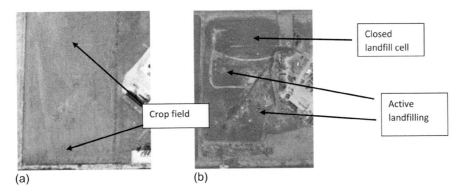

FIGURE 6.15 (a) Aerial photograph showing farmland. (Photograph by Daniel T. Rogers.) (b) Earlier aerial photograph showing landfilling activities. (Photograph by Daniel T. Rogers.)

Historical aerial photographs are effective sources of information, and often help with environmental investigations. In urban areas especially, aerial photographs from several sources are readily available. These sources include:

- Private local companies specializing in aerial photography
- Private national companies specializing in aerial photography
- Local and state historical societies
- Local and state agencies
- Utility companies
- Local companies
- Federal agencies such as USEPA, USGS, Soil Conservation Service, National Forest Service, Bureau of Land Management, National Park Service.

Figure 6.15a,b demonstrate how historical aerial photographs can help identify

RECs. On the left, Figure 6.15a shows a particular property as farmland with no identifiable REC. Analysis of an aerial photograph taken a few years earlier of the same property (Figure 6.17b to the right) indicates the property was used as a landfill.

6.7 OTHER ENVIRONMENTAL LAWS

Now that we have covered some of the most important environmental laws that concern themselves with controlling pollution, we shall briefly discuss the remainder of laws that have been enacted through time that have an environmental focus. The following are the acts that we shall briefly discuss in the sections below:

- Antiquities Act of 1906
- Atomic Energy Act of 1954
- Endangered Species Act of 1973
- Energy Policy Act of 2005
- Fish and Wildlife Coordination Act of 1968
- Food Quality Protection Act of 1996
- Lacey Act of 1900
- Marine Mammal Protection Act 2015
- Mineral Leasing Act of 1920
- National Historic Preservation Act of 1966
- National Parks Act of 1916
- Nuclear Waste Policy Act of 1982
- Oil Spill Prevention Act of 1990

- Atomic Energy Act of 1946
- Coastal Zone Management Act of 1972
- Energy Policy Act of 1992
- Federal Power Act of 1935
- Fisheries Conservation and Management Act of 1976
- Global Climate Protection Act of 1987
- Marine Protection Act of 1972
- Migratory Bird Treaty Act of 1916
- National Forest Management Act of 1976
- Surface Mining Control and Reclamation Act of 1977
- Noise Control Act of 1972
- Ocean Dumping Act of 1988
- Wild and Scenic Rivers Act of 1968

6.7.1 Lacey Act of 1900

The Lacey Act of 1900 is a conservation law that prohibits trade in wildlife, fish, and plants that have been illegally taken, possessed, transported, or sold. Essentially, the Lacey Act made poaching illegal. The Act also regulated the introduction of birds and other animals to places where they have never existed. This is now commonly referred to as invasive species. The Lacey Act is still active and has been amended several times, with the latest occurring in 2008 (United States Fish and Wildlife Service 2017a).

6.7.2 Antiquities Act of 1906

The Antiquities Act of 1906 gave the President of the United States the power to create national monuments with the purpose to protect historic landmarks, historic and prehistoric structures, and other objects of historic or scientific interest. The Antiquities Act was passed as a direct result of concerns over theft from destruction of archaeological sites and was designed to provide an expeditious means to protect federal lands and resources. President Theodore Roosevelt used the authority in 1906 to establish Devil's Tower in Wyoming as the first national monument and used the act 36 times during his presidency, which is the most for any president. Sixteen of the 19 presidents since have used this authority and have created 151 monuments. Some worthy of note include (United States National Park Service 2017a):

- Grand Canyon National Park
- Grand Teton National Park
- Zion National Park
- Olympic National Park
- Statue of Liberty

Since it has been enacted, hundreds of millions of acres of land have been set aside and protected, including marine environments (United States National Park Service 2017a).

6.7.3 Migratory Bird Treaty Act of 1916

The Migratory Bird Treaty Act of 1916 is an environmental treaty between the United States and Canada. The act makes it unlawful to pursue, hunt, take, capture, kill, or sell birds that appear on the protected list. The act does not discriminate between live or dead birds and also grants full protection to any bird parts including feathers, eggs, and nests. Currently, there are over 800 species of birds on the list (United States Fish and Wildlife Service 2017b).

6.7.4 National Parks Act of 1916

The National Parks Act of 1916 is also known as the Organic Act and established the National Park Service so that management of all the national parks would be conducted under one authority (United States National Park Service 2017b). These areas include a diverse variety of land of national importance and include (United States National Park Service 2017b):

• National parks	• National monuments
• National memorials	• National military parks
• National battlefields	• National historic sites
• National parkways	• National recreation areas
• National seashores	• National scenic riverways
• National scenic trails	• National cemeteries
• National heritage areas	• National scenic rivers

From the time Yellowstone National Park was created in 1872, each national park or other lands deemed of national importance was managed separately. The passage of the National Parks Act of 1916 combined management of all these protection lands and locations to the newly created National Parks Service.

6.7.5 MINERAL LEASING ACT OF 1920

The Mineral Leasing Act of 1920 governs leasing public lands for developing deposits of coal, petroleum, natural gas, and other hydrocarbons, in addition to phosphates, sodium, sulfur, and potassium. Previously, the General Mining Act of 1872 authorized citizens to freely prospect for minerals on public lands and allowed a discoverer to stake claims to both minerals and surrounding lands for development. This open-access policy enabled a major oil rush in the western United States and quickly became too much for the government to handle, thus prompting the passage of the act in 1920 (United States Bureau of Land Management 2017).

The Bureau of Land Management (BLM), a division of the Department of the Interior (DOI), is the principal administrator of the Mineral Leasing Act. BLM evaluates areas for potential development and awards leases based on whoever pays the highest price during a period of competitive bidding.

6.7.6 FEDERAL POWER ACT OF 1935

The Federal Power Act of 1935 grew out of the Federal Water Power Act of 1920 and was established to more effectively coordinate the development of hydroelectric projects in the United States. Prior to this, the authority to develop hydroelectric power was left up to individual states. Previously to passage of the act, competing interests between private and public interest groups and companies inhibited any progress. One group desired to keep shipping and the other desired damming and hydroelectric development while others pursued preservation and others recreation (United States Bureau of Reclamation 2017).

6.7.7 ATOMIC ENERGY ACT OF 1954

The Atomic Energy Act of 1954 amended an earlier effort in 1946 and covered both civilian and military uses of nuclear materials. It covered the development, regulation, and disposal of nuclear materials. The amended act made it possible for private companies to develop nuclear power generation plants, which was not part of the original 1946 act (United States Nuclear Regulatory Commission 2017).

6.7.8 NATIONAL HISTORIC PRESERVATION ACT OF 1966

The National Historic Preservation Act of 1966 was intended to preserve historical and archaeological sites in the United States. The act created the National Register of Historic Places. Although the act was not officially passed until 1966, many landmarks in the United States have been protected, such as George Washington's home Mount Vernon and Thomas Jefferson's home as well. Protection of these historic places did not seem appropriate under the Antiquities Act of 1906. Supporters and historians desired a more specifically tailored act that was specific to historical places (United States Advisory Council on Historic Preservation 2017).

6.7.9 WILD AND SCENIC RIVERS ACT OF 1968

The Wild and Scenic Rivers Act of 1968 was passed to protect rivers in the United States because they possessed scenic, recreational, geologic, fish and wildlife, historic, cultural, and other similar values. Rivers, or portions or sections of a river, that obtain the designation are preserved in their free-flowing condition and are not dammed or otherwise impeded. The act essentially vetoes the licensing of new hydroelectric power plants. Currently, there are over 200 rivers, totaling more than 12,500 miles of river in 38 states and Puerto Rico that have been designated under the act. By comparison, more than 75,000 large dams are located across the United States that have modified an estimated 600,00 miles of rivers in the United States (United States National Park Service 2017c).

Designation as a wild and scenic river is not the same as a national park designation, and generally does not confer the same level of protection as a wilderness area designation, as we shall see in the following sections. However, it does protect the free-flowing nature of a river in federal or non-federal areas and does not alter property rights (United States National Park Service 2017c).

6.7.10   FISH AND WILDLIFE COORDINATION ACT OF 1968

The Fish and Wildlife Coordination Act of 1968 is an amended act that was first passed in 1934 with the purpose of protecting fish and wildlife when federal action resulted in the control or modification of a natural stream or body of water.

The act is implemented by the Department of Interior and is authorized to conduct the following to preserve fish and wildlife (United States Fish and Wildlife Service 2017c):

- Developing, protecting, rearing, and stocking all species of wildlife, resources thereof, and their habitat
- Controlling losses from disease and other causes
- Minimizing damages from overabundant species
- Providing public shooting and fishing areas
- Carrying out other necessary measures
- Conducting surveys and investigations of the wildlife of the public domain
- Accepting donations of land and other contributions from private sources

6.7.11   COASTAL ZONE MANAGEMENT ACT OF 1972

The Coastal Zone Management Act of 1972 was established to preserve, protect, develop, restore, or enhance coastal zones of the United States. This act was in response and a recognition that coastal areas of the United States were under significant stress due to development pressures. The act established the National Coastal Zone Management Program (NCZMP) and the National Estuarine Research Reserve System (NERRS). The state of Washington was the first to adopt the program in 1976. The NCZMP is administered by the National Oceanic and Atmospheric Administration (NOAA). The program is designed to establish a basis for protecting, restoring, and establishing a responsibility in preserving and developing coastal communities and resources where they are under the highest pressures. The area does not just include the coastal areas of the Atlantic, Pacific, or Artic Oceans alone but also includes the Great Lakes and island territories to ensure a healthy and thriving ecosystem for future generations (NOAA 2017a).

6.7.12   MARINE PROTECTION ACT OF 1972

The Marine Protection Act of 1972 was the first act to specifically manage natural resources through an ecosystem approach. The ecosystem-based management approach is an integrated management approach that recognizes the full array of interactions within an ecosystem, including humans, rather than considering single issues, species, or ecosystem services in isolation. The ecosystems approach has been integrated into other marine regulations including the Magnuson-Stevens Fishery Conservation and Management Act and the Marine Mammal Protection Act of 2015. Originally the Marine Protection Act of 1972 set aside marine mammals as untouchable and unapproachable. It is viewed as the first animal rights legislation in the United States. The act was amended through the Magnuson-Stevens Fishery Conservation and Management Act and Marine Mammal Protection Act of 2015.

The act prohibited the act of hunting, killing, capture, or harassment of any marine mammal, including the polar bear. There are some exceptions and include capture of marine mammals for scientific study and Alaska Native subsistence (NOAA 2017b).

6.7.13 ENDANGERED SPECIES ACT OF 1973

The Endangered Species Act of 1973 was designed to protect animal species from extinction from a consequence of human interference from economic growth and development untempered by adequate concern and conservation. The act was established in an attempt to reverse a trend of those identified endangered species from extinction at whatever the cost (United States Fish and Wildlife Service 2017d). The act is administered by the Fish and Wildlife Service (FWS) and the National Oceanic and Atmospheric Administration (NOAA).

Listing status is a sliding criteria that indicates the degree to which a certain species is threatened with extinction. The listing status used in the United States is as follows (United States Fish and Wildlife Service 2017d):

- E = endangered. Endangered is the highest of alerts, meaning the closest to extinction in the United States and generally means any species that is in danger of extinction throughout all or a significant portion of its natural habitat range.
- T = threatened. Threatened is any species which is likely to become endangered within the foreseeable future throughout all or a significant portion of its natural habitat range.
- C = candidate. Candidate is a species that is under consideration due to a drop in population or because of loss of a significant reduction in its natural habitat.

The first list of endangered species in the United States included 14 mammals, 36 birds, 6 reptiles and amphibians, and 22 fish. According to the United States Fish and Wildlife Service (FWS), there are currently a total of 711 animal species on the list and 941 plant species for a total of 1,652 on the endangered species list just in the United States (Fish and Wildlife Service 2017d). Some of the animal species in the United States that are or were on the endangered species list include the following (United States Fish and Wildlife Service 2017d):

• Bald eagle – removed in 2007	• Whooping crane
• California condor	• Kirkland's warbler
• Peregrine falcon – removed in 1999	• Gray wolf
• Mexican wolf	• Red wolf
• Grey whale	• Grizzly bear – removed in 2007
• California southern sea otter	• Florida key deer
• Hawaiian goose	• Virginia big-eared bat
• Black-footed ferret	• Florida grasshopper sparrow

World-wide, there are thousands of species of animals and plants that are in danger of extinction because of direct human involvement that includes development, habitat loss and destruction, poaching, pollution, and other factors. The World Wildlife Fund maintains a general list of those animals that are in danger and also includes an additional category different than that of the United States by the addition of another category called critically endangered. Some of the animals currently listed by the World Wildlife Fund as critically endangered or endangered include (World Wildlife Fund 2017):

- Amur leopard
- Cross River gorilla
- Malayan tiger
- South China tiger
- Sumatran rhino
- Asian elephant
- Bluefin tuna
- Chimpanzee
- Hectors dolphin
- Indochinese tiger
- Sea lion
- Tiger

- Black rhino
- Eastern lowland gorilla
- Sumatran orangutan
- Sumatran elephant
- Sumatran tiger
- Western lowland gorilla
- Blue whale
- Borneo pygmy elephant
- Ganges river dolphin
- North Atlantic right whale
- Snow leopard
- Black spider monkey

- Bornean orangutan
- Javan rhino
- Orangutan
- Mountain gorilla
- Amur tiger
- African elephant
- Bonobo
- Galapagos penguin
- Indian elephant
- Red panda
- Sri Lanka elephant
- African giraffe

The above list is just to name a few of the more familiar animals that are now considered threatened with extinction. Unless major efforts from the world community are conducted immediately, the loss of these animals and countless others are likely inevitable. The list below includes some North American species that have gone extinct just since 1500 AD and include (International Union on the Conservation of Nature [IUCN] 2019):

- Great auk – 1852
- California golden bear – 1922
- Labrador duck – 1878
- Cascade mountain wolf – 1940
- Mexican grizzly bear – 1964
- Sea mink – 1860
- South Rocky mountain wolf – 1935
- Carolina parakeet – 1918
- Heath hen – 1932
- Bachman's warbler – 1988

- Passenger pigeon – 1914
- Ivory-billed woodpecker – 1942
- Eastern cougar – 2011 declared
- Eastern elk – 1887
- Newfoundland wolf – 1911
- Caribbean monk seal – 1952
- Southern California kit fox – 1903
- Dusky seaside sparrow – 1987
- Eskimo curlew – 1981
- Tacoma pocket gopher – 1970

According to the World Wildlife Fund (2017) and the International Union for Conservation of Nature (2019), hundreds of species of animals are now extinct directly due to humans in several different ways including hunting, habitat destruction, poisoning, development, and pollution. According to the United Nations, more than 1 million species of animals and plants are threatened with extinction as a direct result of human activities (United Nations 2019b).

6.7.14 NOISE CONTROL ACT OF 1972

The Noise Control Act of 1972 was established to regulate noise pollution with the intent of protecting human health and minimizing the annoyance of noise to the general public (USEPA 2017ac). The act set standards for establishing noise emission limits for nearly every imaginable source of noise including:

- Automobiles
- Trucks
- Buses
- Aircraft
- Heating, ventilation, and air-conditioning units
- Other equipment
- Major appliances

The act also encouraged municipalities and local governments to mitigate noise pollution in urban planning. In 1982, Congress ended funding to USEPA for noise control and delegated the primary responsibility of addressing noise pollution issues to the state and local governments. However, USEPA still maintains authority to conduct research and publish information on noise and its effects on the public, which includes the effects of noise in environmental impact statements for new developments (USEPA 2017ac).

6.7.15   Fisheries Conservation and Management Act of 1976

The Fisheries Conservation and Management Act of 1976 is also known as the Magnuson Fishery Conservation and Management Act and the Sustainable Fisheries Act. The act was established to (United States Fish and Wildlife Service 2017e):

- Establish national fish hatchery conservation and management standards
- Provide for the sustained participation of fishery-dependent communities and minimize economic impacts to those communities
- Modify operation of established fishery management councils
- Mandate action to identify overfished species
- Mandate action to rebuild fishery populations to those species that have been overfished
- Establish guidelines to identify essential fish habitat
- Establish a fishing capacity reduction program
- Conduct research on fishery management, conservation, and economic characteristics of fisheries

6.7.16   National Forest Management Act of 1976

The National Forest Management Act of 1976 was enacted to establish guidelines for the management of renewable resources on national forest land. The act specifically addresses timber harvesting within national forests. The purpose of the act was to protect national forests from permanent damage from excessive logging and clear-cutting practices (United States Forest Service 2017). Throughout the history of the United States there has been numerous legal battles and conflicts between environmental groups and the timber industry.

The demand for timber for home building, farming, and other uses in the United States has always seemed to be high and has placed significant pressure on the environment through habitat destruction. As mentioned in the beginning of this chapter, William Penn was aware of the importance of healthy forests in 1681 when he required that for every 5 acres of forest cleared that one acre remain preserved.

One of the more contentious conflicts was over the northern spotted owl in the Pacific Northwest. The northern spotted owl experienced a decrease in population levels that threatened its extinction. A major study was undertaken to evaluate the causes of the population decrease and the results indicated that there were three major threats to the owl population decrease that involved timber harvesting that included (1) variability of birth and death rates through time, (2) loss of genetic variation, and (3) random catastrophes.

The passage of the act required the National Forest Service to conduct an inventory of all national forests and to use a systematic and interdisciplinary approach to resource management which also included monitoring biological effects of timber harvesting (United States Forest Service 2017). Although deforestation in the United States has not been as much as we will discover in other countries of the world in the next chapter, the decrease is still significant. In 1620, approximately 4.1 million square kilometers of the United States was forest. The low point was in 1926 at 2.9 million square kilometers. Since 1926, the amount of forested land in the United States is 3.0 million square kilometers (United States Department of Agriculture 2014).

FIGURE 6.16   Amount of forest in the United States in 1620. (From United States Department of Agriculture. U.S. Forest Resource Facts and Historical Trends. 2014. www.fia.fs.fed.us/library/brochures/docs/2012/ForestFacts_1952-2012_English.pdf.) (Accessed May 4, 2019.)

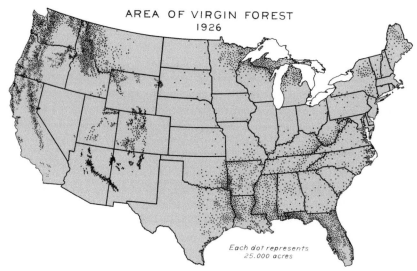

FIGURE 6.17   Amount of forest in the United States in 1926. (From United States Department of Agriculture. U.S. Forest Resource Facts and Historical Trends. www.fia.fs.fed.us/library/brochures/docs/2012/ForestFacts_1952-2012_English.pdf.) (Accessed May 4, 2019.)

Figure 6.16 shows the amount of forest in the United States in 1620 and Figure 6.17 shows the amount in 1926.

6.7.17   SURFACE MINING CONTROL AND RECLAMATION ACT OF 1977

The Surface Mining Control and Reclamation Act of 1977 was enacted for the purpose of addressing the environmental effects of coal mining. The act created two programs that included:

- Regulating active coal mines
- Reclaiming abandoned mine lands

The act also created the Office of Surface Mining within the United States Department of Interior to ensure consistency and fair application of the requirements under the act (United States Office of Surface Mining 2017).

The act was developed because of environmental concerns with the effects of surface coal mining or strip mining. Coal has been mined in the United States since the 1740s but surface or strip mining did not become widespread until the 1930s and by the mid-1970s accounted for more than 60% of coal mining. The act established the following:

- Uniform standards of performance. This provision standardized what was required nationwide and eliminated confusion and differences in regulations from state to state.
- Permitting. The act required that each proposed mine obtain a permit that described:
  - Pre-mining environmental conditions and land use
  - Proposed mining activities and area
  - Post mining reclamation
  - How the mine will achieve act requirements
  - How the land will be used following completion of mining activities
- Bonding. The act required that mining companies post a bond sufficient to cover the cost of reclaiming the mined land.
- Inspection and enforcement. The act permitted inspections and enforcement of violations by the federal government.
- Land restrictions. The act prohibited mining activities within the national parks and wilderness areas. It also allowed for citizen challenges to proposed mining developments.

The act also created a reclamation fund to assist in funding reclamation for mines closed before the passage of the act. Some funds are used to pay for emergencies such as landslides, land subsidence, fires, and conduct high priority cleanups (United States Office of Surface Mining 2017).

6.7.18 Nuclear Waste Policy Act of 1982

The Nuclear Waste Policy Act of 1982 established a program with the intention of providing a safe and permanent location for the disposal of highly radioactive wastes. During the 40 years before enactment of the law, there was no regulatory framework in place for disposal of nuclear wastes generated in the United States (Nuclear Regulatory Commission [NRC] 2017).

Some of the waste that was generated had a half-life of more than a million years and was temporarily stored in various locations and types of containers. Most of the wastes generated were as a result of manufacturing nuclear weapons. Approximately 77 million gallons of nuclear waste had been generated and was being stored in South Carolina, Washington, and Idaho. In addition, there were 82 nuclear power plants that produced electricity and also generated nuclear waste in the United States in 1982. The nuclear waste at the nuclear power plants consisted of spent fuel rods which were stored in water at the reactor sites and many of the plants were in danger of running out of storage space (NRC 2017).

The Nuclear Waste Policy Act created a timetable and procedure for establishing a permanent repository for high-level radioactive waste by the mid-1990s and provided temporary storage for those sites that were running out of space. The USEPA was directed to establish public health and safety standards for potential releases of radioactive materials and the NRC was required to promulgate regulations that covered construction, operation, and closure of repository location. Generators of nuclear waste were to be charged a fee to cover the costs of disposal of nuclear wastes (NRC 2017). Another provision of the act required the Secretary of Energy to issue guidelines for selecting sites of two permanent underground nuclear waste repositories. The Department of Energy (DOE) studied sites in Washington near the Hanford Nuclear Reservation, in Nevada near the nuclear testing site, and sites in Utah, Texas, Louisiana, and Mississippi. Other sites were studied along the east coast of the United States but were quickly ruled out as possible storage locations (NRC 2017).

In 2002, Yucca Mountain was selected as the only site to be characterized as the permanent repository for all the nation's nuclear waste. However, the state of Nevada used its veto power and rejected the recommendation and challenged the recommendation in court. The United States Court of Appeals upheld the state of Nevada's appeal, ruling that USEPA's 10,000 year compliance period for site selection was too short and was not consistent with the National Academy of Sciences' recommended compliance period of 1 million years.

USEPA subsequently revised the standard to 1 million years. At issue is selecting a disposal site that will safely contain the nation's nuclear waste for 1 million years. A license application selecting the Yucca Mountain location was submitted in 2008 and is still under review by the NRC. As of 2010, several lawsuits have been filed in federal courts to contest the legality of the selection process and proposed location of Yucca Mountain as the selected permanent repository site (NRC 2017; USEPA 2017ad). In the meantime, nuclear waste is being stored at various locations in temporary containers.

6.7.19   Global Climate Protection Act of 1987

The Global Climate Protection Act of 1987 delegated responsibility to the USEPA to develop and propose a coordinated national policy on global climate change. In a report to Congress in 1991, USEPA stated that in order to address global climate change, the following elements must be addressed (USEPA 1991b):

- Scientific research on climate change to reduce uncertainties with respect to operation of the climate system and to improve understanding of the impacts of human activities on future climate conditions and responses
- Assessment of environmental, social, and economic impacts of climate change
- Evaluation of policy options and practices to limit, mitigate, or adapt to climate change, including assessment of effectiveness and social and economic impacts
- Development and implementation of feasible and cost-effective policies and practices

USEPA stated in its report to Congress in 1991 that the nature of human existence on the Earth has changed dramatically during the last century. The world population has increased threefold. However, this was in 1991; now in 2017 (26 years later), the human population is estimated at over 7 billion people as opposed to just over 5 billion in 1991 (United States Census Bureau 2019). USEPA (1991b) further states that industrial production has increased by a factor of 50 and the world economy has increased by a factor of 20. In all, USEPA concludes that human activity is changing the features of Earth and the chemistry of the atmosphere. Since the act was passed in 1987, USEPA has supported and produced hundreds of reports and academic journal articles that have been valuable in building our understanding of causes, impacts, and potential solutions to climate change. The CAA gave USEPA the authority to regulate emissions from power plants and other large sources of greenhouse gases, and now requires facilities that are large emitters of greenhouse gases to report emissions to USEPA (USEPA 2017ae).

USEPA has developed and implemented voluntary programs to cut greenhouse gas emissions through the following (USEPA 2017ae):

- Natural Gas STAR, which is a program to limit methane emissions
- Coalbed Methane Outreach Program to encourage mine owners and operators to capture methane rather than allow its escape in the atmosphere
- Environmental stewardship partnership programs to address the most potent greenhouse gases emitted from the aluminum, semiconductor, refrigerant, power, and magnesium industries
- Energy Star Program, which is a voluntary program to assist businesses and individuals in protecting the climate through energy efficiency

Currently, reductions of greenhouse gas emissions are voluntary and Congress has yet to pass legislation to cut greenhouse gas emissions. USEPA (1991b) stated that in order to effect real change in reducing the effects of global climate change from greenhouse gas emissions that it must be through international cooperation. Thus far, efforts to reduce greenhouse gas emissions under an enforceable international treaty have failed, in part, due to language that limits greenhouse gas emissions on developed countries such as the United States but does not limit greenhouse gas emission on developing nations, such as China and India (USEPA 2017ae).

6.7.20 OCEAN DUMPING ACT OF 1988

The Ocean Dumping Act of 1988 amended the Marine Protection, Research, and Sanctuaries Act of 1972 to prohibit the dumping of sewage sludge and industrial waste into the ocean after 1991 (USEPA 2017af). Before 1972, the ocean was in effect a dumping ground that included sewerage, sludges, heavy metals, organic chemical wastes, industrial waste, and radioactive waste (USEPA 2017af). The act prohibits the dumping of most any waste into the ocean including (USEPA 2017af):

- High-level radioactive waste
- Radiological, chemical, and biological warfare agents
- Persistent inert synthetic or natural materials which may float or remain in suspension in the ocean
- Sewerage sludge
- Medical wastes
- Industrial wastes containing the following:
    - Organohalogen compounds
    - Mercury and mercury compounds
    - Cadmium and cadmium compounds
    - Oil of any kind or in any form
    - Known carcinogens, mutagens, or teratogens

Figure 6.18 shows an example of dumping liquid waste in the ocean before the act. Figure 6.19 shows an example of a barge with garbage near New York City to be dumped in the ocean.

FIGURE 6.18 Example of dumping liquids in the ocean. (From United States Environmental Protection Agency (USEPA) 2017af. Summary of the Oil Pollution Act of 1990. 2017af. https://epa.gov/oil-pollution-act.) (Accessed April 4, 2017.)

FIGURE 6.19   Example of solid waste dumping in the ocean. (United States Environmental Protection Agency (USEPA) 2017af. Summary of the Oil Pollution Act of 1990. 2017af. https://epa.gov/oil-pollution-act.) (Accessed April 4, 2017.)

Passage of the act in 1988 was motivated in part by the garbage barges that would dump garbage from New York City into the ocean, as pictured in Figures 6.20 and 6.21.

6.7.21   Oil Spill Prevention Act of 1990

The Oil Spill Prevention Act of 1990 was enacted in large part due to the March 24, 1989, Exxon Valdez oil spill in Prince Williams Sound off the Alaskan shore, which spilled 11 million gallons of crude oil. At the time, it was the largest oil spill in the history of the United States. Cleanup of the oil was mainly achieved through mechanical means and included (USEPA 2017ag):

FIGURE 6.20   Exxon Valdez oil spill. (From United States Environmental Protection Agency (USEPA). USEPA Exxon Valdez Spill Profile. 2017ag. www.epa.gov/emergency-response/exxon-valdez-spill-profile.) (Accessed April 4, 2017.)

FIGURE 6.21 Example of affected wildlife from the Exxon Valdez oil spill. (From United States Environmental Protection Agency (USEPA). USEPA Exxon Valdez Spill Profile. 2017ag. www.epa.gov/eme rgency-response/exxon-valdez-spill-profile.) (Accessed April 4, 2017.)

- Applying oil booms to concentrate the oil and pump the oil into containers aboard recovery ships
- Washing shorelines
- Physically recovering oil from shorelines and beaches
- Excavating shorelines and beaches

Burning the oil and applying a dispersant were both attempted but had little or no effect. Billions of dollars were spent recovering oil and cleaning tens of miles of shoreline and saving and cleaning affected wildlife (USEPA 2017ag; NOAA 2017c). Figure 6.20 shows the Exxon Valdez in Prince Williams Sound after the spill occurred and recovery efforts were underway (USEPA 2017ag). Figure 6.21 shows a photograph of some affected wildlife (USEPA 2017ag). Immediate effects of the spill included estimated deaths from 100,000 to as high as 250,000 seabirds, nearly 3,000 sea otters, 300 harbor seals, nearly 250 bald eagles, 22 orcas, and an unknown number of salmon and herring (NOAA 2017c).

The Oil Spill Prevention Act resulted in instrumental changes in oil production, transportation, and distribution that included the following (USEPA 2017ag; 2018c):

- Establishing a fund for future cleanup of spills
- Increasing spill response preparedness
- Increasing reliability in mapping and location of ships
- Improving ship designs
- Instituting measures and controls to minimize spills during loading and unloading
- Increasing financial assurance requirements

Another oil spill worthy of note is the Deepwater Horizon oil spill in 2010. On April 20, 2010, the offshore drill rig Deepwater Horizon was drilling a well 41 miles off the southeast coast of Louisiana in water that was approximately 5,000 feet deep when an explosion occurred that led to the destruction of the drilling rig and spilled millions of gallons of oil into the Gulf of Mexico. Figure 6.22 shows the Deepwater Horizon engulfed in flames after the explosion.

An accurate estimate on the amount of oil released from the Deepwater Horizon oil spill in 2010 vary widely from just a few million gallons to much more. Cleanup, however, required billions of dollars. The majority of the cleanup has been completed but monitoring continues (NOAA 2017d).

6.7.22   ENERGY POLICY ACT OF 1992 AND 2005

The Energy Policy Act of 1992 was enacted to sets goals and mandates to increase clean energy use, improve overall energy efficiency, and lessen the nation's dependence on imported energy. The

FIGURE 6.22 Deepwater Horizon oil spill in the Gulf of Mexico. (From United States Coast Guard. U.S. Coast Guard Responds to Deepwater Horizon. New Orleans, Louisiana. 2017. www.uscg.gov/deepwater-horizon-images.) (Accessed April 4, 2017.)

FIGURE 6.23 Wind farm in Washington State. (Photograph by Daniel T. Rogers.)

act set standards for many different sectors of the economy including establishing standards for buildings, utilities, equipment standards, renewable energy, and alternative fuels (National Energy Institute 2005). The act was amended in 2005 to provide tax incentives for alternative energy development such as wind energy (see Figure 6.23). The act also made it easier for a new technology to expand oil fracking by exempting waste fluids generated from the Clean Air Act, Clean Water Act, and the Safe Drinking Water Act, and CERCLA (National Energy Institute 2005).

An unforeseen disadvantage to wind turbines is that they kill birds, especially raptors, migratory birds, and bats. Some estimates range from 250,000 to several million birds are killed annually in the United States and this will likely increase as more wind turbines are installed and old ones are replaced with wind turbines that are much larger and kill even more birds because of increased reach and spin speed of the turbines (Eveleth 2013; Curry 2017). However, vertical wind turbines are much more bird friendly because they are more compact and can be installed in urban areas (see Figure 6.24).

FIGURE 6.24 Vertical wind turbine in London, England. (Photograph by Daniel T. Rogers.)

6.7.23 Food Quality Protection Act of 1996

The Food Quality Protection Act (FQPA) was enacted to standardize the way USEPA would manage the use of pesticides and amended FIFRA and FFDCA. It mandated a health-based standard for pesticides used in foods, provided for special protection for babies and infants, streamlined the approval process of safe pesticides, established incentives for the creation of safer pesticides, and required pesticide registrations remain current. The FQPA established a new safety standard that must be applied to all food commodities, that being "reasonable certainty of no harm." In addition, USEPA was required to consider this new risk standard as it applies to babies and infants. The FQPA required that re-testing of all existing pesticide tolerance levels be conducted within 10 years and that USEPA must account for "aggregate risk" and "cumulative exposure" to pesticides with similar mechanisms of toxicity.

As a result of the FQPA, USEPA banned methyl parathion and azinphos methyl because of the risks posed by these two pesticides to children. In 2000, USEPA banned an additional pesticide, chloropyrifis, which was common in agriculture, household cleaners, and commercial pest control products, because of concerns about children's health (USEPA 2017ah). FQPA further raised debate over using clinical studies of pesticides on humans. In 2004, the National Academy of Science released a report supporting the use of clinical studies under very strict regulations, citing that the benefit outweighed the risk to the individual. In 2005, USEPA adopted the Human Studies Regulation that allows human studies, with the exception of pregnant women or children, and mandate a strict ethical code (USEPA 2017ai).

6.7.24 Marine Mammal Protection Act of 2015

The Marine Mammal Protection Act of 2015 amended the earlier 1972 Act in that it required that nations exporting fish and fish products to the United States are required to demonstrate that harvesting fish does not also kill or injure marine mammals in excess of standards established under the act. The act essentially leveled the playing field for domestic and international fishing interests. To comply, other nations that sold fish or fish products to the United States could adopt marine mammal conservation techniques acceptable or in compliance with United States standards or demonstrate compliance using or developing other acceptable methods or standards (NOAA 2017e).

6.8 ENFORCEMENT

In addition to the Polluters Pay Principle described under CERCLA, enforcing environmental laws is a central part of USEPA's strategic plan to protect human health and the environment. To ensure that environmental regulations are taken seriously by the regulated community, the USEPA has established an enforcement policy and has published several guidance documents that provide detail so that enforcement measures are applied equally and consistently (USEPA 2018e; 2018f). In addition, USEPA has a top priority to protect whole communities disproportionately affected by pollution through what is termed the environmental justice network (USEPA 2018f).

In large part, the reason why the United States has been so successful at improving the environment and protecting human health has been because USEPA has an effective enforcement program. Purposeful acts that harm the environment have been greatly reduced in large part because of education and the threat of enforcement. No longer is there a mentality that "It's not a violation until I get caught." In fact, it may be a good time to rethink enforcement policies that have not changed in 30 years.

There are generally two levels of enforcement penalties in the United States, which are civil and criminal (USEPA 2018g). The criminal enforcement program at USEPA was established in 1982 and was granted full law enforcement authority by Congress in 1988 and employs special agents and investigators, forensic scientists and technicians, lawyers, and support staff. Civil enforcement arises most often through an environmental violation. Civil enforcement does not take into consideration what the responsible party knew about the law or regulation that was violated. Environmental criminal liability is triggered through some level of intent (USEPA 2018g). A simple explanation of the difference between a civil or criminal violation is that a criminal act is characterized as a "knowing violation" where a person or company is aware of the facts that create the violation. A conscious and informed action brought about the violation. In contrast, a civil violation may be caused by an accident or mistake (USEPA 2018f; 2018g).

Types of enforcement results from a civil violation include settlements, civil penalties, injunctive relief, and supplemental environmental projects (SEPs) (USEPA 2018f).

To evaluate the level of seriousness of an environmental violation, one must answer three basic questions:

1. Who first discovered the environmental violation? For the most part, environmental regulations are based on self reporting. It is generally a worse situation if the regulatory authority discovers a violation, especially one that should have been self reported.
2. Did the violation cause harm? It is generally a worse situation if the violation caused harm to human health or the environment. A permit violation or exceedance is generally considered a situation that has the potential to cause harm.
3. Was the violation a purposeful act? It will always be a worse situation if the violation was a purposeful act.

6.9 IMPROVING ENVIRONMENTAL REGULATIONS OF THE UNITED STATES

Now that this chapter has nearly come to a close we can reflect back and see how complex and seemingly comprehensive environmental regulations in the United States have become since the USEPA was created nearly 50 years ago. However, as mentioned in the introduction in this chapter, there are areas where existing environmental regulations must improve and still other areas that are in need of regulations where either none exist or if they do exist, are clearly inadequate. The following sections highlight areas where environmental regulations in the United States can and should improve.

6.9.1 MODIFY ENVIRONMENTAL ENFORCEMENT EMPHASIS AND POLICY

At this point, it is not apparent that a need for modifying USEPA's approach to enforcement is in need of change, especially since it has been seemingly effective up to this point. However, it's time to move beyond a litigation heavy and punishment-based enforcement model and enter into more of a reform- and repair-based enforcement policy that emphasizes sustainability.

We will learn in the next chapter that some countries are more advanced compared to the United States in how they have chosen to organize and implement their environmental regulations and where they focus internal resources. In addition, the governments of some countries have different attitudes and relationships with industry. The regulatory agencies of some countries have chosen to form working partnerships with industry with the intent to work more closely together to solve complex environmental issues rather than an adversarial approach which much too often results in polarizing points of view and expensive litigation.

The result of working together saves time, effort, and resources and in the long run has resulted in the fact that some countries are better at protecting human health and the environment than the United States.

6.9.2 SCIENCE-BASED REGULATION IMPROVEMENTS

We discovered in earlier chapters that contaminants behave differently and are more prone to releases in different geologic, hydrogeologic, geographic, and climatic regions. For instance, regulations specific to geologic risk factors such as earthquakes (e.g., California) or climate change (eastern seaboard and the coast along the Gulf of Mexico) should better reflect those areas where there is the highest risk.

This is demonstrated by the fact that two employees from a chemical plant in Texas were indicted for environmental crimes following Hurricane Harvey in 2017 that dumped up to 50 inches of rain on the region. Climate change was credited with exacerbating and increasing rainfall amounts. A United States Chemical Safety Board report stated that chemicals at the plant had to be kept frozen to avoid bursting into flames. But temperatures rose after floodwaters knocked out the plant's main power and the backup. Even though employees went to extremes and at personal risk to prevent harm, the chemicals exploded and caused a fire that burned for days, releasing a toxic chemical cloud. The compliant stated that plant management should have been better prepared for incidents that occur as a result of climate change (Texas Commission on Environmental Quality 2018).

6.9.3 AGRICULTURE

Agriculture is now the biggest polluter in the United States and the world (United Nations 2016). The United Nations estimates that most land pollution is caused by agricultural activities, such as grazing, use of pesticides, fertilizers, irrigation, plowing, and confined animal facilities (United Nations 2016). Agricultural activities release significant amounts of greenhouse gases from plowing, decaying vegetation, and from livestock. Fugitive emissions and erosion are also significant and release significant amounts of chemicals into the air and the water (see Figure 6.25). In addition, we have also learned that fertilizers are a significant source of pollution to the oceans of the world.

Since environmental regulations have been so successful at reducing the harmful effects from industry in the United States, it is now time to reduce harmful effects on the environment from agriculture.

FIGURE 6.25 Fugitive dust emissions from farm equipment. (From United States Environmental Protection Agency. Agriculture 101. 2019l. www.epa.gov/sites/production/files/2015-07/documents/ag_101_agricultur e_us_epa_0.pdf.) (Accessed April 24, 2019.)

6.9.4 Urban Air Quality

The Clean Air Act has made a major impact at improving the air quality throughout the United States, especially with respect to automobile exhaust, ozone, particulate matter, and the other criteria pollutants. According to USEPA (2018b), compared to emissions in 1980, carbon monoxide emissions have been reduced by 72%, lead by 99%, nitrogen oxides by 61%, volatile organic compounds by 54%, particulate matter (PM10) by 61% and, sulfur dioxide by 89% (see Figure 6.26). Improvements on exhaust from diesel trucks and buses have been slowed and are especially noticeable in urban areas. In addition, other sources of particulate matter are of high concern in urban areas as well and needs improvement. This need is in part the result of an increasing population and that urban areas continue to grow as population centers that will inevitably increase automobile congestion and emissions if not addressed as quickly as possible.

As we shall see in the next chapter, even though the United States is in need of improving urban air pollution, many locations in the rest of the world are far worse when it comes to urban air pollution. Actions by some countries (i.e., United Kingdom) include a carbon tax placed on vehicles driven in urban areas with high pollution levels to discourage unnecessary travel.

6.9.5 Banning Harmful Chemicals

Let's examine the regulations of the United States for just a moment from an overview perspective. The majority of environmental regulations focus on wastes that are generated and whether a facility has a permit. They are most concerned, and for good reason, on air emissions, water discharge, and solid and hazardous waste generation and disposal. The reasoning is that these are the primary

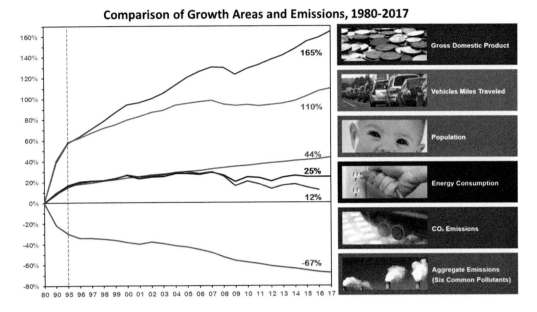

FIGURE 6.26 Reduction of air emissions since 1980. (From United States Environmental Protection Agency (USEPA). Air Quality – 2018 National Summary. 2018b. www.epa.gov/air-trends/air-quality-natio nal-summary.) (Accessed December 24, 2018.)

pathways for harmful substances to migrate from a point of generation to a point of human exposure where they may cause harm to us or the environment. In general, the regulations do not restrict what substances can be used at a facility. TSCA is the set of regulations that has this authority and to date TCSA has very few chemicals that have been banned in the United States.

After reading this chapter, you may have wondered and even asked yourself a question. Why didn't we know about the harm certain chemicals have on humans and the environment before they are widely used? This is a very good question and it happens to be a mistake our society has made time and time again. It happened with DDT, PCBs, asbestos, MTBE, and many other compounds that at first were even advertised as a miracle chemical (i.e., DDT and PCBs) or a miracle substance (i.e., asbestos). But as time marched on and these chemicals or substances were widely used, evidence began to surface that these substances had unintentional and unknown side effects that had a negative effect on human health, the environment, or both. As we explained in Chapter 2, sometimes there is a latency period, sometimes 20 or more years after exposure, that a symptom or negative effect surfaces.

As we have discussed earlier in this chapter, there are over 80,000 chemicals in the United States currently being used. USEPA has been successful in restricting only nine chemicals in its history of over 40 years since promulgated in 1976 and none since 1984. Those substances currently restricted include:

- Polychlorinated biphenyls (PCBs)
- Chlorofluorocarbons (CFCs)
- Dioxin
- Asbestos
- Hexavalent chromium
- Four nitrite compounds that are either:
  - Mixed mono and diamides of an organic acid
  - Triethanolanime salt of a substituted organic acid
  - Triethanolanime salt of tricarboxylic acid
  - Tricarboxylic acid

In June 2016, TSCA was amended significantly and gave USEPA additional authority and responsibility to proactively evaluate the risks of all chemical substances. USEPA has identified ten chemicals as high priority and are currently conducting risk evaluations. They include:

1. 1,4-Dioxane	6. 1-Bromopropane
2. Asbestos	7. Carbon Tetrachloride
3. Cyclic Aliphatic Bromide Cluster	8. Methylene Chloride
4. N-methylpyrrolidone	9. Pigment Violet 29
5. Tetrachloroethylene	10. Trichloroethylene

Analysis of the list above include four of substances that belong to the same chemical group called chlorinated volatile organic compounds (CVOCs), and are also known as chlorinated solvents. Those include carbon tetrachloride, methylene chloride, tetrachloroethylene, and trichloroethylene. In addition, 1,4-Dioxane is sometimes associated with the breakdown of tetrachloroethylene and trichlorethylene, so perhaps there may be as many as five in the CVOC family of chemicals. Therefore, of the ten chemicals USEPA has initially placed in its high priority list, half or more belong or are associated with the same group of compounds, CVOCs or chlorinated solvents. As we mentioned in Chapter 2, it should come as no surprise that having chlorine attached to an organic molecule sometimes has a negative impact on us and the environment if released.

There is perhaps negative feedback that has occurred by not banning more chemicals in the United States, which has resulted in providing a false sense of security concerning chemical use. This sense of false security has developed an attitude of reliance that the USEPA will protect us, after all that is their job, right? Many users may say that if a particular chemical was harmful, it would have been banned. This is where we can improve but have not because TSCA was not strong enough when enacted and the burden of scientific proof required to ban a chemical is too high and takes too long to establish. This is evident when we examine some of our example sites in this chapter that include Woburn, Massachusetts, Toms River, New Jersey, and the Hinckley Site in California.

6.9.6 EMERGING CONTAMINANTS

Emerging contaminants are chemicals or substances that have been recently discovered in the environment and, in some instances, have caused harm or have impacted resources to the extent that they can no longer be used for fear of health risks. Emerging contaminants include many compounds we discussed in Chapter 2, including PFAS, many pharmaceutical compounds, 1,4-dioxane, and others. Many of these emerging compounds are now being detected in the environment and have not been adequately evaluated. This highlights a need for improved prevention measures and the need for improving handling and disposal practices.

6.9.7 INVASIVE SPECIES

Some consider invasive species a type of emerging contaminant. However, for the purposes of this book we treat them separately because there are so many invasive species and their behavior in the environment is so different compared to chemicals and other substances because their mode of migration is through breeding. Invasive species have been present for the last few hundred years but environmental damage is increasing and in need of further regulatory protections.

6.9.8 IMPROVING POLLUTION PREVENTION

The Pollution Prevention Act (PPA) and Oil Spill Prevention Act (OSPA), both of 1990, do not get enough credit in the history of environmental regulations of the United States. The PPA and OSPA were the first real efforts to turn from regulating the environment at the "end of the pipe" to promoting the prevention of pollution from entering the environment. One of the driving factors behind the PPA and OSPA was the realization of how much it cost to clean up the environment after a release.

It was a sobering lesson that just one incident could cause so much harm, cost billions of dollars, and in some instances could not be cleaned up to acceptable levels because technologies simply did not exist. It was during this time that USEPA and many others discovered that sometimes you can throw money at a contaminated site or incident and make it go away and then there are those where all the money and technology in the world would not be able to fix the problem before enormous harm was a reality. Therefore, out of this realization, USEPA began to focus efforts on pollution prevention, in part because it made common sense in that it's proactive at environmental protection and lowers response cost much like fire prevention. As we shall see in the next chapter, it was out of these and similar efforts in the European Union that the term sustainability took hold.

6.9.9 Urban Land Use

One important item that should be added to pollution prevention efforts is urban land use. There is much that we can accomplish at protecting human health and the environment by not locating high-risk activities in areas that are especially sensitive to pollution.

Currently, urban planning for industrial activities are near transportation routes and are generally not located near residential areas. This is a good practice. However, there is often a complete disregard for evaluating geologic and hydrogeologic risk factors that can greatly influence the migration of contaminants through air, water, and soil that ultimately lead to human and ecological exposure from these areas. The examples for Woburn, Massachusetts, Hinkley site in California, Toms River, Love Canal, and Times Beach are just a few of hundreds of examples that highlight this risk but has not been adequately addressed with effective legislation for land use planning.

6.9.10 Noise Pollution

An additional type of pollution we have not addressed in this chapter is noise pollution. The reason is that noise pollution is largely not regulated in the United States at a federal level. However, as we shall see in the next chapter, noise pollution is a regulated form of pollution in many countries of the world, including the European Union, Mexico, and even China.

6.9.11 Household Waste

Some local efforts have been initiated at educating the public on the potential dangers of some households wastes that include cleaners, solvents, electrical equipment, waste paints, oil and grease, and many other substances. However, much more effort and incentives are needed. The landfills where much of the waste from households are disposed are not as adequately constructed to prevent releases to groundwater from many of the substances that are located within the majority of municipal landfills. In addition, septic tanks are an additional source of pollution to the ground and groundwater and highlight the need for expanding local POTWs so that this untreated waste is not directly discharged into the environment.

6.9.12 Nonpoint Source Pollution

As described in the summary of the Clean Water Act, nonpoint source pollution is in need of regulation, especially from rural and agricultural land, to minimize the build-up and discharge of pollutants that include fertilizers, pesticides, herbicides, bacteria, and other pollutants into the waters of the United States, which causes massive marine deaths in the Gulf of Mexico and perhaps other coastal locations (e.g., Florida). This will be more difficult to regulate in part because it affects areas that are not within the geographical boundary of the United States. The source of much of this pollution originates from nonpoint sources located in rural and agricultural land within the interior portions of the country.

Environmental regulations concentrate on urban land, which only accounts for 2.5% of developed land in the United States (see Section 6.2). Other areas that include crop land, range land, and managed forest account for 75% of developed land in the United States and is generally not regulated by the USEPA. USEPA has been successful at improving the environmental quality in urban areas to such a degree that pollution effects from nonpoint sources, which account for 75% of other developed land in the United States, is much more noticeable and in need of regulation and protection.

6.9.13 SUSTAINABILITY

The United States is in need of a clearer picture of what it means to be sustainable. To many, it's a buzz word light on specific goals or even a basic plan for achieving sustainability. As we shall discover in future chapters in the book, the European Union and other countries have established a clearer picture and have set goals for sustainability. When we discuss sustainability later in this book, we will discuss the need to move beyond reactive measures to proactive sustainability measures to improve our environment in the future. If there are delays in enacting sustainability measures, it may be too late. A whole chapter is dedicated to this discussion point.

6.9.14 CONSUMER PRODUCTS AND PACKAGING

Plastic has become one of the more important current pollution issues and was the focus of Earth Day 2018 (Earth Day 2018). Plastic has many valuable uses. However, humans have become addicted to single-use of disposable plastic. Around the world, one million plastic drinking bottles are purchased every minute and up to 5 trillion single-use plastic bags are used world-wide every year. In total, half of all plastic produced is designed to be used only once and then thrown away. Plastic waste is now so common in the natural environment that some scientists have even suggested that plastic could serve as a geological indicator of the Anthropocene Era (United Nations 2018). From the 1950s through the 1970s, only a small amount of plastic was produced and was not a significant issue. By the 1990s plastic waste had more than tripled. In the early 2000s, plastic output increased more than in the previous 40 years. As of the end of 2018, more than 300 million metric tons of plastic waste is generated per year, which is equivalent to the weight of the entire human population (United Nations 2018). Although plastic containers and packaging are nearly impossible to regulate by USEPA, there is a clear need to address this issue.

6.9.15 SUMMARY OF IMPROVING OR NEED OF NEW REGULATIONS

The sections highlighted above describe the majority of subject areas where environmental regulations in the United States are either in need of improvement or where new regulations are needed. For reference purposes and to list those areas that have the greatest need for new or improved environmental regulations, the following is provided:

- Modify Environmental Enforcement Emphasis and Policy
- Science-based improvements to environmental regulations to account for regional risks and climate change
- Agriculture
- Pesticides and herbicides
- Fertilizers
- Septic tanks

- Invasive species
- Urban planning and land use
- Residential and household waste
- Urban air
- Further or new restrictions on harmful chemicals, especially those that are persistent and mobile in the environment
- Emerging contaminants, such as those described in Chapter 2
- Noise
- Plastic
- Sustainability

Many of the items listed above fall into the exceptions scenario that we first touched on and explained in the introduction of this chapter. The main theme with the exceptions is that environmental regulations should address all identified risks for a given practice, situation, or exposure scenario. It makes little sense if human health and the environment are protected from nine out of ten exposure risks when the remaining risk factor causes harm. This example is realized when considering the lack of regulations of the items listed above. Another item worthy of discussion is that there are political factors and areas where there is overlapping jurisdictions that may have negative outcomes that must be addressed when discussing the need for further environmental regulations in the United States. For instance, the United States Department of Agriculture would almost certainly interject an interest in environmental regulation of farm land. In addition, the Department of Interior would also be interested in sustainability regulations as it pertains to national forests and parks and environmental regulations on rangeland.

This will become more evident as we discuss the challenges that are faced by many of the countries of the world in the next chapter.

6.10 SUMMARY OF ENVIRONMENTAL REGULATIONS OF THE UNITED STATES

After reading this chapter, you should have a much greater understanding of just how much environmental regulation we have in the United States and why. It should also be apparent that staying in compliance is a complex undertaking requiring much skill and expertise.

There is much we can be proud of when we examine the improvements of our air quality, water quality, and cleanup of sites of environmental contamination, which now number in the tens of thousands of sites across the United States. Looking back over the sections of different environmental regulations, you may have noticed that much of the legislation was oftentimes a reaction to an incident or environmental tragedy. Laws and regulations originated from many circumstances because of unintentional actions that ended up making quite a mess and causing significant harm to human health, the environment, or both. Some of the incidences that we have touched upon include the Exxon Valdez spill in 1990, Deepwater Horizon in 2010, Love Canal, Bhopal India in 1984, Valley of Drums, and Toms River to the killing and extinction of animal species that also nearly included the American bison, California condor, grizzly bear, and numerous others. However, this is a reactionary approach and not a proactive approach. When we discuss sustainability later in this book, we will discuss the need to move beyond reactive measures to proactive sustainability measures to improve our environment in the future. If there are delays in enacting sustainability measures, it may be too late.

The amount and complexity of environmental laws and regulations in the United States is due to the fact that we are an industrialized society and consume enormous amounts of energy and goods and our hunger for those goods and energy continues to rise as we continue to improve our standard of living. The United States produces large volumes of waste of all types and as our population continues to increase, the amount of waste produced will also continue to increase.

These central facts combined with the dynamic and innate complexities of the natural world add an almost infinite number of negative outcomes in a world with over 7 billion people all wanting to improve their standard of living. The fact is, if we do not have effective and comprehensive environmental regulation, we simply threaten our existence and numerous other species as we are slowly devoured in our own garbage.

Our advances in science have created many technologies that have improved our standard of living and in some circumstances have improved our relationship with nature. But our technological advancements have also created many unintended negative side effects that question the benefit of some technologies. An example was presented in Chapter 2 and involves the creation of synthetic compounds that nature has difficulty breaking down. Some of these compounds include polychlorinated biphenyls, chlorinated hydrocarbons, and many pesticides, including DDT, which began the environmental movement in the United States in earnest over 50 years ago. This is to say nothing about plastics which we all should know about since we commonly use and discard huge amounts of plastic in our everyday life.

Although efforts are underway to reduce our energy needs and our veracious hunger for consuming nature, one thing seems certain, our thirst for consumer goods and energy needs will continue to grow, and as our population increases our energy needs will increase even faster. This means that environmental regulation will have to grow with us in volume and complexity as we consume more and more of Earth's resources and replace it with our waste.

The next chapter will focus on environmental regulations and pollution world-wide. We will focus on industrialized nations and a few developing nations with large populations including India and China. We will not be examining countries with what are considered extreme hardships that include Somalia, Syria, Afghanistan, Iran, and others.

We will evaluate how countries' environmental laws compare to the United States. Some findings might be rather surprising in that it will become apparent that a few countries are more advanced at protecting human health and the environment than the United States. Surprising still might be the impression that some countries either lack environmental laws or the will to enforce the environmental laws they have due to other overriding social or economic challenges. As we shall see in the next chapter, it is not enough just to have environmental regulations. Many other factors are required to have an effective environmental regulatory framework that actually accomplishes that ultimate goal of protecting human health and the environment.

REFERENCES

American Academy of Pediatrics. 2011. Policy Statement – Chemical-Management Policy: Prioritizing Children's Health. *Journal of the American Academy of Pediatrics*. Vol. 127. No. 5. pp. 983–990.

American Society for Testing Materials. 2005. *Standard Practice for Environmental Site Assessments*. E1527-05. ASTM, West Philadelphia, PA.

Bates, C.G. and Ciment, J. 2013. *Encyclopedia of Global Social Issues*. M.E. Sharpe Publishers. New York. 1450p.

Beck, E.C. 1979. The Love Canal Tragedy. *Journal of the Environmental Protection Agency*. Washington, DC. www.epa.gov/epa/aboutepa/love-canal-tragedy.html. (accessed March 29, 2017).

Brownwell, F.W. and Zeugin, Z.B. 1991. *Clean Air Handbook*. Government Institutes, Inc. Rockville, Maryland. 318p.

California Department of Toxic Substances and Control (DTSC). 2013. *Chemical Look-Up Table*. Technical Memorandum. Sacramento, California. 5p.

California Department of Toxic Substance and Control (DTSC). 2017. www.dtsc.ccelern.csus.edu/wasteclass. (accessed March 24, 2017).

California Environmental Protection Agency (CalEPA). 2001. *Storm Water Sampling Guidance Document*. California Stormwater Task Force. Sacramento, California. 30p.

California Environmental Protection Agency (CalEPA). 2005. *Use of Human Health Screening Levels (CHHSLs) in Evaluation of Contaminated Properties*. Sacramento, California. 65p.

California Environmental Protection Agency (CalEPA). 2017. *Air Emission Background Information*. www.arb.ca.gov/ei/general.htm. (accessed March 3, 2017).

California Regional Water Quality Control Board. 2018. *PG & E Hinkley Chromium Cleanup*. www.waterboards.co.gov/lahontan/water_issues/projects/pge/. (accessed December 9, 2018).

Carson, R. 1962. *Silent Spring.* Houghton Mifflin, Boston, MA.

Carlisle, E. 2000. *The Gulf of Mexico Dead Zone and Red Tides.* The Louisiana Environment. Tulane University. New Orleans, LA. 6p.

Conn, R.L., Leng, M.L., and Solga, J.R. 1984. *Pesticide Regulatory Handbook.* Executive Enterprises Publishing Company. New York. 450p.

Cornell University Law. 2017. *Resource Conservation and Recovery Act (RCRA) Overview.* www.law.cornell.edu/wex/rcra. (accessed March 23, 2017).

Curry, A. 2017. *Will Newer Wind Turbines Mean Fewer Bird Deaths. The National Geographic.* www.nationalgeographic.com/news/energy/2014/04/140427altamont-pass. (accessed April 5, 2017).

Earth Day. 2018. Earth Day Network Campaign: End Plastic Pollution. https://earthday.or/campaigns/plastic-campaign. (accessed December 24, 2018).

Eccleston, Charles S. 2008. *NEPA and Environmental Planning: Tools, Techniques, and Approaches for Practitioners.* CRC Press. New York. 432p.

Eckerman, I. 2005. *The Bhopal Saga – Causes and Consequences of the World's Largest Industrial Disaster.* Universities Press. London, UK. 284p.

Environmental Education Associates. 2019. The Famous Benjamin Franklin Letter on Lead Poisoning. http://environmentaleducation.com/wp-content/uploads/userfiles/Ben%20Franklin%20Letter%20on%20EEA%281%29.pdf. (accessed March 3, 2019).

Eveleth, R. 2013. *How Many Birds to Wind Turbines Really Kill?* Smithsonian Magazine. Washington, DC. www.smithsoniation.com/smart-new/how-many-birds-do-wind-turbines-kill. (accessed April 5, 2017).

Hahn, R.W. 1994. United States Environmental Policy: Past, Present and Future. *Natural Resource Journal.* John F. Kennedy School of Government, Harvard University. Cambridge, MA. Vol. 34. No 1. pp. 306–348.

Haynes, W. 1954. *American Chemical Industry – A History.* Vols. I–IV. Van Nostrand Publishers. New York.

Illinois Environmental Protection Agency (IEPA). 2017. *Tiered Approach to Corrective Action Objectives (TACO).* IEPA. Springfield, IL. 284p.

International Union on the Conservation of Nature (IUCN). 2019. *Extinct Species in North America since 1500 AD.* www.iucn.org. (accessed February 20, 2019).

Journal of Roman Studies. 2019. *Fresh Water in Roman Law: Rights and Policy. Vol. 107.* www.cambridge.org/core/journals/journal-of-roman-studies/article/fresh-water-in-roman-law-rights-and-policy/548B1C559B3D6ACEDF50C4576DD14603/core-reader#fns01. (accessed March 3, 2019).

Kaufman, M.M., Rogers, D.T., and Murray, K.S. 2011. *Urban Watersheds: Geology, Contamination, and Sustainability.* CRC Press. New York. 583p.

Lazarus, R. 2004. *The Making of Environmental Law.* Cambridge Press. Cambridge, UK. 367p.

Malakoff, D. 1998. Death by Suffocation in the Gulf of Mexico. *Journal of Science.* Vol. 281. pp. 190–192.

Markell, D. 2014. An Overview of TSCA, Its History and Key Underlying Assumptions, and Its Place in Environmental Regulation. New Directions in Environmental Law. *Washington University Journal of Law and Policy.* Vol. 32. No. 448. pp. 333–376.

Michigan Department of Environmental Quality. 2016. *Cleanup Criteria and Screening Levels Development and Application.* www.michigan.gov/documents/deq/deq-rrd-chem-CleanupCriteriaTSD_527410_7.pdf. (accessed February 17, 2019).

National Aeronautics and Space Administration (NASA). 2019. *Largest-Ever Ozone Hole over Antarctica.* https://visibleearth.nasa.gov/view.php?id=54991. (accessed March 3, 2019).

National Energy Institute. 2005. *Summary of the Energy Policy Act of 1992 and 2005.* www.nei.org/summary-of-energy-policy-act. (accessed April 4, 2017).

National Law Review. 2016. *What's New About the Revised TSCA – Toxic Substance Control Act.* www.natlawreview.com/article/tsca. (accessed March 22, 2017).

National Oceanic and Atmospheric Administration (NOAA). 2017a. *Coastal Zone Management Act.* www.noaa.gov/coastal-zone-management-act.html. (accessed April 3, 2017).

National Oceanic and Atmospheric Administration (NOAA). 2017b. *Marine Mammal Protection Act.* https://noaa.gov/marine-mammal-protection-act. (accessed April 17, 2017).

National Oceanic and Atmospheric Administration (NOAA). 2017c. *Exxon Valdez Oil Spill.* Office of Response and Restoration. www.noaa.gov/exxon-valdez-oil-spill-restoration. (accessed April 4, 2017).

National Oceanic and Atmospheric Administration (NOAA). 2017d. *Deepwater Horizon Oil Spill.* Office of Response and Restoration. https://noaa.gov/deepwater-horizon-oil-spill. (accessed April 4, 2017).

National Oceanic and Atmospheric Administration (NOAA). 2017e. *Summary of the Marine Mammal Protection Act of 2015.* https://noaa.gov/Marine-mammal-protection-act-summary. (accessed April 4, 2017).

National Oceanic and Atmospheric Administration (NOAA). 2018. *Gulf of Mexico Harmful Algal Bloom Bulletin. Southwest Florida Region.* https://go.usa.gov/xn9g2. (accessed December 11, 2018).

National Oceanic and Atmospheric Administration (NOAA). 2019. *Koppen-Geiger Climate Changes.* https://sos.noaa.gov/datasets/koppen-geiger-climate-changes-1901-2100/. (accessed February 16, 2019).

New Jersey Department of Environmental Protection (NJDEP). 2008. *Introduction to Site-Specific Impact to Ground Water Soil Remediation Standards Guidance Document.* Trenton, NJ. 10p. www.nj.gov/dep/srp/guidance/rs/igw. (accessed May 5, 2017).

New Jersey Department of Environmental Protection (NJDEP). 2015. *Site Remediation Program: Remediation Standards.* www.nj.gov/dep/srp/guidance. (accessed May 4, 2017).

New Jersey Department of Environmental Protection (NJDEP). 2017. *Soil Cleanup Criteria.* www.nj.gov.srp/guidance/scc. (accessed May 5, 2017).

Nuclear Regulatory Commission. 2017. *Nuclear Waste Policy Act of 1982.* www.nrc.gov/NWPA. (accessed April 4, 2017).

Ohio Environmental Protection Agency (Ohio EPA). 2014. *Land Disposal Restrictions (An Overview).* Division of Materials and Waste Management. Ohio EPA. Columbus, OH. 3p.

Ohio Environmental Protection Agency (Ohio EPA). 2019. *Air Permit Exemptions.* www.ohio.gov/portal/41/sb/publications/airpermit/exemptions.pdf. (accessed February 19, 2019).

Pennsylvania Historical Association. 2019. *The* Vision of William Penn. http://explorepahistory.com/story.php?storyId=1-9-3&chapter=1. (accessed March 3, 2019).

Rogers, D.T. 1992. The Importance of Site Observation and Follow-up Environmental Site Assessment – A Case Study. *In*: Stanley, A. editor. *Groundwater Management Book 12*, pp. 563–573. Dublin, OH: National Groundwater Association.

San Francisco Chronicle. 1996. *PG&E to Pay 333 Million in Pollution Suit.* www.sfgate.com/article/PG-E-to-Pay-333-million-in-pollution-suit-3303933.php. (accessed December 9, 2018).

Schierow, L.J. 2013. *The Toxic Substance Control Act (TSCA): A Summary of the Act and Its Major Requirements.* Congressional Research Service (CRS) Report to Congress. Washington, DC. 16p.

Smithsonian. 2019. *Air Pollution Goes Back Further than You Think.* www.smithsonianmag.com/science-nature/air-pollution-goes-back-way-further-you-think-180957716/. (accessed March 3, 2019).

Sun, M. 1983. Missouri's Costly Dioxin Lesson. *Science.* Vol. 219. pp. 367–369.

Tarlock, D.A. 2001. *History of Environmental Law.* Encyclopedia of Life Support Systems. University of Chicago Press. Chicago, IL. 1450p.

Texas Commission on Environmental Quality (TCEQ). 2018. *Hurricane Harvey Response: Arkema Chemical Plant.* Crosby, TX. www.tceq.gov/respone/hurricanes/arkema-facility-response. (accessed December 23, 2018).

United Nations. 2016. *Human Development Report.* United Nations Development Programme. www.hdr.undr.org/sites/2017. (accessed September 16, 2017).

United Nations (UN). 2017a. *Environmental Principles and Concepts Derived from the 1972 Stockholm Conference and the 1992 Rio Declaration on Environmental Law and Training.* www.UNEP.org. (accessed February 25, 2017).

United Nations. 2017b. *World Population through Time.* www.unitednations.org/population. (accessed March 4, 2017).

United Nations. 2018. *The Facts on Plastic Pollution.* www.unenvironment.org/beat-plastic-pollution. (accessed December 24, 2018).

United Nations. 2019a. The United Nations Environment Programme. www.unenvironment.org/environment-you. (accessed March 3, 2019).

United Nations. 2019b. *Intergovernmental Science-Policy Platform on Biodiversity and Ecosystems Services (IPBES) Summary for Policymakers of the Methodological Assessment Report of the IPBES Scenarios and Models of Biodiversity and Ecosystems Services.* www.ipbes.net/system/tdf/downloads/pdf/spm_deliverable_3c_scenarios_20161124.pdf?file=1&type=node&id=15245. (accessed May 9, 2019).

United States Advisory Council on Historic Preservation. 2017. *National History Preservation Act.* www.achp.gov. (accessed April 3, 2017).

United States Army Corps of Engineers. 2017. *Overview of Wetlands Regulation in the United States.* www.saw.usace.army.mil/wetlands. (accessed March 15, 2017).

United States Bureau of Land Management (BLM). 2017. *Summary of the Mineral Leasing Act.* .www.blm.gov/mineral-leasing-act.html. (accessed April 3, 2017).

United States Bureau of Reclamation. 2017. *Summary of the Federal Power Act.* www.usbr.gov/federal-power-act/html. (accessed April 3, 2017).

United States Census Bureau 2019. *United States Population through Time and World Populations Clock.* www.UScensus.gov/USpopluation. (accessed February 20, 2019).

United States Central Intelligence Agency. 2019. *World Fact Book. Map of North America.* www.cia.gov/library/publications/the-world-factbook/attachments/images/large/north_america-physical.jpg?154714 5652. (accessed February 16, 2019).

United States Coast Guard. 2017. *U. S. Coast Guard Responds to Deepwater Horizon.* New Orleans, LA. www.uscg.gov/deepwater-horizon-images. (accessed April 4, 2017).

United States Department of Agriculture. 2011. *Land Use in the United States.* www.ers.usda.gov/data-products/major-land-use. (accessed December 8, 2018).

United States Department of Agriculture. 2014. *United States Forest Resource Facts and Historical Trends.* www.fia.fs.fed.us/library/brochures/docs/2012/ForestFacts_1952-2012_English.pdf. (accessed May 4, 2019).

United States Department of Commerce. 2018. *Principle Federal Economic Indicators.* Bureau of Economic Analysis. www.bea.gov. (accessed December 8, 2018).

United States Department of Labor. 2017. *History of the Occupational Safety and Health Administration.* www.osha.gov./history. (accessed February 28, 2017).

United States Department of Transportation. 2013. *Federal Motor Carrier Safety Administration (FMCSA) Placard Requirements.* www.fmcsa.dot.gov. (accessed March 22, 2017).

United States Department of Transportation. 2017. *Summary of the Federal Hazardous Material Transportation Act.* Office of Hazardous Materials Safety. www.USDOT.gov./HMTA. (accessed March 18, 2017).

United States Environmental Protection Agency (USEPA). 1988. *Superfund Record of Decision: Reich Farms,* Toms River, NJ. Office of Emergency and Remedial Response, Washington, DC. EPA/ROD/R-02-88/070. 108p.

United States Environmental Protection Agency (USEPA). 1989a. *Superfund Record of Decision: Wells G&H, Woburn, Massachusetts.* Office of Emergency and Remedial Response, Washington, DC. EPA/ROD/R-01-89/036. 85p.

United States Environmental Protection Agency (USEPA). 1989b. *Superfund Record of Decision: Ciba-Geigy,* Toms River, NJ. Office of Emergency and Remedial Response, Washington, DC. EPA/ROD/R-02-89/076. 120p.

United States Environmental Protection Agency (USEPA). 1991a. *Land Disposal Restrictions: Summary of Requirements.* USEPA Office of Solid Waste and Emergency Response. Washington, DC. 86p.

United States Environmental Protection Agency (USEPA). 1991b. *U.S. Efforts to Address Global Climate Change: Report to Congress.* United States Department of State. Washington, DC. 87p.

United States Environmental Protection Agency (USEPA). 1992. *Lead and Copper Rule.* www.epa.gov/dwreginfo/lead-and-copper-rule. (accessed March 3, 2019).

United States Environmental Protection Agency (USEPA). 2004. *EPA Removes Love Canal from Superfund List.* www.epa.gov./admpress/love-canal. (accessed March 29, 2017).

United States Environmental Protection Agency (USEPA). 2005a. *Introduction to Hazardous Waste Identification (40CFR Parts 261).* USEPA Office of Solid Waste and Emergency Response (5305 W). EPA530-K-05-012. Washington, DC. 30p.

United States Environmental Protection Agency (USEPA). 2005b. *Standards and Practice for All Appropriate Inquiries.* 40 Code of Federal Regulation (CFR), Part 312. US Government Printing Office, Washington, DC.

United States Environmental Protection Agency. 2006. *Methyl Tertiary Butyl Ether (MTBE).* https://archive.epa.gov/mtbe/html/faq.html. (accessed March 3, 2019).

United States Environmental Protection Agency. 2007a. *Guide to the Clean Air Act.* Office of Air Quality Planning and Standards. Research Triangle Park, NC. 24p.

United States Environmental Protection Agency. 2007b. *The Plain English Guide to the Clean Air Act.* Office of Air Quality Planning and Standards. Research Triangle Park, NC. 26p.

United States Environmental Protection Agency. 2008. *Acid Rain.* www.epa.gov/acidrain. (accessed March 21, 2017).

United States Environmental Protection Agency. 2009. *Industrial Stormwater Monitoring and Sampling Guide.* EPA 832-B-09-003. Washington, DC. 49p.

United States Environmental Protection Agency. 2010. *National Pollution Discharge Elimination System Permit Writer's Manual.* United States Environmental Protection Agency. Office of Water. Washington, DC. 320p.

United States Environmental Protection Agency. 2011. *PSD and Title V Permitting Guidance for Greenhouse Gases.* Office of Air and Radiation. Washington, DC. 94p.

United States Environmental Protection Agency. 2012. *Spill Prevention, Control and Countermeasures Plan (SPCC) Program*. United States Environmental Protection Agency (USEPA). Office of Emergency Management. Washington, DC. www.epa.gov/emergencies. (accessed March 15, 2017).

United States Environmental Protection Agency. 2013. *USEPA Greenhouse Gas Reporting Program*. https://epa.gov/green-house-gas-reporting. (accessed March 3, 2017).

United States Environmental Protection Agency. 2014. *The Toxic Substance Control Act: History and Implementation*. USEPA, Washington, DC. www.epa.gov/tsca/summary. (accessed March 22, 2017).

United States Environmental Protection Agency. 2015a. *Acid Rain Emission Reductions*. USEPA Archive. www.epa.gov/acid-rain-reductions. (accessed March 22, 2017).

United States Environmental Protection Agency, USEPA. 2015b. *List of Lists*. EPA 55-B-15-001. Office of Solid Waste and Emergency Response. Washington, DC. 125p.

United States Environmental Protection Agency. 2016a. *Improved Vehicle Emissions Guide*. www.epa.gov/vehicleemissions. (accessed March 4, 2017).

United States Environmental Protection Agency. 2016b. *Flint Drinking Water Documents*. www.epa.gov/flint/flint-drinking-water-documents. (accessed March 3, 2019).

United States Environmental Protection Agency. 2017a. *National Environmental Protection Act*. www.epa.gov/NEPA. (accessed February 23, 2017).

United States Environmental Protection Agency. 2017b. *Summary of Environmental Regulations of the United States*. https://epa.gov/summary-environmental-laws. (accessed February 24, 2017).

United States Environmental Protection Agency. 2017c. *Pesticide Laws and Regulations*. www.epa.gov./pesticide. (accessed February 27, 2017).

United States Environmental Protection Agency. 2017d. *Summary of the Clean Air Act*. www.epa.gov/laws-regulations/summary-clean-air-act. (accessed March 1, 2017).

United States Environmental Protection Agency (USEPA). 2017e. *NAAQS Table*. www.epa.gov/criteria-air-pollutants/naaqs-table. (accessed March 3, 2017).

United States Environmental Protections Agency (USEPA). 2017f. *Non-attainment Areas of the United States*. www.jpg.gov/nonattainmentmap. (accessed March 2, 2017).

United States Environmental Protection Agency (USEPA). 2017g. *Hazardous Air Pollutants (HAPs)*. www.epa.gov/haps. (accessed February 20, 2019).

United States Environmental Protection Agency (USEPA). 2017h. *Title V Operating Permits*. www.epa.gov.title-v-operating-permits. (accessed March 3, 2017).

United States Environmental Protection Agency (USEPA). 2017i. *National Emission Standards for Hazardous Air Pollutants (NESHAPs)*. www.epa.gov/NESHAPs. (accessed February 20, 2017).

United States Environmental Protection Agency (USEPA). 2017j. *Area Sources of Urban Air Toxics*. www.epa.gov/urban-air-toxics/area-sources-urban-air-toxics. (accessed February 20, 2017).

United States Environmental Protection Agency (USEPA). 2017k. *Endangerment and Cause or Contribute Findings for Greenhouse Gases. Section 202(a) of the Clean Air* Act. www.epa.gov.climatechange/endangerment. (accessed March 3, 2017).

United States Environmental Protection Agency (USEPA). 2017l. *Summary of the Safe Drinking Water Act*. www.epa.gov/laws-regulations/summary-safe-drinking-water-act. (accessed March 4, 2017).

United States Environmental Protection Agency. 2017m. *Drinking Water Standards*. www.epa.gov/safedrinkingwater. (accessed March 15, 2017).

United States Environmental Protection Agency. 2017n. *Hazardous Materials Transportation* Act. www.epa.gov/HMTA/overview. (accessed March 21, 2017).

United States Environmental Protection Agency. 2017o. *Summary of the Toxic Substance Control Act*. https://epa.gov/laws-regulations/summary-toxic-substances-control-act. (accessed March 22, 2017).

United States Environmental Protection Agency. 2017p. *Summary of the Resource Conservation and Recovery Act*. www.epa.gov/rcra/summary. (accessed March 23, 2017).

United States Environmental Protection Agency (USEPA). 2017q. *Criteria for the Definition of Solid Waste and Solid and Hazardous Waste Exclusions*. .www.epa.gov/criteria-definition-solid-waste. (accessed March 30, 2017).

United States Environmental Protection Agency (USEPA). 2017r. *USEPA Office of Solid Waste. SW-846 Test Methods Manual*. USEPA. Washington, DC. 1, 357p. www.USEPA.gov/sw-846. (accessed March 24, 2017).

United States Environmental Protection Agency (USEPA). 2017s. *Resource Conservation and Recovery Act Listed and Characteristic Hazardous Wastes*. Office of Solid Waste. Washington, DC. www.epa.gov/hazarodus-wastes. (accessed March 26, 2017).

United States Environmental Protection Agency (USEPA). 2017t. *Used Oil Regulations Under the Resource Conservation and Recovery Act.* Office of Solid Waste. Washington, DC. www.epa.gov.used-oil. (accessed March 26, 2017).

United States Environmental Protection Agency (USEPA). 2017u. *CERCLA Sites in the United States.* www. epa.gov/CERCLA-sites. (accessed March 26, 2017).

United States Environmental Protection Agency (USEPA). 2017v. Times Beach, *Missouri Archive.* www.epa. gove/times-beach. (accessed March 27, 2017).

United States Environmental Protection Agency (USEPA). 2017w. *Love Canal Tragedy.* www.epa.gov/history/love-canal. (accessed March 29, 2017).

United States Environmental Protection Agency (USEPA). 2017x. *Superfund Site Profile: Valley of Drums, Kentucky.* www.epa.gov/valley-of-drums-profile. (accessed March 29, 2017).

United States Environmental Protection Agency (USEPA). 2017y. *EPA to Update Community on Toms River, New Jersey Superfund Site.* https://archive.epa.gov/epa/newsreleases/epa-update-community-toms-river-nj-superfund. (accessed December 9, 2018).

United States Environmental Protection Agency (USEPA). 2017z. *Superfund Amendments and Reauthorization Act (SARA).* www.epa.gov/SARA. (accessed March 31, 2017).

United States Environmental Protection Agency (USEPA). 2017aa. *Emergency Planning and Community-Right-to-Know Act (EPCRA).* www.epa.gov/EPCRA. (accessed March 31, 2017).

United States Environmental Protection Agency (USEPA). 2017ab. *Pollution Prevention Act.* www.epa.gov/pollution-prevention-act. (accessed April 3, 2017).

United States Environmental Protection Agency (USEPA). 2017ac. *Summary of the Noise Control Act.* www. epa.gov/noise-control-act. (accessed April 3, 2017).

United States Environmental Protection Agency (USEPA). 2017ad. *Summary of the Nuclear Waste Policy Act.* www.epa.gov/laws-regulations/summary-nuclear-waste-policy-act. (accessed April 4, 2017).

United States Environmental Protection Agency (USEPA). 2017ae. *Global Climate Change Act.* www.epa. gov/GCCA. (accessed April 4, 2017).

United States Environmental Protection Agency (USEPA). 2017af. *Summary of the Oil Pollution Act of 1990.* https://epa.gov/oil-pollution-act. (accessed April 4, 2017).

United States Environmental Protection Agency (USEPA). 2017ag. *USEPA Exxon Valdez Spill Profile.* www. epa.gov/emergency-response/exxon-valdez-spill-profile. (accessed April 4, 2017).

United States Environmental Protection Agency (USEPA). 2017ah. *Summary of the Food Quality Protection Act.* www.epa.gov/fqpa-summary. (accessed April 4, 2017).

United States Environmental Protection Agency (USEPA). 2017ai. *Addressing Major Public Policy Concerns: Accomplishments Under the Food Quality and Protection Act.* www.epa.gov/fqpa/policy-concerns. (accessed April 4, 2017).

United States Environmental Protection Agency (USEPA). 2018a. *Regulations for Greenhouse Gas (GHG) Emissions.* www.epa.gov/regulations-emissions-vehicles-and-engines/regulations-greenhouse-gas-ghg-emissions. (accessed December 24, 2018).

United States Environmental Protection Agency (USEPA). 2018b. *Air Quality – 2018 National Summary.* www.epa.gov/air-trends/air-quality-national-summary. (accessed December 24, 2018).

United States Environmental Protection Agency (USEPA). 2018c. *Summary of the Clean Water Act.* https://epa.gov/laws-regulations/summary-clean-water-act. (accessed December 18, 2018).

United States Environmental Protections Agency. 2018d. *Basic Information about Nonpoint Source (NPS) Pollution.* www.epa.gov/nps/basic-inofmration-about-nonpoint-source-nps-pollution. (accessed December 11, 2018).

United States Environmental Protection Agency. 2018e. *Enforcement Policy, Guidance and Publications.* www.epa.gov/enforcement/enforement-policy-guidance-publications. (accessed December 11, 2018).

United States Environmental Protection Agency. 2018f. *Enforcement Policy, Guidance and Publications.* www.epa.gov/enforcement/criminal-enforement-overview. (accessed December 11, 2018).

United States Environmental Protection Agency. 2018g. *Basic Information in Enforcement.* www.epa.gov/enforcement/basic-information-enforement. (accessed December 11, 2018).

United States Environmental Protection Agency. 2018h. *Oil Spills and Preparedness Regulations.* www.epa.gov/oil-spills-preventions-and-preparedness-regulations. (accessed December 11, 2018).

United States Environmental Protection Agency. 2019a. *Laws and Regulations.* www.epa.gov/laws-regulations/regulations. (accessed February 17, 2019).

United States Environmental Protection Agency. 2019b. *Generally Available Control Technologies.* www.epa.gov/airtoxics/112dpg.html. (accessed February 17, 2019).

United States Environmental Protection Agency (USEPA). 2019c. *National Recommended Water Quality Criteria-Human Health Criteria Table.* www.epa.gov/wqc/national-recommended-water-quality-criteria-human-health-criteria-table. (accessed February 17, 2019).

United States Environmental Protection Agency (USEPA). 2019d. *Drinking Water Contaminants – Standards and Regulations.* www.epa.gov/dwstandardsregulations. (accessed February 17, 2019).

United States Environmental Protection Agency (USEPA). 2019e. *Defining Hazardous Waste.* www.epa.gov/hw/defining-hazardous-waste-listed-characteristic-and-mixed-radiological-wastes. (accessed February 17, 2019).

United States Environmental Protection Agency (USEPA). 2019f. *Hazardous Waste Manifests.* www.epa.gov/hwgenerators/uniform-hazardous-waste-manifest-instructions-sample-form-and-continuation-sheet. (accessed February 20, 2019).

United States Environmental Protections Agency (USEPA). 2019g. *CERCLA.* www.epa.gov.superfund. (accessed February 20, 2019).

United States Environmental Protection Agency (USEPA). 2019h. *Regional Screening Levels.* www.epa.gov/RSLs. (accessed February 20, 2019).

United States Environmental Protection Agency. 2019i. *Superfund Enforcement: 35 Years of Protecting Communities and the Environment.* www.epa.gov/enforcement/superfund-enforcement-35-years-protecting-communities-and-environment. (accessed February 17, 2019).

United States Environmental Protection Agency. 2019j. *Innocent Landowner Defense.* www.epa.gov/enforcement/innocent-landowner. (accessed February 20, 2019).

United States Environmental Protection Agency. 2019k. *Brownfields.* www.epa.gov/brownfields/all-appropriate-inquiry. (accessed February 20, 2019).

United States Environmental Protection Agency. 2019l. *Agriculture 101.* www.epa.gov/sites/production/files/2015-07/documents/ag_101_agriculture_us_epa_0.pdf. (accessed April 24, 2019).

United States Fish and Wildlife Service. 2017a. *Summary of the Lacey* Act. https://.fws.gov/us-conservation-laws/lacey-act-html. (accessed April 3, 2017).

United States Fish and Wildlife Service. 2017b. *Summary of the Migratory Bird Treaty* Act. www.nps.gov/migratory-bird-act. (accessed April 3, 2017).

United States Fish and Wildlife Service. 2017c. *Fish and Wildlife Coordination Act.* www.fws.gov/fish-and-wildlife-coordination-act.html. (accessed April 3, 2017).

United States Fish and Wildlife Service. 2017d. *Endangered Species Act.* www.fws.gov/endangered/laws-policies/. (accessed February 20, 2019).

United States Fish and Wildlife Service. 2017e. *Fishery Conservation and Management Act of 1976.* www.fws.gov/FCMA. (accessed April 3, 2017).

United States Forest Service. 2017. *Forest Conservation and Management* Act. www.nfs.gov/FCMA. (accessed April 3, 2017).

United States Forest Service. 2018. *Forest Products Cut and Sold from the National Forests and Grasslands.* www.fs.fed.us/forestmanagement/products/cut-sold/index.shmtl. (accessed December 8, 2018).

United States National Park Service. 2017a. *Summary of the Antiquities Act.* www.nps.gov/history/antiquities-act-html. (accessed April 3, 2017).

United States National Park Service (NPS). 2017b. *History of the National Park Service.* www.nps.gov/history. (accessed April 3, 2017).

United States National Park Service. 2017c. *Wild and Scenic Rivers* Act. www.nps.gov/wild-and-scenic-rivers-act-html. (accessed April 3, 2017).

United States Nuclear Regulatory Commission. 2017. *Summary of the Atomic Energy* Act. www.nrc.gov/atomic-energy-act/html. (accessed April 3, 2017).

United States Occupational Safety and Health Administration. 2017. www.osha.gov. (accessed February 14, 2017).

United States Office of Surface Mining and Reclamation. 2017. *Surface Mining Control and Reclamation* Act. www.osmra.gov. (accessed April 3, 2017).

United States Research Archives. 2017. www.UnitedStates.research.archives.gov (accessed February 24, 2017).

Vallianatos, E.G. 2014. *Poison Spring: The Secret History of Pollution and the EPA.* Bloomsbury Press. New York. 284p.

Wisconsin Department of Natural Resources. 2017. *Sample Stormwater Pollution Prevention Plan (SWPPP).* www.dnr.wi.gov/topic/stormwater/documents/sampleSWPPP.pdf. (accessed March 17, 2017).

World Wildlife Fund (WWF). 2017. *Species List of Endangered, Vulnerable, and Threatened Animals.* www.wwf.org/species/endangered-list. (accessed April 3, 2017).

7 Environmental Regulations and Pollution of the World

7.1 INTRODUCTION

In this chapter we will explore the environmental regulations and pollution of many countries and regions of the world. The chapter is organized by continent and then by country. We will start with North America, then move to Europe, Asia, South America, Africa, Oceania, and then conclude with Antarctica and the oceans of the world. Table 7.1 lists each country by continent that will be examined. To keep the scope of this book manageable, we have to limit the number of countries to be evaluated. In addition, we will focus on regulations that primarily address pollution of the air, water, and land. We will concentrate on industrialized nations and a few developing nations with large populations including India and China. This is because many who will be interested in environmental regulations in countries other than the United States will likely be interested in the regulations of the countries chosen to be highlighted. The countries that will be evaluated account for 70% of the human population and 75% of the land area on Earth.

To widen our perspective and grasp a more comprehensive and educated view, we will also evaluate countries that are important or unique to all humans. These countries will provide us with special insight into specific issues, situations, and problem-solving challenges due to a host of complex circumstances that some countries are facing. These countries have special challenges that, the combination of which, are unlike other countries and will require us to explore potential reasons behind some of the challenges faced by these nations as a result of social concerns, political unrest, lack of action, economic, geographic, geological or other factors that result in poor or excellent environmental performance. Some of these countries include Australia, New Zealand, Tanzania, Egypt, Brazil, Argentina, Korea, Peru, Malaysia, and Chile.

We will focus our discussion of each country on the fundamentals introduced in Chapter 1, which include:

- Protect the air
- Protect the water
- Protect the land

To address each of the above-listed fundamentals, we will examine how each country addresses air pollution, water pollution, solid and hazardous waste, and remediation of polluted sites. We will examine how most countries address legacy pollution issues through a discussion of the framework that each country has developed to investigate and remediate historical sites of environmental contamination. This means that we must also examine the significant historical sites of contamination in each country because the experiences and lessons learned from these sites have influenced and uniquely shaped each country's choices in developing environmental regulations perhaps more than most other factors.

We will compare and contrast how these countries' environmental laws differ and see how they compare to the United States. Some findings might be rather surprising in that it will become apparent that some countries are more advanced compared to the United States in how they have chosen to organize and implement their environmental regulations and where they focus internal

TABLE 7.1

Countries and Areas in Which Environmental Regulations Are Examined

North America	Europe
Canada	Austria
Mexico	Belgium
South America	Bulgaria
Argentina	Croatia
Brazil	Cyprus
Peru	Czech Republic
Chile	Denmark
Asia	Estonia
China	Finland
India	France
Japan	Germany
Saudi Arabia	Greece
South Korea	Hungary
Indonesia	Ireland
Malaysia	Italy
Oceania	Latvia
Australia	Lithuania
New Zealand	Luxemburg
Africa	Malta
Egypt	Netherlands
Kenya	Norway
Tanzania	Poland
South Africa	Portugal
Antarctica	Romania
The Oceans	Russia
	Slovakia
	Slovenia
	Spain
	Sweden
	Switzerland
	Turkey
	United Kingdom

resources. In addition, the governments of some countries have different attitudes and relationships with industry. The regulatory agencies of some countries have chosen to form working partnerships with industry with the intent to work more closely together to solve complex environmental issues rather than an advisory type approach which much too often results in polarizing points of view. The result of working together saves time, effort, and resources and in the long run has resulted in the fact that some countries are better at protecting human health and the environment than the United States.

Surprising still might be that some countries either lack key environmental laws or the will to enforce the environmental laws they have due to other social or economic challenges or the political realities some countries face. One overriding difficulty many developing countries face is lack of infrastructure, training, and education to properly implement their own environmental regulations. Another overriding factor is climate and climate change, which in some instances has greatly increased the negative effects of pollution on the environment and humans. Figure 7.1 is a map that shows the climate regions of the world that we will refer to throughout this chapter (NOAA 2018a).

Lastly, we will examine Antarctica and the world's oceans. Antarctica is a continent that has no permanent human settlements. As a first thought you might ask yourself why should we examine Antarctica, there should not be any significant environmental issues in Antarctica, correct? To most of you the answer will be humbling and perhaps even disturbing. We shall have to wade through the information and data presented in this chapter to begin to understand and appreciate the depth and magnitude of the challenges that are confronting humans from an environmental perspective and that are very evident even in Antarctica. We will also briefly discuss the world's oceans which we will learn has been the great dumping ground for human waste for centuries.

To start our journey of environmental regulations and pollution of the world, let's first look at some global assessments that have been implemented by the United Nations (UN) and the World Health Organization (WHO) that describe the current status of pollution and human health on Earth.

7.2   GLOBAL ASSESSMENTS AND STANDARDS

Pollution of the planet encompasses numerous aspects of the land, air, and water. However, there are some of what are called sinks or final resting places of pollution and the largest sink on Earth are the oceans. In fact, in the geological sciences, there is a common reference that states "whatever is created eventually ends up in the ocean." Other sinks include the land and in some instances the atmosphere.

International organizations such as the WHO and the UN have begun to assess, on a global scale, the quality of fresh water and the air on Earth. A global assessment of the land has not yet been completed. Both have published guidelines for assessing, improving, or maintaining acceptable air and water quality standards as far back as 1983 (WHO 2013; United Nations 2016a; 2016b). According to WHO (2013; 2015a; 2015b) and the United Nations (2016a; 2016b), approximately 2.8 billion humans lack access to basic sanitation and improved drinking water. While this number is improving, it still represents nearly 30% of the planet's human population.

In addition, according to WHO (2018a), nine out of ten people on Earth breathe air that is considered unacceptable because it contains high levels of pollution. The United Nations estimates that most land pollution is caused by agricultural activities, such as grazing, use of pesticides, fertilizers, irrigation, plowing, and confined animal facilities (United Nations 2016c).

Significant effort was made to include data tables for air pollution emission standards, water quality, remediation goals, and others for every chemical compound for every country evaluated. However, this was just not possible to keep this book manageable. Therefore, only select tables listing specific chemicals are provided and website addresses are provided for those not included.

7.2.1   Air

According to the World Health Organization, nine out of ten people on Earth breathe air containing high levels of pollution. Updated estimates indicate that nearly 7 million deaths occur each year from breathing unhealthy air (WHO 2018a). Most of the deaths occur in low to middle-income countries in Asia and Africa. The most problematic pollutant is fine particulate matter, which are particles less than 2.5 microns and penetrate deep into the lungs and cause diseases that include (WHO 2018a):

• Stroke	• Heart disease	• Lung cancer
• Pulmonary diseases	• Respiratory infection	• Pneumonia

The major sources of air pollution world-wide include (WHO 2018a):

- Inefficient use of energy in households
- Industry
- Agriculture
- Transportation
- Use of coal

The inefficient use of energy in households is directly related to access to clean cooking fuels, which is the case for an estimated population of 3 billion people world-wide (WHO 2018a). The most significant types of air pollutants world-wide include (WHO 2018a):

• Particulate matter (PM10)	• Fine particulate matter (PM2.5)
• Carbon monoxide (CO)	• Ozone (O_3)
• Sulfate	• Nitrates
• Black carbon	• Lead
• Volatile organic compounds (VOCs)	• Benzo(a)pyrene
• Benzene	

The World Health Organization in 2017 has ranked the top ten most air-polluted cities in the world as the following (WHO 2018a):

1. Cairo, Egypt	2. Delhi, India
3. Beijing, China	4. Moscow, Russia
5. Istanbul, Turkey	6. Guangzhou, China
7. Shanghai, China	8. Buenos Aires, Argentina
9. Paris, France	10. Los Angeles, United States

The World Health Organization published ambient air quality guidelines in 2005 for ozone, particulate matter, sulfur dioxide, nitrogen dioxide, and carbon monoxide (see Table 7.2).

The guideline limits listed in Table 7.2 are not enforceable and are based on human health risk posed by air pollutants. WHO examined 37 air pollutants but has only set guidelines for the five listed in Table 7.2. WHO includes guidance for countries in the form of statistical modeling related to exposure and risk for those compounds that are of concern but did not establish a guideline. This is due to many factors that relate to exposure, risk, geography, and other factors at specific locations (WHO 2014).

Many cities throughout the world exceed WHO guidelines for air quality. Figure 7.2 shows the major cities of the world that exceed the WHO guidelines for particulate matter (PM2.5) on an annual basis (WHO 2016a). As you can see from examining Figure 7.2, most cities in the United States, Australia, Canada, and Europe do not exceed the WHO guidelines. However, many cities in southern Asia, central Africa, and Mexico City exceed the WHO guidelines. The most air-polluted city in the United States ranks tenth worst in the world and is Los Angeles, California.

TABLE 7.2
WHO Ambient Air Guidelines

Parameter	Units of Measure	Guideline Limit
Ozone (O_3)	Parts per billion	50
Particulate Matter (PM10)	Micrograms per cubic meter	25
Sulphur dioxide (SO_2)	24 hour, parts per billion	8
Nitrogen dioxide (NO_2)	Yearly, parts per million	21
Carbon Monoxide (CO)	8 hour, parts per million	9

Source: World Health Organization. Ambient Air Pollution Database May 2014. www.who.org/ambient-air-database (accessed September 19, 2017.) and World Health Organization. World Health Organization Global Urban Ambient Air Pollution Database. 2016a. www.who.int/phe/health_to pics/outdoorair/database.htm. (accessed October 28, 2018.)

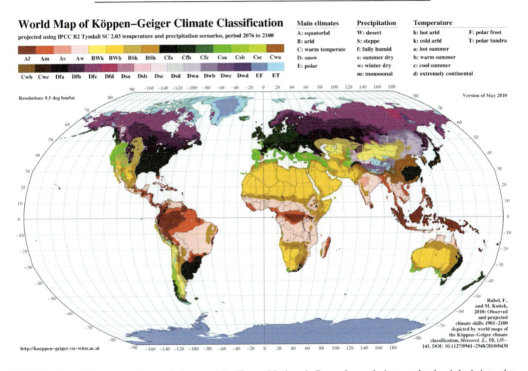

FIGURE 7.1 Climate regions of the world. (From National Oceanic and Atmospheric Administration. Koppen–Geiger Climate Changes. 2018a. https://sos.noaa.gov/datasets/koppen-geiger-climate-changes.) (Accessed February 10, 2019.)

The US Department of State has an air quality monitoring program for many large cities of the world that is updated daily and is termed an Air Quality Index (AQI) (US Department of State 2018a). The US Department of State air quality guidelines for the air quality index for fine particulate matter (PM2.5) are listed in Table 7.3.

7.2.2 WATER

Global driving forces for issues related to water include climate change, increasing water scarcity, population growth, demographic changes, and urbanization. These driving forces are expected to

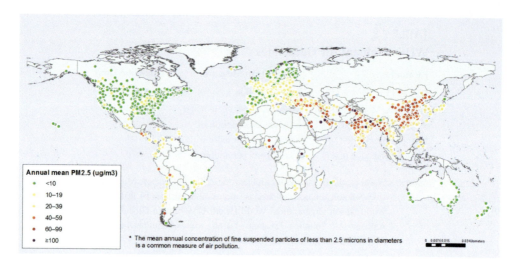

FIGURE 7.2 Annual mean PM2.5 at select cities world-wide. (From World Health Organization. World Health Organization Global Urban Ambient Air Pollution Database. 2016a. www.who.int/phe/health_topics/outdoorair/database.htm.) (Accessed October 28, 2018.)

place additional stress on fresh water supplies and sanitation systems and services. As climate change scenarios become increasingly reliable, existing infrastructure will need to be adapted and planning of new systems and services will need to be updated. Extreme weather conditions are also reflected in the increased frequency and intensity of natural disasters and will become the new normal (WHO 2013). Figure 7.2 shows the climate regions of the world.

According to the United Nations (2015a; 2015b), the most prevalent water quality problem globally is eutrophication, which is a result of high-nutrient loads (mainly phosphorus and nitrogen). The most significant sources of water pollution on Earth is agriculture. Other sources include domestic sewerage, industrial effluents, and atmospheric inputs from fossil-fueled burning and wildfires. A significant issue related to water is sanitation and the spread of disease through untreated water with human excrement and from livestock and wildlife, called zoonotic pathogens (WHO 2013). As stated above, the WHO (2013; 2015a; 2015b) and the United Nations (2016a; 2016b), estimate that 2.8 billion people do not have access to proper sanitation, while Baum et al. (2013) estimates that the actual number could be as high as 4.1 billion because most evaluations do not account for all biological disease-causing pathogens.

The WHO lists the top ten most water polluted cities in the world as the following (WHO 2018a):

1. Sumgavit, Azerbaijan	2. Linfen, China
3. Tianjing, China	4. Sukinda, India
5. Vapi, India	6. La Oroya, Peru
7. Dzerzhinsk, Russia	8. Norilsk, Russia
9. Chernobyl, Russia	10. Kabwa, Zambia

Of the countries we will evaluate, only one city is located in a country that we will not evaluate, that being Zambia. Of the countries we will evaluate, four are located in the former Soviet Union, two are located in China, two in India, and one in Peru. Contaminants that have affected the most water polluted cities in the world include (WHO 2018a):

TABLE 7.3

Air Quality Index (US Department of State 2018a)

AQI for PM2.5	Value (ug/m³)	Health Effects Statement	Cautionary Statement
Good	0–50	Little or no risk	None
Moderate	51–100	Unusually sensitive individuals may experience respiratory symptoms	Active children and adults, and people with respiratory disease, such as asthma, should limit prolonged outdoor exertion
Sensitive Groups	101–150	Increased aggravation of heart and lung disease and premature mortality in persons with cardiopulmonary disease and the elderly; significant increase in respiratory effects in the general population	Active children and adults, and people with respiratory disease, such as asthma, should avoid prolonged outdoor exertion; everyone else, especially children, should limit prolonged outdoor exertion
Unhealthy	151–200	Increased aggravation of heart and lung disease and premature mortality in persons with cardiopulmonary disease and the elderly; significant increase in respiratory effects in the general population	Active children and adults, and people with respiratory disease, such as asthma, should avoid all outdoor exertion; everyone else, especially children, should limit outdoor exertion
Very Unhealthy	201–300	Significant aggravation of heart and lung disease and premature mortality in persons with cardio-pulmonary disease and the elderly; significant increase in respiratory effects in the general population	Active children and adults, and people with respiratory disease, such as asthma, should avoid all outdoor exertion; everyone else, especially children, should limit outdoor exertion
Hazardous	301–500	Serious aggravation of heart and lung disease and premature mortality in persons with cardio-pulmonary disease and the elderly; serious risk of respiratory effects in the general population	Everyone should avoid all outdoor exertion
Beyond Index	501 and greater	Extremely high levels and health risk	Take immediate steps to reduce exposure

ug/m³ = micrograms per cubic meter

Source: US Department of State. US Department of State Air Quality Monitoring Program. 2018a. www.stateair.net/web/post/1/5.html. (accessed October 1, 2018.)

- Volatile organic compounds (VOCs)
- Oils and fuel
- Radioactive materials
- Garbage
- Zoonotic pathogens
- Biological contaminants including *E. coli*
- Pesticides
- Fertilizers
- Plastic
- Other pathogens

In 2011, the WHO published minimum standards for water quality. The WHO Water Quality Standards are intended as a benchmark for those countries that have not established standards of their own and are considered minimum standards (WHO 2015a). Table 7.4 lists the WHO Water Quality Standards.

The WHO guidelines for drinking water are not mandatory limits. They are intended to provide a scientific point of departure for the development of national or regional numerical drinking water standards for those countries that do not have standards or may have their own standards but are

TABLE 7.4
WHO Water Quality Standards

Parameter	Standard (ug/l)	Parameter	Standard (ug/l)	Parameter	Standard (ug/l)
Acrylamide	0.5	Alachlor	20	Aldicarb	10
Aldrin and dieldrin	0.03	Antimony	20	Arsenic	10
Atrazine	100	Barium	700	Benzene	10
Benzo(a)pyrene	0.7	Boron	2,400	Bromate	10
Bromodichloromethane	60	Bromoform	100	Cadmium	3
Carbon tetrachloride	4	Carbofuran	7	Chlorate	700
Chlordane	0.2	Chlorine	5,000	Chlorite	700
Chloroform	300	Chlorotoluron	30	Chlorpyrifos	30
Chromium	50	Copper	2,000	Cyanzine	0.6
2,4-D	30	2,4-DB	90	DDT	1
Dibromoacetonrile	20	1,2,3-Chloropropane	1	1,2-Dibromoethane	0.4
Dibromochloromethane	100	Dichloroacetate	50	Dichloroacetonitrile	20
1,2-Dichlorobenzene	1,000	1,2-Dichloroethane	30	1,2-Dichloroethene	5-
1,4-Dichlorobenzene	300	Dichloromethane	20	Dichloroprop	20
1,2-Dichloroporpane	40	Dimethoate	6	1,4-Dioxane	50
1,3-Dichloropropene	20	Edetic Acid	600	Endrin	0.6
Di(ethylhexyl)phthalate	8	Epichlorohydrin	0.4	Ethylbenzene	300
Fenoprop	9	Fluoride	1,500	Hydroxyatrazine	200
Hexachlorobutadiene	0.6	Isoproturon	9	Lead	10
Lindane	2	MCPA	2	Mecoprop	10
Mercury	6	Methoxychlor	20	Metolachor	10
Microcystin-LR	1	Molinate	6	Monochloroacetate	20
Nickel	70	Nitrate (as NO_3^-)	50,000	Nitrite (as NO_2^-)	3,000
Nitrilotriacetate acid	200	Pendimethalin	20	Pentachlorophenol	9
N-Nitrodimethylamine	0.1	Selenium	40	Simazine	2
Sodium dichloroisocyanurate	40,000	Styrene	20	2,4,5-T	9
Terbuthylazine	7	Tetrachloroethene	40	Simazine	2
Toluene	700	Trichloroacetate	200	Trichloroethene	20
Trifluralin	0.02	Trihalomethanes	1	Uranium	30
Vinyl chloride	0.3	Xylenes	500		

ug/l = microgram per liter

Source: WHO. Guidelines for Drinking-water Quality. Geneva, Switzerland. 516 pages. 2015a.

incomplete or, in some cases, may not be satisfactory. The WHO guidelines were established using the following criteria (WHO 2015a):

- The probability of exposure
- Chemical concentration likely to present a risk of observable health risks
- Evidence of health effects and relative ease of control of the different sources of exposure

The WHO lists many other compounds that they do not provide numerical values in the guideline due to mitigating factors such as geography, environmental factors, social, cultural, economic, dietary, and other conditions affecting potential exposure. Therefore, WHO recommends that it may be necessary to undertake a drinking water assessment before establishing standards. Chemical compounds and microbes not listed in Table 7.4 but, in many instances, should be evaluated on a case by case basis after a drinking water assessment is completed, include the following (WHO 2015a; 2015b):

• *E. coli* and other microbes	• Amitraz	• Chlorobenzilate
• Chlorothalonil	• Cypermethrin	• Deltramethrin
• Diazinon	• Dinoseb	• Ethylene thiourea
• Fenamiphos	• Formothion	• MCPB
• Hexachlorocyclohexanes	• Methamidophos	• Oxamyl
• Phorate	• Propoxur	• Pyridate
• Quintozene	• Toxaphene	• Triazophos
• Tributyltin oxide	• Trichlorfon	• Aluminum
• Ammonia	• Asbestos	• Bentazone
• Beryllium	• Bromide	• Chloride
• Bromochloroacetate	• Carbaryl	• Chloral hydrate
• *Bacillus thuringiensis*	• Chlorine dioxide	• Chloroacetones
• 2-Chlorophenol	• Chloropicrin	• Cyanide
• Cyanogen chloride	• Dialkyltins	• Dibromoacetate
• 1,1-Dichloroethene	• 1,1-Dichloroethane	• Dichloramine
• 1,3-Dichlorobenzene	• 2,4-Dichlorophenol	• Diquat
• 1,3-Dichloropropane	• Diflubenzuron	• Endosulfan
• Di(2-Ethylhexyl)adipate	• Fenitrothion	• Fluoranthene
• Formaldehyde	• Glyphosate	• Hardness
• Hexachlorobenzene	• Heptachlor	• Hydrogen sulfide
• Inorganic tin	• Iodine	• Iron
• Malathion	• Manganese	• Methoprene
• Methyl parathion	• MTBE	• Molybdenum
• Monobromoacetate	• Monochlorobenzene	• MX
• Nitrobenzene	• Navaluron	• Parathion
• Permethrin	• pH	• 2-Phenylphenol
• Petroleum products	• Pirimiphos-methyl	• Potassium
• Propanil	• Pyriproxyfen	• Silver
• Sodium	• Spinosad	• Sulfate
• 1,1,1-Trichloroethane	• TDS	• Trichloramine
• Trichloroactonitrile	• Trichlorobenzenes	• Zinc

7.2.3 LAND

Land pollution is defined as degradation or even destruction of the Earth's surface, soil, rock, or subsurface as a result of anthropogenic activities. The expansion of housing developments, businesses, industry, infrastructure, and agriculture all in response to an increasing population over the last several decades accounts for humans modifying over 50% of the Earth's topsoil (United Nations 2016c). Land also becomes contaminated from the migration of contaminants through the soil column or when contaminants change media such as those carried by the atmosphere perhaps hundreds or even thousands of miles and are deposited on the land or water perhaps in a different country.

The surface of the Earth is approximately 510,100,000 square kilometers (Coble et al. 1987). Approximately 70% is covered by water or 361,800,000 square kilometers and 30% is land or 148,300,000 square kilometers. The amount of land on Earth that is farmed is estimated at 40% of 59,320,000 square kilometers (United Nations 2016c). This is a huge number especially since the amount of urbanized land on Earth is estimated to be only 2.7% and is home to roughly half the current human population (United Nations 2016c). The amount of land currently used for farming becomes an even larger value since 10.4% of Earth's land or 15,600,000 square kilometers is covered by ice and is not farmable, 20% or 29,660,000 square kilometers is mountainous, and another 20% or 29,660,000 square kilometers is desert and has little or no topsoil (Coble et al. 1987). When factoring out those land areas not suitable for farming, only 7% of land remains available on Earth.

The World Health Organization and United Nations (2012) have not established guidelines for soil or groundwater. However, most countries that have not developed standards for the remediation of land that includes, soil, sediment, or groundwater commonly refer to a set of standards established by the Dutch in 1987 and were revised in 1994, 2001, and 2015 and are commonly referred to as the Dutch Intervention Values (DIVs) (Lijzen et al. 2001; Netherlands Ministry of Infrastructure and the Environment 2015).

The DIVs are based on potential risks to human health and ecosystems and are technically evaluated. The values presented usually have two integers for each chemical listed. One is a target level, which is considered optimal and likely will result in no or limited health or ecological effect and the other is a level or concentration above which some intervention is necessary to lower risk posed by site-specific exposure pathways.

The DIVs have been used world-wide as a benchmark for cleanup of contaminated sites in many areas of the world where cleanup values have not been established. In addition, even if there are established values, the Dutch Intervention Values are used for comparison purposes. Therefore, in many instances, the Dutch Intervention Values represent the benchmark for cleanup standards world-wide.

Table 7.5 presents a partial list of Dutch Intervention Values. A complete list can be viewed at www.sanaterre.com/guidelines/dutch.html. (Lijzen et al. 2001; Netherlands Ministry of Infrastructure and the Environment 2015)

7.2.4 Summary of Global Assessments and Standards

It may not be clearly noticeable at this point, but there is a general lack of enforcement of international standards for dealing with pollution. As we shall see during our journey through the environmental regulations of select countries of the world, many countries struggle with meeting existing standards, in part, because enforcement is either absent or substandard. This is significant because, as we already know, pollution does not respect political boundaries.

We will now turn our attention from global standards to those of select countries. We shall begin with North America. Since the United States was covered in its own chapter we will start with a short introduction of North America and then describe the environmental regulations and challenges dealing with pollution in Canada and Mexico. We will then move to Africa, Asia, Oceania, South America, and conclude with Antarctica and the Oceans.

7.3 NORTH AMERICA

North America is a continent entirely within the Northern Hemisphere and almost entirely within the Western Hemisphere. North America covers an area of approximately 24,709,000 square kilometers (9,540,000 square miles) and covers 4.8% of the total surface area of Earth and 16.5% of Earth's land surface (Marsh and Kaufman 2015). Figure 7.3 shows North America (United States Central Intelligence Agency 2018).

North America is the third largest continent on Earth, after Asia and Africa, and is fourth in population after Asia, Africa, and Europe. As of 2013, the estimated population of North America was 565 million people in 23 independent states or about 7.5% of the total human population on Earth (United Nations 2017a). The United States at 327 million, Mexico at 130 million, and Canada at 33.4 million make up the majority (approximately 85%) of the human population of North America. Estimates of human habitation of North America range from approximately 40,000 to 17,000 years ago. Together with South America, this age range of human habitation is rather recent compared with other continents, excluding Antarctica (United Nations 2017a).

TABLE 7.5

Dutch Intervention Values for Select Compounds for Soil and Groundwater

Compound	Soil		Groundwater	
	Target Value mg/kg	Intervention Value mg/kg	Target Value ug/l	Intervention Value ug/l
Arsenic (As)	29.0	55.0	10	60
Barium (Ba)	160	625	50	625
Cadmium (Cd)	0.8	12	0.4	6
Chromium (Cr)	100.0	380	1	30
Copper (Cu)	36.0	190	15	75
Nickel (Ni)	35.0	210	15	75
Lead (Pb)	85.0	530	15	75
Mercury (Hg)	0.3	10.0	0.05	0.3
Silver (Ag)	None	15	None	40
Selenium (Se)	0.7	100	None	40
Zinc (Zn)	140	720	65	800
Chloride	None	None	100,000	None
Cyanide	None	20	5	1,500
Cyanide (complex)	None	50	10	1,500
Thiocyanate	None	20	None	1,500
Benzene	None	1.1	0.2	30
Ethyl benzene	None	110	4	150
Toluene	None	320	7	1,000
Xylenes (sum)	None	17	0.2	70
Styrene (vinylbenzene)	None	86	6	300
Phenol	None	14	0.2	2,000
Cresols (sum)	None	13	0.2	200
Naphthalene	None	None	0.01	70
Phenanthrene	None	None	0.003	5
Anthracene	None	None	0.0007	5
Vinyl chloride	None	0.1	0.01	5
Dichloromethane	None	3.9	0.01	1,000
Trichloroethene (TCE)	None	2.5	24	500
Tetrachloroethene (PCE)	None	8.8	0.01	40
Pentachlorophenol	None	12	0.04	3
Polychlorinated biphenyl (PCB) (sum)	None	1	0.01	0.01
Monochloroanilines (sum)	None	50	None	30
Dioxin (sum)	None	0.00018	None	None
Chloronaphthalene (sum)	None	23	None	6
Chlordane	None	4	0.00002	0.2
DDT (sum)	None	1.7	None	None
DDE (sum)	None	2.3	None	None
DDD (sum)	None	34	None	None
DDT/DDE/DDD sum	None	None	0.000004	None
Aldrin	None	0.32	0.000009	None
Dieldrin	None	None	0.0001	None
Lindane	None	1.2	0.009	None
Atrazine	None	7.1	0.00029	150

(*Continued*)

TABLE 7.5   (CONTINUED)
Dutch Intervention Values for Select Compounds for Soil and Groundwater

Compound	Soil		Groundwater	
	Target Value mg/kg	Intervention Value mg/kg	Target Value ug/l	Intervention Value ug/l
Carbofuran	None	0.017	0.0009	100
Asbestos	None	100	None	None

mg/kg = milligram per kilogram
ug/l = microgram per liter

Source: Netherlands Ministry of Infrastructure and the Environment. 2015. Dutch Pollutant Standards. 2015. www. 0government.nl/ministry-of-infrastructure-and-the-environment. or at www.sanaterre.com/guidelines/dutch.html. (accessed February 17, 2019.).

The North American Free Trade Agreement (NAFTA) addresses the issues of environmental protection of the three participating countries – the United States, Canada, and Mexico. However, it leaves establishing environmental rules and standards to the three participating countries, with two exceptions (US Department of State 2017):

1. They are to comply with existing treaties between themselves
2. They are not to reduce their environmental standards as a means of prompting investment in business

As we shall see in the following sections, the three countries that primarily comprise North America that we will examine may have similar environmental regulations, but they differ in practice significantly due to social concerns, corruption, lack of enforcement, and physical geography.

7.3.1 CANADA

Canada is the second-largest country in land area on Earth (9,984,670 square kilometers) with significant natural resources including timber, petroleum, and minerals (see Figure 7.3). The estimated population of Canada in 2009 was 33,487,208, which is about 10% of the population of the United States. It's a significant trade partner of the United States; in fact, approximately 75% of exported goods from Canada are to the United States and the majority of the population of Canada live within 200 kilometers of the US-Canadian border (Canadian Government 2017). Canada closely resembles the United States in its free market–oriented economic system and pattern of industrial productions. Canada and the United States also share several industries (Canadian government 2017).

Since the early 1970s, provincial governments have adopted several environmental laws and regulations pertaining to the protection or enhancement of the environment. These laws and regulations require duties and obligations on individuals or companies that cannot be ignored and which, in the event of non-compliance or of an environmental occurrence, release, or incident, can potentially lead to individual or company liability not only under the environmental laws but also in civil or common law context. In addition, the same can be said for the laws that have been adopted at the federal level (Canadian Government 2017).

Canada was given the right to self-govern from England in 1849 and officially became its own country in 1867. The Constitution Act of 1867 described which level of government (federal or provincial) can be done within its jurisdiction. However, environmentally related issues had not yet been contemplated. As a result, legislative and enforcement authority over the environment is split between the federal and provincial governments, who each have adopted their own set of rules

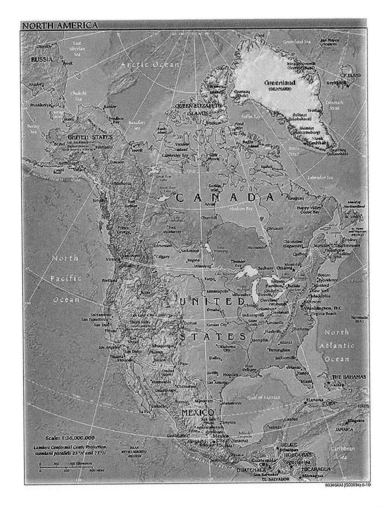

FIGURE 7.3 Map of North America. (From United States Central Intelligence Agency. 2018. The World Fact Book. 2018. www.cia.gov/library/publications/the-world-factbook.gos/NA.html.) (Accessed October 10, 2018.)

(Canadian Government 2017). This has led to an extra level of complexity with each province being different, similar to that of the United States.

The federal government has exclusive jurisdiction over criminal law, coastal and inland fisheries, navigation, federal works and laws for the peace, order, and good government. The provinces have jurisdiction over property and civil rights as well as matters of a local and private nature. Provinces can also delegate many environmental powers to municipalities. As a result of this delegation of power, municipalities regulate matters such as noise, nuisances, pesticides, sewers, and land use planning. As an example, the provincial government of Quebec has delegated responsibility for air emissions and wastewater discharges to the sewer system and waterways within its boundary to the Montreal Metropolitan Community (Canadian Government 2017).

7.3.1.1 Environmental Regulatory Overview of Canada

The main environmental laws at the federal level in Canada include (Environment Canada 2016):

- Canadian Environmental Protection Act (CEPA) of 1999
- Fisheries Act of 2012

- Transportation of Dangerous Goods Act
- Species at Risk Act
- Canadian Environmental Assessment Act

The provinces and territories of Canada have adopted some form of the environmental legislation at the federal level. The following is a general list of each province and dominant environmental law (Environment Canada 2016; Canadian Environmental Protection Act of 1999):

- Province of Alberta, the main environmental regulations are included in the Environmental Protection and Enhancement Act (Province of Alberta 2017)
- Province British Columbia, it's the Environmental Management Act (Province of British Columbia 2017)
- Province of Manitoba, it's the Environment Act (Province of Manitoba 2017)
- Province of New Brunswick, it's the Clean Environment Act (Province of New Brunswick 2017)
- Provinces of Newfoundland and Labrador, it's the Environment Act (Province of Newfoundland and Labrador 2017)
- Northwestern Territory, it's the Environmental Protection Act (Northwest Territories 2017)
- Province of Nova Scotia, it's the Environment Act (Province of Nova Scotia 2017)
- Province of Ontario, there is the Environmental Protection Act, Ontario Water Resources Act, Safe Drinking Water Act, Endangered Species Act, and the Environmental Assessment Act, which constitutes the majority of legislation in Ontario (Province of Ontario 2017)
- Province of Quebec, the main environmental legislation is called the Environmental Quality Act under which a series of subsections address hazardous materials, biomedical wastes, residual materials, air quality, and contaminated sites (Province of Quebec 2017)
- Province of Saskatchewan, it's the Environmental Management and Protection Act (Province of Saskatchewan 2017)
- Yukon Territory, it's the Environment Act (Province Yukon Territory 2017)

In Canada, the purpose of environmental law is to promote protection of the environment and particularly in the context of sustainable development. In Canada, "environment" or "natural environment" is defined as the air, land, water (including groundwater), and all other external conditions or influences under which humans, animals, and plants live or are developed or have dynamic relations. A "contaminant" is defined as any solid, liquid, gas, odor, heat, sound, vibration, radiation, or any combination, resulting directly or indirectly from human activities (Canadian Environmental Protection Act 1999). Environmental laws in Canada create a prohibition against contaminating the environment and establish duties to protect the quality of the environment and obligations associated with permits and approvals that allow impacts (discharges) to the environment, but in a controlled and regulated manner in order to prevent harm to the environment (Environment Canada 2016). Failure to comply may result in sanctions, fines, and administrative orders, much the same as in the United States.

7.3.1.2 Air

Air in Canada is regulated under the Canadian Environmental Protection Act of 1999. Canada regulates stationary and mobile sources such as internal combustion engines. In addition, Canada and the United States have entered into a transboundary air agreement in 2003 to identify and address transboundary air pollution (Canadian Government 2019a; 2019b). This action was emphasized and was a priority for Canada because of the realization that much of the air pollution in Canada was generated in the United States.

As in the United States, Canadian regulations on air pollution have concentrated on particulate matter (PM), ozone (O_3), nitrogen dioxide (NO_x), carbon dioxide (CO_2), sulfur dioxide (SO_2), and volatile organic compounds (VOCs). Since 2000, each of the pollutants listed above have decreased, some greater than 50%. Currently, the air in Canada in urban locations is generally considered of good quality (WHO 2016a; US Department of State 2018a).

As in the United States, any emissions from a stationary source likely requires permitting. The first step is to file a "notice of intent," which is similar to a construction permit in the United States, if the emissions originate from an industrial source from industries such as oil and gas and other petroleum-related industry, metal melting and smelting, pulp and paper, iron and steel, potash, cement lime, chemicals, fertilizers, and manufacturing (Canadian Government 2019c).

7.3.1.3 Water

In Canada, enforceable limits for drinking water are called guidelines and were established by the Federal-Provincial-Territorial Committee and are updated as needed (Health Canada 2017). The drinking water limits for Canada are listed at www.healthcanada.gc.ca/waterquality (Health Canada 2017).

Comparing the Canadian drinking water criteria with that of the United States one can immediately conclude that the criteria are nearly the same as the United States. However, one size does not fit all in that there are slight differences. For example, arsenic and lead have slightly lower limits in Canada than in the United States.

According to the Canadian Government (2019d), 150 billion liters of untreated wastewater is discharged into the environment daily in Canada and represents a real threat to the environment. Wastewater effluent is regulated under the Canadian Fisheries Act, which was enacted in 2012. The Fisheries Act set mandatory minimum effluent quality standards that, if necessary, must be achieved through secondary water treatment before being discharged. The wastewater regulations apply to owners and operators of wastewater systems that collect, or are designed to collect, 100 cubic meters or more of influent per day and that discharge to surface water. Key elements of the act require monitoring, sampling, and analysis, and reporting on effluent quality and quantity (Canadian Government 2019d). Canadian federal wastewater limits are listed at https://laws-lois.ju stice.gc.ca.eng/regulations/SOR-2012-139/fulltext.html (Canadian Government 2019d).

Due to extreme differences in geography and activity, a risk assessment may be required to establish site-specific discharge limits depending on the receiving environment. Risk assessments often include initial characterization or the effluent and considers the characteristics of the receiving environment and mixing that occurs in an allocated mixing zone (Canadian Council of Ministers of the Environmental 2019). This process differs from that of the United States in that risk assessments for wastewater are rarely conducted, perhaps because there is much less wilderness in the United States compared to the northern parts of Canada. An additional water issue in Canada is stormwater. Many municipalities in Canada have difficulties dealing with stormwater combined with sewer discharges. Stormwater discharges, as in the United States, tend to overwhelm water treatment plants resulting in the mass discharge of untreated wastewater to surface water bodies (Canada Water 2019).

7.3.1.4 Solid and Hazardous Waste

In Canada, hazardous waste and hazardous recyclable materials are defined as those that are flammable, corrosive, or are inherently toxic and are regulated under the Canadian Environmental Protection Act of 1999. Canada acknowledges that many household items may be hazardous that include batteries, computer and other electrical equipment, cleaners, paints, pesticides, herbicides, and other items. Providing a separate definition for hazardous waste and hazardous recyclable material provides regulators with more flexibility in managing waste (Canadian Environmental Protection Act 1999). The Canadian Environmental Protection Act (1999) includes authority to:

- Set criteria to assess the environmentally sound management of wastes and hazardous recyclable materials and refuse to permit import or export if criteria are not met

- Require exporters of hazardous waste to submit export-reduction plans
- Regulate the export and import of prescribed non-hazardous wastes for final disposal
- Control inter-provincial movements of hazardous wastes and hazardous recyclable materials

The Canadian Environmental Protection Act (1999) lists nine different hazardous waste classes that include:

1. Explosives	2. Gases	3. Flammable liquids
4. Flammable solids	5. Oxidizers	6. Toxics and infectious
7. Radioactive	8. Corrosives	9. Miscellaneous

The basic approach relies on the nine different hazardous waste classes and analytical tests such as the Toxic Characteristic Leaching Procedure (TCLP), which is commonly applied in the United States to evaluate whether a waste should be classified as hazardous (Canadian Environmental Protection Act 1999).

7.3.1.5 Remediation Standards

Similar to the United States, Canada has developed remediation standards for many chemical compounds. As in the United States, the federal government of Canada has established guidelines so that each province or territory could develop remediation criteria for specific chemical compounds based on toxicity, exposure assumptions, and land use. Remediation standards for the province of Ontario are listed at www.ontario.ca/page/soil-ground-water-and-sediment-standards-use-under -part-xv1-environmental-protection-act (Province of Ontario Canada 2011).

Perhaps the most interesting and most significant difference between the Canadian remediation goals and the United States is that Canada has established cleanup criteria for agricultural soil while the United States has not. In most circumstances, agricultural land is exempt from environmental regulations under United States Environmental Protection Agency (USEPA) in the United States.

In the United States, circumstances where agricultural land is included in environmental assessments is when there is a proposed land-use change and the agricultural land is then developed as residential, commercial, or industrial. It is under this circumstance that the land is evaluated and often investigated for the presence of contamination. The detection of contamination may fall within a gray zone as to whether any detected contamination is the result of agricultural activities which may be exempt or from other activities such as open dumping, chemical spills, leaking tanks, or other activities not directly associated with agricultural activities.

7.3.1.6 Summary of Environmental Regulations of Canada

Canada's population is 10% of the United States but has larger land area and is located north of the continental United States. These geographic and demographic factors have influenced environmental laws within Canada and also how they are applied. This is, in part, due to the northern portions of Canada's sensitivity to pollution and human habitation and that much of Canada's natural resources including oil and natural gas, forest products, and minerals are located in the northern portions of Canada. Canada has to deal with many pollution issues, namely air and water, caused by the United States since Canada shares its southern border with the United States. This is because pollution does not respect political borders and is a subject that will repeatedly appear in this chapter.

In general, the regulations of Canada are robust and have been in place and enforced long enough to be effective. Canada also has vast areas that are relatively pristine from industrial and urban development leaving much of the country with limited human-caused environmental impacts. However, having large areas of uninhabited land also places stress on providing infrastructure and pollution prevention activities by increasing the cost caused by distance, isolation, and remoteness.

7.3.2 Mexico

The official name of what we commonly refer to as Mexico is the United Mexican States (OECD 2015). It's the world's eighth-largest country by land size at 1,972 million square kilometers with a population estimated at 120 million as of 2015, which is approximately 35% of the population of the United States and almost four times that of Canada (see Figure 7.3). Historically, Mexico has received criticism for not instituting effective environmental laws and regulations. Once Mexico developed laws and regulations addressing growing environmental concerns, it was again criticized for not enforcing its own laws (OECD 2015). Over the past 20 years, that has changed. As we shall see, Mexico has made significant strides in improving the protection of human health and the environment. In fact, Mexico has streamlined much of the requirements for industry into a single operating permit, which is more than we can say the United States has accomplished. However, Mexico still has a long way to go in two major areas of the environment, that being water and air. As we shall see in the next few sections, availability of safe drinking water and improving air quality, especially in and around Mexico City, remain key challenges for Mexico.

7.3.2.1 Environmental Regulatory Overview of Mexico

Environmental regulations are present in Mexico and are patterned after the United States, specifically California. The federal government of Mexico, through the Secretariat of the Environment and Natural Resources (SEMARNAT), has sole jurisdiction over acts that affect two or more states, acts that include hazardous waste, and procedures for the protection and control of acts that can cause environmental harm or serious emergencies to the environment. SEMARNAT's main roles include (OECD 2015):

- Develop environmental policy
- Enforce environmental policy
- Assist in urban planning
- Develop rules and technical standards for the environment
- Grant or deny license
- Authorization of permits
- Decide on environmental impact studies
- Offer opinions and assist states with environmental programs

SEMARNAT enforces the laws, regulations, standards, rulings, programs, and limitations issued through the National Environment Institute and the Federal Attorney Generalship of Environmental Protection (PROFEPA). Environmental laws in Mexico address air, water, hazardous waste, pollutants, pesticides, and toxic substances, along with environmental protection, natural resource protection, environmental impact statements, risk evaluation and determination, ecological zoning and sanctions (OECD 2015).

7.3.2.2 Air

In 1992, the United Nations labeled Mexico City air as the most polluted on Earth (Mage et al. 1996). The major air pollution sources that have been blamed for causing much of Mexico City's air pollution come from car exhaust. However, Mexico City's unique geography, located in a mountain valley, where atmospheric inversions are frequent and result in trapping air contaminants near the ground surface, is also a contributing factor (Yip and Madl 2002; Marsh and Grossa 2012). Mexico City is home to more than 21 million people, making it the most populated city in the western hemisphere and accounts for approximately 20% of Mexico's total population (World Population Review 2017). Figure 7.4 is a visual demonstration of the poor air quality of Mexico City.

Air quality in Mexico City is representative of the rest of Mexico in that automobile exhaust combined to a lesser extent of industrial emissions combined with its geography present challenges

FIGURE 7.4   Air quality in Mexico City as evidenced by the haze. (Photograph by Daniel T Rogers.)

to improving overall air quality. Soon after the United Nations declared that Mexico City had some of the most polluted air and a discovery that breathing polluted air contributes to infant mortality in the urbanized world, air quality regulations were initially enacted in Mexico (Mage et al. 1996; Loomis et al. 1999). Currently, ambient air in Mexico is regulated at the federal level by SAMARNAT. In 1993, SAMARNAT issued a series of official Mexican standards or normas (NOM), that regulate maximum concentrations of environmental contaminants in the air. These include eight criteria pollutants, which are the following (Mexico Environmental and Natural Resource Ministry 2017):

- Carbon monoxide (CO)
- Ozone (O_3)
- Nitrox dioxide (NO_2)
- Sulfur dioxide (SO_2)
- PM2.5 (particulate matter less than 2.5 microns)
- PM10 (particulate matter less than 10 microns)
- Total suspended particulate matter (TSP)
- Lead (Pb)

The most significant air pollutants in Mexico are ozone (O_3), sulfur dioxide (SO_2), nitrogen oxides (NOx), volatile organic hydrocarbons (VOCs), and carbon monoxide (CO), most of which all originate from the incomplete combustion of fossil fuels (Mage et al. 1996; United Nations 2016b). Ambient air quality standards for criteria pollutants in Mexico are listed at www.MexicoAirQualityStandards (United Nations 2016b).

In Mexico, there are seven general categories of industry where rules are consistently applied within each category. The seven categories include (Mexico Environmental and Natural Resource Ministry 2017):

1. Federal public activities
2. Water works, general communications networks, and oil, gas, and carbon transportation activities

3. Chemical and petrochemical plants, iron and steel mills, paper factories, sugar refiners, drink manufacturers, cement producers, automotive parts makers, and the generation and transmission of electricity activities
4. Mineral and non-mineral exploration, extraction, treatment, and refining activities
5. Federal touristic development activities
6. Treatment, storage, and disposal plants of hazardous waste and activities, including nuclear waste
7. Exploitation of slow regenerating vegetation in forest and tropical jungle activities

7.3.2.3   Water

In general, Mexico has poor water quality. According to the OECD (2015), Mexico's water resources are some of the most degraded in the world, especially in the developed cities of Mexico City, Guadalajara, and Monterrey and in the northern portion of the country which is more arid than the southern portion. Despite recent improvements and as of 2015, only 14% of Mexico's population are connected to a sewerage system that treats wastewater (OECD 2015).

In addition, the quality of treatment in many areas is not adequate and the effluent discharge in many cases does not meet fundamental water quality standards (OECD 2015). Figures 7.5 and 7.6 are examples of poor surface water quality near Mexico City.

Water regulations have been passed but the lack of infrastructure and economic funds for building wastewater treatment plants, associated piping, education, and enforcement of regulations is far from being realized for the majority of Mexico's population (OECD 2015). In 1988, then President Carlos Salinas created the Mexico National Water Commission (CONAGUA). At the beginning, CONAGUA was given the task of defining federal policies to strengthen service providers through technical assistance and financial resources. Progress has been slow and in 2015, Mexico presented a new general water law that proposed the transfer of several functions from both federal and state levels to newly created institutions at the watershed level. The new law also designated that each individual has the right to 50 liters of water per day.

Mexico published drinking water criteria in 1996 and 1997 and are listed at www.Mexico.nom-003-SEMARNAT-1997 (Mexico Environmental and Natural Resource Ministry 1997).

FIGURE 7.5   Surface water showing poor water quality. (Photograph by Daniel T. Rogers.)

FIGURE 7.6 Surface water showing poor water quality and discharge of untreated water. (Photograph by Daniel T. Rogers.)

Many of the substances with water quality limits are higher than in the United States and water quality standards only exist for a few of the more common compounds. In large part, this is due to internal struggles within Mexico to improve water quality in a nation where surface water and groundwater are often already contaminated (OECD 2015). Stormwater regulations in Mexico do not yet exist but perhaps will be forthcoming in the not too distant future. According to Mexico National Water Commission, named CONAGUA (2019), 63% of water used in Mexico was from surface water and 37% was from groundwater. Overexploitation of groundwater has resulted in surface subsidence in many locations, especially in Mexico City (refer to Figure 7.7).

FIGURE 7.7 Example of land subsidence in Mexico City due to excessive groundwater withdrawal. (From USGS. 2017. Examples of Land Subsidence in Mexico City. 2017. www.USGS.gov/edu/gallery/landsubsid ence-learning.html.) (Accessed August 30, 2017.)

Today, CONAGUA has focused on the priorities at the national level that were set through a development plan with the aim of attaining the following (Mexico Environmental and Natural Resource Ministry 2016):

- Improving water productivity in agriculture
- Improving access and quality to water supply and sanitation
- Supporting an integrated and sustainable water resource management in basins and aquifers
- Improving the technical, administrative, and financial development of water resources
- Consolidating user and society participation
- Identifying and reducing risk with respect to meteorological phenomena
- Evaluating the effects of climate change to the water cycle
- Creating a culture of compliance

The future challenges that Mexico faces in order to improve water quality include (Mexico Environmental and Natural Resource Ministry 2016):

- Infrastructure – currently lack of infrastructure including water treatment and wastewater treatment and a reliable delivery system is such a large impediment that the government has historically not even attempted to improve its infrastructure on a large scale
- Groundwater overexploitation – this has resulted in land subsidence in many urban areas
- Land subsidence – primarily the result of groundwater overexploitation, which has resulted in impeding the structural integrity of many urban buildings
- Flooding – as urban areas continue to expand, more land will be covered with impervious surface that will continue to exacerbate flooding and also will have a negative effect on increased land subsidence because of reduced groundwater recharge
- Increasing urbanization – continued development without infrastructure construction before development continue to exacerbate the already poor distribution system, especially in large urban areas
- Basic water quality – currently, one of the immediate health concerns are water-borne disease from untreated potable water that in some cases contain fecal matter
- Intermittent supply – this has resulted in individual residences to collect rain water from roofs during rain events and storing water in black plastic storage tanks installed on rooftops
- Limited wastewater treatment – continues to increase the risk of water-borne diseases, especially in larger urban areas
- Water use for irrigation – currently agriculture is the largest consumer of water in Mexico.
- Inefficient use of urban water – this encompasses many different items that include faulty or leaking delivery system and open sewers
- Limited cost recovery – due to lack of infrastructure, some delivery systems to individual users do not account for how much was used due to lack of water meters

The issues identified in this section may seem dire or even hopeless but, as we shall see later in this book, investment in manufacturing and slow but determined progress has made a big difference in improving water quality in Mexico over the last 20 years (OECD 2015).

7.3.2.4   Solid and Hazardous Waste

In Mexico, the definition of hazardous waste is very similar to that of a characteristic waste in the United States, and is defined as a substance that is explosive, ignitable, corrosive, chemically reactive, or is toxic (directly or indirectly) to plants, animals, or humans. Air, water, and solid waste and hazardous waste are all handled similar to the United States (i.e., California in particular), with one exception and that being stormwater (Basurto and Soza 2007; Rodriguez et al. 2015; Mexico

Environmental and Natural Resource Ministry 2019a). Within many Mexican states, stormwater is of little or no concern due to dry conditions, especially in the northern and west regions of the country. However, that will likely change soon.

Permits, authorizations, and documentation likely required at manufacturing facilities in Mexico include the following (Basurto and Soza 2007; Rodriguez et al. 2015; Mexico Environmental and Natural Resource Ministry 2019a):

- Operating licensed
- Wastewater discharge registration
- Hazardous waste generator's manifest
- Monthly log of hazardous waste generation
- Ecological waybills for the incineration and/or exportation of hazardous materials and wastes
- Semi-annual report on hazardous wastes sent to recycling, treatment, or rinal disposition
- Accidental hazardous waste spill manifest
- Delivery, transport, and receipt of hazardous waste manifests
- Environmental impact studies

In Mexico, the operating license is a key document/permit and must be obtained before operations begin. However, in some instances this may be difficult to obtain far in advance since many items may change that will affect the details in the license that render the license invalid even before operations begin (Mexico Environmental and Natural Resource Ministry 2019a). In addition, if any changes are made within the facility, the operating license may no longer be valid and must be amended, much as it is in the United States and as we shall see, as it is in most of the world.

In the United States, for instance, with an air permit, if a modification in operations is made that increases or has the potential to increase emissions, then a permit modification is necessary and in some circumstances a new permit may be required, that being a construction permit. This same principle applies in most other countries of the world.

In Mexico, and except in relation to the petroleum and petrochemical industries or those treatment facilities of hazardous waste, a single operating license may be obtained through a single application. The single operating license integrates the various permits, licenses, and authorizations for matters of environmental impact, water service, emissions of atmospheric contaminants, and generation and treatment of hazardous waste (Mexico Environmental and Natural Resource Ministry 2019a). Obtaining a single license that covers all environmental matters is very different than how environmental permits are handled in the United States. A single operating license or permit that covers all aspects of potential environmental impact a facility may have upon the environment is far more manageable in most all aspects than having separate permits for every source. With the exception of Title V permits for air emissions, the United States issues permits separately.

7.3.2.5 Remediation

In 2004, Mexico passed what is termed "Mexico's General Law for Prevention and Integral Management of Wastes," commonly referred to as Mexico's Waste Law. The passage of this law significantly advanced Mexico's efforts to develop a detailed regulatory framework for dealing with legacy sites and the discovery of new or unknown sites of environmental contamination (Mexico Environmental and Natural Resource Ministry 2019b). It mirrors the United States Brownfields Revitalization Act and All Appropriate Inquiry. Figures 7.8 to 7.10 show some of the challenges of contaminated sites in Mexico and mirror many aspects of similar issues that were faced in the United States in the early 1970s, that being uncontrolled and unregulated waste disposal that resulted in high contaminant levels in soil and groundwater and also affected surface water quality. Figure 7.8 shows uncontrolled landfilling, Figure 7.9 shows an area of drum disposal, and Figure 7.10 shows an area where solid wastes were openly burned, which potentially resulted in contributing to poor air quality.

FIGURE 7.8 Example of uncontrolled surface waste disposal. (Photograph by Daniel T. Rogers.)

FIGURE 7.9 Example of improper waste disposal. (Photograph by Daniel T. Rogers.)

FIGURE 7.10 Example of open burning of solid waste. (Photograph by Daniel T. Rogers.)

The Mexican Waste Law created strict liability against owners and possessors (e.g., operators) of a contaminated site. Previously, parties who caused the contamination were held responsible for cleanup. The Mexican Waste Law holds that one does not have to have caused the contamination to be held liable for its cleanup. To protect new landowners, the law mandates disclosure of known contamination by hazardous materials or wastes from owners to potential third-party buyers or tenants (Mexico Environmental and Natural Resource Ministry 2019b).

In addition, the Mexican Waste Law does not allow for the transfer of a site contaminated with hazardous materials or wastes without express authorization from Mexico's environmental ministry. Lastly, the environmental ministry or SEMARNAT will not grant approval until the contamination is remediated or until an agreed plan has been approved by all parties. The impact of the Mexican Waste Law and its regulation has been realized in situations where properties contain known or suspected contamination (Mexico Environmental and Natural Resource Ministry 2019b). As one can imagine, potential purchasers are reluctant to take title of contaminated property without recourse to prior studies that characterize the potential contaminate issue(s) and define the responsibility and costs associated with addressing the contamination. Sellers, on the other hand, are reluctant to conduct environmental investigations because of the potential to discover contamination that may require notice to SEMARNAT and that may cost significant time and resources to address.

The most common method to achieve cleanup objectives in Mexico is to simply excavate the contaminated media, usually soil, and dispose of the material in a landfill, if possible. The Mexico Environmental and Natural Resource Ministry (2004; 2005; 2015) has published remedial objectives at www.Mexicana.nom-147-SEMARNAT/ss-2003, www.Mexicana.nom-138-SEMARNAT/ss-2003, and www.Mexicana.nom-133-SEMARNAT/ss.2003 (Mexico Environmental and Natural Resources Ministry 2004; 2005; and 2015).

Cleanup levels for soil have been divided into groups and are based on exposure scenarios and land use similar to the United States and other countries. The categories include residential, agricultural, commercial, and industrial. Occasionally, some of the groups are combined under one heading. Note that in Mexico agricultural land is specifically noted and has applicable cleanup objectives unlike in the United States (Mexico Environmental and Natural Resource Ministry 2019b).

The objective of remediation in Mexico is much the same throughout the world and that is to either eliminate or reduce contaminant concentrations or to control levels of contamination through institutional controls or other acceptable methods such that the risk posed by the presence of the contamination no longer puts in danger human health or adversely affects the environment. Remedial efforts in Mexico should accomplish the following (Mexico Environmental and Natural Resource Ministry 2019b):

- Permanently reduce contaminant concentrations
- Reduce contaminant bioavailability, solubility, or both
- Avoid contaminant dispersion in the environment
- Establish institutional controls, if necessary

There are two regulatory agencies that are responsible for overseeing compliance with the investigation and remediation of contaminated sites in Mexico and they are the Procuraduria Federal de Proteccion al Ambiente (PROFEPA), which is a regulatory division of SEMARNAT, and the Health Secretariat, which is involved in cases of risk to human health (Mexico Environmental and Natural Resource Ministry 2019b).

7.3.2.6 Summary of Environmental Regulations in Mexico

Mexico has made significant progress in environmental regulations in the last two decades but lacks significant infrastructure, such as wastewater treatment in Mexico City and other locations. Other significant impediments include corruption and lack of political priorities on issues pertaining to the environment. Mexico also struggles with consistent enforcement of environmental regulations in parts of the country.

7.4 EUROPE

We will discuss the environmental laws and regulations of Europe, first through the 28 member European Union (EU) and then move on to other European countries that are not members of the EU but need to be covered. These other countries include Russia, Norway, Switzerland, and Turkey. We will discuss the United Kingdom (UK) as part of the EU since the UK is still operating environmentally as if it were still an EU member (see Figure 7.11).

7.4.1 EUROPEAN UNION COUNTRIES

The EU is a collection of 28 countries but soon to be 27 with the recent planned exit of the UK. The EU represents 4,324,773 square kilometers in area, the seventh largest in the world with a population of 507,416, 607 residents as of a 2014 census (European Union 2019).

7.4.1.1 European Union Environmental Regulatory Overview

The EU is considered by many to have the most extensive and strictest environmental laws of any international organization. The EU's environmental legislation addressed issues such as (EEA 2019a):

- Acid rain
- Thinning of the ozone layer
- Air quality
- Noise pollution
- Solid and hazardous waste
- Water pollution
- Sustainable energy

The Institute for European Environmental Policy (2019) estimated that the body of EU environmental laws total over 500 directives, regulations, and decisions. The 1972 Paris Summit meeting of heads of state and government of the European Economic Community (EEC) is often used to pinpoint the beginning of the EU's environmental policy making and declarations (EEA 2019a).

Countries of the EU include (EEA 2019):

• Austria	• Belgium	• Bulgaria	• Croatia
• Cyprus	• Czech Republic	• Denmark	• Estonia
• Finland	• France	• Germany	• France
• Hungary	• Ireland	• Italy	• Latvia
• Lithuania	• Luxembourg	• Malta	• Netherlands
• Poland	• Portugal	• Romania	• Slovakia
• Slovenia	• Spain	• Sweden	• United Kingdom

The EU is composed of two major institutional groups. The first group is composed of the following (EEA 2019a):

- European Commission, which initiates and implements EU law and represents the driving force and is the executive body
- Council for the EU, which provides legislative approval to laws from the commission and represents the governments of each of the member states
- Court of Justice, which ensures compliance with the law and serves as an appellate judicial body when questions of EU law arise
- Court of Auditors, which controls the EU budget

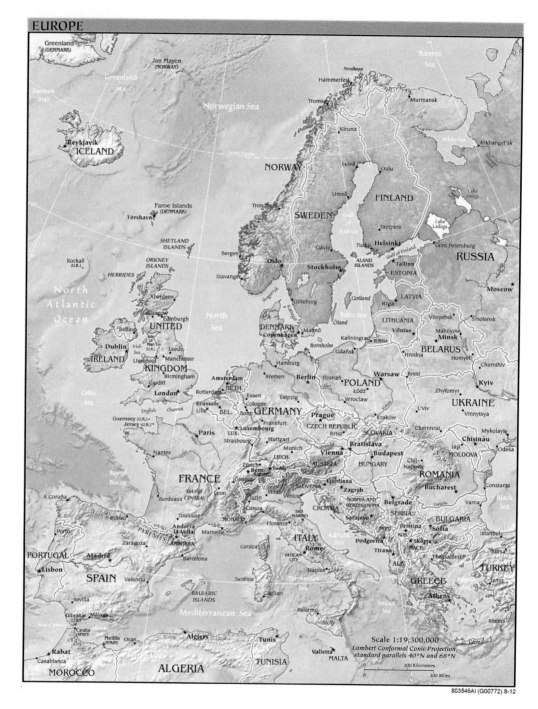

FIGURE 7.11   Map of Europe. (From United States Central Intelligence Agency. 2018. The World Fact Book. 2018. www.cia.gov/library/publications/the-world-factbook.gos/NA.html.) (Accessed October 10, 2018.)

The second group consists of the following (EEA 2019a):

- European Economic and Social Committee, which expresses the opinions of organized civil society on economic and social issues
- Committee of the Regions, which represents the local and regional authorities in Europe
- European Central Bank, which is responsible for monetary policy and managing the currency of the EU, the Euro
- European Ombudsman, which investigates EU institutions and bodies
- European Investment Bank, which assists in EU objectives by financing government projects

The EU has three forms of binding legislation as part of its institutional framework that includes (EEA 2019a):

1. Directives. Directives are binding on the member states to which they are addressed regarding the results to be achieved. The member states, however, may choose how to bring about those results. A directive does not take effect until a member state passes national legislation to implement the provisions of European law, which it must do within 2 years of the date of passage of any specific directive. A directive normally enters into force on the date specified in the directive or on the twentieth day after publication in the EU's Official Journal. Directives are the most frequently used EU laws.
2. Regulations. Regulations are binding in their entirety and apply directly to all member states, similar to national laws. They are stronger and much less common than directives. In addition, regulations go into effect in the member states immediately, without national implementing legislation. Approximately 10% of EU laws are regulations. They supersede any conflicting national laws and are not allowed to be transposed into national law, even if the law is identical to the regulation. Regulations are binding the day they come into force, which like directives, is either specified in the document itself or is the twentieth day after publication in the EU Official Journal.
3. Decisions. Decisions are binding in the entirety on those to whom they are addressed. Decisions are individual legislative acts that differ from directives or regulations in that they are specific in nature and are used to specify detailed administrative requirements or update technical aspects of regulations or directives. They may be addressed to a certain government, an enterprise, or to even an individual.

Other forms of legislation include recommendations and opinions, which are not binding but are expected to be taken into account when decisions are made. In addition, resolutions are nonbinding statements by which the Council of Ministers expresses a political commitment to a specific objective. Often what happens is that after some time passes, usually a few years, a resolution gives way to either a directive or a regulation (EEA 2019a).

The European Commission, which is the executive of the EU, consists of one commissioner from each member state and is headed by a president who serves a 5-year term. Commission members are fully independent and are not permitted to take instructions from the government with which they originate (EEA 2019a).

The responsibilities of the European Commission include (EEA 2019a):

- Guardianship of treaties and the initiation of infringement proceedings against member states and others who disobey the treaties and other community law
- Negotiation of international agreements
- Adoption of technical measures to implement legislation adopted by the Council

- Matters relating to the environment, education, health, consumer affairs, the development of trans-European networks (TENS), research and development of policy, culture, and economic and monetary union
- Preparing the groundwork for incorporating former Soviet bloc countries into the EU, identifying areas where these groups can align their policies with those of the EU, and a procedure of implementation
- To initiate legislation only in areas the EU is better placed than individual member states to take effective action, thus taking the principle of "subsidiarity" into account
- Taking action when necessary against those who do not respect their treaty obligations by perhaps bring them before the European Court of Justice when compliance is not voluntary
- Managing policies and negotiating international trade and cooperation agreements
- Acting as a mediator between conflicting interests of member states
- Sustaining, managing, and developing agricultural and regional development policies
- Developing cooperation with other European nations not members of the EU and countries in Africa, Caribbean, and the Pacific Rim
- Promoting research and technological developmental programs

The EEA plays an important role in EU environmental policy by providing the commission with information used in setting EU environmental policy. The EEA aims to support sustainable development, which helps to improve Europe's environment through the provision of timely, targeted, relevant, and reliable information to policy making agents and the public. The EEA provides research and clarifies Europe's environmental issues and challenges and employs a whole range of tools to assist policymakers and the public to address environmental quality and sustainable development issues. In effect, the EEA is separate from other institutions, and is charged with providing objective information and analysis. Other duties of the EEA is to provide objective information and analysis including a catalog of data sources and projects and a prototype called the European Information and Observation Network (EINETICS), which is a computerized network for environmental data (EEA 2019a).

Current EEA members include the 28 member nations plus Bulgaria, Romania, Turkey, Iceland, Norway, and Liechtenstein. A membership agreement has been entered into with Switzerland, Albania, Bosnia and Herzegovina, Croatia, former Yugoslav Republic of Macedonia, and Serbia and Montenegro have all applied for EEA membership. EEA's priorities include (EEA 2019a):

• Air quality	• The states of the soil
• Flora and Fauna	• Biotopes
• Land use	• Natural resources
• Waste management	• Noise emissions
• Coastal protections	• Hazardous chemicals

The need for EU environmental protection has grown through time as industrial and population growth has occurred. The Single European Act of 1987 provided the first expressed legal basis for EU environmental policy and is outlined in four basic principles that include (EEA 2019a):

1. Applying the principle of "polluters pay," which is similar to the policy adopted in the United States under the passage of CERCLA in 1980
2. Pollution prevention, which is similar to the Pollution Prevention Act of the United States, enacted 3 years later

3. Rectify environmental degradation at the source, when possible
4. Integration of environmental protection into other EU policies

7.4.1.2   Air

In Europe, emissions of many air pollutants have decreased substantially over the past few decades. However, because of its dense population and number of mobile and stationary sources of air pollution, concentrations of many air pollutants are still too high and poor air quality persists. A significant proportion of Europe's population live in cities where the poor air quality is generally located and includes ozone, nitrogen oxides, and particulate matter. According to the EEA, approximately 90% of Europe's city inhabitants are exposed to air quality levels considered harmful to health (EEA 2019b).

Air emissions management in the EU has recently focused on greenhouse gas emissions from both mobile and stationary source, much like what is presently conducted in the United States but a bit more strict within the EU. The EU regulates greenhouse gases from mobile sources such as automobiles, diesel trucks, agricultural machinery, motorcycles, and stationary sources, industry, residential and household and other sources of greenhouse gases such as tools, aircraft, and seagoing ships (EEA 2019b).

For air emissions, the EU also regulates the equivalent of criteria pollutants in the United States which include ozone, sulfur dioxide, nitrogen oxides, particulate matter, carbon monoxide, and lead. European legislation on air quality is built on certain principles. Each member state is divided into zones and agglomerations. The member state is to then conduct a study within each zone or agglomeration using measurements, modeling, and other empirical techniques. Where air quality levels are exceeded or elevated, the member state is then to prepare a plan or program to ensure compliance in the future. In addition, air quality data is to be shared with the general public (EEA 2019c). The EEA regulates air much the same way as USEPA regulates air in the United States. For example, EEA regulates the same air contaminants and requires zones or agglomerations that exceed the standards and objectives to prepare a plan for obtaining attainment, which is similar to the United States where USEPA requires state regulatory agencies to prepare State Implementation Plans (SIPs) in areas of non-attainment (EEA 2019b).

EEA air pollutant standards are listed at https://ec.europa/environment/air/quality/standards.htm (EEU 2019c).

In a 2009 directive, the EU established an emission trading system (ETS) as the principle method to combat climate change. It's a key instrument for delivering Europe's portion of reduction commitments under the Paris agreement of the United Nations Framework Convention on Climate Change. Since initial formation in 2005, it has evolved to become a key tool for the cost-effective reduction of industrial GHG emissions by giving companies the flexibility to make investments where they deliver the biggest gains (EEA 2019b).

The ETS works on the "cap and trade" principle on the total amount of certain GHGs, such as carbon dioxide, nitrous oxide, and perfluorinated compounds. An estimated 11,000 facilities are subject to the ETS, which include mainly heavy, energy-using sites in the power generation sector and some large manufacturing operations, and some aircraft operators. The EU ETS applies to 31 countries, those of the 28 member states and Iceland, Liechtenstein, and Norway. The overall volume of GHGs that can be emitted each year by facilities subject to the directive is set at the EU cap at the Convention on Climate Change and the cap is reduced each year. In addition, facilities are also required to have an approved monitoring plan for monitoring and reporting annual emissions. Currently, this program covers approximately 45% of GHG emissions estimated for the EU (EEA 2019b).

For industrial facilities, air permitting in the EU takes a holistic approach that differs from the United States. Air permits issued to industrial facilities are usually issued under the Industrial

Emissions Directive (IED) and include conditions to prevent and control all environmental impacts from a facility taking into account its full environmental performance, which are (EEA 2019d):

• Air emissions	• Water use
• Water discharge	• Land use
• Generation of solid and hazardous waste	• Use of raw materials
• Energy use and efficiency	• Noise prevention
• Accident rate and prevention	• Odors
• Site closure	• Site restoration

The EU is of the opinion that this integrated air permitting approach is best for the environment because it takes into account the whole environment and not a portion and is therefore viewed as more effective, protective, and efficient. The primary responsibility for implementation of the IED rests with each member state. In turn, these governments assign the responsibility to a department or agency within the respective country to deliver the practical implementation activities of the directive. As an example, the UK has delegated the responsibility of implementation to the environment agencies within the four component parts of the country, that being England, Wales, Scotland, and Ireland. These four agencies are empowered by legislation that transposes the IED into national law (EEA 2019d).

7.4.1.3   Water

The EU regulates water as drinking water and as all other water, that being surface water, groundwater, coastal and marine waters, under the following directives:

- Drinking Water Directive
- Water Framework Directive
- Marine Strategy Directive
- Birds and Habitats Directives
- Floods Directive

The Drinking Water Directive had its beginning in 1975 and then was culminated in 1980 in setting binding water quality targets for drinking water. The EEA drinking water quality standards are listed at https://ec.europa/drinkingwater (EEA 2019e).

The Water Framework Directive was first developed and passed in 1975 and culminated into its final form in 1980 along with the Drinking Water Directive. The Water Framework Directive established binding targets for surface water including lakes, streams, and rivers. It also established standards for fish waters, shellfish waters, bathing waters, and groundwater. At first, the priority was to target those surface water bodies that were sources of drinking water (EEA 2019f).

An innovative method employed by the EEA was the concept of water management through a watershed or river basin approach. This concept has been especially challenging in the past because many river basins within the EU cross international borders, such as the Rhine River. However, from a scientific and even a practical point of view, regulating a river basin is much more effective because it is a single geographical and hydrogeological unit (EEA 2019f).

The EU also has what are termed "Additional Monitoring Parameters" that are equivalent in the United States to Secondary Drinking Water Standards. These limits are listed at https://ec.europa/drinkingwater (EEA 2019e).

In the United States, environmental protection on a watershed wide basis is perhaps not a new concept but one that is not widely used for environmental protection. The Delaware River watershed is within six different states and was protected and regulated under an executive order signed by John F. Kennedy during his presidency in the early 1060s (Delaware River

Watershed Initiative 2019). The Colorado watershed has also been regulated, predominantly by the Bureau of Land Management within the Department of Interior, for its hydroelectric capacity in the Hoover Dam and Glen Canyon Dam and for its water diversion practices through canals and aqueducts for agricultural purposes and potable water sources, namely Las Vegas (USGS 2019). The watershed or river basin approach includes two evaluation criteria in which each river basin is evaluated; one is the general requirement for ecological protection and the other is minimum chemical standards. Within these two criteria falls protection of aquatic ecology, protection of unique and valuable habitats, protection of drinking water resources, and protection of bathing water.

A good ecological protection standard is evaluated not through a rigorous standard but on a relative scale because ecological conditions vary widely within the EU. However, a good chemical protection status is defined in terms of compliance with all the quality standards established for chemical substances at the European level. These standards ensure at least a minimum chemical quality, especially with respect to very toxic substances. Under the directive, the EEA handles groundwater differently. The EEA takes the approach for groundwater that there should not be any impacts to groundwater at all, except from naturally occurring, undisturbed sources. For this reason, the EEA has not set chemical standards for groundwater, because according to EEA, setting chemical standards then establishes the concept that there are allowed levels of pollution for groundwater (EEA 2019f; 2019g).

However, a few standards have been established for groundwater and include nitrates, pesticides, and biocides, which generally originated from agricultural use and not from industry. As an additional level of protection for groundwater resources, the EEA has required that there will be no direct discharges of any wastewater to groundwater (EEA 2019g). This is a different approach to that within the United States, even California. For instance, Class V injection wells are allowed in California, which are essentially what is termed elsewhere in the United States as "Dry Wells." In addition, many western states have what are termed "infiltration basins" where stormwater runoff is collected and is then allowed to migrate vertically to groundwater to recharge groundwater without treatment.

Europe treats groundwater as a hidden resource which is quantitatively much more significant than surface water because, in part, it represents 97% of fresh water on Earth that is not ice (EEA 2019g). In addition, groundwater recharges many surface water sources and is mostly responsible for base flow in many river systems. Groundwater migrates or moves much more slowly than surface water, which then means that it is more susceptible to becoming contaminated from anthropogenic activities because of the longer migration time frame and hence more opportunity to be environmentally degraded. What this means in many circumstances is that pollution released many decades ago may still threaten groundwater supplies, not only currently but for perhaps centuries in the future.

7.4.1.4 Solid and Hazardous Waste

Europe is densely populated and nowhere more that this becomes evident than in the EEA's approach to solid and hazardous waste. The EEA's waste management approach is heavily reliant on reuse and recycling (EEA 2019h). Over the last 20 years, European countries have shifted focus from disposal methods to prevention and recycling, and also examining methods to reduce packaging waste. The EEA has set a target of recycling at 65% of all waste by 2030 (EEA 2019h).

The EEA (2019i) regulates generators of hazardous waste through a "cradle to grave" method much like the United States. The classification of wastes into hazardous or non-hazardous waste is based on methods of generation and characteristics, much like the United States, that include whether the waste material is explosive, flammable, corrosive, toxic, infectious, a gas, oxidizer, or radioactive (EEA 2019i). Hazardous waste limits are listed at http://ec.eurpoa.eu/environment/was te/hazardous_index.html (EEA 2019i).

7.4.1.5   Remediation

In the EU, land pollution has affected an estimated 2.5 to perhaps as high as 3.5 million sites (EEA 2019j). Cleaning up these sites will take decades and cost on average from 0.5% to 2% of Gross domestic product (GDP) per year as estimated by the EEA (2019h). Therefore, Europe seems to have similar challenges with contaminated land as the United States. Figure 7.12 shows contamination to soil from landfilling that occurred during World War I in France. In the EU, contaminated land is treated as two groupings, land and groundwater. The EEA lists cleanup criteria for many chemical compounds as target levels and intervention values. These values are commonly referred to as Dutch Intervention Values (DIVs) (Netherlands Ministry of Infrastructure and the Environment 2015).

Contaminants present above the intervention value require further evaluation or in many instances may require a remedial action. Table 7.5 lists both the target and intervention values for select compounds in soil and groundwater. Soil values are presented as dry weight and are calculated for a "standard soil" containing 10% organic matter. The values presented in the table for heavy metals in groundwater reflect the standard for groundwater that is considered shallow or less than 10 meters (Netherlands Ministry of Infrastructure and the Environment 2015).

The soil intervention values indicate when the functional properties of the soil for humans, plants, and animals is considered seriously impaired or threatened. A concentration of a single chemical compound at or exceeding its respective intervention value is considered a serious case of soil or groundwater contamination if the affected volume exceeds 25 cubic meters of soil or 100 cubic meters of groundwater (Netherlands Ministry of Infrastructure and the Environment 2015). A comparison of the DIVs to cleanup levels in the United States indicated that in some instances the cleanup criterion for individual chemical compounds is lower in the EU than in the United States (i.e., vinyl chloride) (Netherlands Ministry of Infrastructure and the Environment 2015).

7.4.1.6   Summary of Environmental Regulations of the European Union

Environmental regulations of the European Union are considered by many to be the most comprehensive set of environmental regulations currently known to the world. In what started at the Paris Summit in 1972 in establishing the EU's environmental policy and basing its regulatory framework upon that of the United States Environmental Protection Agency, it has now become the system that sets the example. No more is this evident than in the field of sustainability where we shall see examples of the EU's commitment to environmental stewardship.

FIGURE 7.12 Impacted soil from landfilling during World War I in France. (Photograph by Daniel T. Rogers.)

7.4.2 RUSSIA

Russia is the largest country in the world, encompassing much of eastern Europe and the whole northern portion of Asia. The total land area of Russia is 17,098,246 square kilometers (6,601,670 square miles). Due to its size and geographic distribution (see Figure 7.11), Russia has many different climate zones that include tundra, coniferous forest, mixed and broadleaf forest, grassland, and semi-desert (Blinnikov 2011). Russia occupies 11% of the world's land surface and stretches through 11 time zones and has 40 UNESCO biosphere reserves. The population of Russia as of 2016 was estimated at 143.4 million. Interestingly, the population of Russia in 1991 was estimated to be 148.5 million (United Nations 2017a; 2017b).

According to the United Nations (2017a), the population of Russia has stabilized and has started to increase since 2015. Russia's largest city by population is its capital, Moscow, with a population of 11.5 million, followed by Saint Petersburg with a population of 4.8 million (United Nations 2017b). For centuries, Russia has built its economy on its vast and rich natural resources that include large regions of wilderness and timber, and mineral and energy resources and fresh water (Newell and Henry 2017).

The discussion of the environmental regulations of Russia will be more in depth than most other countries. This is because Russia is a very large country with enormous natural resources, particularly in oil and gas, forestry products, and minerals (much of which have not been fully developed), and significant biodiversity. How Russia decides to implement and enforce its environmental regulations will likely have either future positive or negative implications for the rest of the planet.

7.4.2.1 Environmental Regulatory Overview and History of Russia

Russia's relationship with the environment has historically not been considered good, with many troubling concerns that have been a carry-over from its Soviet days. From the aftermath of the Chernobyl nuclear disaster (in what is the Ukraine today), the drying up of the Aral Sea, unchecked industrial emissions and discharges, and its dependence on mining, oil, and gas, Russia has generated plenty of attention from environmental organizations world-wide (Brain 2016).

According to the OECD (2019), environmental policies and regulations in Russia continue to suffer from an implementation gap that continues to this day. In addition, many regulatory agencies have undergone multiple reorganizations and continue to be fragmented, which has sometimes brought many agencies to the brink of institutional paralysis.

Environmental policy in Russia strives to achieve a balance between protecting the natural world and economic development. Although this objective may sound pleasing, it has been difficult to achieve in practice or actually begin especially since these ideologies are often times in direct conflict. There are numerous environmental issues in Russia. Many of these issues have been attributed to policies of the former Soviet Union, which was a period of time when officials within the government felt that pollution control was an unnecessary hindrance to economic development and industrialization (OECD 2019). As a result, 40% of Russia's territory began experiencing symptoms of significant ecological stress by the 1990s that have been traced in large part back to a number of environmental issues that include (National Intelligence Council 2012):

- Deforestation
- Energy irresponsibility
- Unchecked disposal of hazardous wastes
- Surface mining
- Improper nuclear waste disposal

For most of the Soviet period, the task of environmental protection was fragmented and shared by more than 15 ministries, each of which was responsible for a particular economic sector, but many environmental issues had overlap with other sectors that resulted in confusion over who was responsible and who had the actual power to affect change and ultimately begin to protect human health and the environment (Henry and Douhovnikoff 2008).

The Chernobyl nuclear accident in 1986 and Mikhail Gorbachev's glasnost, or policy of openness, were catalysts for change in Russia and opened the doors for environmentalism in Russia for the first time. In addition, it also gave environmental watch groups and the international community some insight and data on just how large the disregard for the environment had been during the days of the Soviet Union (Henry and Douhovnikoff 2008).

In 1988, the Russian State Committee on Environmental Protection was established and gave the committee the authority to conduct environmental reviews for all new government projects (Russian Federal Law on Environmental Protection 2002). Environmental protection gained further popularity and importance with the collapse of the Soviet Union. One of the first laws passed by the newly formed Russian Federation was the 1991 Federal Act on the Protection of the Natural Environment. Russia also announced that it had committed itself to the principle of sustainable development in the early 1990s (Henry and Douhovnikoff 2008).

Along the lines of openness, it was announced in 1993 through a commission on the environment chaired by Aleksei Yablokov, who at the time was President Boris Yeltsin's advisor on the environment, that the Soviet Union had disposed of 2.5 million curies of radioactive waste in the Sea of Japan starting in 1965 (Henry and Douhovnikoff 2008).

From 1991 to 2000, environmental protection officials were largely ineffective because of difficult conditions due to lack of resources, bureaucratic infighting, and lack of real authority to effect change. Intense lobbying by industrial groups eroded environmental protection over time. Other significant issues included widespread corruption, lack of funding, constant reorganization, lack of clear legal direction and laws, and pressure for economic development. Confronted with these obstacles, it was no surprise that the environmental movement in the newly formed Russian Federation failed. Since 1995, the little power that the environment authorities had gradually disappeared. In 1996, President Yeltsin reduced that status of the Ministry of Environment to committee status and in 2000 the new president Vladimir Putin dissolved the Committee of Environment and the Federal Forest Service and passed on the responsibility to the Ministry of Natural Resources (MNR). The motivation behind this action was to encourage exploitation of Russia's natural resources in order to ignite economic development due to the Russian financial crisis of 1998 (OECD 2019; Henry and Douhovnikoff 2008).

In Russia, environmental management relies on a set of command and control, economic, and information instruments. Russia began setting hygiene standards in 1922, at the very start of the Soviet era. Most of the environmental quality standards in Russia are a carryover from the Soviet Union. Since environmental protection in Russia was passed to the MNR in 2000 by presidential decry, the MNR operates as the executive authority and is responsible for the following (Newell and Henry 2017):

- Public policy and statutory regulation for the study, use, renewal, and conservation of natural resources including hunting, hydrometeorology, environmental monitoring, and pollution control
- Public environmental policy and statutory regulation as it pertains to production and waste management, conservation and state environmental assessments
- Ensuring compliance with international treaties on environmental matters
- Working with other federal agencies, states, and local governmental authorities on environmental matters

Russia is a signatory to the following select international conventions (OECD 2019):

- 1972 Convention on the International Regulations for Preventing Collisions at Sea
- 1972 Convention on the Prevention of Marine Pollution by Dumping of Wastes and Other Matter

- 1973 Convention to Regulate International Trade in Endangered Species of Wild Flora and Fauna
- 1974 International Convention for the Safety of Life at Sea
- 1978 Convention for the Prevention of Pollution at Sea
- 1979 United Nations Convention on Long-range Trans-boundary Air Pollution
- 1982 United Nations Convention on the Law of the Sea
- 1985 Vienna Convention for the Protection of the Ozone Layer
- 1987 Montreal Protocol on Substances that Deplete the Ozone Layer
- 1989 Basel Convention on the Trans-boundary Movements of Hazardous Wastes and Disposal
- 1992 Convention on the Protection of the Black Sea from Pollution
- 1992 Convention on Biological Diversity
- 1992 United Nations Framework Convention on Climate Change
- 1992 Convention on the Protection and Use of Trans-boundary Watercourses and International Lakes
- 1997 Kyoto Protocol on Climate Change
- 2001 Stockholm Convention on Persistent Organic Pollutants

Facilities subject to federal environmental supervision are those facilities that exceed one or more of the following (OECD 2019):

- A waste disposal site that accepts 10,000 tons of hazardous waste or more
- A facility that discharges 15 million cubic meters of wastewater per year into a water body
- A facility that emits 500 tons a year or more of polluting substances into the ambient air

Environmental laws in Russia cover eight main topics that include (OECD 2019):

- Air emission management
- Chemical management
- Water management
- Waste management
- Safety management
- Environmental impact assessment
- Energy management
- Dangerous goods – hazardous materials management

Each of the above eight main topics are similar to the United States and other countries of the world.

7.4.2.2 Air

Responsibility for issuing air emission permits in Russia is divided between two agencies. The Federal Environmental, Industrial and Nuclear Supervision Service (Rostechnadzor) is responsible for the development and approval of the procedure and methods for determining air emission limits, and issuing air emission permits for radioactive sources. The Federal Service for Supervision of Natural Resources (Rosprirodnadzor) is responsible for the development and approval of the procedure and methods for determining air emission limits from non-radioactive sources. Federal standards require that Rostechnadzor and Rosprirodnadzor must include the quantity of each polluting substance to be emitted from each source.

Rostechnadzor and Rosprirodnadzor officials have the right to enter all facilities where any air pollution sources are present to evaluate compliance with emission permits and review any relevant documents including testing and laboratory results, information logs or other related material. Rostechnadzor officials also have the right to issue violation notices and fines for sources that are not in compliance with permit terms. In 2009, the Ministry of Natural Resources and Environment set out a plan to develop three orders aimed at completing the regulatory framework for air protection that remain in the process of being developed and include (Russian Ministry of Natural Resources 2017):

1. Developing procedures and methods for determining standards for maximum permissible and temporarily agreed emissions in ambient air
2. Creating a list of pollutants that are subject to regulation
3. Providing reliable sources of information for the management of air protection

After the regulatory framework has been completed, the ministry is expected to begin work on calculating environmental damage caused by air pollution and creating an inventory of the sources of air pollution that have a measurable negative impact on air quality.

7.4.2.3 Water

Water pollution is one of Russia's most significant environmental issues (World Water 2017; Russian Ministry of Natural Resources 2019). The most significant include the following:

- A significant portion of the Russian population do not have access to safe drinking water
- Evidence of oil pollution is visible in most urban areas
- An estimated 40% of surface water sources do not meet basic environmental standards

The Federal Agency for Water Resources is responsible for enforcing laws related to water resources and is under the jurisdiction of the Ministry of Natural Resources of the Russian Federation. While water pollution from industrial sources has diminished because of a decline in manufacturing, municipal wastes and nuclear contamination are constant threats to important water supply sources in Russia (Russian Ministry of Natural Resources 2019).

Evidently, the practice of discharge of untreated wastewater and sewerage into rivers and the ocean in Russia is relatively common (United Nations 2017c). It is estimated that more than 70% of all the wastewater from municipal sources is untreated. The issue of lack of wastewater treatment from mining operations is especially acute (Yablokov 2016).

In Russia, any industrial facility that discharges wastewater is required to obtain a permit. Wastewater discharge permits are issued by Rostechnadzor for a fee. The fees for wastewater discharge are reduced for those facilities that treat the wastewater before it's discharged into a water body (Russian Ministry of Natural Resources 2019).

Facilities accepting, transporting, and treating wastewater using a centralized water supply and discharge systems are required to monitor the quality and properties of the wastewater discharged. There are also discharge limits established in individual permits that are issued for polluting substances and microorganisms (Russian Ministry of Natural Resources 2019). This is similar to the EU and NPDES permits in the United States. Specific limits on wastewater discharge are established locally on a case by case basis.

Under a Russian Presidential Decree (No. 177, 2003) and Article 63 of the Federal Law on Environmental Protection, Russia implemented a unified environmental monitoring system that includes monitoring of water bodies, internal sea waters, and the Baikal lake system, along with wild animal resources, state of the subsoil resources, air quality, soil quality, etc. Monitoring is organized and performed by the Federal Water Resources Agency with the purpose of providing data on the improvement of water quality (Russian Ministry of Natural Resources 2019). This is similar to research and monitoring conducted in the United States by USEPA and USGS and other United States federal and international organizations and even state agencies.

The main specific priorities under the program include (Russian Ministry of Natural Resources 2019):

- To provide water resources needed by the population and the economy
- To prevent and eliminate floods and other harmful effects to water
- To ensure the security of hydro-electrical installations
- To protect bodies of water from pollution
- To develop a system for monitoring and forecasting the state of water resources and to provide information in water resources
- To improve the system of state management and conservation of water resources
- To provide a regulatory and scientific and technical basis for the water supply

7.4.2.4 Solid and Hazardous Waste

In Russia, the law that addresses waste management was enacted in 1998 and is termed Industrial and Consumption Waste and is Russian Federal Law No. 89-FZ, 1998 (Russian Ministry of Natural Resources 2019). Solid waste generation has increased significantly as the Russian Federation became an independent nation and as many Russian residents adopted a western-style consumption style. The vast majority of solid waste in Russia is disposed in landfills with a small percentage being incinerated.

Russia has begun to emphasis pollution prevention initiatives through reducing and minimizing the amounts of waste, particularly industrial waste during the production process. Many facilities are required to employ best management practices to reduce the amount of industrial wastes generated. As in the United States, Russia requires each facility to keep records on the generation and disposal of industrial wastes. In addition, Russia requires industrial facilities to obtain a permit to generate hazardous wastes and sets limits on the volumes of wastes that can be generated and sets analytical limits on wastes similar to that of the United States and European Union (Russian Federal Law on Environmental Protection 2002; Russian Ministry of Natural Resources 2019). The definition of hazardous waste in Russia is similar to the United States and the rest of the developed world and is defined as waste containing harmful substances with hazardous properties making it flammable, toxic, explosive, or reactive and also includes substances containing potentially infectious agents or those representing a direct or indirect danger to human health or the environment either alone or in combination with other substances.

Solid waste in Russia is also classified into five environmental hazard classes according to Russian Federal Law 7-FZ, 2002 (Russian Ministry of Natural Resources 2019):

- Class I – Very hazardous
- Class II – Highly hazardous
- Class III – Moderately hazardous
- Class IV – Low hazard
- Class V – Virtually harmless

Handling and transportation of hazardous waste is regulated under Russian Federal Law 89-FZ, 1998. All hazardous wastes must be properly classified on the basis of their composition and characteristics. The owner and generator of the waste is responsible for assuming the cost for proper handling of the waste.

7.4.2.5 Environmental Impact Assessment

Environmental impact assessments are generally required under Article 32 of the Russian Federal Law No. 7-FZ for projects if there is a potential that operation of the project could have an adverse effect on the environment directly or indirectly. The procedure and nature of the work to be done in order to conduct the Environmental Impact Assessment (EIA), as well as the necessary documentation, depend on a multitude of factors similar to that of many other countries and the United States. The EIA process in Russia involves several stages that are as follows (OECD 2019; Artic Center 2017):

- The facility must first prepare all documentation containing a general description of the proposed project including its purpose and alternatives and submit them to the applicable regulatory authority
- Inform the general public of the project including the location and nature of the project by using publications or announcements in the central or regional press
- Evaluate all environmental risks
- Evaluate and analyze the state of the territory on which the project is intended to be located
- Develop any measures that may be required to restore the environment
- Calculate the consumption of natural resources that may be necessary to the project

The legal or physical person responsible for the project must certify and approve the final version of the EIA before submitting it together with all necessary documents to the state environmental expert committee for consideration. Regulations are in effect since 2010 that require environmental audits of the progress and impact of the project to determine whether the project meets the requirements set forth in EIA and Russian environmental protections laws (OECD 2019).

7.4.2.6   Summary of Environmental Regulations and Protection in Russia

Russia is a country that has vast resources but has not performed well when it has come to environmental protection. Russia appears to have a robust set of environmental regulations but has difficulty in following their own rules. Not unlike other parts of the world, including the United States, Russia has had its share of environmental incidents that have caused huge negative impacts on human health and the environment. However, what may be different in Russia more than any other country examined in the world is that the occurrence of environmental incidents causing great harm has not resulted in change for the positive.

The Russian environmental movement which began in the late 1980s and into the early 1990s under Gorbachev's reforms have for the most part crumbled due to economic hardship and political instability largely by the current administration. Currently, the Putin administration has labeled many internal environmental groups as "anti-Russian" and have used aggressive tactics to quash openness when the subject matter is focused on environmental protection (Newell and Henry 2017). This reality together with corruption, poor environmental enforcement, general lack of environmental awareness and care indicate that environmental protection will not improve in the short term. According to Newell and Henry (2017), Russia's most significant environmental challenge is the illegal and unregulated use of its natural resources.

7.4.3   Norway

Norway is located on the Scandinavian Peninsula and also includes the island of Jan Mayen and the archipelago of Svalbard. Norway occupies a total area of 385,252 square kilometers (148,747 square miles) and as of January 2017 has a population of 5,258,317 (United Nations 2017a). Norway is bordered by Finland and Russia to the northeast and Sweden to the east, the Atlantic Ocean to the west, the Skagerrak to the south with Denmark beyond the strait.

Norway maintains a combination of market economy and a Nordic welfare model that includes universal health care and a comprehensive social security system. Norway has extensive natural resource reserves that include petroleum, natural gas, minerals, timber, seafood, fresh water, and hydropower. Norway's petroleum industry accounts for approximately 25% of Norway's GDP which, on a per capita basis, is the world's largest producer of oil outside the Middle East (United Nations 2017a). In July 2013, the Norwegian Climate and Pollution Agency and the Norwegian Directorate for Nature Management merged into what is now called the Norwegian Environment Agency (NEA), which is under the direction of the Ministry of Climate and Environment.

7.4.3.1 Environmental Regulatory Overview of Norway

The Norwegian Environment Agency has been assigned key tasks to achieve national objectives that include (Norwegian Environment Agency 2019a):

- A stable climate and strengthened adaptability
- Biodiverse forests
- Unspoiled mountain landscapes
- Rich and varied wetlands
- An unpolluted environment
- An active outdoor lifestyle
- Well managed cultural landscapes
- Living seas and coasts
- Healthy rivers and lakes
- Effective waste management and recycling
- Clean air and less noise pollution

Norway has set goals to be achieved by 2020 that include (Norwegian Environment Agency 2019a):

- Pollution of any type will not cause injury to health or environmental damage
- Releases of substances that are hazardous to health or the environment will be eliminated
- The growth of wastes will be less than economic growth
- Wastes will be fully used for recycling and energy recovery
- To ensure safe air quality
- Noise will be reduced by 10% of 1999 values

7.4.3.2 Air

Norwegian Environment Agency (2019a) has set mean annual air quality goals of 20 ug/m³ for particulate matter (PM10), 8mg/m³ for fine particulate matter (PM2.5), and 40 mg/m³ for nitrogen dioxide (NO_2). The value for fine particulate matter (PM2.5) is lower than the value set by the EEA of 25 ug/m³. The values for PM10 and nitrogen dioxide are the same as the EEA.

Although Norway is committed to improving its own air quality by reducing emissions, Norway has committed resources and funding for the United Nations and World Bank Pollution Management and Environmental Health (PMEH) program that focuses on improving the air quality in the major urban areas in other countries including China, Egypt, India, Nigeria, and South Africa. In addition, Norway has provided funding to address indoor air quality by providing clean cookstoves in these same countries mentioned above (Norwegian Environment Agency 2019a).

7.4.3.3 Water

Water management in Norway is divided between several different ministries that include (N. orwegian Environment Agency 2019a):

- Health and care services
- Environment
- Petroleum and energy
- Local government
- Regional development

Surface water accounts for 90% of water use in Norway while groundwater accounts for the remaining 10%. Norway generally has good water quality and treats 96% of wastewater before being discharged. The main sources of water contamination in Norway include agriculture, salmon farming, and the 4% of untreated wastewater. Norway also struggles with stormwater from its urban areas, which tends to overwhelm its treatment plants and increases flooding potential (Norwegian Environment Agency 2019a).

7.4.3.4 Solid and Hazardous Waste

Not unlike much of the EU, Norway struggles with solid waste and limited landfill space. Norway also incinerates a portion of its solid waste for energy recovery (Norwegian Environment Agency 2019b). Norway recycles up to 97% of its plastic, which is higher than any other nation measured. The recycling effort is successful because there is an environmental tax on plastic producers and an incentive for citizens to recycle, which is equivalent of up to $0.15 to $0.30 rebate per container depending on size (Norwegian Environment Agency 2019b).

In Norway, hazardous waste is defined as a waste that has the potential to cause serious pollution or involve risk of injury to people or animals. Solid and hazardous waste in Norway is handled using the EU standards (Norwegian Environment Agency 2019b).

7.4.3.5 Remediation

Norway has not ignored impacts to the environment that have resulted from oil and gas, forestry, mining, agriculture, shipping, and other activities. Given that most of Norway's inhabitants reside along its coastal areas and is located in a far northern latitude with significant cold-weather climate patterns all place additional difficulties in addressing its environmental remediation efforts.

Norway follows risk assessment guidelines for evaluating and remediating sites of environmental contamination and has established remediation standards for different types of land use. Remediation standards in Norway are in some instances much lower than the United States (Norwegian Environment Agency 2019c). For example, the soil value for lead in Norway is set at 60 mg/kg, whereas an acceptable level for lead in the United States may be as high as 400 mg/kg depending on the exposure route (Norwegian Environment Agency 2019c).

Remediation standards are listed at the Norwegian Environmental Agency (2019c) at www.m iljodirektoratet.no/old/klif/publikasjoner/andre/1691/ya1691.pdf.

7.4.3.6 Summary of Environmental Regulations of Norway

In Norway, sustainability is a fundamental principle for all development. The government's strategy is based on the principles of equitable distribution, international solidarity, the precautionary principle, polluters-pay principle, and the principle of common commitment. In part, this is achieved through an awareness and value of ecosystems and their connection with sustainability. Adequate knowledge and study of the condition of ecosystems is a necessary precondition for good nature management (Norwegian Environment Agency 2019a).

Sustainability also includes efforts to address climate change where Norway has been divided into sectors with each sector having its own climate change action plan with established targets for reducing greenhouse gases and integrating the action plan into specific sustainability targets (Norwegian Environment Agency 2019a).

7.4.4 Switzerland

Switzerland is officially known as the Swiss Federation. Switzerland is located in west-central Europe. Switzerland has a land area of 41,285 square kilometers (15,940 square miles) and is a land-locked country that is located in the Alps and Swiss Plateau. As of 2016, the population of Switzerland was 8,401,201. The human density of Switzerland is 195 people per square kilometer, which is very dense considering that much of the country is uninhabitable because of the mountainous terrain. For comparison purposes, the United States has a density of 33 humans per square kilometer (United Nations 2017a). Human occupation in Switzerland dates back nearly 150,000 years. Evidence of farming dates to 5,300 BC (United Nations 2016c).

Switzerland is one of the most developed countries on Earth. Switzerland has the highest nominal wealth per adult and the eighth-highest GDP per capita. Switzerland ranks near the top of many metrics of national performance including government transparency, civil liberties, quality of life, economic competitiveness, and human development. Zurich and Geneva consistently each rank among the top cities in the world with respect to quality of life (Bowers 2011).

7.4.4.1 Environmental Regulatory Overview of Switzerland

Switzerland is associated with lakes, mountains, and clean air along with a good quality of life. The natural environment is an integral part of Switzerland's identity. Along these lines, Switzerland has been proactive at (OECD 2016):

- Protecting natural resources, especially forests
- Higher density urban planning
- Reducing carbon dioxide and other greenhouse gas emissions
- Preserving water quality
- Maintaining biodiversity
- Improving air quality
- Preserving soil
- Cleaning up contaminated sites

Switzerland is under intense environmental pressure due to its population density and amount of land surface that is uninhabitable because of its mountainous nature (see Figure 7.13). During the 1970s and 1980s, ambitious environmental policies were implemented that included substantial government funding in promoting environmental awareness due to recognizing the environmental decline of land areas within the country, notably its forests (EEA 2015a).

Switzerland has designed and implemented pollution abatement policies with ambitious objectives. Most of these objectives have been achieved with remarkable success and include air pollution emission rates among the lowest in Europe and very high levels of wastewater infrastructure and in waste management facilities. This success was realized by means of a very active regulatory approach combined with rigorous enforcement, strong support from the public, and a considerable financial effort. The environmental policy applied in Switzerland rests on an extremely comprehensive body of federal environmental regulations. Their enactment is not without careful forethought

FIGURE 7.13   Switzerland's mountainous landscape. (Photograph by Daniel T. Rogers.)

and public input, active research, and referendums that are voted upon by the citizens on major environmental policy issues (OECD 2016).

Although Switzerland spends approximately 1.7% of its GDP on environmental issues, which is high compared to other European countries, much work remains to be accomplished which includes (OECD 2016):

- Meeting air pollution target for NOx, VOCs, and ozone
- Maintaining and upgrading wastewater treatment infrastructure
- Upgrading solid waste management infrastructure
- Cleaning up contaminated sites
- Source identification and cleanup of nonpoint source water pollution
- Continuing to restore forests and natural areas

7.4.4.2 Air

Air emissions in Switzerland are regulated under the Switzerland Environmental Protection Law, which was first enacted in 1983 and has been subsequently amended (Switzerland Environmental Protection Act 2019). The Environmental Protection Law sets ambient air quality standards very similar to the EEA. The Environmental Protection Act established the following (Switzerland Environmental Protection Act 2019):

- Maximum emission values
- Regulations on construction and equipment
- Traffic and operating regulations
- Heat insulation in buildings requirements
- Motor fuel regulations

According to the Environmental Protection Act, Switzerland's ambient air limits must be set such that any air pollution (Switzerland Environmental Protection Act 2019):

- Does not endanger people, animals, plants, their biological communities, and habitats
- Does not seriously affect the well-being of the population
- Does not damage buildings
- Does not harm soil fertility, vegetation, or waters

Switzerland's air quality has improved since the mid-1980s when the Environmental Protection Act was enacted. Ambient air quality standards can be viewed at www.admin.ch/opc/en/classified -compilation/19830267/index.html (Switzerland Environmental Protection Act 2019).

7.4.4.3 Water

In Switzerland, water is managed by the Federal Office of Public Health (FOPH) and the Federal Office of the Environment (FOEN). The FOPH is concerned with epidemics and infectious disease, food safety, and the safety of drinking water. FOEN is concerned with long-term conservation and maintaining natural resources. Switzerland's water quality standards are listed at: www.unece.org/ fileadmin/DAM/env/water/Protocol/reports/pdf (Switzerland Environmental Protection Act 2019).

In Switzerland, 80% of drinking water is obtained from springs while the remaining 20% is obtained from surface water sources. All potential households in Switzerland are connected to a central sewerage treatment plant or a decentralized treatment system (Switzerland Office for the Environment 2019). This represents the highest percentage of water treatment of any country that we will evaluate.

Challenges facing Switzerland with respect to water include agricultural runoff that contain fertilizers, herbicides, and pesticides. In addition, stormwater runoff is another challenge especially during very wet periods. In addition, Switzerland also has to deal with pollution originating from other countries through atmospheric deposition on land and surface water bodies (Switzerland Office for the Environment 2019).

7.4.4.4   Solid and Hazardous Waste

Solid and hazardous waste is regulated under the Switzerland Environmental Protection Act (2019). Since Switzerland is a small mountainous country, landfill space is at a premium. Therefore, Switzerland is strict on the amounts and types of wastes that are permitted and is dedicated to efforts of waste avoidance. For instance, products intended for once-only or short-term use may be prohibited because of limited landfill space. In addition, Switzerland requires any manufacturers to avoid any waste generation, especially hazardous waste that does not readily degrade under natural conditions, and to treat any waste so that it contains as little bound carbon as possible and is insoluble (Switzerland Office for the Environment 2019).

To discourage the generation of waste, Switzerland imposes high costs on generators of solid and hazardous waste and under certain circumstances requires prepaid disposal fees (Switzerland Office for the Environment 2019).

7.4.4.5   Remediation

Remediation of contaminated sites in Switzerland is regulated under the Switzerland Environmental Protection Act (2019). Switzerland has adopted the "polluters pay principle" similar to the United States. The methodology for evaluating potential contaminated sites uses the risk assessment approach similar to the EU and United States. Switzerland also requires that an environmental impact study be conducted for any new proposed development and also requires that all such reports be made public (Switzerland Office for the Environment 2019).

7.4.4.6   Summary of Environmental Regulations of Switzerland

Switzerland's environmental regulations are very comprehensive and also show a commitment from the government and residents. Switzerland has realized that it must be strict on environmental pollution matters since the country is small, mountainous, and heavily populated. Luckily, Switzerland does not have an industrial base as large as many other nearby countries but does have to deal with the challenges of pollution from agricultural sources and urbanization.

7.4.5   TURKEY

Turkey is a transcontinental country in Eurasia. Portions of Turkey are located within what is considered Asia and southeastern Europe. For the purposes of this book, we will treat Turkey as a part of Europe (see Figure 7.11). Turkey is bordered by eight countries that include: Greece and Bulgaria to the northwest; Georgia to the northeast; Armenia, Iran, and Azerbaijani to the east, and Iraq and Syria to the south. Turkey is also bordered by seas on three sides that include: the Aegean Sea to the west; the Black Sea to the north; and the Mediterranean Sea to the south. Ankara is the capital but Istanbul is the largest city in the country. Turkey covers 783,356 square kilometers (302,455 square miles) and has a population of 79,814,871 as of 2016 that equals a human density of 102 per square kilometer (United Nations 2016c; Turkish Statistical Institute 2017).

Turkey is a charter member of the United Nations, an early member of the North Atlantic Treaty Organization (NATO) and the OECD. Turkey's growing economy and diplomatic initiatives have led to its recognition as a regional power while its location has given it geopolitical and strategic importance throughout history (Porter 2009). Turkey faces some significant environmental issues that include (Anderson 2011; European Environment Agency 2013):

- Conservation of its biodiversity
- Air pollution

- Water pollution
- Greenhouse gases
- Land degradation

7.4.5.1 Environmental Regulatory Overview of Turkey

Current environmental laws in Turkey are similar to the EU and the United States and cover air pollution, water pollution, and land pollution and target improving human health and protecting its biodiversity. First enacted on August 11, 1983, the objective was to (Turkey Ministry of Environment and Urbanization 2018a; 2018b):

> improve the environment which is the common asset of all Turkish citizens; make better use of, and preserve the land and natural resources in rural and urban locations; prevent water, land and air pollution; by preserving vegetative and livestock assets and natural and historical richness, organize all arrangements and precautions for improving and securing health, civilization and life conditions of present and future generations in conformity with economical and social development objectives, and based on certain legal and technical principles

However, Turkey is facing significant economic hardships that greatly influence its ability to address its environmental issues. Therefore, Turkey faces significant challenges in improving environmental quality short term and long term if it does not conduct control measures to improve and protect human health and the environment. This is especially acute since the biodiversity of plant life in Turkey is so significant (Anderson 2011; European Environment Agency 2013; Environmental Health 2017). There are more than 3,000 endemic plant species in Turkey, making it one of the richest biodiversity areas on Earth. This is due to Turkey's wide variety of habitats and unique position between three continents and three seas (Anderson 2011). In addition, Turkey has not made significant progress at remediating sites of environmental contamination and remediation standards are either lacking or not enforced (EEA 2013).

7.4.5.2 Air

Air pollution is particularly significant in Turkey, especially in urban areas, and is especially acute in Istanbul, Ankara, Erzurum, and Bursa and consistently ranks among the worst air in the world. In the winter months, the combustion of heating fuels combined with automobile exhaust make for especially high levels of particulate matter (EEA 2014; Environmental Health 2017). In fact, according to the European Environment Agency, 97.2% of the urban population in Turkey is exposed to unhealthy levels of particulate matter.

Turkey is currently experiencing significant environmental deterioration of air quality due to combustion of fossil fuels, predominantly from automobiles in urban areas, and lack of enforcement and resources to employ effective air pollution control technologies at power plants and industrial facilities.

The root cause of some of the environmental issues facing Turkey are the overriding social and economic factors that Turkey is currently facing that inhibit enforcement of current environmental laws and regulations (European Environment Agency 2015b).

7.4.5.3 Water

As with air pollution, Turkey faces significant challenges with water pollution of various types including drinking water, surface water, groundwater, and the marine environment. Approximately 45,000 ships pass through the narrow Bosporus Straits every year and fears of a significant incident are real. Fears of an environmental incident have increased since the Turkish Environmental Protection Ministry has limited internal resources to oversee and inspect shipments of hazardous materials (Turkish Ministry of Environmental and Urbanization 2019).

Turkey does not have significant surface water resources, which indicates that if there is a water shortage, Turkey may not be prepared. In addition, water quality and availability of water is poor even for industry. Water quality is also significantly degraded, especially from agricultural fertilizers, pesticides and herbicides (EEA 2015b).

7.4.5.4 Solid and Hazardous Waste

Rapid growth in Turkey has resulted in the generation of significant amounts of waste materials. Turkey has not kept up the pace at establishing the required infrastructure for solid waste management. For example, in 2015 it was estimated that over 80% of the solid waste generated in Turkey was not landfilled properly and was discarded by other means (Akkoyunio et al. 2017). In addition, there are over 2,000 open dump sites scattered throughout the country (Berkun et al. 2011).

7.4.5.5 Remediation

As stated above, Turkey has environmental laws in place but currently lack enforcement abilities due to other overriding social and economic factors. Land degradation is also severely stressed from uncontrolled development leading to unchecked erosion, which has also affected water quality. Turkey is also facing significant environmental deterioration of land due to overuse, lack of urban green space, poor urban planning, overdevelopment, deforestation, population growth. and drought (EEA 2015b).

7.4.5.6 Summary of Environmental Regulations in Turkey

Turkey is facing some of the most significant environmental deterioration compared to other countries we have examined thus far. The main causes appear to be centered on its economic and financial hardships, which have also caused a significant degree of political unrest. This indicates that if the human population at large is experiencing hardship, the natural environment is likely experiencing hardship and stress at an even larger scale and severity. As mentioned above, this threatens its biodiversity and ecosystems with collapse if improvements are not enacted.

7.5 AFRICA

Africa is known as the location where humans are believed to have evolved nearly 2 million years ago (Smithsonian 2017). Africa is the second-largest continent and second most populous. Africa occupies approximately 30.3 million square kilometers (11.7 million square miles) and covers 6% of Earth's total surface area and 20.6% of Earth's land area (United Nations 2017a). Africa's population in total is approximately 1.22 billion and has an average age of 19.7, which is the youngest average age of any continent (United Nations 2017a) (see Figure 7.14).

Africa has a large diversity of ethnicities, cultures, and languages and has 54 separate identified countries, nine territories, and two independent states with limited or no recognition. Some locations within Africa are developed while much of the rural areas are not (see Figures 7.15 to 7.17). Africa has very diverse environmental climates, economics, historical ties, and governmental systems. This diversity has hindered its development and contributes to its environmental degradation and lack of environmental controls of pollution (Mwambazambi 2010; USEPA 2013).

Africa faces significant environmental degradation as a continent caused by many factors including (Chikanda 2009; University of Michigan 2017):

- Over population
- Lack of pollution controls
- Urban sprawl and unchecked development
- Deforestation
- Invasive species
- Soil degradation and erosion
- Water degradation
- Lack of basic sanitation

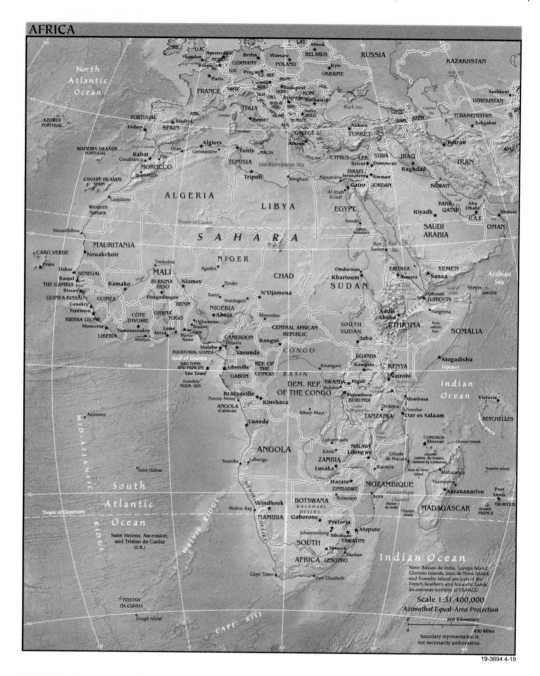

FIGURE 7.14 Map of Africa. (From United Nations 2019. Geospatial Information Section. Map of Africa. www.un.org/Depts/Cartographic/english/htmain.htm. 2019.) (Accessed April 15, 2019.)

- Political unrest and armed conflicts
- Poaching of wildlife, which is significant in most African countries

USEPA has been involved in a collaborative effort to stabilize environmental issues in many parts of Africa. Specifically, Sub-Saharan Africa relating to a growing population and industrial pollution issues that have impacted native population health and wellness, particularly vulnerable populations such as children, the elderly and economically disadvantaged. Areas of focus have centered

FIGURE 7.15 Typical east central Africa living and working conditions. (Photograph by Daniel T. Rogers.)

FIGURE 7.16 Photo demonstrating nomadic livestock herding practices in east central Africa. (Photograph by Daniel T. Rogers.)

FIGURE 7.17 Urban area in Africa lacking adequate building safety and basic sanitation. (Photograph by Daniel T. Rogers.)

on air quality, water quality, and reducing exposure to toxic chemicals (USEPA 2018a). USEPA has not been involved significantly or at all in collaborating with northern countries of Africa above or within the Saharan Desert for numerous reasons, some of which include territorial disputes and political unrest. Specific programs where USEPA is assisting Sub-Saharan countries in Africa include (USEPA 2018a):

- Air quality management
- Safe drinking water practices
- Lead paint abatement
- Reducing and managing methane emissions
- Improving environmental governance
- Improving sanitation
- Improving hazardous waste management

7.5.1 SOUTH AFRICA

South Africa is located at the southern tip of the African continent and occupies 1,220,813 square kilometers, making it the twenty-fifth largest country by size. It is bordered to the north by Namibia, Botswana, Zimbabwe, and Mozambique. Swaziland and Lesotho are two countries that are located within South Africa. The Atlantic Ocean is located to the west and southwest and the Indian Ocean is located to the east and southeast. The population of South Africa as estimated in 2018 exceeds 57 million (Statistics South Africa 2018).

South Africa is largely located in a dry climatic region of Africa with most of its western region located in a semi-desert. Rainfall increases toward the east and falls primarily in summer. Recently, the Cape Town region of South Africa has experienced drought conditions that have now threatened its potable water supply. The central cause of the drought has been blamed on significant increases in population, which now exceeds 4 million, and on climate change (Welch 2018). Figure 7.18 shows Cape Town.

7.5.1.1 Environmental Regulatory Overview of South Africa

South African environmental law attempts to protect and conserve the environment of South Africa. South African environmental law encompasses natural resource conservation and utilization,

FIGURE 7.18 Cape Town, South Africa. (Photograph by Daniel T. Rogers.)

as well as land-use planning and development, and enforcement. In South Africa, the National Environmental Management Act (NEMA) of 1998 forms the framework for the protection of human health and the environment. The South Africa National Environmental Management Act covers environmental subjects that include (South African Government Gazette 2018):

• Sustainable development	• Intergenerational equity
• Environmental justice	• Environmental rights
• Public trust	• Precautionary principle
• Preventative principle	• Polluters-pay principle
• Local-level governance	• Common differentiated responsibility

Although some names may be different, NEMA is very similar to that which we are accustomed to in the United States and most other countries. Especially to what is described as (South Africa Department of Environmental Affairs 2018a):

- Preventive principle, which relates to the treatment or capture of contaminants before they are emitted or released into the environment (i.e., installation and operation of air pollution control equipment). This is the fundamental notion of NEMA regulating generation, treatment, storage, and disposal of hazardous waste, and the use of pesticides.
- Precautionary principle is designed such that lack of full scientific certainty shall not be used as a reason for postponing measures to prevent environmental degradation.
- Polluters-pay principle, which is analogous to Superfund in the United States.
- Environmental justice is intended to safeguard unfair environmental discrimination toward any person or persons.
- Environmental rights does not mean to imply that the environment has rights but is intended to address the rights of the individual to an environment that is safeguarded.
- Local-level governance requires that decisions affecting local municipalities should be made by local municipalities.

NEMA defines the "environment" as the surroundings within which humans exist and include (South African Department of Environmental Affairs 2018b):

- The land
- The water
- The atmosphere of the Earth
- Microorganisms
- Plant life
- Animal life
- Any part or combination of the items listed above and the interrelationships among and between them
- The physical, chemical, aesthetic, and cultural properties and conditions that influence human health and well-being

Furthermore, NEMA goes on to also define the environment as:

The aggregate of surrounding objects, conditions and influences that impact the life and habits of man or any other organism or collection of organisms

NEMA addresses the following distinct but inter-related areas of general concern and include (South African Department of Environmental Affairs 2018b):

- Land-use planning and development
- Resource conservation and utilization
- Waste management and pollution control

One of the main purposes of NEMA was to set up a national environmental management system that would outline procedures for cooperative governance within governmental agencies such that the act does not impose overly burdensome requirements on the private sector (South Africa National Environmental Management Act 1998). Other prominent environmental legislative laws in South Africa include (South Africa Department of Environmental Affairs 2018b):

- Air Quality Act No. 39 of 2004
- Protected Areas Act No. 31 of 2004
- Protected Areas Amendment Act No. 31 of 2004
- Biodiversity Act No. 72 of 2005
- Clean Development Regulations Act of 2004
- Protected Areas Act No. 57 of 2004
- Environmental Conservation Act No. 50 of 2004
- Marine Living Resources Act of 2004
- Biodiversity Act: Threatened or Protected Species Amendment Act of 2011
- Environmental Impact Assessment Regulations of 2010
- Environmental Management Regulations of 2010

7.5.1.2   Air Quality Standards

South Africa ambient air quality standards were enacted in 2004 and cover compounds such as sulfur dioxide, nitrogen dioxide, particulate matter (PM10), ozone, benzene, lead, and carbon monoxide (South Africa Department of Environmental Affairs 2009a). Limits are similar to the United States and the EU and are listed at https://environment.gov.sa/legislative/acts/regulations.

7.5.1.3   Water Quality Standards

South Africa established water quality standards for drinking water and other potable uses in 1996 and does not include water that is sold as a beverage or in swimming pools (South Africa Department of Environment 1996). Water quality standards are similar to the United States and the EU and are listed at https://environment.gov.sa/legislative/acts/regulations.

7.5.1.4   Waste Management

In South Africa, solid and hazardous waste are differentiated and are regulated by the National Environmental Waste Act No. 59 of 2009 (South African Department of Environmental Affairs 2009b). Implementation guidelines were published in 2012 (South Africa Department of Environment 2012). Hazardous wastes are also termed priority wastes in South Africa. Solid and hazardous waste management in South Africa relies heavily on landfill disposal of wastes.

According to the most recent information available, South Africa produced approximately 108 million metric tons of solid waste in 2011, of which 98 million metric tons were disposed in landfills. The remaining 10 million metric tons (or just under 10%) was recorded as being recycled. The largest volumes of waste are produced by the industrial and mining sectors (South Africa Department of Environmental Affairs 2012).

In South Africa, hazardous waste is defined as (South Africa Department of Environmental Affairs 2012):

Any waste that contains organic or inorganic elements or compounds that may, owing to the inherent physical, chemical or toxicological characteristics of that waste, have a detrimental impact on health and the environment

Hazardous wastes are classified according to the following criteria (South Africa Department of Environmental Affairs 2012):

- Reactive
- Corrosive
- Flammable
- Characteristic
- Explosive
- Gases
- Radioactive
- Organic (for halogenated organic wastes)
- Infectious
- Miscellaneous dangerous substance (examples of wastes in this category may include asbestos, dry ice, and other environmentally hazardous substances that do not fall into any of the above-listed categories)

Classifying solid wastes by characteristic in South Africa is conducted by comparing the concentration of contaminants within the waste in two ways: leachable fraction and total concentration. If the concentration of the specific substance exceeds either or both criteria then it is considered hazardous. Leachable concentration is to be determined by collecting a representative sample of the waste material and analyzing the waste using the toxic characteristic leaching procedure similar to that outlined by the USEPA. Compounds and concentration limits which are similar to the United States, EU, and Australia are provided at https://environment.gov.sa/legislative/acts/regulations (South Africa Department of Environmental Affairs 2012).

Generators of the waste must select the potential chemical contaminants that are either known to exist or may be present within the waste material. This may require knowledge of site activities, site history, or the processes which created the waste. Generators must be able to justify the chemical contaminants selected for analysis and keep records of that decision for 3 years. If chemical contents of a waste are unknown, then it is generally recommended to conduct a comprehensive analysis which may include (South Africa Department of Environment 2012):

- Volatile organic compounds
- Polycyclic aromatic hydrocarbons
- Semi-volatile organic compounds
- Phenols
- Polychlorinated biphenyls
- Heavy metals

7.5.1.5 Remediation Standards

Remediation standards for South Africa were established in 2013 (South Africa Department of Environmental Affairs 2013). South Africa remediation standards are divided into categories using the following definitions:

- Contaminant is defined as any substance present in an environmental medium at concentrations that exceed natural background concentrations
- Informal residential means an unplanned settlement on land which has not been proclaimed as a residential and consists mainly of makeshift structure(s) not erected according to approved architectural plans
- Remediation is the management of a contaminated site to prevent, minimize, or mitigate damage to human health or the environment

- Soil Screening Value 1 are soil quality values that are protective of both human health and eco-toxicological risk for multi-exposure pathways, and is inclusive of migration to a water source of either surface water or groundwater
- Soil Screening Value 2 are soil quality values that are protective of risk to human health in the absence of a water resource or ecological exposure
- Standard residential means a settlement that is formally proclaimed and serviced, and generally developed with formal permanent structures including land parcels

Remediation values are similar to the United States and EU and are listed at https://environment. gov.sa/legislative/acts/regulations.

Groundwater remediation values in South Africa follow either surface water criteria or the Dutch Intervention Values described earlier in this chapter.

The variations in the climate of South Africa allows for a wide variety of crops that range from tropical fruit to corn and tree plantations. This in turn has led to extensive use of pesticides, herbicides, and fungicides to the point where South Africa is one of the largest importers of these types of chemicals in all of Africa (Quinn et al. 2011). Estimates on the annual total pesticide use in South Africa exceed 2,800 metric tons. However, remediation standards for most of these chemicals is lacking.

South Africa depends heavily on groundwater resources but due to drought in many areas, the resource is experiencing stress from overexploitation and many wells are now dry, especially in the Cape Town region. Currently, dam levels are at 60% capacity and there is a water use limit placed on the population of 50 liters of water per day (South Africa Department of Environmental Affairs 2018b).

7.5.1.6 Summary of South Africa Environmental Regulations

Compared to the United States, environmental regulations in South Africa have only recently become enacted in that the National Environmental Management Act is only 20 years old and many of the detailed regulations for air, water, and solid and hazardous waste are generally 10 years old. The framework of the regulations are predominantly based on USEPA. This may be true from a framework point of view but many of the details are similar to the European Union and Australia. In fact, as we shall discover when we are evaluating Australia, the solid waste classification in South Africa is nearly identical to Australia.

The environmental regulations of South Africa appear to be very robust and comprehensive, even when comparing them to the United States or the European Union. However, they are still rather recent and given the political and financial pressures, and drought in South Africa, implementation and delays in enforcement have hindered and slowed progress. Evidence of this is that recycling efforts have not seen significant progress and remain at only 10% (South Africa Department of Environmental Affairs 2012). Another very real disadvantage within South Africa is lack of sufficient infrastructure due to geography in that it is rather isolated from other developed countries and therefore must import technologies and equipment over a longer distance.

Finally, climate change appears to have impacted South Africa as most significantly realized with the ongoing drought and the city of Cape Town experiencing a water shortage that might be more severe than any developed city on Earth at the moment. The impact is clear but the response and long-term plan to address climate change is not yet certain.

7.5.2 KENYA

Kenya is located in east-central Africa on the Indian Ocean between Somalia and Tanzania. Other countries that share a border with Kenya include Ethiopia, South Sudan, and Uganda (see Figure 7.14). Kenya is 582,650 square kilometers in size and has an estimated population of 48 million (Kenya National Bureau of Statistics 2018). Nairobi is the largest city in Kenya and is considered the tenth largest city in Africa with an estimated population of 6.5 million residents (Kenya National Bureau of Statistics 2018). The name originates from the Maasai phrase Enkare Nyrobi, which translates to "cool water," and is a reference to the Nairobi River which flows through the city. Central and western Kenya are located in the Kenyan Rift Valley, characterized by mountains and volcanoes, some of which are considered active. Kenya's highest peak, Mount Kenya, which exceeds 5,700 meters in elevation, is located in this region (Kenya Geological Society 2018).

The population of Kenya has increased significantly over the last 50 years, especially Nairobi, where the population has increased from 0.5 million to 6.5 million. This has placed pressure on providing adequate infrastructure to support a population increase of that magnitude that quickly (Kenya National Bureau of Statistics 2018). The increase in population has also placed stress on the natural environment in Kenya and include water and air pollution from urban and industrial areas and agriculture, deforestation, pesticide and herbicide use, erosion, desertification, and poaching of wildlife (Kenya National Environment Management Authority 2018). Approximately 8% of the landmass of Kenya are contained with 22 national parks and 28 national reserves (Kenya Wildlife Service 2018). Kenya's economy is dominated by agriculture followed by manufacturing, much of which is agriculture related. Kenya is the banking capital of central Africa and tourism is also very significant to the economy (Kenya National Bureau of Statistics 2018).

7.5.2.1   Kenya Environmental Regulations

Environmental regulations were enacted starting in 2006 and now include several acts that cover the following environmental areas (Kenya National Environmental Management Authority 2018):

Air quality	Domestic water
Industrial water discharge	Wetlands
Noise	Solid waste
Hazardous waste	Environmental impact assessments
Chemical regulations	Waste transport
Controlled substances	Coastal protection
Biodiversity	Land development

7.5.2.2   Air

Air pollution in Kenya is most significant near its three largest cities – Nairobi, Mombasa, and Kisumu, which account for near 30% of its population (Kenya National Bureau of Statistics 2018; Kinny et al. 2011). Fugitive emissions are also prevalent since many roads in Kenya are dirt or gravel. Although environmental regulations in Kenya are considered robust, as we shall see in this chapter, air regulations are considered weak not because of a need for additional regulation, but due to implementation and enforcement of existing laws and lack of cooperation between government ministries in Kenya (Barczewski 2013).

Air quality regulations were enacted in 2009 titled The Environmental Management and Co-Ordination Air Quality Regulations. The Air Quality Act of 2009 defined air pollution as meaning any change in the composition of air caused by air pollutants (Kenya Air Quality Act 2009). The Air Quality Act of 2009 sets limits for the following sources:

- Industry and other stationary sources
- Mobile sources, such as motor vehicles, and required vehicle emission testing
- Occupational air quality limits
- Fugitive sources
- Particulate emissions from demolition sites

- Open burning
- Cross-border air pollution
- Exposure to hazardous substances
- Effects of stockpiling of material
- Emissions from waste incinerators

The Kenya Air Quality Act of 2009 required owners or operators of industrial sites to obtain a permit or license to emit air pollutants and required each facility to conduct a detailed study that includes:

- Initial emission assessment report
- Preliminary assessment of stationary sources
- Atmospheric impact report
- Monitoring records
- Notification of excessive emissions
- Stack emission recording and reporting
- Continuous monitoring system requirements

The Air Quality Act of 2009 also set ambient air quality tolerance limits similar to the United States and EU and are listed at https://nema.go.ke/NEMA/airqualityact (Kenya Air Quality Act 2009).

7.5.2.3 Water

Water quality regulations were enacted in 2006 with the Environmental Management and Co-Ordination Water Quality Act (Kenya Water Quality Act 2006). The Water Quality Act regulates drinking water, protects the sources of drinking water, industrial use and effluent discharge, and agricultural and recreational use. Water quality standards for drinking or domestic use and effluent discharge are similar to the United States and EU and are located at https://nema.go.ke/NEMA/waterqualityact (Kenya Water Quality Act of 2006).

7.5.2.4 Solid and Hazardous Waste

Management of solid and hazardous waste in Kenya is regulated by the Environmental Management and Co-Ordination Waste Management Act of 2006 (Kenya Waste Management Act 2006). The act covers topics that include:

- Solid waste
- Hazardous waste
- Industrial waste
- Pesticides and toxic substances
- Biomedical wastes
- Radioactive wastes

The act outlines the responsibilities of the generator, transporter, and disposal facility, permitting, licensing, transportation, and environmental audit procedures. It also outlines requirements for evaluation of an environmental impact assessment, training, labeling, packaging, segregation, monitoring, and classification (Kenya Waste Management Act 2006).

Hazardous waste determination in Kenya is by content and percentage of certain chemicals considered hazardous by their very nature. For example, waste that contain the following are considered hazardous in Kenya (Kenya Waste Management Act 2006):

- Radio-nuclides
- Medical waste

- Pharmaceutical, drugs, or medicines
- Biocides, germicides, herbicides, insecticides, fungicides
- Wood preserving chemicals
- Organic solvent waste
- Heat treatment and tempering wastes containing cyanide
- Mineral oil waste
- PCBs
- Wastes from inks, dyes, pigments, and paints
- Waste chemicals from research, development, or teaching
- Explosives
- Wastes containing metal carbonyls, beryllium, hexavalent chrome, copper, zinc, arsenic, selenium, cadmium, antimony, tellurium, mercury, thallium, lead, fluorine, phosphorus, phenol, ethers, and cyanide at 1% or more by weight
- Waste at a pH of less than 2
- Waste at a pH greater than 11.5
- Wastes containing asbestos
- Halogenated organic solvents at 0.1% or more by weight
- Any congener of polychlorinated dibenzo-furan
- Any congener of polychlorinated dibenzo-p-dioxin

Other wastes that are considered hazardous not listed above include wastes that are (Kenya Waste Management Act 2006):

• Flammable	• Explosive	• Combustible
• Oxidizers	• Organic peroxides	• Toxic or poisonous
• Infectious	• Corrosive	• Eco-toxic
• Toxic gas	• Persistent wastes	• Leachate
• Radioactive	• Carcinogens	• Medical waste

7.5.2.5 Remediation Standards

Remediation standards in Kenya were first enacted in 2003 and was amended in 2009 and are termed the Environmental Impact and Assessment and Audit Act (Kenya Impact Assessment and Audit Act 2003). The act requires that an environmental impact assessment be conducted before any land is developed or re-developed and must consider the following at a minimum:

- Environmental, social, cultural, economic, and legal considerations
- Identify environmental impacts and scale of impacts
- Identify and analyze alternative
- Develop an environmental management plan
- Consult with the regulatory agency on a regular basis
- Seek public comment
- Hold at least three public meetings
- Prepare a detailed report

After regulatory review and acceptance, a license will be issued. Following development, environmental audits and monitoring shall be conducted at intervals outlined in the license.

The environmental impact assessment is a detailed environmental study similar to that in the United States and usually includes extensive investigation and testing for the presence of contamination and if discovered, the nature and extent of impacts must be defined. Following the completion of testing, the assessment is required to evaluate whether there are any risks posed by the presence of contamination and lower those exposure risk to an acceptable level, if necessary (Kenya Impact Assessment and Audit Act 2003).

7.5.2.6 Summary of Environmental Regulations of Kenya

Kenya does, in fact, have a very robust set of environmental regulations in place. The regulations have not been enacted for very long, most are less than 15-years old. So it should come as no surprise that the current major obstacle is implementation and enforcement along with cooperation between ministries responsible for the environmental interpretation and enforcement. Kenya air and water environmental regulations are similar to that of the United States and the European Union. However, the solid and hazardous waste regulation are unique to Kenya and are strict. This is likely due to a combination of factors that include Kenya's reliance on its national parks and biodiversity for tourism, lack of a significant industrial base, and perhaps disincentives for additional industrial development. An example of the strict solid and hazardous waste regulations in Kenya is that Kenya considers any plastic waste to be hazardous.

7.5.3 TANZANIA

Tanzania is located in east-central Africa on the Indian Ocean located along the eastern border of the country. Kenya is located immediately to the north and Mozambique, Malawi, and Zambia are located along the southern border. The Democratic Republic of Congo, Burundi, and Rwanda are located immediately to the east (see Figure 7.14).

Tanzania has an estimated population of slightly more than 57 million and is 947,300 square kilometers in size (Tanzania Bureau of Statistics 2018). Tanzania's highest elevation is Mount Kilimanjaro at an elevation of 5,895 meters. There are numerous national parks, conservation areas, and game reserves within Tanzania including the Serengeti National Park, Ngorongoro Crater Conservation Area, and the Selous Game Reserve (Tanzania National Bureau of Statistics 2018). Lake Victoria, Africa's largest lake, is located in the northwest portion of the country. Northern and central portions of Tanzania are mountainous and are part of the African Rift Valley. Most of Tanzania's population is located in the northern portion of the country and the eastern border with the Indian Ocean (Tanzania National Bureau of Statistics 2018).

The largest city in Tanzania, which was the former capital of the country, is Dar es Sallaam with an estimated population of over 5.5 million. The capital of Tanzania is now Dodoma. Dodoma has an estimated population of greater than 2.2 million and includes outlying areas (Tanzania National Bureau of Statistics 2018). The population of Tanzania has increased significantly since 1963 when the population was estimated at 11 million. This has placed pressure on providing adequate infrastructure to support a population increase of that magnitude (Tanzania National Bureau of Statistics 2018). Tanzania has the second-largest economy in central Africa, second to Kenya. The economy is dominated by agriculture, which employs approximately 50% of the workforce. Approximately one-third of the population of Tanzania live at or below the poverty level (Tanzania National Bureau of Statistics 2018).

7.5.3.1 Environmental Regulatory Overview of Tanzania

Not unlike Kenya, Tanzania's increase in population has placed stress on the natural environment and include water and air pollution, deforestation, pesticide, and herbicide use, erosion, desertification, and poaching of wildlife. Water-borne illnesses, such as malaria and cholera, account for over half of the diseases affecting the population (Tanzania National Bureau of Statistics 2018).

7.5.3.2 Air

Air quality in Tanzania is regulated under the Environmental Management Air Quality Act of 2007. The Tanzania Air Quality Act established ambient air quality standards and emission standards for motor vehicles (Tanzania Air Quality Act 2007). In addition, Tanzania has emission standards for cement plants. Cement plants in Tanzania are numerous since the population has greatly increased over the past few decades and construction for housing and commerce has placed a large demand for building materials. The objective of the Tanzania Air Quality Act of 2007 was to:

1. Set baselines parameters on air quality
2. Enforce minimum air quality standards
3. Encourage effective and optimal air pollution control equipment and procedures
4. Ensure protection of human health and the environment

To achieve the objectives, the Tanzania Air Quality Act of 2007 conducted the following:

- Established criterion and procedures for measuring ambient air quality
- Established minimum ambient air quality standards
- Established emission standards and limits for various sources of air pollution
- Prescribed stack heights
- Prescribed criteria and guidelines for air pollution controls for mobile and stationary sources of air pollutants
- Established occupational air quality standards

The Air Quality Act requires owners or operators of air pollutant emission sources to obtain a permit and to document operations of air pollution control equipment to ensure optimal performance. Tanzania Ambient Air Quality Standards are similar to the United States and EU and are listed at www.parliament.go.tz/acts-list-air-quality (Tanzania Air Quality Act 2007).

7.5.3.3 Water

Water quality varies significantly in Tanzania. Urban areas generally have better water quality than rural areas. The latest estimate from 2015 is that 26 million people or approximately half the population lack access to at least basic water. Wastewater treatment is also a challenge in Tanzania. Of the 20 urban locations within the country, 11 provide some water treatment. Within the urban areas where wastewater treatment is possible, the range in actual hookups to sanitation range from 4% to 45% of households that are actually connected to sewers (Tanzania Ministry of Water and Irrigation 2015). Water quality in Tanzania is regulated through the Tanzania Water Resources Act (2009) and the Water Supply and Sanitation Act (2009). Tanzania drinking water standards for select compounds are similar to the United States and EU and are located at www. tanzania.go.tz/egov_uploads/documents/tanzania (Tanzania National Environmental Standards Compendium 2018).

7.5.3.4 Solid and Hazardous Waste

Solid and hazardous waste is regulated through the Tanzania Environmental Management Act of 2008, more commonly referred to as the Hazardous Waste Control and Management Regulations (2008). The Hazardous Waste Control and Management Regulations cover the following topics:

- Classification procedures
- Packaging
- Labeling
- Handling
- Transporting
- Storage
- Permitting

In Tanzania, wastes are considered hazardous if they have the following (Tanzania Environmental Management Act 2008)

- Characteristics such as explosive, flammable, corrosive, reactive or have the potential to produce a leachate considered toxic
- Are generated from certain types of operations such as:
  - Medical waste
  - Pharmaceuticals, drugs, or medicines
  - Biocide production
  - Wood-preserving chemicals
  - Organic solvent production
  - Heat treatment and tempering operations that use cyanide
  - Used mineral oils
  - Oil and water mixtures
  - Waste containing PCBs
  - Ink, dye, pigment, paint, lacquer, and varnish operations
  - Waste from resins, latex, plasticizers, glues, and adhesives
  - Waste chemical substances from research, development, or teaching
  - Residues from waste treatment operations
- Waste having the following constituents:
  - Metals such as beryllium, hexavalent chromium, copper, zinc, arsenic, selenium, cadmium, antimony, tellurium, mercury, thallium, lead
  - Inorganic cyanides
  - Asbestos
  - Phenols
  - Halogenated organic solvents
  - Ethers
  - Any congener of polychlorinated dibenzo-furan
  - Any congener of polychlorinated dibenzo-p-dioxin

7.5.3.5   Remediation Standards

Soil quality standards were established in Tanzania in 2007 with the Environmental Management Act that established maximum allowable concentrations of many compounds in soil. Groundwater standards have not been established. The intention of the Environmental Management Act of 2007 was to make clear that any intentional disposal of solid refuse or putrid solid matter onto the ground is prohibited (Tanzania Environmental Management Act 2007).

7.5.3.6   Summary of Environmental Regulations of Tanzania

The environmental regulations in Tanzania are similar to that of its neighbor Kenya. Differences between Kenya and Tanzania focus on the fact that Kenya is more developed and is further along at implementation of its regulations. Tanzania appears to struggle with providing basic sanitation and supplying water to its population. This is largely due to the lack of basic infrastructure within the country. In addition, approximately one-third of the population of Tanzania are considered living below the poverty level. Therefore, it should come as no surprise that Tanzania is also struggling with maintaining its natural environment and the health of its residents.

7.5.4   Egypt

Egypt is located in northern Africa and has an estimated population of 99 million and a land area of 1,001,449 square kilometers (United Nations 2017a). The Mediterranean Sea forms Egypt's northern border, Libya is located to the west, Sudan to the south, and Jordan and Saudi Arabia to the east (see Figure 7.14). Egypt is the third most populous country in Africa behind Nigeria and Ethiopia. Approximately 95% of the population of Egypt lives along the Nile River, the Nile Delta, or along the Suez Canal. These regions are among the most densely populated areas of the world. An estimated 75% of the population of Egypt are under the age of 25, making it one of the most youthful country populations in the world (United Nations 2018a). Cairo, located along the Nile River, is the largest city in Egypt with an estimated metropolitan population of over 20 million residents (United Nations Populations Programme 2018b).

7.5.4.1 Environmental Regulatory Overview of Egypt

In Egypt, the environment is regulated by the Ministry of Environment. Environmental standards in Egypt were first enacted in 1994 and were amended in 2009 and are simply known and referred to as The Environmental Law and covers protection of the land, air, water, and marine environment (Egypt Environmental Affairs Agency 2018).

7.5.4.2 Air

Air pollution in the urban area of Egypt is considered serious and decreases life expectancy by more than 2 years (World Bank 2013a). Air quality in Cairo is from 10 to 100 times greater than acceptable world standards. The source of much of the air pollution in Cairo is attributed to automobile exhaust from the more than 2 million cars that are on the roads daily. Excessive air pollution has also accelerated deterioration of many ancient landmarks and relics of Egypt's history. Additional sources are from industry and frequent dust storms from desert regions that surround the city (Marey et al. 2010). The Egypt Environmental Law of 2009 addresses air pollution to ensure that total pollution emitted by all sources in any area is within permissible limits (Egypt Environmental Affairs Agency 2018). Egyptian ambient air quality standards are similar to the United States and EU and are listed at www.eeaa.gov.eg/en-us/laws/envlaw.aspx (Egypt Environmental Affairs Agency 2018).

The Environmental Law addresses the following (Egypt Environmental Affairs Agency 2018):

- Open burning of any kind including agricultural fields, garbage, solid waste, and other types of unauthorized burning
- Excessive motor vehicle exhaust
- Application of pesticides and herbicides
- Properly maintaining equipment and machines emitting air pollutants
- Providing appropriate and adequate ventilation
- Prohibiting smoking in public transport areas
- Limiting radioactivity in ambient air
- Regulating ozone-depleting substances

To further address the air quality in the Cairo area specifically, governmental authorities have adopted various measures to address the air quality issues including (Egypt Environmental Affairs Agency 2018):

- Constructing better roads with overpasses to keep traffic moving
- Constructing satellite cities to relieve inner-city congestion
- Improving mass transit
- Providing incentives for taxis and industry to convert to natural gas

However, the overall air quality has not improved and has actually worsened because of several factors that include (The World Bank 2013a):

- Increased population
- Increased development

- Increased agricultural activities, primarily rice farming
- Ineffectiveness of traffic pattern management
- Increased motor vehicles
- Under capacity of mass transit
- Lack of enforcement of emission standards

7.5.4.3   Water

The main source of water in Egypt is the Nile River with an estimated current extraction rate of 56 billion cubic meters of freshwater every year and represents 97% of the total volume of freshwater used each year in Egypt. There are water shortages further away from the Nile River especially in rural areas. This is not a surprise since much of Egypt is considered a desert. Major improvements in providing piped water supply to residents has been realized in that nearly 100% of residents in urban areas and 93% of residents in rural areas have piped water available. However, only about 50% of the population are connected to sanitary sewers (World Health Organization 2018b; 2018c). The Egyptian Environmental Law of 2009 addressed mainly the marine environment along Egypt's northern coast and the area along the Suez Canal. Drinking water is regulated by the Egyptian Ministry of Health beginning in 1995 and focused on protecting the source of drinking water, that being the Nile River (World Health Organization 2018b). Egyptian water quality standards are similar to the United States and EU and are listed at www.who.int/countries/egy/en (World Health Organization 2018b).

7.5.4.4   Solid and Hazardous Waste

Solid and hazardous waste is regulated through The Environmental Law of 2009 and administrated through the Egypt Environmental Affairs Agency (2018). In Egypt, all solid and hazardous wastes are regulated and include (Egypt Environmental Affairs Agency 2018):

- No wastes shall be disposed of in unlicensed locations
- All treatment of wastes must be conducted under an applicable permit and license
- No wastes shall be imported
- Appropriate health and safety measures for all those who handle or ship wastes
- All waste generators shall be registered

The Egyptian solid and hazardous waste regulations include (Egypt Environmental Affairs Agency 2018):

• Transportation	• Labeling	• Manifesting
• Storage	• Disposal	• Characterization
• Licensing	• Operations	• Packaging
• Treatment	• Disposal	

In Egypt, a hazardous waste is defined as (Egypt Environmental Affairs Agency 2018):

Wastes of activities and processes or their ashes that maintain their harmful properties and have no subsequent original or substitutive uses

Unlike most countries of the world, no characteristics of hazardous wastes, processes, waste streams, or constituents of wastes have been identified or defined (Ramadam and Nadim 2014).

7.5.4.5   Remediation

Egypt requires any new proposed development conduct an environmental impact assessment (EIS) which must be conducted in accordance to the elements, designs, specifications, bases and pollutant loads determined by the Egyptian Environmental Affairs Agency (Egypt Environmental Affairs Agency 2018). The focus of the EIA is not on current soil or groundwater quality but only addresses future potential impacts.

7.5.4.6 Summary of Egyptian Environmental Regulations

Egypt is a very urbanized country with very dense population centers. Together with an increasing population, this has placed enormous stress on the health of its population (World Health Organization 2018b). Air pollution, sanitation, remediation of contaminated sites, and deterioration of ancient historical relics are of immediate and pressing concern. Near future progress on solving environmental issues do not look promising due to recent political strife and violence within the country.

7.6 ASIA

Asia is the largest and most populous continent. The land area of Asia is estimated at 43,608,000 square kilometers and represents approximately 29.4% of Earth's land surface. Asia's estimated population is greater than 4 billion and represents approximately 56% of the world's human population. China and India account for most of the population of Asia with a combined population of 2.7 billion. Figure 7.19 shows the continent of Asia. Countries within the continent of Asia that we will evaluate include China, India, Japan, South Korea, Saudi Arabia, Malaysia, and Indonesia. We have already evaluated Russia, most of which lies within Asia, but was evaluated as part of Europe. This is because the western part of Russia, including its capital of Moscow, is located within the continent of Europe.

7.6.1 CHINA

China is located in southeast Asia and is the third-largest country by area in the world behind Russia and Canada with a land area of 9,596,960 square kilometers. The geography of China is perhaps the widest ranging of any country on Earth. As an example, China's lowest point is Turpan Pendi at 154 meters below mean sea level, which is one of the lowest places on Earth. In contrast, China's highest point is Mount Everest, the highest point on Earth at 8,848 meters above mean sea level. Due to the extremes in elevation differences, China's climatic range is from tropical along the southeastern portion of the country to subarctic along the western boundary. China's estimated population now exceeds 1.42 billion, making it the most populous country on Earth (United Nations 2017a). Most of China's population is located along its eastern coastline near the Yellow Sea and the East China Sea. The two largest cities in China include Shanghai and China's capital, Beijing. Shanghai has an estimated population of 24.5 million and Beijing has an estimated population of 21 million (United Nations 2017a). Figure 7.20 shows Shanghai. Figure 7.21 is an example of rapid urbanization in China through high rise building construction.

China's economy is now considered the largest in the world at $23.12 trillion, followed by the European Union at $19.9 trillion, and then the United States at $19.3 trillion. However, China is still a relatively poor country in terms of its standard of living at only $15,600 per person compared to the United States at $59,500 per person based on total GDP. This is staggering to imagine that China's economy is larger than that of the United States but will likely continue to grow significantly in the future. China has built its economy on relatively cheap labor for manufacturing. China's largest trading partner is the United States (US Department of State 2017; United States Department of Commerce 2018; United States Census Bureau 2018).

7.6.1.1 Environmental Regulatory Overview of China

China faces significant environmental issues due to its huge population, rapid growth, and relatively low standard of living. Those environmental issues include air and water pollution, water

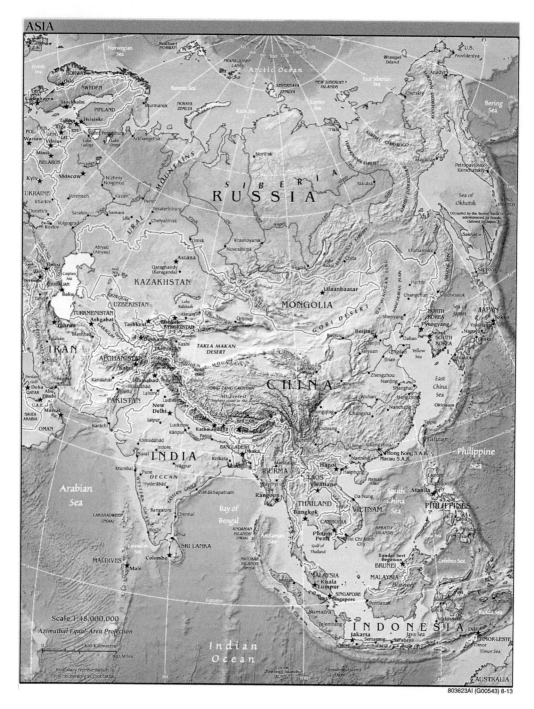

FIGURE 7.19 Map of Asia. (From United States Central Intelligence Agency. 2018. The World Fact Book. 2018. www.cia.gov/library/publications/the-world-factbook.gos/NA.html.) (Accessed October 10, 2018.)

shortages in the northern part of the country, deforestation, sanitation, soil and groundwater impacts, soil erosion, and desertification. Much of the air pollution is due to its heavy dependence on coal, the rapid increase in the number of motor vehicles, especially in its urban regions, and industry (USEPA 2018b). Figure 7.22 is an example of poor air quality along the Great Wall of China near Beijing.

FIGURE 7.20   Downtown Shanghai. (Photograph by Daniel T. Rogers.)

FIGURE 7.21   Example of rapid urbanization in China. (Photograph by Daniel T. Rogers.)

China is a party to many international environmental treaties, protocols, and agreements including signing the Kyoto Protocol to reduce carbon emissions. Others include the following (United Nations 2018c):

- United National Convention to Combat Desertification
- Marine dumping
- Ozone layer protection
- Endangered species
- International Tropical Timber Agreement
- International Convention for the Regulation of Whaling
- Ship pollution
- Wetlands
- Convention on biological diversity

FIGURE 7.22   Air pollution along the Great Wall of China near Beijing. (Photograph by Daniel T. Rogers.)

- Climate change
- Antarctic environmental protocol
- Nuclear Test Ban Treaty

The United States Environmental Protection Agency (USEPA) and China's Ministry of Ecology and Environment (MEE) have been collaborating for several years to strengthen and expand China's capabilities at mitigating pollution that now presents significant challenges to human health and the environment in China (USEPA 2018b). Objectives of the collaboration have been to improve many of China's environmental protection programs including the following (USEPA 2018b):

- Improving air quality
- Reducing water pollution
- Preventing exposure to chemicals and toxics
- Remediating soil
- Treating hazardous waste
- Disposal of hazardous and solid waste
- Improving environmental enforcement
- Promoting environmental compliance
- Improving existing environmental laws
- Enhancing environmental education
- Chemical management

USEPA has also been collaborating with China's Ministry of Science and Technology on joint research to better access air emission sources and impacts on China to better develop mitigation practices and to enhance sustainability. Areas of research have included (USEPA 2018b):

- Soil remediation
- Groundwater remediation
- Water sustainability
- Toxicology
- Motor vehicle emissions
- Clean cooking methods
- Air pollution monitoring and assessment

7.6.1.2 Air

According to the WHO (2016b), 16 of the world's 20 cities with the worst air quality are located in China. Contributors to the poor air quality include:

- Motor vehicle exhaust
- Industry
- Combustion of coal, much of which is high in sulfur

Coal is considered the most significant source of air pollution in China. China generates 80% of its electricity from coal generating plants. China burns 6 million tons of coal per day (USEPA 2018b). Motor vehicles also represent a significant source of air pollution in China. It is estimated that China will have 400 million vehicles by 2030, which by any calculation means that motor vehicle pollutants will continue to be a significant issue in the future. Just as in other countries, the main components of the air pollution in China include (USEPA 2018b):

- Particulate matter (PM10)
- Fine particulate matter (PM2.5)
- Sulfur dioxide
- Nitrogen oxide
- Nitrogen dioxide
- Ozone

China leads the world in premature deaths caused by air pollution at 1.2 million per year according to the WHO (2016b) and the National Geographic Society (2017).

China first promulgated ambient air standards in 1978 (Siddiqi and Chong-Xian 1984). Article II of the law adopted in 1978 states:

The State protects the environment and natural resources and prevents and eliminates pollution and other hazards to the public.

From passage of the law, China developed ambient air quality standards that were announced in 1982. There are three divisions in the ambient air standards in China that include (Siddiqi and Chong-Xian 1984):

- Class I: This level represents the ideal standard, at which there is no direct or indirect harmful effect of any type on humans or ecosystems, even under long-term exposure.
- Class II: This level is set at what is believed to be threshold concentrations for effects on sensitive plants, and above which chronic effects on humans may become noticeable. This level is set as the standard in areas that are urban residential, commercial, cultural, or rural.
- Class III: This is the level considered necessary to protect people from acute or chronic poisoning, and to protect animals and plants (not including sensitive plants).

During the 2008 Olympics, China instituted a mandate that shut down factories 3 weeks prior to the 2008 Summer Olympics and placed restrictions on the use of motor vehicles. The result was a measurable positive effect on the health of people in Beijing (Rich et al. 2010). In 2015, China formed the Clean Air Alliance of China (CAAC 2015) to formulate recommendations to improve ambient air quality in China. China is now beginning to address the challenge of improving its air quality with the formation of the CAAC (USEPA 2018b).

7.6.1.3 Water

Water pollution in China is as important as air pollution but is less widely known. China suffers from water shortages and severe water pollution. The foundation behind China's water shortage and pollution is its huge population combined with rapid industrialization and a historical lack of environmental oversight and enforcement (United Nations 2013a).

Along with the challenges of water pollution, sanitation is also a significant contributor to its poor water quality. Approximately 100 million Chinese do not have access to a reliable water source and an estimated 460 million Chinese do not have access to basic sanitation. Although these estimates have improved greatly in recent decades, China still faces significant challenges (World Health Organization 2015b). According to the China Research Academy of Environmental Sciences (2013), both surface water and groundwater are heavily polluted in many areas of China. This is exacerbated by the uneven distribution of groundwater and surface water resources across China due to its physical geology. Northern portions of China account for approximately 65% of the land area but only have 30% of the water resources.

In contrast, the southern portion of China which accounts for 35% of the land area has 70% of the surface and groundwater resources. This uneven distribution has led to over extraction and exploitation of groundwater in northern China, which has led to a decrease in water levels in some groundwater supply wells of more than 100 meters in some cases (China Research Academy of Environmental Sciences 2013). Causes of water pollution in China have been identified as the following (China Research Academy of Environmental Sciences 2013):

- Discharge and infiltration of untreated urban stormwater runoff
- Discharge and infiltration of untreated industrial discharges
- Runoff and infiltration from areas of soil contamination
- Runoff and infiltration from agricultural areas
- Lack of groundwater protection legislation
- Lack of awareness

In response to the degradation of both surface and groundwater, the Chinese Ministry of Environment has launched a series of research efforts to understand both surface water and groundwater impacts so that by the year 2020 surface water and groundwater impacts will be identified and risks reduced (China Research Academy of Environmental Sciences 2013). Figures 7.23 and 7.24 are examples of improving water quality in China. Figure 7.22 is a photo of water discharge before 2013 that was untreated sanitary and industrial discharge. Figure 7.23 is a photo taken from the same location

FIGURE 7.23 Untreated sanitary and industrial wastewater. (Photograph by Daniel T. Rogers.)

FIGURE 7.24 Sanitary and industrial wastewater discharge following installation of a wastewater treatment system. (Photograph by Daniel T. Rogers.)

after a wastewater treatment system was installed resulting in greatly improved water quality that meets or exceeds water quality standards.

As stated earlier in this section, China faces challenges in supplying clean drinking water to its population. Approximately 60% of China's population obtains its drinking water from surface water bodies that mainly include rivers, and 40% obtains its drinking water from groundwater. The majority of groundwater withdrawal for drinking water occurs in the northern regions of China (World Health Organization 2015b). China established drinking water criteria on July 1, 2012, by adopting the World Health Organization Guidelines for drinking water (see Table 7.4) (World Health Organization 2015a).

7.6.1.4 Solid and Hazardous Waste

Efforts to regulate solid and hazardous waste in China have been plagued by numerous significant social, economic, and political challenges. In 2007, the first national census of hazardous waste in China estimated that China produced 45 million metric tons of hazardous waste but only 10 million metric tons were produced from industry (Wencong 2012). However, much information was lacking that includes (Wencong 2012):

- Types of waste
- How the wastes were generated
- How the wastes were transport
- Where the wastes were disposed

The first law addressing solid waste in China was passed in 1995 and was titled Law of the People's Republic of China on the Prevention and Control of Environmental Pollution by Solid Waste (China Ministry of Ecology and Environment [CMEE] 2018). As part of China's solid waste law, a national catalog of hazardous waste was developed to identify and classify solid wastes as either hazardous or not hazardous. In China, as in most other countries, a waste is hazardous if the waste is any of the following (CMEE 2018):

• Corrosive	• Ignitable	• Flammable
• Medical waste	• Toxic	• Reactive
• Explosive	• Pesticide	• Herbicide

Other types of wastes considered hazardous in China include:

• Electronic wastes	• Fluorescent light bulbs
• Waste Hg-containing thermometers	• Waste mineral oils
• Waste solvents	• Waste film and photographs
• Waste Ni-Cd battery cells	• Disinfectants
• Waste paints	

Exempted from the law in China are wastes generated by households in daily life, which is similar to that of the United States (CMEE 2018). Until recently, China did not have a reliable infrastructure for disposal of hazardous wastes. Therefore, many industrial sites stored wastes until a satisfactory transport method and disposal site was established nearby that was in compliance with China's solid waste law. However, historically solid wastes were disposed of anywhere convenient including farm fields, next to ponds or lakes, or wastes were simply burned. Figure 7.25 is an example of burning waste. Figure 7.26 is an example of disposal of wastes near a pond. China is currently examining and will likely revise its solid and hazardous waste regulations sometime in 2019 (CMEE 2018).

7.6.1.5   Remediation

In 2014, the China Ministry of Environmental Protection announced five standards of environmental protection (MEP Notice No. 14 of 2014) that provided a framework for future regulatory development of soil and groundwater pollution prevention and remediation (MEP 2014). The five standards included (MEP 2014):

1. Technical Guidelines for Environmental Site Investigation
2. Technical Guidelines for Environmental Site Monitoring
3. Technical Guidelines for Risk Assess of Contaminated Sites
4. Technical Guidelines for Site Soil Remediation
5. Definition of Terms of Contaminated Sites

On August 31, 2018, China passed a new law on soil pollution and control. The new law went into effect on January 1, 2019. The law stipulates that the China Ministry of Environmental Protection

FIGURE 7.25   Historical burning of solid waste in China. (Photograph by Daniel T. Rogers.)

FIGURE 7.26   Historical disposal of solid waste in a pond. (Photograph by Daniel T. Rogers.)

establish soil remediation standards and conduct a census on the condition of soils nation-wide every 10 years (China Ministry of Environmental Protection 2018). The law strengthens the responsibilities of governments and the responsibilities and liabilities of polluters. In addition, the law also require the polluters of farmland to develop rehabilitation plans, provide them the appropriate government authority, and to carry out the plans.

Until the China Ministry of Environmental Protection adopts cleanup values for pollutants of its own, the DIVs are commonly used for comparison purposes to evaluate whether soil, sediment, and groundwater require cleanup and to what extent.

7.6.1.6   Summary of Environmental Regulations in China

Environmental regulations in China are similar to that of the United States and European Union with two major exceptions:

1. Requirement that each facility obtain one operating permit that covers all operating conditions including environmental aspects
2. Many regulations are either rather new (i.e., air and water) or are currently under development (i.e., land)

China is a rapidly developing nation with significant air, water, sanitation, and land pollution issues. Together with its large population, China faces significant future challenges. However, China now has the largest economy in the world and will likely continue to grow and prosper from this development. In addition, China has made major strides at developing an environmental framework that can hopefully deal with improving the quality of its air, water, and land from pollution in the future. The combination of a strong and growing economy with a robust set of environmental regulations and political will means that it is likely that China will continue to improve environmentally at a fast pace.

7.6.2   Japan

Japan is an island nation comprised of four main islands and thousands of smaller islands just to the east of the western coast of Asia in the Pacific Ocean (see Figure 7.19). The population of Japan is estimated at just over 127 million (United Nations 2017a). The capital of Japan is Tokyo, which has a population of 36 million residents and is considered the largest urban center in the world (United Nations 2017a). According to the United Nations population projects, Japan's population is not expected to grow over the next 10 years. In fact, Japan's population is expected to decrease by more than 2 million in the next 10 years (United Nations 2017a).

Japan is a highly developed country with the fourth-largest economy in the world behind China, the United States, and the European Union. Japan has an estimated GDP of over $5 trillion annually (United Nations 2017a). Japan is ranked the sixty-first largest country with a land area of 377,973 square kilometers and has 29,751 kilometers of coastline (United Nations 2017a). The population density of Japan is approximately 348 individuals per square kilometer, which is much higher than the world-wide average of 51 and ten times that of the United States, which is 33 individuals per square kilometer (United Nations 2017a).

Japan experiences natural disasters in the form of volcanic eruptions, landslides, tsunamis, and typhoons that can lead to significant environmental concerns and loss of life. For example, an earthquake in March 2011 damaged more than 270,000 buildings and was also responsible for the generation of an estimated 24 million metric tons of waste debris. In addition, the earthquake and subsequent tsunami damaged the Fukushima Nuclear Power Plant, which released significant quantities of radiation that had large aerial impacts and will have lasting effects for decades (Japan Ministry of the Environment 2018a).

Japan's economy is primarily the production of motor vehicles, electronics, industrial tools, steel, and other metals. Japan also has some food production that is comprised of rice, sugar beets, fruits and vegetables, fish, and beef (US Department of State 2018b). Japan has undergone rapid industrialization since the end of World War II in 1945. The United States and Japan have a very close relationship on matters including (US Department of State 2018b):

- Development assistance
- Environment
- Women's empowerment
- Infectious disease
- Space exploration
- International diplomatic initiatives
- Health
- Resource protection
- Science and technology
- Medicine
- Education

7.6.2.1 Environmental Regulatory Overview of Japan

Environmental regulations in Japan are some of the strictest in the world and are the responsibility of the Japan Ministry of the Environment. Environmental regulations in Japan had their beginning in 1967 and were significantly revised in 1994 as Japan was experiencing significant industrialization (Japan Ministry of the Environment 2018b). Japan has significant environmental issues related to air, water, and land. Disposal of solid waste is of particular concern since Japan has relatively little land compared to other countries and has a large population with a high density and generates large amounts of solid waste (Japan Ministry of the Environment 2018a).

Japan itself points to an incident that occurred in 1896 at the Ashio Copper Mine as the beginning of the environmental movement in Japan. A massive flood from torrential rains occurred in the region of the mine in September 1896 and caused not only the mine to flood but also the Watarase, Tone, and Edo Rivers to overflow their banks. Damage from the flood nearly destroyed 136 towns in the path of the flood and contaminated the entire affected area with heavy metals, including many agricultural areas that effectively poisoned the soil and thousands of residents in the path of the flood or who ate food from farms within the affected area (Kichira and Sugai 1986; Japan Ministry of the Environment 2018b). This incident is considered a natural disaster triggered by human actions that impacted an entire region of Japan because of a general lack of appreciation for the environment that included deforestation, poor mining practices, little or no drainage control, and lack of knowl-edge of potential environmental impact from mining wastes. This incident highlighted the need to place more emphasis on safety and environmental integrity and started Japan's first mass-citizens movement in Japanese history (Japan Ministry of the Environment 2018b).

7.6.2.2 Air

Since the 1980s, Japan has improved its air quality significantly, especially the reduction of sulfur dioxide, which has been reduced by 78%. Air exhaust, especially from diesel engines, has received particular attention in Japan where approximately 80% of the nitrogen oxides and particulate mat-ter originate from diesel engines (Japan Ministry of the Environment 2018b). In addition, in 1986, Japan produced nearly 120,000 metric tons of chlorofluorohydrocarbons (CFCs), or ozone-depleting substances, and now produces virtually none (Japan Ministry of the Environment 2018b).

An additional concern of air pollution in Japan is air pollution originating in China that affects the air quality in Japan. Sources of pollution in Japan believed to originate in China include photo-chemical air pollutants that react with sunlight and produce smog and higher ozone levels, nitrogen oxide, acid snow and acid rain, and hydrocarbons (Japan Ministry of the Environment 2018c). As a result of the significant air quality issues experienced in the 1960s and 1970s during a rapid indus-trial development phase, Japan enacted some of the most strict air pollution control laws in the world (Japan Ministry of the Environment 2018d). Ambient air quality standards for Japan are at www.env.go.jp/en/air/aql/aq.html (Japan Ministry of the Environment 2018d).

7.6.2.3 Water

The Ashio Mine incident is often mentioned when discussing water pollution in Japan (Japan Ministry of the Environment 2018d). Japan has made significant progress over the past few decades at improving water quality and the reduction of water pollution, especially heavy metals, as a result of regulations on industrial wastewater. However, other pollutants in water such as volatile organic compounds and other organics have not experienced as much improvement, especially in urban riv-ers and streams and inland lakes and reservoirs (Japan Ministry of the Environment 2018e).

Pollution of fresh water is not the only type of water pollution that Japan faces. Pollution in the ocean waters surrounding Japan has also been a significant issue, especially since Japan consumes high volumes of fish and other foods that have their origin in the marine environment. Mercury in the form of methylmercury discharged into Minamata Bay for over 30 years from nearby industries contaminated both marine life and residents who relied on the marine environment of Minamate Bay as a source of food (Japan Fact Sheet 2018). Other examples of anthropogenic deterioration of water resources in Japan include (Japan Ministry of the Environment 2018e):

- Loss of wetlands
- Construction of dams
- Lining riverbanks with cement
- Straightening river channels

Water discharge regulations in Japan are similar to that of the United States and many developed countries of the world in that Japan requires every facility that discharges industrial wastewater to obtain a permit (Japan Ministry of the Environment 2018f). However, the water quality stan-dards are strict and inspections and enforcement of regulations are usually frequent and strict as well. Japan water quality standards are at www.env.go.jp/en/pollution/issues (Japan Ministry of the Environment 2018f).

7.6.2.4 Solid and Hazardous Waste

Solid and hazardous waste in Japan, along with air and water, is perhaps the most important environ-mental issue in Japan. Being a highly urban and industrialized country with a high relative population and density, which is exacerbated by the fact that Japan is a relatively small country, has placed enormous pressure on Japan to control its solid and hazardous waste (Japan Ministry of the Environment 2018g). Currently, Japan estimates that it produces approximately 50 million metric tons of solid and hazardous waste per year. As a measure to limit the amount of wastes that are disposed of in landfills, Japan incinerates up to 80% of its solid waste and has an aggressive recycling effort for paper, plastic, and metals. As part of recycling efforts, Japan also has aggressive programs in place to reuse material and reduce the amount of waste generated (Japan Ministry of the Environment 2018g).

Of the 50 million metric tons of solid and hazardous waste generated in Japan, approximately 42 million is from industry and the remainder is municipal waste (Japan Ministry of the Environment 2018g). Japan requires specialized air and water pollution control equipment on all incinerators used to burn solid waste. Japan measures dioxin levels in ambient air and water because of its significant use of incinerators used to burn solid wastes and the levels of dioxin have been significantly reduced by well over 95% since 1998 (Japan Ministry of the Environment 2018h). Japan has developed contaminant limits on the leachate from residual waste following the incineration process that is composed of what is termed cinder dust, ash, sludge, and slag, and are located at www.env.go.jp/en/soil/leachate.html (Japan Ministry of the Environment 2018i).

7.6.2.5 Remediation

During the post-war period, Japan faced the reality of dealing with millions and millions of metric tons of waste and urban waste land. At the time, these wastes were dumped into the ocean and rivers and piled up in the open, causing plagues of flies and mosquitoes and the spread of infectious diseases. The Japanese government made several attempts at passing laws such as the Public Cleansing Act of 1954 and the Act on Emergency Measures of 1963 concerning the development and living environment, and the Basic Act for Environmental Pollution Control was enacted in 1967 (Japan Ministry of the Environment 2018h). None of these laws adequately addressed the issue of soil pollution.

It wasn't until the enactment of the Japan Soil Contamination Countermeasures Act of 2002, which had the objective to formulate measures to evaluate the degree to which land in Japan was impacted by contaminants and to provide methods and criteria to prevent further deterioration and restore impacted locations to protect human health, did Japan make significant progress at restoring and improving its land environment (Japan Ministry of the Environment 2018h). Cleanup levels for soil in Japan are listed at www.env.go.jp/en/soil/leachate.html (Japan Ministry of the Environment 2018j).

The general conditions under the Soil Contamination Countermeasures Act of 2002 are similar to that of the United States in that Japan requires that those who are responsible for environmental restoration are those who have polluted the environment. This is under a Japanese provision in the regulations termed the polluters pay principle.

Japan is heavily dependent on surface water as its primary source of drinking water. In fact, 89% of Japan's drinking water is from surface water sources and 11% is from groundwater. Since Japan is densely populated and a relatively small country by land area and is an island nation, providing a reliable source of drinking water and preventing fresh water sources from contamination is a very high priority (Japan Ministry of the Environment 2018a). Therefore, Japan has an array of over 4,000 monitoring wells located throughout the country to monitor groundwater and thousands of surface water monitoring points to monitor surface water quality. The purpose of the well network is to act as an early warning system so that potential threats to water quality are addressed at the earliest stage possible (Japan Ministry of the Environment 2018k).

In addition, Japan has strict soil cleanup values and monitoring points near potential sources of contamination such as landfills and heavily industrialized areas (Japan Ministry of the Environment 2018k). Lastly, Japan has undertaken several water conservation and recycling efforts (Japan Ministry of the Environment 2018k). Groundwater quality values are at www.env.go.jp/en/spcl/html (Japan Ministry of the Environment 2018k). In Japan, soil quality values are the same as groundwater quality values. This is done to protect the groundwater sources (Japan Ministry of the Environment 2018k).

7.6.2.6 Summary of Environmental Regulations of Japan

Japan is one of the most developed nations on Earth and as a result has experienced significant environmental degradation of its air, water, and land. The contributing factors to environmental degradation in Japan can be traced to wars, lack of knowledge, deforestation, erosion and drainage control, rapid industrializations, and lack of effective pollution control measures. In addition, Japan's geography and geology have greatly contributed to its environmental degradation and loss of life from volcanic eruptions, landslides, tsunamis, earthquakes, and typhoons.

However, Japan now has one of the most comprehensive and effective environmental regulatory structures and has improved its air, water, and soil quality. To maintain its quality of life and to sustain its population, Japan must continue to develop methods to continue to protect its air, water, and land. Japan has a regulatory structure similar to that of the United States and air, water, and soil quality criteria are strict and consistent from one media to the next. Japan's inspection and enforcement activities are significant and robust compared to other countries.

7.6.3 INDIA

India is the second most populated country on Earth with an estimated population of over 1.2 billion and is second in population only to China. India is the seventh-largest country on Earth with a total area of 3,287,263 square kilometers (see Figure 7.19). The largest river in India is the Ganges, which is 2,525 kilometers in length. India has varied geography, temperature, and precipitation ranges. India experiences monsoons in the summer months that often flood many urban areas. India's lowest elevation is–2.2 meters and its highest elevation is 8,586 meters (Kanchenjunga), the third-highest peak on Earth in the Himalayan mountains along its northern border. India shares borders with Bangladesh, China, Pakistan, Nepal, Myanmar, Bhutan, and Afghanistan. Southern portions of India are located along the coastline of the Indian Ocean, which is more than 7,500 kilometers in length (Marsh and Kaufman 2015).

India faces significant pollution issues of its air, water, and land. This becomes difficult to manage because of its large human population, limited financial resources, lack of infrastructure, and current low technological capabilities. However, between 1984 and 2010 India has made significant progress in addressing environmental issues and improving environmental quality (World Bank 2011). This is reflected in India's increase in life expectancy, which was 41 years in 1940 compared to 68 in 2016 (World Bank 2016).

India is one of the top agricultural countries of the world and because of this India uses large quantities of water for irrigation. Approximately 20% of its population does not have access to usable water and 21% of disease in India can be traced back to unsafe water (WHO 2018d). The Ganges River is considered especially polluted with untreated sewerage discharge to the river at several locations. There is a wide disparity in the standard of living in India, perhaps more than most any other country of the world. The average per person GDP in India is $3,600 (World Bank 2016).

7.6.3.1 Environmental Regulatory Overview of India

One of the most significant environmental incidents in India occurred on December 3, 1984, when more than 36,000 kilograms of methyl isocyanate leaked from a Union Carbide pesticide manufacturing plant in Bhopal, India. At least 3,800 people were immediately killed and thousands more were injured (Broughton 2005). The disaster highlighted the need for environmental and safety standards, preventative and engineering strategies, and disaster preparedness.

This incident, perhaps more than any other environmental incident in India, changed the course of environmental regulations in India and in many other parts of the world, including the United States. The disaster also demonstrated that a seemingly local incident of the release of contaminants into the environment was intimately tied to global market economic markets (Broughton 2005). In addition, this incident and many others that have occurred historically throughout the world, many of which we have mentioned and discussed in some detail, highlight the need for consistent environmental regulation and enforcement globally.

7.6.3.2 Air

The major sources of air pollution in India are from burning of fuelwood and garbage in cook stoves, burning of biomass and dried livestock waste, burning garbage, automobile exhaust, and ineffective air pollution control equipment at industrial facilities. According to the US Department of State (2018a), Delhi, the capital city of India, with its estimated population of 19 million residents, earned the distinction of being the most polluted city on Earth, as air quality for fine particulate matter (PM2.5) reached a level of 1,010 . ug/m^3. This is especially disturbing when comparing the detected level with the WHO standard for PM2.5, which is 25 ug/m^3.

In 1981, India enacted the Air Prevention and Control of Pollution Act that had an objective of improving air quality through prevention, control, and abatement. The Air Prevention and Control of Pollution Act has six main purposes that include:

- To advise the central government on any matter concerning the improvement of the quality of the air and the prevention,. control, and abatement of air pollution
- To plan and cause to be executed a nation-wide program for the prevention, control, and abatement of air pollution
- To provide technical assistance and guidance to the State Pollution Control Board
- To carry out and sponsor investigation and research related to prevention, control, and abatement of air pollution
- To collect, compile, and publish technical and statistical data related to air pollution
- To develop ambient air quality standards

Ambient air quality standards for India are located at www.cpcb.nic.in/upload/latest_final_air_standard.pdf (India Central Pollution Control Board 2009).

The Air Pollution Control Act of 1981 required that each manufacturing facility evaluate whether there were any air emissions from its operations and if so, whether a permit was necessary. In general, manufacturing facilities with any air emissions likely require a permit (India Ministry of Environment, Forestry, and Climate Change 2018a).

7.6.3.3 Water

The enactment of the India Water Pollution Control Act of 1974, as amended, prohibited any person or entity from knowingly causing or permitting any poisonous, noxious, or pollutant matter to enter any stream, well, or sewer (India Ministry of Environment, Forestry, and Climate Change 2018b). In 1998, the act was amended to include effluent limits on industrial discharge of wastewater and required industrial plants to obtain permits to discharge wastewater (India Ministry of Environment, Forestry, and Climate Change 2018c). Drinking water quality and effluent standards in India are at www.environmentallawsofindia.com/water/pollution/act (India Ministry of Environment, Forestry, and Climate Change 2018c).

While India has established criteria for drinking water and industrial effluent, enforcement and infrastructure within India stand as the major impediments to improving overall water quality (Agarwal 2015).

7.6.3.4 Solid and Hazardous Waste

In India, as in most other countries, a solid waste is considered hazardous if it has any of the following characteristics (India Ministry of Environment, Forestry, and Climate Change 2018d):

• Toxic	• Radioactive
• Ignitable	• Reactive
• Corrosive	• Explosive

Similar to that of the United States, India requires as part of its solid and hazardous waste regulations that a detailed evaluation, termed Environmental Impact Assessment (EIA), of any facility that treats, stores, or is the final disposal location for hazardous wastes. The EIA evaluates all aspects of potential impact that the facility may pose including (India Ministry of Environment, Forestry, and Climate Change 2018d):

- Surface and groundwater
- Wetlands
- Forested areas
- Nearby urban areas
- Noise
- Wildlife
- Marine life
- Invertebrates
- Other potential sensitive habitats

For compliance purposes in India, each facility that generates a solid waste must follow the recommended steps (India Ministry of Environment, Forestry, and Climate Change 2018d):

- Identify the point of generation of the waste
- Collect data on the process that generated the waste
- Characterize the waste pertaining to the properties of ignitability, corrosivity, flammability, reactivity, and toxicity
- Evaluate whether the waste was generated under any of the 18 listed waste categories
- Evaluate whether the waste can be recycled, reused, or is considered a solid waste requiring disposal, or a hazardous waste requiring disposal at a licensed TSDF
- Identify the appropriate site for disposal
- Arrange for proper transportation of the waste from a licensed transporter

7.6.3.5   Remediation

In India, remediation standards for cleanup of contaminated sites are generally administered through the local or provincial governmental authority and following the procedures outlined in conducting an Environmental Impact Assessment (EIA) (India Ministry of Environment, Forestry, and Climate Change 2018e). The process of conducting an EIA in India is nearly identical to that of the United States, EU, and many other nations. The process includes following the fundamental steps (India Ministry of Environment, Forestry, and Climate Change 2018e):

- Screening: Considered the initial stage and addresses whether there is a need to conduct an EIA.
- Scoping: Once it's established that an EIA is necessary, the scoping stage is conducted and determines all relevant environmental concerns
- Public Consultation: Public consultation provides the opportunity for input from any persons of interested stakeholders to provide input through the EIA process
- Investigation: This stage may involve several steps or iterations before the data set is complete and concerns are well understood
- Appraisal: This stage is generally considered a data gap analysis and can either accept data, reject data, or require that more data be collected
- Risk Evaluation: This stage then takes all the data and then conducts a project-specific health risk assessment. The standards used to conduct a health risk assessment are those guidelines established by USEPA or the European Union

- Remediation goals: After each of the stages listed above are completed, remediation goals with specific targets are established

7.6.3.6 Summary of Environmental Regulations of India

India has a large disparity of distribution of wealth and standard of living. The average standard of living for the residents of India is less than one-twentieth (1/20) that of the United States. Together with its large population of well over 1 billion, lack of infrastructure, and other social and political struggles, the environmental pressures are, at times, overwhelming. However, despite these huge challenges, India has made progress at improving its environment and life expectancy since the Bhopal disaster in late 1984.

India has established a robust set of environmental regulations that compare adequately to the United States, EU, and other rather developed nations. However, lack of infrastructure and enforcement along with social and economic factors discussed above highlight the need for continued improvement.

7.6.4 SOUTH KOREA

South Korea is located on the Korean Peninsula in northeast Asia. Since 1950, Korea has been divided in half at the 38th parallel as a result of the Korean War. The Amnok River forms the border of North Korea, China, and Russia. The Yellow Sea is located to the east, the East China Sea and Korea Strait is to the south, and the Sea of Japan is located to the east (see Figure 7.19). Korea is approximately 20% of the size of California or about 100,000 square kilometers (United Nations 2012). The devastation caused by the Korean War in the early 1950s is still apparent today. The two Korea's have evolved from a common cultural and historical base into two very different societies with radically dissimilar political and economic systems (Asia Society 2018). North Korea is heavily influenced by Chinese and Soviet/Russian culture and South Korea is heavily influenced by western culture. Today, the population of North Korea is approximately 25 million while the population of South Korea is just over 51 million (United Nations 2017a).

South Korea is mountainous along its western boundary and is much less mountainous along its eastern and southern parts. In fact, South Korea is considered 70% mountainous. Rivers tend to flow from east to west with a general descending topography toward the west and south. South Korea has a total land area of 97,230 square kilometers with a population density of approximately 526 people per square kilometer (United Nations 2016c). Since the cease fire with North Korea in 1953, South Korea has essentially been rebuilt. The capital of South Korea is Seoul, which is located in the northwestern portion of the country. Seoul has a population of greater than 10 million residents and more than 25 million in the metropolitan proper.

South Korea, which was primarily an agrarian society before the Korean War with 75% of its population living in rural areas and farming small plots of land, has been completely transformed to an industrial and urban nation. South Korea now has 83% of its population living in urban areas. Since much of the population in Korea is urban, land use is an intermixing of industry, residential and agriculture, which is very different than the United States. This presents a challenge in applying environmental regulations evenly. However, the cost and time spent on transportation is much lower because many people live near where they work, and food is grown nearby (United Nations 2012). Figure 7.27 shows an example of the intermixed land use between industrial, residential, agricultural, and recreational.

The standard of living in Korea per person is just over $35,000, which is more than twice that of China but is well below that of the United States. South Korea experienced rapid growth in the 1970s and remains under rapid development. The GDP of South Korea is $1.69 trillion, which is approximately 12 times less than the United States. As a result of this development, the natural environment suffered deterioration of the air, water, and land and resulted in the passage of numerous environmental laws (United Nations 2012).

FIGURE 7.27 Intermixed land use in Korea, recreational (bottom), agriculture and industry (middle), and residential (upper). (Photograph by Daniel R. Rogers.)

7.6.4.1 Environmental Regulatory Overview of South Korea

The United States Environmental Protection Agency (USEPA) is heavily engaged in collaborating with Korea as a result of the Korea-United States Free Trade Agreement (USEPA 2018c).

Through this mechanism, the USEPA and the Korean Ministry of the Environment and partner agencies in both countries have been cooperating to strengthen (USEPA 2018c):

- Environmental governance
- Air quality
- Water quality
- Reduce waste
- Recycling
- Reusable energy
- Protections against toxic pollutants

7.6.4.2 Air

Air quality in South Korea has been improving but is still a major environmental issue, especially in its capital city of Seoul. As with Japan, South Korea experiences increased air pollution from China in addition to its own sources of pollution (Korea Ministry of the Environment 2018a). The major air pollutants in South Korea should include (USEPA 2018c):

- Fine particulate matter (PM2.5)
- Particulate matter (PM10)
- Carbon monoxide
- Nitrogen dioxide
- Ozone
- Lead
- Smog
- Hydrocarbons

- Benzene
- Other volatile organic compounds (VOCs)

South Korea set air quality standards for each of the pollutants listed above in 1983 and set the standard for benzene in 2010 (Korea Ministry of the Environment 2018a).

The Korean Ministry of the Environment has been measuring trends in air pollutants within the country for 15 years and has measured a decreasing trend for lead and particulate matter (PM10). However, sulfur dioxide, nitrogen dioxide, and ozone have either been steady or have increased in concentration in ambient air (Korea Ministry of the Environment 2018a). Ambient air quality standards are similar to the United States and EU and can be viewed at www.eng.me.go.kr/eng/web/index.do?menuld/252 (Korea Ministry of the Environment 2018a).

7.6.4.3 Water

South Korea has an aggressive and comprehensive water quality monitoring program with 1,476 water quality monitoring stations throughout the nation, which is something learned from Japan. The monitoring stations are located along rivers (697 stations), lakes and marshes (185 stations), agricultural areas (474 stations), and 120 other monitoring stations at other strategic locations. Water is being monitored for dissolved oxygen, total organic carbon, pH and other parameters including volatile organic compounds (VOCs), biochemical oxygen demand, *E.coli*, phosphorus, and nitrogen. Groundwater is also monitored twice yearly for 20 different analytes including those listed above at 2,499 locations (Korea Ministry of the Environment 2018b).

Surface water quality has improved by approximately 30% since measurements began in 1997. However, approximately 15% of the surface water regions do not meet water quality objectives. One particular issue with surface water is eutrophication with 33 of the 49 lakes being monitored showing medium levels of contaminants that are attributed to cause eutrophication (Korea Ministry of the Environment 2018b).

Water quality for groundwater in Korea is better when compared to surface water. A major concern for groundwater is properly abandoning wells that are no longer in service. Approximately 43,000 wells have been located and properly abandoned. South Korea developed its water quality criteria based on the USEPA method that uses the Integrated Risk Information System (IRIS), which is based on human health and ecological risks but does not consider mobility or persistence.

Drinking water is supplied to 98.1% of the Korean population as of 2012 (Korea Ministry of the Environment 2018b). Drinking water quality criteria are similar to the United States and EU and are located at www.me.go.kr/eng/web/index.do (Korea Ministry of the Environment 2018b).

In Korea, it is required to obtain an industrial discharge permit for any facility that discharges wastewater. Facilities that propose discharge of specific hazardous substances are not permitted to be located near source water protection areas and designated lakes, which is also the case in the United States in what is called well-head protection areas (Korea Ministry of the Environment 2018c). Similar to that of the United States, Korea calculates and manages allowable pollutant load to surface water from water treatment plants under what is termed Total Water Pollution Load Management System (TPLMS). The target and allowable pollutant load is specific to each river watershed and is determined through the nation-wide water quality monitoring program (Korea Ministry of the Environment (2018d).

Effluent standards are calculated for each permit issued and fees are collected if the facility discharges any of the following compounds (Korea Ministry of the Environment 2018e):

1. Organic substances
2. Suspended solids
3. Cadmium and its compounds
4. Cyanide
5. Organo-phosphoric compounds
6. Lead and its compounds

7. Hexavalent chromium compounds
8. Arsenic and its compounds
9. Mercury and its compounds
10. PCBs
11. Copper and its compounds
12. Chrome and its compounds
13. Phenols
14. Trichloroacetatic ethylene
15. Tetrachloroethylene
16. Manganese and its compounds
17. Total nitrogen
18. Total phosphorous

7.6.4.4 Solid and Hazardous Waste

The Waste Control Act of 2016 governs the disposal of solid and hazardous waste in Korea and amends previous versions of 2007, 2011, and 2015. The purpose of the Waste Control Act of 2016 is to contribute to environmental conservation and the enhancement of the people's standard of living by minimizing the production of wastes and disposing of generated wastes in an environmentally friendly manner (Korea Ministry of the Environment 2018f). Innate in the stated purpose above is the desire to minimize the amount of wastes generated and to improve the quality of its disposal so as not to significantly impact the environment.

The Waste Control Act of 2016 states the following requirements (Korea Ministry of the Environment 2018f):

- Every business entity shall reduce the generation of wastes to the maximum extent possible by improving the manufacturing process, etc., of products and minimize the discharge of wastes by recycling his/her own wastes
- Every person shall take prior appropriate measures with respect to the discharge of wastes to prevent any harm to the environment or the health of the resident
- Waste treatment shall be properly managed in a manner that reduces their quantities and degree of hazard or otherwise is consistent with environmental conservation and the protection of the people's health
- Any person who causes environmental pollution by discharging wastes shall be responsible for restoring the affected environment and bear the expenses incurred in restoring the damage caused by such pollution
- To the extent possible, wastes originated in the Republic of Korea shall be disposed of within the Republic of Korea and the importation of wastes shall be restrained
- Wastes shall be recycled rather than incinerated or buried in order to contribute to the improvement of resource productivity

Korea instituted a tracking system as part of its waste management system in 1999 to track the transportation of hazardous wastes (Korea Ministry of the Environment 2018g).

In Korea, a waste is hazardous if it is a listed waste and is any of the following (Korea Ministry of the Environment 2018f):

• Explosive	• Flammable
• Radioactive	• Infectious
• Medical waste	• Reactive
• Corrosive	• Characteristic waste

A waste is a characteristic hazardous waste if a leachate test indicates a concentration of the listed compounds exceeding concentrations listed at www.law.go.kr/DRF/law/wastecontrolact (Korea Ministry of the Environment 2018f).

Listed hazardous wastes in Korea are similar to other nations and include the following (Korea Ministry of the Environment 2018f):

- Waste pesticides or herbicides
- Electronic wastes
- Waste thermometers that contain mercury
- Waste oils
- Waste solvents
- Medical wastes
- Wastewater treatment sludges
- Paint waste
- Waste from pigments, glue, varnish, and printing ink

In efforts to reduce solid wastes, Korea has a vigorous recycling program that also includes food wastes. As stated by the Korean government, increased living standards have resulted in excessive convenience, the use of disposable items, and over-packaged products that cause a waste of resources and generate unnecessary waste and bring about negative impacts to the environment. To combat this, the Korean Ministry of the Environment began a recycling program in 1994 and continues to improve the program through the following (Korea Ministry of the Environment 2018h):

- Reduction of plastics
- Reduction of packaging waste
- Reuse of paper cups and bags
- Introduction of farm produce, green packaging guidelines, and safety regulations
- Developing technologies for flexible packaging and paper containers
- Recycling of metals
- Recycling of wood products

7.6.4.5   Remediation

Korea has established a nearly identical law as in the United States for dealing with legacy contamination of land, groundwater, and surface water – the Korean Liability, Compensation, and Relief System for Damages from Environmental Pollution Act of 2016 (Korea Ministry of the Environment 2018h). Soil and groundwater cleanup values are at www.eng/me/go.kr/eng/web/law/Actno.166922 (Korea Ministry of the Environment 2018h).

In an effort to be proactive and avoid soil and groundwater contamination, the Korea Ministry of the Environment has identified 22,868 specific facilities that are subject to soil and groundwater monitoring that include (Korea Ministry of the Environment 2018i):

- Gasoline service stations
- Certain industrial plants
- Petroleum storage sites
- Abandoned metal, coal, and asbestos mines
- Hazardous waste storage and disposal facilities
- Chemical plants
- Treatment plants
- Power generating plants
- Other ecologically sensitive areas

Each site must conduct tests to ensure that no soil or groundwater is present every 1 to 5 years (Korea Ministry of the Environment 2018j). This type of proactive approach to environmental issues is more advanced than most any country in the world including the United States.

Through this program, Korea has identified, investigated, and remediated several sites and continues to be proactive to restore its environment (Korea Ministry of the Environment 2018j).

7.6.4.6 Summary of Environmental Regulations of Korea

Korea has essentially rebuilt its country in the last 70 years due to war. This has provided an opportunity to re-construct a country with less political and social impediments and more opportunity for ease of modernization.

Korea perhaps has the most complete set of environmental regulations that we have examined thus far, even more comprehensive than the United States and the European Union. This is reflected in the fact that Korea regulates its farmland as strict as other locations such as residential and industrial properties. In addition, Korea has robust recycling laws that involve all citizens, not just industry or commercial businesses. In addition, from a remediation point of view, Korea monitors its high-risk locations for the presence of contamination as an early warning system and also closely monitors is surface water and groundwater resources as well.

Korea has a standard of living of just over $35,000 per person, which is well below the average in the United States of $59,500. The population of Korea slightly exceeds 50 million residents, which is approximately 7 times less than the United States. However, Korea is only around 100,000 square kilometers in size, which is approximately 100 times smaller in land area than the United States. This means that Korea has a rather large population for its overall land size.

The economic, geographical, and demographical statistical facts listed above along with its environmental regulations hold insight as to why Korea has such a complete environmental system with close monitoring. With its low relative land area and high population, Korea is forced to protect its environment and must be efficient in its use and care of natural resources. This puts pressure on its economy, but Korea has found an effective balance especially since its standard of living is just twice that of China.

7.6.5 SAUDI ARABIA

Saudi Arabia is located in westernmost Asia and occupies the majority of the Arabian Peninsula. The total land area of Saudi Arabia is 2,149,690 square kilometers which makes it the twelfth largest country in the world (see Figure 7.19). The population of Saudi Arabia is approximately 30 million (US Department of State 2018c). Saudi Arabia is bordered by Jordan and Iraq to the north, Kuwait to the northeast, Qatar, Bahrain, and the United Arab Emirates to the east, Oman to the southeast and Yemen to the south. It is separated from Israel and Egypt by the Gulf of Aqaba.

Oil was discovered in Saudi Arabia in 1938 and has since developed into the largest oil producer and exporter of oil in the world, controlling the second largest oil reserve and eighth-largest gas reserve (US Department of State 2018c). Oil production has provided Saudi Arabia with economic prosperity and substantial political leverage in the region. The political system in Saudi Arabia is considered a monarchy and has no free elections (US Department of State 2018c).

Saudi Arabia's geography is dominated by the Arabian Desert, associated with semi-desert and several mountain ranges. There are very few lakes and no permanent rivers. Average summer temperatures can exceed 45°C (113°F) and have a temperature of 54°C (129°F). In summer, the temperature rarely drops below freezing. The largest city in Saudi Arabia is Riyadh, which has an estimated population of 6.5 million residents. The GDP of Saudi Arabia is estimated at 646 billion and per capita income is estimated at nearly $21,000 per citizen (United Nations 2002; 2004; 2018d).

Since Saudi Arabia is essentially a country located in a desert, water is a precious resource. More than 50% of fresh water for the country is from desalinization plants located near the Red Sea. The remainder originates from groundwater and some surface water located in the western mountains of the country (United Nations 2018d). The remainder originates from groundwater and some surface water located in the western mountains of the country (United Nations 2018d).

7.6.5.1   Environmental Regulatory Overview of Saudi Arabia

Environmental issues that Saudi Arabia face are closely tied to its oil industry, which accounts for most of its economy. Environmental issues in Saudi Arabia are viewed towards the development of the country's oil and gas reserves. Saudi Arabia first passed the General Environmental Regulations and Rules for Implementation in 2001 and formerly adopted environmental laws for ambient air, drinking water, wastewater discharge, solid and hazardous waste, emissions from mobile sources, noise, and pollution prevention and preparedness in 2012. The Presidency of Meteorology and Environment (PME) is responsible for implementing environmental regulations in Saudi Arabia (Chakibi 2013).

7.6.5.2   Air

Air pollution in Saudi Arabia is generally confined to its petroleum and related industries and motor vehicle exhaust. In addition, frequent sand and dust storms also significantly affect ambient air quality (United Nations 2018d). Ambient air standards for Saudi Arabia are similar to the United States and EU and are listed at www.pme.gov.sa (Saudi Arabia Presidency of Meteorology and Environment 2012).

7.6.5.3   Water

Data on water pollution in Saudi Arabia is limited but is believed to be influenced by the petroleum and related industries. Drinking water in Saudi Arabia is supplied by the government at almost no cost to its citizens. Saudi Arabia essentially has adopted water quality criteria established by the World Health Organization (Al-Omran et al. 2014; World Health Organization 2015a). Saudi Arabia wastewater discharge limits are similar to the United States and EU and are listed at www.pme.gov. sa (Saudi Arabia Presidency of Meteorology and Environment 2012).

7.6.5.4   Solid and Hazardous Waste

Saudi Arabia defines wastes as substances which have been discarded or neglected and can no longer be put to beneficial use. Saudi Arabia defines hazardous waste as a type of waste that has characteristics that render them hazardous to human health and include waste that are (Saudi Arabia Presidency of Meteorology and Environment 2012):

• Toxic	• Flammable	• Explosive	• Radioactive
• Corrosive	• Reactive	• Infectious	• Listed wastes

Saudi Arabia also has a hazardous waste category as listed wastes that include any wastes generated by the following (Saudi Arabia Presidency of Meteorology and Environment 2012):

- Medical wastes
- Biocide or phyto-pharmaceutical processes
- Wood preservatives
- Waste mineral oil
- Wastes from heat treatment and steel tempering containing cyanides
- Waste oil
- PCBs
- Waste paints, ink dyes, pigments, lacquers, and varnish
- Waste resins, latex, plasticizers, glues, and adhesives

- Wastes generated from research and development, and teaching arising from activities known to be harmful or suspected to be harmful to humans
- Waste from photographic chemicals
- Wastes from metal treatment processes

Saudi Arabia also considers wastes as hazardous if they contain any of the following constituents (Saudi Arabia Presidency of Meteorology and Environment 2012):

• Metal carbonyls	• Beryllium	• Hexavalent chromium
• Copper	• Zinc	• Arsenic
• Selenium	• Cadmium	• Antimony
• Tellurium	• Mercury	• Thallium
• Lead	• Inorganic fluorine	• Inorganic cyanide
• Acidic solutions	• Basic solutions	• Organic phosphorus
• Organic cyanide	• Asbestos	• Phenols
• Ether compounds	• Halogenated organics	• Organic solvents
• Organic halogens	• Dibenzo-furans	• Dibenzo-p-dioxins

7.6.5.5 Remediation

In Saudi Arabia, new developments or expansions of any facility require that the owner conduct a comprehensive Environmental Impact Assessment (EIA). The process begins with notification of the presidency of meteorology and environment and retaining a qualified consultant approved by the regulatory agency. Each project is graded as Category I, II, or III by the regulatory authority according to the potential environmental impact, with Category I as the least amount of potential impact and Category III as the most potential environmental impact (Saudi Arabia Presidency of Meteorology and Environment 2012). Each category must address the following (Saudi Arabia Presidency of Meteorology and Environment 2012):

- Air quality
- The marine and coastal environment
- Surface and underground water
- Flora and fauna
- Land use
- Urban development
- General scenic view

After the EIA is conducted, a risk evaluation is conducted to evaluate whether there are any unreasonable risks posed by the presence of contamination discovered during the investigation process. Cleanup criterion are set by the regulatory authority based on the EIA and risk evaluation (Saudi Arabia Presidency of Meteorology and Environment 2012).

7.6.5.6 Summary of Environmental Regulations of Saudi Arabia

Saudi Arabia's published environmental laws are comprehensive and in line with other developed countries. However, information on enforcement and contaminated site remediation is not available and is not readily shared with the outside world. Therefore, it's difficult to evaluate the current status of environmental compliance and sustainability. However, Saudi Arabia does face serious environmental issues related to air and especially water because Saudi Arabia is a desert country and water is scarce and can't afford to be polluted.

7.6.6 INDONESIA

Indonesia is the largest country in Southeast Asia and is made up of 17,508 islands, of which 6,000 are uninhabited, that are spread between the Indian and Pacific Oceans that link the continent of Asia with Australia (see Figure 7.19). Indonesia has five major islands that include Sumatra, Java, Kalimantan, Sulawesi, and Irian Jaya (United Nations 2018e). The land area of Indonesia is characterized as tropical rainforest. Since most of the soil originates from volcanic activity, the soils in Indonesia are considered very fertile. The total land area of Indonesia is 1,904,569 square kilometers and ranks fourteenth largest in the world. Indonesia shares borders with Malaysia, Papua New Guinea, and East Timor (United Nations 2018e).

The population in Indonesia is estimated at 261 million residents, making it the fourth most populous nation in the world. The population density of Indonesia is 147 individuals per square kilometer, which is roughly 5 times more densely populated than the United States. Indonesia is home to over 500 languages and dialects. The capital of Indonesia is Jakarta with an estimated population of 9.6 million residents. The gross domestic product of Indonesia is $687 billion and the per capita standard of living is $3,000, which is approximately 20 times less than the United States. Trading partners with Indonesia include Singapore, Japan, and the United States (United Nations 2018e). The significant environmental issues on land in Indonesia include deforestation, air pollution, acid rain, and surface water pollution. Air pollution is especially significant in urban areas, especially Jakarta. The surrounding marine environment have been heavily impacted by poorly planned coastal developments, overfishing, and ocean acidification, which has placed 95% of Indonesian reefs under serious threat (United Nations 2018e; Conservation International 2018).

7.6.6.1 Environmental Regulatory Overview of Indonesia

In Indonesia, the Ministry of Environment is responsible for administering and enforcing environmental laws and is also responsible for national environmental policy and planning (Indonesia Ministry of Environment 2018a). The Indonesia Ministry of Environment is collaborating with USEPA to improve air quality, water quality, disposal of wastes, and cleanup sites of environmental contamination in Indonesia (USEPA 2018d).

7.6.6.2 Air

When evaluating air pollution in Indonesia, one must look no further than Jakarta, its capital city of nearly 10 million residents, as the most polluted city in Indonesia. Jakarta routinely has air pollution by fine particulate matter (PM2.5) of 150 ug/m^3 or greater, which is ranked in the World Health Organization Category Ambient Air Quality Index as unhealthy (see Table 7.2 and Figure 3.1) (World Health Organization 2018a). The primary source of the PM2.5 is vehicle exhaust. According to the Indonesia Ministry of Environment, the typical air-polluting industries in Indonesia are manufacturing, mining, energy, oil and gas, and vehicle exhaust (Indonesia Ministry of Environment 2018a). Indonesia ambient air quality standards are listed at www.moe.indonesia.org/ambientairstandards (Indonesia Ministry of Environment 2018b).

In 2009, Indonesia enacted a vehicle emission standard that requires emission tests of each vehicle and also set a phase out of the use of leaded gasoline and set a 500 part per million maximum level for sulfur in diesel fuel (Indonesia Ministry of Environment 2018b). Indonesia also enacted a permit process that sets site specific emission standards at industrial facilities that emit pollutants to the atmosphere. The permits require that appropriate air pollution control equipment be installed for each air emission source and that all air pollution control equipment be properly maintained (Indonesia Ministry of Environment 2018b). Burning of garbage and burning forest land for agriculture preparation in Indonesia has historically been a systemic problem. In 2010, the Ministry of Environment started a program to begin to reduce and control this type of air pollution, which often creates a visible and noticeable haze, sometimes regionally (Indonesia Ministry of Environment 2018b).

7.6.6.3 Water

The World Health Organization estimated that 59% of residents have access to sanitation and approximately 30% have access to safe drinking water (WHO 2018e). Indonesia laws that govern water were first enacted in 1974. Responsible agencies for use of water in Indonesia are spread across many different government ministries including but not limited to:

- Ministry of Forestry
- Ministry of Finance
- Ministry of Mining
- Ministry of Agriculture
- Ministry of Environment
- Ministry of Commerce
- Ministry of Energy
- Ministry of Industry

Since the government departments responsible for water are spread across several ministries, no one ministry accepts responsibility because of competing interests (World Health Organization 2018e).

In general, drinking water quality in Indonesia does not meet national standards on two separate levels (World Health Organization 2018e):

- Indonesia does not supply enough water to its citizens, demand is greater than what can be supplied
- The water quality is not achieved because of pollution

Drinking water quality in Jakarta is especially impacted with poor water quality and scarce availability since 81% of the city's water supply comes from the Citarum River, which is heavily polluted with toilet activities, domestic wastewater, trash disposal and other sources of pollution (Water Environment Partnership in Asia 2018).

Indonesia lacks basic infrastructure and financial resources to supply safe water to its citizens in the form of (Choudhary 2017):

- Wastewater treatment is in some places non-existent
- Basic sanitation to prevent human excrement from contaminating water supplies
- Supplying water filters to citizens
- Pipelines are inadequate
- Demand for water is estimated at 50% of current need

Groundwater in much of Indonesia is also impacted by pollution and is not considered an alternative source of safe drinking water (Choudhary 2017).

Water pollution in Indonesia has been traced to health risk and disease that include (Choudhary 2017):

- Gastro-enteritis
- Typhoid
- Cholera
- Paratyphoid
- Dysentery
- Diarrhea

Boiling water as water treatment is also not recommended because it does not remove chemical pollutants (Choudhary 2017). Water quality standards for Indonesia are listed at www.aecon.org. indonesia (Indonesia Ministry of Environment 2018a).

7.6.6.4   Solid and Hazardous Waste

Management of waste in Indonesia is a mounting challenge because of a large population of over 250 million people, lack of infrastructure, and lack of financial resources. This Ministry of Environment published a new regulatory framework in 2014 in an attempt to streamline the regulations and make existing regulations stricter (Indonesia Ministry of Environment 2018a).

The revised solid waste regulations also attempt to improve waste disposal methods, specifically to increase the amount of solid waste disposed in licensed landfills instead of disposed in streets, rivers, and streams, and open burning of waste. Previous to the new solid waste regulations, solid waste was disposed in the following manner, as measured in 2008 (Lokahita 2017):

- 68% in landfills
- 16% small scale incineration
- 7% composting
- 5% open burning
- 4% dumped in rivers and streams

Since the regulations were updated, Indonesia has made improvements but these improvements in many circumstances have been met with a negative impact because of a rapid increase in population, which overshadows many of the attempted improvements (Lokahita 2017).

7.6.6.5   Remediation

Remediation standards for contaminated sites in Indonesia generally rely on the Dutch Intervention Values list in Table 7.5 and USEPA criteria discussed in Chapter 2. The USEPA is collaborating with the Ministry of Environment of Indonesia to assist and educate Indonesia so that at some point they will be able to build a framework of regulations to address legacy sites of environmental contamination. USEPA is currently collaborating with Indonesia on four sites and is working together with the Ministry of Environment to improve emergency response and cleanup procedures (USEPA 2018d).

7.6.6.6   Summary of Environmental Regulations in Indonesia

As stated by USEPA, Indonesia is a key factor in the global environmental arena, with significant importance in ecological resources but with significant challenges to protect and restore its environment (USEPA 2018d). Lack of infrastructure, large population, lack of technology, lack of financial resources, and lack of effective regulations and enforcement have all contributed to an environmental deterioration in Indonesia that has led to:

- Poor water quality
- Poor air quality
- Poor waste disposal practices
- Deterioration of the marine environment and reef system
- Poor health
- Disease
- Extinction of species

7.6.7   MALAYSIA

Located near the equator, Malaysia is a country of two parts in southeastern Asia (see Figure 7.19). The land area of Malaysia is 330,603 square kilometers, which ranks sixty-sixth in size. Malaysia is bordered by Thailand to the north and Indonesia to the south. Malaysia is a wet and tropical climate and averages approximately 3 meters of rain annually. The population of Malaysia slightly exceeds

31 million residents, has a GDP of 314.5 billion, and a per capita income of nearly $10,000. The capital of Malaysia is Kuala Lumpur with its population of nearly 2 million residents located on the western peninsula region south of Thailand (United Nations 2017a).

7.6.7.1 Environmental Regulatory Overview of Malaysia

Environmental issues in Malaysia include air pollution, water pollution from raw sewage from humans and livestock, and industrial sources, deforestation, and loss of habitat. Industrial sources of water pollution originate from Malaysia's main industries, which are tin mining, natural rubber productions, and palm oil production (United Nations 2018f).

7.6.7.2 Air

Sources of air pollution in Malaysia include open burning of waste, vehicle exhaust, and industry (Malaysia Department of Environment 2015). In 2015, the Malaysia Department of Environment developed new ambient air quality criteria. Ambient air quality standards for Malaysia are at www.doe.gov.my/portelv1/info-umum/english-ambient-air-standards (Malaysia Department of Environment 2018a).

In general, Kuala Lumpur has the poorest air quality in Malaysia largely due to fine particulate matter (PM2.5) (Malaysia Department of Environment 2018b). However, even though the air quality of Kuala Lumpur is often considered poor, it still is routinely much better than Jakarta, Indonesia (United Nations 2018f).

7.6.7.3 Water

Sources of water pollution in Malaysia include sewerage which directly discharges to waterways, garbage and other solid waste dumped into rivers and streams, discharge from livestock farms including pig and chicken farms, and industrial discharge. Water quality is especially poor in urban areas (Malaysia Department of Environment 2018c). Water quality standards for drinking water in Malaysia are listed at www.doe.gov.my/waterquality (Malaysia Department of Environment 2018c). Malaysia sewerage and industrial effluent discharge standards are listed at www.doe.gov.my/discharge-standards/malayasia.htm (Malaysia Department of Environment 2018d).

On occasion, water polluted with ammonia has shut down water treatment plants cutting fresh water to hundreds of thousands of residents in Malaysia. Water supplies are routinely restored in stages after the pollution passes by the water treatment intake locations. Sources of the ammonia have been traced to animal feed lots, specifically pig and chicken farms (United Nations 2018f).

7.6.7.4 Solid and Hazardous Waste

Not unlike Indonesia, Malaysia struggles with appropriate disposal of solid and hazardous waste. Although the problem is not at the magnitude as present in Indonesia because Malaysia's population is only 12% of Indonesia's, the trends and habits are similar (United Nations 2018e; United Nations 2018f).

7.6.7.5 Remediation

Malaysia has set priorities for improving and restoring their environment but achieving success will require changing cultural norms, time, and foreign investment. Currently, Malaysia is focusing on air and water pollution because those are the two environmental issues that are the highest priority for improving human health (Malaysia Department of Environment 2015; United Nations 2018f). Efforts to improve air quality include (Malaysia Department of Environment 2018a):

- Upgrading transportation infrastructure to improve traffic flow and reduce vehicle emissions
- Restricting emissions from power generating plants and industry through a comprehensive permitting process
- Eliminating the open burning of garbage and solid waste

Efforts to improve water quality include (Malaysia Department of Environment 2018a):

* Improve zoning and land use planning
* Upgrade treatment plants
* Require industrial, agriculture, and livestock to pretreat water before discharge

7.6.7.6 Summary of Environmental Regulations of Malaysia

Malaysia suffers from much of the same environmental issues as Indonesia. Those issues of particular concern include urban air pollution, rapid increases in population, providing clean water, sanitation, soil and groundwater contamination, deforestation, and erosion.

7.7 OCEANIA

Oceania is a region made up of thousands of islands in the Central and South Pacifica Ocean (See Figure 7.28). The two largest land masses that make up Oceania include Australia and the microcontinent of New Zealand. Oceania also includes three island regions that include Melanesia, Micronesia, and Polynesia (National Geographic Society 2018a).

From a geographic perspective, Oceania is divided into three region types that include (National Geographic Society 2018a):

* Continental Type – this type includes Australia and New Zealand
* High Islands – Also called volcanic islands, which includes Melanesia, Mount Yasur, and Vanuatu
* Low Islands – Also called coral islands and includes Micronesia and Polynesia

We will evaluate the environmental regulations of Australia and New Zealand located within Oceania, which represent the two largest land masses.

7.7.1 AUSTRALIA

Australia is the sixth-largest country of the world with a land mass of 7,686,850 square kilometers. Australia is generally considered a desert. However, there is a small area located along its northeastern coast that is rainforest (see Figure 7.29) (United Nations 1997). Australia's population is estimated at 24.9 million with most of the population living along the eastern coast and the western city of Perth (United Nations 2017a). Australia's largest city is Sydney, located in the southeastern portion of the country adjacent to the Pacific Ocean. Australia has no land borders with other countries. The Indian Ocean is located along its western shore and the Pacific Ocean is located along its eastern shore (United Nations 1997).

The average per capita income in Australia is estimated at $67,442, which ranks second highest in the world just behind Switzerland (World Bank 2018a). The gross domestic product of Australia is estimated at $1.69 trillion and ranks twenty-seventh in the world. Economic drivers in Australia include mining and tourism along with lesser amounts of farming and livestock. Mining in Australia consists of coal, natural gas, gold, aluminum, iron, and other heavy metals, and diamonds and opal (World Bank 2018a).

From an environmental perspective, Australia is perhaps best known for the Great Barrier Reef, which is comprised of over 3,000 individual reef systems and coral cays and hundreds of tropical islands along its northwestern coast (United Nations 2018g). The Great Barrier Reef contains the largest collection of coral reefs in the world, with 400 different types of coral and 1,500 species of fish. The Great Barrier Reef was designated a United Nations World Heritage Site (UNESCO) in 1981 and covers an area of 348,000 square kilometers (United Nations 2018g).

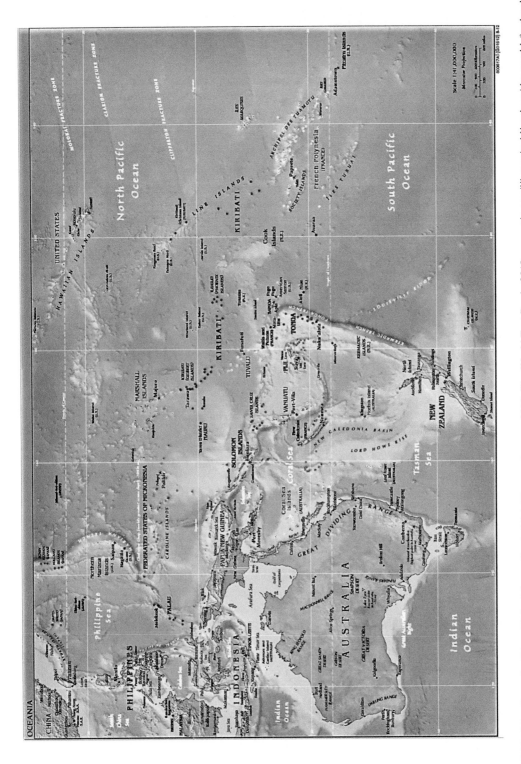

FIGURE 7.28 Oceania. (From United States Central Intelligence Agency. 2018. The World Fact Book. 2018. www.cia.gov/library/publications/the-world-factbook.gos /NA.html.) (Accessed October 10, 2018.)

FIGURE 7.29 Australia's northeast coastline near Cairns. (Photograph by Daniel T. Rogers.)

7.7.1.1 Environmental Regulatory Overview of Australia

Environmental issues in Australia are wide ranging and can be summarized as the following (Australia Department of Environment and Energy 2018a; The National Geographic Society 2018b):

- Invasive species – specifically, the introduction of the rabbit
- Endangered species – including the Tasmanian devil and southern cassowary
- Extinction – more than 50 species of birds, mammals, and marsupials have gone extinct, including the Tasmanian wolf
- Ocean dumping
- Land degradation – caused by land clearing, over grazing, and forest clearing
- Soil erosion – from poor land-use practices
- Poor water quality – from poor industry practices and land use planning
- Salinization – from poor water quality irrigation techniques
- Urbanization
- Pollution due to mining, including heavy metals, gems, and coal
- Climate change – coral reef in the Great Barrier Reef system are dying off from warmer ocean waters as a result of climate change

7.7.1.2 Air

Australia does not have the level or severity of air pollution that is currently experienced by most other countries. However, Australia does have reduced air quality in several urban locations, especially along its east coast (United Nations 2018a). In addition, since Australia does not share a common land boundary with other countries, and most other countries are at a rather large distance, it does not significantly experience air pollution migration from other countries (Australia Department of Environmental and Energy 2018a).

In Australia, carbon monoxide, lead, nitrogen oxide, particulate matter, ozone, and sulfur dioxide are regulated and are termed criteria pollutants as they are in the United States. The two criteria pollutants of greatest concern in Australia are particulate matter and ozone (Australia Department of Environmental and Energy 2018b). Ozone is of particular concern in urban areas and particulate matter is a concern in urban areas and in rural areas of Australia. Rural Australia has a particularly high level of particulate matter because the interior portions of Australia are desert with little or no vegetation and most of the roads are not paved. Therefore, any wind can create particulate matter and road traffic in many instances creates large amounts of particulate matter.

In June 1998, the Australia National Environment Protection Council (NEPC) established Australia's first ambient air quality standards. The NEPC is a statutory body with law-making powers that were established under the National Environmental Protection Act of 1994. The Air NEPM set standards for six pollutants that are listed at www.env.gov.au/protection/air-quality-standards (Australia Department of Environment and Energy 2018c).

NEPC is also gathering data on five other pollutants it termed "Air Toxics" and are known as "Air Toxics NEPM" to set limits in the future. The Air Toxics NEPM include (Australia Department of Environment and Energy 2018c):

- Benzene
- Formaldehyde
- Benzo(a)pyrene
- Toluene
- Xylenes

Emissions standards for individual facilities are not set by NEPC but rather are set by individual states and territories within Australia, which is similar to that of the United States, and are set under a permit process (Australia Department of Environment and Energy 2018c).

7.7.1.3 Water

Water pollution in Australia, including the marine environment, has historically originated from four main sources that include (Australia Department of Environmental and Energy 2018d; Coolaustralia.org 2018a):

- Dumping of rubbish, sewerage, and industrial waste in the ocean
- Mining waste discharged into rivers and on the land
- Industry
- Urbanization

The King River in Tassie is considered the most polluted river in Australia. For nearly 70 years, it was used as a sewer and dumping ground for mining waste from a nearby copper mine. The river is so acidic that nothing can live in the river water (Coolaustralia.org 2018a).

Australia set drinking water quality standards in 2011 and were updated in 2018 (Australia National Health and Medical Research Council 2018). Australia perhaps has the most comprehensive drinking water quality management system of any country we will evaluate. Australia regulates drinking water in many of the same aspects as the United States but continues to update and monitor the quality of the water supplied to the public in a comprehensive manner. Australia evaluates microbial quality of water through evaluating the risk of disease from the following water pathogens (Australia National Health and Medical Research Council 2018):

- Bacterial pathogens
- Protozoa
- Viruses
- Helminths
- Cyanobacteria

Australia also comprehensively evaluates the physical quality, chemical quality, radiological quality, and treatment chemicals of drinking water on a routine basis to ensure the best water quality possible. Water quality limits are listed at https://nhmrc.gov.au (Australia National Health and Medical Research Council 2018). Effluent discharge of wastewater from industrial sites to any receiving water in Australia, including the marine environment, is regulated through each state or territory using a site-specific risk assessment approach that is based on internationally accepted risk assessment approaches (Australia National Health and Medical Research Council 2018). Discharge from ships is regulated by the Australia Maritime Safety Authority under separate permits and regulations (Australian Maritime Safety Authority 2018).

7.7.1.4 Solid and Hazardous Waste

In 2009, Australia set goals for reducing the amount of wastes generated nation-wide. The goal was to create less waste and improve resources (Australia Department of Environment 2009). The need to revise its national policy and improve regulations was because Australia saw an increase in the amount of waste generated nation-wide by 34.7% over a 4 year period from 2002 to 2006 and the amount of hazardous waste doubled during this same period of time (Australia Department of Environment 2009).

Since 2009, Australia is now recycling approximately 60% of its solid waste. The waste generated per person in Australia in 2006 was estimated at 2.783 metric tons per person and was estimated at 2.705 metric tons in 2015, which represents a slight decrease nation-wide in the amount of solid waste produced (Australia Department of Environment 2016).

In Australia, a solid waste is defined as material or products that are unwanted or have been discarded, rejected, or abandoned. Wastes typically arise from three dominant waste streams that include (Australia Department of Environment and Energy 2018e; 2018f):

- Domestic and municipal
- Industrial
- Construction and demolition

Determining whether a solid waste is hazardous or not in Australia is similar to that of the United States and many other countries. In Australia, a solid waste is hazardous if it is a per-listed waste or has the following characteristics (Australia Department of Environment 2018e; 2018f):

• Explosive	• Flammable
• Ignitable	• Gas
• Oxidizing agents	• Corrosive
• Reactive	• Infectious
• Medical	• Toxic
• Radioactive	• Characteristic

A waste is a characteristic hazardous waste if a leachate test of the waste identifies any of the compounds above those listed at www.env.gov.au/protection/waste-class.pdf (Australia Department of Environment 2018e). Soil, sediment, and water remediation values for select compounds and pathways are listed at www.environment.gov.au/site-cleanup.pdf (Australia Department of Environment and Conservation 2010).

7.7.1.5 Remediation

Cleanup or restoration of agricultural land in Australia is perhaps the most significant issue, not industry or urbanization. Approximately 50% of agricultural land in Australia is considered severely degraded because of rising salinity from irrigation, erosion, and vegetative loss. The result has been a reduced capacity for crops, which in turn then requires more land for agriculture and continued land degradation. This cycle of agriculture kills farmland, contaminates drinking water, and destroys natural ecosystems if it continues without drastic restoration efforts and modified farming techniques (Coolaustralia.org 2018b).

7.7.1.6 Summary of Environmental Regulations of Australia

Australia has a comprehensive set of environmental regulations and has an enforcement system and culture that is very engaged in protecting human health and the environment. Australia has some disadvantages when it comes to its dry climate, poor soil, and desert geography in that water is more a precious resource than other parts of the world since it is of limited quantity. Therefore, as we discussed earlier in this section, Australia must protect its water from contamination. In addition, off Australia's northeast coast is the largest reef system in the world, the Great Barrier Reef. The Great Barrier Reef is experiencing an unprecedented die-off of coral due to a warming of ocean water believed to be directly linked to climate change.

7.7.2 New Zealand

New Zealand is an island country 1,200 kilometers southeast of the coast of Australia in the South Pacific Ocean. New Zealand is comprised of nearly 600 islands, two of which are large and make up most of the land mass and are called the North Island and the South Island (see Figure 7.28).

New Zealand is comprised of 267,710 square kilometers of land making it the seventy-fifth largest country in the world. The population of New Zealand is 4.8 million. New Zealand's largest city is Auckland with an estimated population of 1.6 million (United Nations 2018h). The Northern Island of New Zealand is characterized as volcanic in nature and the Southern Island is characterized as mountainous in nature. New Zealand's economy is considered the fifty-third largest in the world when measured by gross domestic product of US$200.7 billion in 2016. Per capita income of New Zealand is $37,860 as measured in 2016. New Zealand's economy is considered one of the most globalized and depends greatly upon international trade. Major industries in New Zealand are aluminum mining, food processing, metal fabrication, wood, and paper products. Trading partners include Australia, the United States, China, Japan, Germany, and South Korea (United Nations 2018h).

7.7.2.1 Environmental Regulatory Overview of New Zealand

Environmental issues facing New Zealand include deforestation, soil erosion, loss of habitat, and surface and groundwater quality degradation. Much of the development in New Zealand has been for farming, which has impacted the environment perhaps more than any type of human activity, including industrial impacts. New Zealand actively is involved in sustainability issues and has openly embraced efforts to deal with climate change beginning as early as 1991 (New Zealand Ministry for the Environment 2018a). In New Zealand, the Resources Management Act of 1991 is the main law that regulates discharges into the environment, including air and water. Much of the responsibility for managing discharges and permits has been handed down to district and regional councils. Therefore, just as in the United States, there are differences from district to district in how these issues are handled but the basic goals, objectives, and specifics are the same (New Zealand Ministry for the Environment 2018a).

7.7.2.2 Air

Air quality in New Zealand is similar to that of Australia in that it is most degraded in urban areas but is better than many other urban areas of the world (United Nations 2018a). New Zealand's footprint for greenhouse gases is unique in that the majority of greenhouse gases originate from agriculture and not industry (New Zealand Ministry for the Environment 2018b). In fact, New Zealand's industry only accounts for a small percentage, less than 10% of total emissions. Two main sources of greenhouse gases originate from agriculture and include actual farming and livestock, especially the dairy industry which produces large quantities of methane from cattle (New Zealand Ministry for the Environment 2018b).

New Zealand's ambient air quality standards were first enacted in 1994 and then were updated in 2002 (New Zealand Ministry of the Environment 2018c). The ambient air quality standards are listed at www.gw.govt.nz/form-5a (New Zealand Ministry for the Environment 2018c). As with many other countries, air permits are necessary for any industrial emitting source. Permit applications are comprehensive and typically include stack testing or meeting other performance criteria and extensive record keeping to document compliance. Permit limits are low if emitting sources include any compounds listed at www.gw.govt.nz/form-5a (New Zealand Ministry for the Environment 2018c).

7.7.2.3 Water

Water quality in New Zealand rivers has been monitored since 1989 and by international standards is of good quality. However, there are signs of declining quality from increased concentrations of nitrogen from anthropogenic sources, namely agriculture (New Zealand Ministry of the Environment 2018d). Water quality standards in New Zealand are listed at www.mfe.nz/publications/drinking-water (New Zealand Ministry of the Environment 2018e).

Agriculture is the largest land use in the lowlands of New Zealand, particularly dairy and livestock, which has been linked to increased concentrations of nitrogen in soil, surface water, and groundwater. Availability of water has also declined due to increased dairy farming which uses more water in New Zealand than any other activity (New Zealand Ministry of the Environment 2018f). Wastewater discharges in New Zealand require what is termed a resource consent from the district regulatory authority. The resource consent is a discharge permit that makes the discharge subject to facility-specific discharge conditions that take into account the quality of the wastewater and the quality of the receiving body of water (New Zealand Ministry for the Environment 2018f).

7.7.2.4 Solid and Hazardous Waste

In New Zealand, a waste is considered hazardous if it presents some degree of physical, chemical, or biological hazard to people or the environment (New Zealand Ministry for the Environment 2018g). Hazardous waste can be gases, liquids, or solids that are:

Explosive	Flammable
Infectious	Ignitable
Oxidizing agents	Corrosive
Reactive	Medical
Toxic	Eco-toxic
Radioactive	Characteristic

Similar to many other countries, hazardous wastes are tracked from the generator to the transporter to the final disposal site using a manifest system (New Zealand Ministry for the Environment 2018g).

New Zealand has developed a waste list or what is termed the "L-Code," which is a method that reflects typical waste streams in New Zealand. The L-Code provides guidance on identifying wastes in a consistent manner and serves as the basis for record-keeping systems, especially for hazardous wastes. The L-Code is as follows (New Zealand Ministry for the Environment 2018h):

1. Wastes resulting from exploration, mining quarrying, and physical and chemical treatment of minerals
2. Waste from agriculture, horticulture, aquaculture, forestry, hunting and fishing, food preparation and processing

3. Waste from wood processing and the production of panels and furniture, pulp, paper, and cardboard
4. Wastes from leather, fur, and textile industries
5. Wastes from petroleum refining, natural gas purification, and pyrolytic treatment of coal
6. Wastes from inorganic chemical processes
7. Wastes from organic chemical processes
8. Wastes from the manufacture, formulation, supply, and use of coating (paints, varnishes, and vitreous enamels), adhesives, sealants, and printing inks
9. Wastes from the photographic industry
10. Wastes from thermal processes
11. Wastes from chemical surface treatment and coating of metals and other material: non-ferrous hydro-metallurgy
12. Wastes from shaping and physical and mechanical surface treatment of metals and plastics
13. Oil wastes and wastes of liquid fuels
14. Waste organic solvents, refrigerants, and propellants
15. Waste packaging; adsorbents, wiping cloths, filter materials and protective clothing not otherwise specified
16. Wastes not otherwise specified
17. Construction and demolition wastes (including excavated soil from contaminated sites)
18. Wastes from human or animal health care and/or related research
19. Wastes from waste management facilities, offsite wastewater treatment plants, and preparation of drinking water and water use for industrial applications
20. Municipal wastes (household waste and similar commercial, industrial, and institutional wastes) including separately collected fractions

New Zealand has an extensive reduce, reuse, recycle program and also generates 80% of its electricity from renewable resources that include wind, hydro, and geothermal. The program for reducing waste follows a circular economy approach that is based on three principles, which include (New Zealand Ministry for the Environment 2018i):

- Design out waste and pollution
- Keep products and material in use
- Regenerate natural systems

7.7.2.5 Remediation

A Contaminated Sites Remediation Fund (CSRF) was established as part of the Resource Management Act of 1991 that supports regional councils to achieve goals for contaminated land management. Investigation and remediation is conducted using a prioritizing method to determine which sites are funded under the program. The New Zealand Ministry for the Environment assesses each application and selects ten sites that are of the highest priority for funding. Once a site is remediated to an acceptable risk, they are removed from the list and replaced with a new site in need (New Zealand Ministry for the Environment 2018j).

New Zealand has developed detailed and comprehensive guidelines for conducting environmental investigations at sites of potential concern with the objective to achieve a nationally consistent approach to evaluate whether a risk to human health and environment exists (New Zealand Ministry for the Environment 2018k). New Zealand uses a Risk Screening System (RSS) that guides the investigation so that each site is investigated consistently and to appropriately rank a site by risks posed to protect human health and the environment. New Zealand utilizes the USEPA Integrated Risk Information System (IRIS) and the United States Agency for Toxic Substance and Disease Registry to develop minimum risk levels (New Zealand Ministry for the Environment 2018l).

7.7.2.6 Summary of Environmental Regulations of New Zealand

The data certainly demonstrates that New Zealand has a comprehensive, effective, and forward-thinking environmental regulation program and a population that is engaged in protecting its air, water, and land. This is evident in New Zealand's efforts to reduce the amount of solid waste that is generated, and that 80% of its electricity is generated from renewable resources. Challenges for New Zealand include striking a balance with its agricultural activities and environmental degradation of its surface and groundwater and continued population increase and urbanization.

7.8 SOUTH AMERICA

South America forms the southern landmass of the Americas, generally considered as south of the Panama Canal (see Figure 7.30). South America is composed of three basic types of terrain: the Andes Mountains that span nearly the entire western boundary, the interior Amazon River Basin, and the Guiana Highlands in the east (United Nations 2017a). South America has a very diverse climate with one of the wettest in Columbia and Chile to the driest desert in the world, the Atacama (United Nations 2018i; 2018j; 2018k).

South America is composed of 12 countries and has a total estimated population of 427 million and represents only 5.6% of the total human population on Earth. The land area of South America is 17,840,000 square kilometers, making it the fourth largest continent behind Asia, Africa, and North America. The population density of South America is 21.4 per square kilometer (United Nations 2017a). Most people live along the eastern and western coast of the continent.

7.8.1 Argentina

Argentina is bordered by the Atlantic Ocean to the east, Andes Mountains and Chile to the west, Bolivia and Paraguay to the north and Brazil and Uruguay to the northeast (see Figure 7.30). It is the eighth-largest country in the world with 2,780,400 million square kilometers of land. Argentina has a population of 42.1 million, making it the thirty-second most populous country. The population density of Argentina is 16.2 people per square kilometer. The largest city in Argentina is Buenos Aires, which is also the capital of the country and has a population of over 15 million (United Nations 2018i). Argentina has a diverse economy that consists of manufacturing, agriculture, forestry, fishing, construction, transport, and utilities. The gross domestic product of Argentina was over $639 billion in 2017, which translates into a per capita income of $3,500 per individual, which is well below that of many other nations (World Bank 2018b).

7.8.1.1 Environmental Regulatory Overview of Argentina

Environmental issues facing Argentine include air pollution, water pollution, deforestation, salinity, and lack of environmental controls and enforcement (United Nations 2018i).

7.8.1.2 Air

Air pollution in Argentina is most acute in its capital city of Buenos Aires. However, there is very limited data on current and historical air pollution levels in Buenos Aires and throughout Argentina. Argentina has established ambient air quality standards but they have not been implemented for the following (Argentina Secretariat of the Environment 2018a):

- Carbon monoxide
- Nitrogen dioxide
- Sulfur dioxide
- Ozone

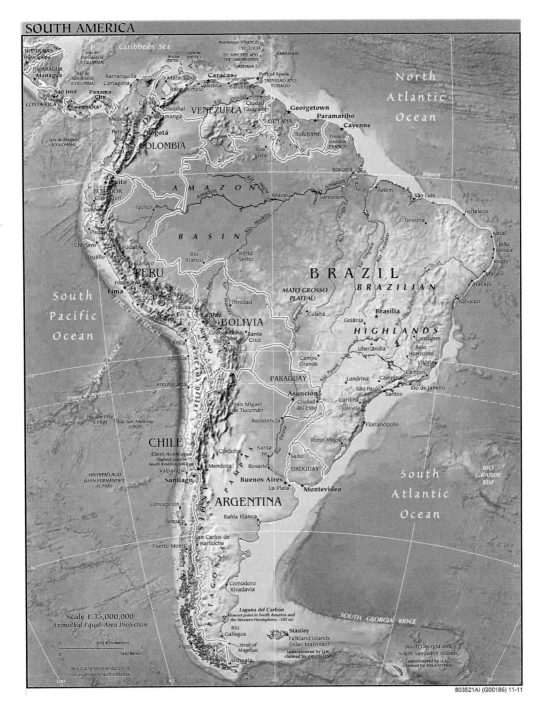

FIGURE 7.30 Map of South America. (From United States Central Intelligence Agency. 2018. The World Fact Book. 2018. www.cia.gov/library/publications/the-world-factbook.gos/NA.html.) (Accessed October 10, 2018.)

- Particulate matter (PM10)
- Fine particulate matter (PM2.5)

Argentina ambient air standards are listed at www.fmed.uba.ar/depto/toxico/plaguicidas/20.284 (Argentina Secretariat of the Environment 2018a).

7.8.1.3 Water

The environmental management of water bodies under federal jurisdiction in Argentina are regulated under Law 25.688 (Argentina Secretariat for the Environment 2018b). Drinking water in urban areas of Argentina is generally regarded as safe. However, up to an estimated 20% of drinking water sources in rural areas is not considered safe (United Nations 2018i).

Argentina is struggling with cleanup of the 64-kilometer long Matanzas-Riachuelo River which flows through Argentina's most populous city of Buenos Aires and is considered one of the most polluted rivers on Earth. The river has seen contamination for over 200 years from agriculture, livestock, tanneries, and industrial pollutants from over 15,000 industrial facilities that have discharged to the river and are located within the river drainage basin (World Bank 2013b). Cleanup of the river was to begin in 2018 with assistance from the World Bank, Argentina Secretariat for the Environment and Sustainable Development, and several municipalities along and near the river at a cost of several billion US dollars (World Bank 2018c). Cleanup is expected to continue for decades before being complete. The situation along the Matanzas-Riachuelo River is but one example of the challenges that plague the development, implementation, and enforcement of water resources management in Argentina. Management and changing water management techniques in Argentina are further hindered because multiple institutions at the national, provincial, and river basin level have jurisdiction (World Bank 2018d).

Identified issues that must be addressed include (World Bank 2000; 2018d):

- Incomplete or outdated legal and regulatory framework
- Limited capacity in water management at the central and provincial levels
- Outdated procedures for water resources planning
- Lack of integrated national water resources information system
- Deficient water resources monitoring system
- Serious water pollution problems
- High risk of flooding in urban and rural areas
- Lack of flood risk reduction strategies
- Lack of appropriate incentives for conservation and pollution reduction

7.8.1.4 Solid and Hazardous Waste

Solid and hazardous waste is regulated in Argentina by Law 24.051 and addresses the generation, transport, handling, treatment, and final disposal of hazardous waste. However, specific details on waste characterization, infrastructure, and management structure on a national level, such as the management of water resources, is lacking. In Argentina, a waste is hazardous if it can damage living beings or contaminate land, air, water, or the environment. Any facility that generates a hazardous waste must register with the National Registry of Hazardous Waste Generators and Operators, which is required to operate any business that generates a hazardous waste (Argentina Secretariat for the Environment 2018c). Just as in the United States and several countries of the world, the hazardous waste tracking system in Argentina is a "Cradle to Grave" system (Argentina Secretariat for the Environment 2018c).

7.8.1.5 Remediation

As with water and solid and hazardous waste, remediation standards on a national basis are lacking in Argentina and are plagued by several impediments that include overlapping jurisdictions, competing interests, lack of reliable cleanup standards, and lack of infrastructure (United Nations 2018i).

7.8.1.6 Summary of Environmental Regulations of Argentina

Argentina is a rather large country for its population. Argentina's significant environmental issues appear to be concentrated in and immediately around its largest city of Buenos Aires and involve air, water, and land pollution. Investigation and cleanup of one of the world's most polluted rivers will likely be a benchmark achievement or failure for Argentina's future at protecting human health and the environment. Cleanup of a 64-kilometer river in an urban environment in any country, including the United States, is a monumental task and is perhaps too much to ask for a country that is struggling with social and economic issues as well.

7.8.2 BRAZIL

Brazil occupies 48% of the continent of South America and has a total land area of 8,514,215 square kilometers (see Figure 7.30). The only countries larger in land area are Russia, Canada, China, and the United States. Much of the climate of Brazil is tropical, with its southern portion being temperate. The largest river in Brazil, and the second-longest in the world, is the Amazon River, with a length of 6,992 kilometers (United Nations 2018j).

Brazil has the eighth largest economy in the world with a gross domestic product of $2.138 trillion. Brazil has the second-largest economy in the Americas behind the United States. Significant natural resources include gold, uranium, iron, and timber. Main trading partners with Brazil include China, the United States, Argentina, and Japan. Brazil's main industries include textiles, chemicals, cement, lumber, iron ore, tin, steel, aircraft, motor vehicles, and other machinery equipment (United Nations 2018j).

Brazil undertook an ambitious program to reduce its dependence on imported oil, which accounted for 70% of its historical needs. By 2007, Brazil became self-sufficient in oil and is now one of the world's leading producers of hydroelectric power. In fact, hydroelectric power now generates 90% of Brazil's electricity needs. This was accomplished by the Itaipu Dam on the Parana River and the Tucurui Dam in Para in northern Brazil. Brazil is also expanding its use of nuclear power with two nuclear power plants currently in operation and has plans for ten more power plants (World Bank 2018e).

The population of Brazil is estimated at 207 million, making it the second-highest in population in the Americas only behind the United States. The two largest cities in Brazil include Sao Paulo, with an estimated population of 21.5 million and Rio de Janeiro, with a population of 12.8 million. Per capita income of Brazil is approximately $15,500 or nearly one-quarter of the per capita income of the United States and about the same as China (World Bank 2018e).

7.8.2.1 Environmental Regulatory Overview of Brazil

The Amazon rainforest is the largest tropical rainforest in the world spanning 5.5 million square kilometers, 60% of which is located in Brazil. Rainforests world-wide once covered 15% of the land on Earth but due to anthropogenic activities, only 6% are left and are decreasing each year. Current environmentally-related threats to the Amazon rainforest include (National Geographic Society 2018c):

- Deforestation. Deforestation in the Amazon rainforest vary between the nine countries that are within or border the rainforest. Types of deforestation within the Amazon basin include:
    - Clearing the forest by burning it to make room for cattle pastures for large ranches and local farmers who clear smaller, but perhaps more numerous, patches for agriculture.
    - Mining, especially for gold
    - Urbanization from expending cities on the fringe

- Loss of biodiversity. Species lose their habitat or can no longer survive in the small fragments of forest left behind after clearing, especially if the patches of forest are not interconnected.
- Habitat degradation. Fragmentation by road building and other anthropogenic activities lead to habitat degradation that place additional negative pressures on indigenous species.
- Modified global climate. Deforestation releases CO_2 into the atmosphere and also prevents CO_2 removal and the release of oxygen into the atmosphere by the trees that are deforested.
- Interruption or loss of the water cycle. Deforestation reduces the critical water cycling services that trees provide by increasing the amount of erosion and decreasing the amount of evapotranspiration, which in turn disrupts and decreases subsequent rain events.
- Social effects. Deforestation ultimately leads to degradation of the ecosystem and contributes to overall decline of long-term economic balance.
- Logging. Logging is another example of deforestation that destroys habitat.
- Hydroelectric plants. Damming rivers in the Amazon rainforest flood.
- Poaching. Poaching of mammals, birds, reptiles, and even insects lowers the natural population numbers in species and places additional stress on Amazon wildlife.
- Pollution. Anthropogenic activities introduce a plethora of pollutants into the rainforest ecosystem through the air, water, and land that include but are not limited to organic and inorganic compounds, heavy metals, particulate matter, suspended solids, erosion, noise, increased atmospheric carbon dioxide, decreased oxygen, and habitat loss.

7.8.2.2 Air

Air pollution in Brazil is largely due to rapid urbanization and a deforestation technique called "slash and burn" (see Figure 7.31), which is a method where vast tracts of forest land are cut and burned to make way for development. In addition, may smaller towns still burn garbage under a legal framework that allows prescribed burning (see Figures 7.32 and 7.33). Efforts to improve air quality are underway, especially in Brazil's largest cities (USEPA 2018e).

Ambient air quality standards in Brazil are similar to the United States and EU and include the following compounds (Brazil Ministry of the Environment 1990):

- Particulate matter (PM10)
- Fine particulate matter (PM2.5)
- Ozone

FIGURE 7.31   Slash and burn techniques in Brazil. (Photograph by Daniel T. Rogers.)

FIGURE 7.32 Photograph of burning agricultural fields in Brazil. (Photograph by Daniel T. Rogers.)

FIGURE 7.33 Example of burning garbage in Brazil. (Photograph by Daniel T. Rogers.)

- Sulfur dioxide
- Carbon monoxide
- Nitrogen dioxide

Brazil ambient air quality standards are located at www.brazilgovnews.br/standards-on-air-quality -htm (Brazil Ministry of the Environment 1990).

The National Environment Council of Brazil created the ambient air standards in 1990 and established the ambient air quality program (Brazil Ministry of Environment 2018a). Air pollution in urban areas is worst in Sao Paulo and Rio de Janeiro (United Nations 2018a).

Environmental licensing is required in Brazil for any enterprise or activity that uses or may generate pollutants or hazardous substances. In the case of an air permit, the process is similar to the United States, the European Union, China, and other countries. The following is a general outline of the permit process (Brazil Ministry of Environment 2018a):

- Initial application. The initial application is termed a construction permit if it's new construction. The construction permit requires proof of ownership, certification of use or zoning, and proposed architectural or building improvements.

- Habitation permit. Following approval of the construction permit, a habitation permit is the next step
- Registration for the article of association.
- Business license.
- Sanitary permit. Once the sanitary permit is obtained, other required permits for water, water treatment, and waste generation can proceed.

7.8.2.3 Water

Water resources management in Brazil is the responsibility of the National Water Agency. In general, Brazil has vast quantities of fresh water, especially in the northern portions of the country near the Amazon River. Up to 97% of the population has access to clean water. However, water pollution in Brazil from sewerage is acute and is especially so in many of the larger cities, including Rio de Janeiro and Sao Paulo (Brazil Ministry of Environment 2018a).

According to the WHO (2015b), only 15% of sewerage in Brazil is collected and treated before discharge. Not only does the discharge of untreated sanitary waste impact the rivers and other fresh water bodies in Brazil, but it also has a noticeable and significant effect on the ocean and beaches, especially near the large cities of Sao Paulo and Rio de Janeiro. (WHO 2015b). Water quality standards vary from each of the 27 states in Brazil and also within many of the 389 municipalities that supply water.

Some large manufacturing facilities, as in the United States and elsewhere, have installed their own water treatment and wastewater treatment plants in Brazil. Some of these facilities require more improved water quality than the local municipality can provide and others require a more reliable source. These facilities obtain a permit to withdraw water from a water source, usually a nearby river, treat the water before it's used, and then treat the water again before it's discharged back into the river. Figures 7.34 and 7.35 show an example of this process.

7.8.2.4 Solid and Hazardous Waste

Brazil amended its 1998 waste law in 2010 (Brazil Ministry of Environment 2018a) that brought together in one law a set of principles, goals, instruments, targets, and actions toward the integrated management of environmentally sound solid waste management. The Brazilian Waste Law outlines

FIGURE 7.34 Fresh water river intake in Brazil. (Photograph by Daniel T. Rogers.)

FIGURE 7.35 Untreated water (top), treated water (bottom). (Photograph by Daniel T. Rogers.)

concepts and provisions similar to that of the United States and includes (Brazil Ministry of the Environment 2018b):

- Solid waste management
- Recycling
- Reuse
- Reduce
- Sustainability
- Pollution prevention
- Polluters pay principle
- Shared responsibility or cradle to grave concept

Hazardous wastes in Brazil follow that of most other countries and include (Brazil Ministry of the Environment 2018b):

• Explosive	• Ignitable
• Flammable	• Infectious
• Oxidizing	• Toxic
• Corrosive	• Characteristic
• Reactive	• Medical

Radioactive waste is addressed separately (Brazil Ministry of the Environment 2018b). The volumes of hazardous waste generation in Brazil from 2006 to 2015 increased by 30%. The significant challenges for Brazil recently is the management of medical waste and e-waste, both of which appear to need improvement (Macedo and Sant'ana 2015).

7.8.2.5 Remediation

Much like the rest of the world, Brazil uses risk-based standards for remediation. Remediation thus far in Brazil has generally been for its larger sites that pose the highest risk and usually involves water and river systems (Brazil Ministry of Environment 2018a).

7.8.2.6 Summery of Environmental Regulations of Brazil

Brazil is the largest and most populated country in South America and the country that has 60% of the Amazon Rainforest. Brazil's most significant environmental issues are:

- Sanitation, especially in the poor areas within its major cities
- Cleaning up river systems, which is related to sanitation

- Air pollution in larger urban areas
- Protecting the rainforest

The sanitation issue is one that plagues many countries in South America and Mexico, portions of Africa, and Asia. Unfortunately, the longer it takes to finally address the sanitation issue the harder and more costly it becomes and continues to spread disease (United Nations 2015b)

Brazil's environmental regulations seem robust but are not providing the desired results because of several factors that include:

- Lack of financial resources
- Overlapping responsibilities between agencies
- Other social and economic priorities
- Lack of political will and corruption, especially with respect to addressing the larger more complex environmental issues such as protecting the rainforest and sanitation

7.8.3 CHILE

Chile is located along the extreme western boundary of South America. Chile shares land borders with Argentina, Peru, and Bolivia. Chile's western border is the Pacific Ocean (see Figure 7.30).

From north to south Chile extends 4,270 kilometers but averages only 170 kilometers east to west, making Chile a long and narrow nation. The land area of Chile is 756,096 square kilometers. The Andes Mountains run along the entire eastern border of Chile. Chile is prone to earthquakes. In fact, Chile has experienced 28 earthquakes in the 20th century of a magnitude 6.9 on the Richter scale or greater (USGS 2018; United Nations 2018k). Because Chile is so long and narrow in the north-south direction and goes from sea level to nearly 7,000 meters in elevation in the Andes Mountains, the climatic regions vary significantly and often over very short distances.

The population of Chile is 18.4 million. Santiago is Chile's largest city and is the capital with a population of 6.3 million, which represents 33% of Chile's entire population. The gross domestic product of Chile was $247.03 billion in 2016. Per capita income is estimated at $15,059. Chile is considered one of South America's leading economies (World Bank 2015).

7.8.3.1 Environmental Regulatory Overview of Chile

The main statutory environmental framework in Chile is contained in the Environmental Law (No. 19.300, 1994 and as amended in 2010), which introduced the Environmental Impact Assessment System, which set forth the process for obtaining an environmental license. The 1994 law is known as the General Environmental Framework Law, which established the National Environment Commission (CONAMA), which reports directly to the president's office through the Ministry of General Secretariat of the Presidency. CONAMA coordinates government environmental policies, prepares environmental regulations, and fosters integration of environmental concerns in other policies (United Nations 2015b).

The amendments to the 1994 law enacted in 2010 created the Ministry for the Environment. The Ministry for the Environment, which elevated environmental issues to a cabinet-level rank, is charged with assisting the president of the Republic of Chile in the design and implementation of policies, plans, and programs for the protection of the environment, which includes (United States Library of Congress 2010):

- Biological diversity
- Renewable natural resources
- Hydraulic resources
- Sustainable development

- Environmental policy
- Regulatory framework

The Environmental Impact Assessment System requires all investment projects and productive activities to undergo a detailed process to determine the real effects proposed operations would have on the environment (Chile Ministry of the Environment 2018a).

USEPA has been collaborating with Chile since 2004 on issues relating to (USEPA 2017; Chile Ministry of the Environment 2018a):

- Risk management related to waste and contaminated sites
- Environmental enforcement
- Environmental forensics
- Compliance with air pollution standards
- Training
- Technical consultation
- Public participation
- Environmental education

Chile's most significant environmental issues center around deforestation and the subsequent erosion that it causes and air pollution in its capital city of Santiago. The city of Santiago experiences increased air pollution due to automobile exhaust and wood burning that is difficult to abate because the city is surrounded by mountains (Chile Ministry of the Environment 2018a). Many households use wood for fuel because it's more economical than other forms of energy. Other environmental issues in Chile include degraded water quality, especially in and around Santiago, and from mining operations. Unfortunately, most of Chile's mining is in the northern desert region of the country where any water quality issues are heightened because of the relative scarcity of water resources (Chile Ministry of the Environment 2018a).

7.8.3.2 Air

Air pollution in Chile ranges from good to unhealthy (WHO 2016a). Santiago usually experiences the lowest air quality in the country and is larger due to automobile exhaust and wood burning and is exacerbated because the city is surrounded by mountains which inhibits the circulation of air during certain times of year, especially the winter months when colder air is trapped near the surface of the ground (WHO 2016a). However, the buildup of ozone is heightened in the summer months due to increased solar radiation (Diaz-Robles et.al 2011).

Similar to the United States, areas that do not meet ambient air quality standards are designated as non-attainment areas by geographic region. When an area is designated as non-attainment, either an atmospheric prevention plan (APP) or an atmospheric decontamination plan (ADP) is required, which is similar to the state implementation plan (SIP) in the United States (Chile Ministry of the Environment 2018). These plans require that regions of non-attainment lower emissions until attainment is achieved. The effectiveness of the ambient air quality is evident in that the area of Santiago has reduced its sulfur dioxide levels by 77% since 1994. Other contaminants such as ozone have been on the increase mostly due to an increasing population and more automobiles on the roads (Diaz-Robles et al. 2011). Chile's ambient air quality standards are listed at http://portal.mma.gob.cl.aire (Diaz-Robles et al. 2011; Chile Ministry of the Environment 2018b).

7.8.3.3 Water

Compared to other countries of South America, Chile has good water quality and sanitation with nearly 99% of its population with access to treated water and basic sanitation (WHO 2017). The northern portion of the country appears to suffer the worst water quality in Chile for two reasons (United Nations 2013b):

- Water is scarce because it's a desert region.
- Most mining, especially copper, is located in the same region and requires large quantities of water that is obtained from groundwater.

The resulting water contaminated with heavy metals impacts surface water quality quite significantly partially because water is relatively scarce. In contrast, water in southern Chile is considered by some to be the purest fresh water on Earth because of its remoteness and lack of anthropogenic sources of pollution (United Nations 2013b).

Unlike many other countries, water in Chile is largely privatized and has been since 1981. The advantage of privatization is that the quality of water is of better quality. However, the disadvantage is that water is expensive. In addition, it is estimated that prices will increase significantly due to an increasing population and climate change which has negatively impacted water availability through increased temperatures and a general decrease in precipitation levels since 1976 (United Nations 2013b).

7.8.3.4   Solid and Hazardous Waste

Environmental regulations in Chile on solid and hazardous waste are similar to that of the United States and cover the following (Chile Ministry of the Environment 2018c):

- Solid waste
- Hazardous waste
- Storage and handling of hazardous substances
- Transporting solid and hazardous waste
- Disposal

Hazardous wastes in Chile follow that of the United States and include (Chile Ministry of the Environment 2018c):

• Explosive	• Ignitable	• Flammable
• Infectious	• Oxidizing	• Toxic
• Corrosive	• Characteristic	• Reactive
• Medical	• Certain residues	

Certain identified processes that produce waste are also considered hazardous and are similar to listed wastes in the United States. Some of these include the following (Chile Ministry of the Environment 2018c):

• Textiles	• Petroleum	• Solvents
• Used absorbents	• Refrigerants	• Propellants
• Pesticides	• Herbicides	• Adhesives and glues

7.8.3.5   Remediation

Remediation of sites of environmental contamination in Chile has focused on mining and abandoned mines. USEPA has been actively collaborating with the Chile Ministry of the Environment focusing on (USEPA 2017):

- Management of environmental aspects of mining
- Abandoned mine risk evaluation

- Remediation of contaminated mining sites
- Site investigation and characterization
- Conducting risk assessments
- Mapping
- Selection of remedial options
- Enforcement measures at mining sites

Additional remedial focus has been on improving air quality, especially in the urban area of Santiago through a Chilean Ministry of Environment and USEPA partnership designed to improve air quality, protect the climate, and provide public health benefits (USEPA 2017).

7.8.3.6   Summary of Environmental Regulations of Chile

Chile is a small country with a relatively low population located in an extremely diverse climate and geographic region. This tends to complicate regulation of the environment because there are so many different, separate and distinct, and sometimes unrelated geologic and hydrologic environments. Therefore, Chile's geology and geography heavily influence its efforts to protect and improve environmental regulations.

Air quality regulations appear to be the most advanced in Chile compared with water, solid and hazardous waste, and remediation regulations. However, availability of clean water will likely be more of a pressing issue in the future due to climate change.

Chile's environmental regulations first became serious in 1994 and then were revised significantly in 2010. Together with cooperation and collaboration with the United States Environmental Protection Agency (USEPA), it does not appear that environmental matters will largely be ignored in the future. Anticipation is that future environmental regulations in Chile will be very similar to that of the United States.

7.8.4   Peru

Peru is located in western South America. Peru's west coast extends along the Pacific Ocean for 2,414 kilometers. Columbia and Ecuador are located to the north, Brazil to the east, and Bolivia and Chile to the south (see Figure 7.30). Peru covers a land area of 1,279,999 square kilometers, making it the twentieth largest country by area in the world. Peru has an estimated population of 31 million people. The GDP of Peru was $470 billion in 2017, which translates into a per capita income of $13,735. The economy of Peru has significantly improved over the last 20 years. As evidence of strong economic growth, poverty in Peru has been reduced from nearly 60% in 2004 to just below 26% in 2012 (World Bank 2018f).

The Peruvian economy is 43% service (including tourism) and 47% industry and manufacturing. Major exports include metals (copper, zinc, gold, mercury), chemicals, pharmaceuticals, machinery, textiles, and fish meal. Peru ranks second in the world for copper production. Copper accounts for 60% of Peru's exports (World Bank 2018f). The capital of Peru is Lima. It is located along the western coast with the Pacific Ocean. The population of Lima and associated surrounding communities is estimated at over 12 million residents, which represents 40% of the entire population of Peru (World Bank 2018f).

The geography of Peru is rather unique in that the Andean Mountain range, which trends north to south, divides the country from its coastal region to the west from the Amazon Basin to the east (United Nations 2005). The Andean Mountain range acts as a divide in geographical and climatic terms.

7.8.4.1   Environmental Regulatory Overview of Peru

The principle environmental issues of Peru include water pollution, soil erosion, soil pollution, and deforestation. Air pollution is significant in Lima where pollutant concentrations sometimes reach unhealthy levels (WHO 2018f). Peru is susceptible to high levels of soil erosion as deforestation occurs and is often exacerbated especially in mountainous regions with increased erosion. Water pollution and lack of sanitation are also issues, especially in rural area where an estimated 62% of the population have access to clean water and access to sewage is as low as 22% (United Nations 2005).

The USEPA is collaborating with Peru on several environmental issues since 2009 that include air, water, waste, land degradation, remedial technologies and management, and others (USEPA 2018f).

The Peruvian Ministry of Environment is a ministry of the cabinet of Peru created in 2008. Its function is to oversee the environmental sector with the authority to design, establish, and execute government policies concerning the environmental management and strategic development of natural resources (Peru Ministry for the Environment 2018). Peru, by means of Law 30327, established an integrated permitting process titled "Global Environmental Certification" (IntegrAmbiente) that integrates all necessary environmental permits into one permit similar to the process in China (Peru Ministry for the Environment 2018a).

7.8.4.2   Air

The city of La Oroya, located in the Andes Mountains 176 kilometers northeast of Lima, is considered by many to be one of the most polluted cities in the world. The source of the pollution is a copper smelter and associated copper mine which has resulted in significantly elevated levels of lead, copper, zinc, and sulfur dioxide in the air (USEPA 2018f). Ambient air quality standards are similar to the United States and EU and are listed at www.temasactuales.com/assets/pdf/gratis/peraiqual. htm (Peru Ministry for the Environment 2018a).

In 2016, the World Health Organization released a study of air quality which identified Lima as the city with the worst air pollution in South America (WHO 2016a). Given that Lima is a city of more than 12 million residents, is surrounded by mountains, is subject to air inversions which tend to trap pollutants in the low layers of the atmosphere, it is not a big surprise that the air quality is poor.

7.8.4.3   Water

Just as with air pollution, Peru faces water pollution issues associated with its copper mining operations along with degradation of water quality due to deforestation. Access to safe water and sanitation has improved in Peru in recent years but 10% of the population remains without access to safe drinking water and nearly 80% without improved sanitation (WHO 2018f). The Ministry of Housing, Construction, and Sanitation is the governing entity that formulates, approves, executes, and oversees national water and sanitation policy (Peru Ministry of Housing, Construction, and Sanitation 2018).

USEPA is collaborating with the Peruvian government to improve water quality, specifically addressing mining wastes contaminated with various metals, including mercury (USEPA 2018f). Surface water quality in particular has been gradually declining due to mining discharges to surface water that either have not been treated adequately or have not been treated at all. Other anthropogenic sources of surface water pollution are from industry, municipalities, and polluted agricultural runoff (USEPA 2018f). The wastewater discharge standard in Peru are listed at www.gob.pe/vivienda (Peru Ministry of Housing, Construction, and Sanitation 2018).

7.8.4.4   Solid and Hazardous Waste

Solid waste management in Peru was essentially non-existent until the Ministry of the Environment was created in 2008. Environmental concerns in Peru are not considered a national priority, including solid waste management. However, there are only nine officially recognized solid waste landfills in Peru and five are located in and around the capital city of Lima. This means that many parts of the country still rely on informal disposal of solid waste in dumps and other poorly managed and poorly constructed locations (Peru Ministry for the Environment 2018b; Peace Corp 2016).

USEPA is currently collaborating with Peru on many environmental issues including upgrading infrastructure for solid and hazardous waste management (USEPA 2018f).

7.8.4.5 Remediation

Remediation of contaminated sites in Peru is improving but still has a long way to go to be considered effective. USEPA has been assisting Peru in establishing an effective program through the following activities (USEPA 2018f):

- Management of environmental aspects of mining
- Mine risk evaluation
- Remediation of contaminated mining sites
- Site investigation and characterization
- Conducting risk assessments
- Mapping
- Selection of remedial options
- Enforcement measures at mining sites
- Managing mercury

Mercury is utilized at many gold mines in Peru to assist in extracting gold from gold-bearing ore. Releases of mercury are common along with other heavy metals including iron, manganese, lead, copper, and zinc (USEPA 2018f). These heavy metals have not only impacted the soil but have significantly degraded surface water quality and are a significant source of air pollution, especially near major mines, including the city of La Oroya located in the Andes Mountains 176 kilometers northeast of Lima (USEPA 2018f).

7.8.4.6 Summary of Environmental Regulations of Peru

Although Peru does not have a significant population, the environmental concerns due to mining, especially copper and gold, and lack of infrastructure are significant. An additional item of note is that the environment has only become a national priority since 2008 when the Ministry of the Environment was established. Peru is making progress but the environmental damage in erosion, deforestation for farming and mining activities, air pollution, water quality degradation, and the huge future costs of environmental remediation of legacy sites will certainly present Peru with huge economic costs well into the future.

7.9 ANTARCTICA

When evaluating environmental regulations of the world, one must not exclude an entire continent in that discussion. Therefore, we shall and must discuss environmental regulations and pollution in Antarctica. Antarctica is a continent on the bottom of the world with a total land area of 14 million square kilometers with 98% of its surface covered in ice. Antarctica shares no border with any other country (see Figure 7.36).

There are no permanent settlements in Antarctica. In 1959, Antarctica was established as an international area dedicated to science and research. Several nations including Argentina, France, Japan, United Kingdom, and the United States signed the Antarctic Treaty in 1959. The treaty contains 14 articles which establish key elements related to the peaceful international coordination on the continent, some of which include (Antarctic and Southern Ocean Coalition 2018):

- Banning military intervention
- Peaceful scientific research
- International information exchange
- No territorial sovereignty

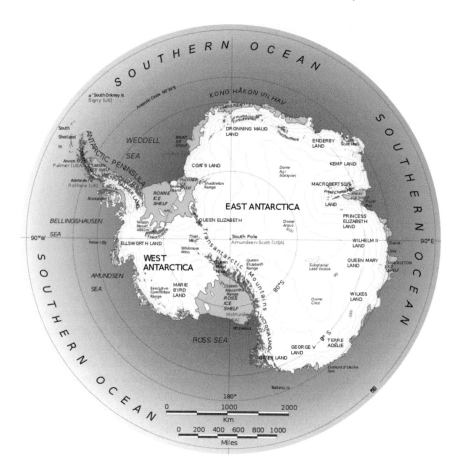

FIGURE 7.36 Map of Antarctica. (From NASA 2018. Map of Antarctica. 2018. https://nasa.gov/images/c ontnet/600827main_24.jpg.) (Accessed November 12, 2018.)

7.9.1 POLLUTION IN ANTARCTICA

Some pollution issues in Antarctica include (McConnell et al. 2014; NASA 2014):

- Ozone depletion
- Increased atmospheric CO_2 concentrations
- Solid and hazardous waste generated from research activities and tourism
- Marine pollution from anthropogenic sources such as spills from ships and cargo vessels
- Litter
- Lead
- Invasive species such as dandelions

7.9.2 ENVIRONMENTAL REGULATIONS OF ANTARCTICA

Environmental regulations in Antarctica do exist and are traced back to the Antarctic Treaty of 1959. All activities in Antarctica must be conducted in accordance with the regulations set forth by the Norwegian Polar Institute. For any activity, the following must be conducted (Norwegian Polar Institute 2018):

- Notification of planned activity
- Preparation of an environmental impact assessment

- Preparation of contingency plans
- Obtaining insurance
- List of experience and equipment
- Preparation of environmental emergency plan
- Waste handling and disposal plan
- Provisions for preventing introduction of alien species
- Preparation of a final report

7.9.3 SUMMARY OF ENVIRONMENTAL REGULATIONS IN ANTARCTICA

It should not come as a surprise after reading this chapter that there are environmental impacts even in Antarctica from anthropogenic sources. This presents an interesting and perhaps conflicting phenomenon to some of you because Antarctica is predominantly devoid of human settlement. It is essentially a wilderness of continental size. So how can Antarctica be polluted and if Antarctica is polluted what does that say about the rest of the Earth. Antarctica is perhaps the best example that pollution does not respect country borders. Evaluating Antarctica has placed a spotlight on the migration of pollution on Earth. One migration pathway that is common for pollution follows Fick's First Law of Molecular diffusion that we covered in Chapter 2 and which is in full display for us to observe and quantify in Antarctica.

7.10 OCEANS

Oceans cover 70% of the surface of the Earth or 360 million square kilometers. In many respects the oceans of the world are the last resting place for most anthropogenic pollution. As an example, an estimated 7 million metric tons of plastic is dumped into the world's oceans every year (See Figure 7.37) (NOAA 2018b). This is just a portion of the pollution that enters the oceans every year. NOAA (2018b) estimates that billions of pounds of trash and other pollutants enter the oceans each year.

7.10.1 OCEAN POLLUTION

The oceans have been the dumping ground for not just our garbage but for some of our most toxic wastes for hundreds of years. The oceans have been treated as the human privy and garbage disposal. Only in the past few decades have the dumping of wastes in the oceans come under some environmental regulations. However, polluting the oceans still continues to this day. The oceans

FIGURE 7.37 Plastic and other garbage on a beach in Hawaii. (From NOAA. Ocean Pollution. 2018b. www.noaa.gov/resource-collections/ocean-pollution.) (Accessed November 10, 2018.).

receive pollution from direct discharge or dumping, from deposition from the atmosphere, from rivers and streams that discharge into the oceans, and directly from the land itself.

The oceans were once believed to be so vast and deep that the effects of pollution would never directly threaten the oceans. This has been proven false. The rate of accumulation of pollution is proportional to human population and increasing energy requirements. The more people, the more pollution ends up in the oceans. The most common types of pollution in the oceans include plastic and other forms of garbage, heavy metals, organic pollutants such as PCBs and oil, radioactive substances and waste, biological pollutants, pesticides, herbicides, fertilizers, heavy metals, and untreated raw sewerage (NOAA 2018b).

Pollution in the oceans does not all originate from direct discharge but also is delivered through the atmosphere from increased concentrations of carbon dioxide in the atmosphere that is partially adsorbed by the oceans and causes acidification of the surface layers (NOAA 2017). Absorption of carbon dioxide by the oceans has been estimated to be as high as 25%, which represents millions of tons of carbon dioxide absorbed by the oceans every year. The additional carbon dioxide absorbed by the oceans reacts with ocean water and forms an acid called carbonic acid and is lowering the pH of the oceans world-wide. In fact, the oceans are acidifying faster than they have in some 300 million years. The effect of the acidification has been linked to bleaching of coral reefs, including the Great Barrier Reef off the northeastern coast of Australia (Natural Resource Defense Council 2018).

An additional form of pollution in the ocean that is not commonly mentioned is noise. Sound waves travel farther and faster in the oceans than they do in the atmosphere. Increased human-induced noise in the ocean is harming numerous marine species because they use sound to navigate, find food, and mate. These crucial activities for marine life survival are having widespread impacts (Natural Resource Defense Council 2018).

7.10.2   Environmental Regulations of the Oceans

The United States has enacted several laws over the last several decades aimed at protecting the Great Lakes and the oceans from pollution. As we learned in Chapter 3, the most significant federal laws addressing waterways were first enacted in the United States as far back as 1899 and include (USEPA 2018g):

- Rivers and Harbors and Refuse Act of 1899
- Coastal Zone Protection Act of 1972
- Fisheries Conservation and Management Act of 1976
- Global Climate Protection Act of 1987
- Marine Protection Act of 1972
- Marine Mammal Protection Act of 2015
- Ocean Dumping Act of 1988
- Oil Spill Protection Act of 1990

7.10.3   Summary of Environmental Regulations of the Oceans

For centuries, the oceans have been subject to the freedom of the seas doctrine, which is a principle recognized in the 17th century that limited national rights and jurisdiction over the oceans to a narrow belt of sea surrounding a nation's coastline that typically is 19.3 kilometers (United Nations 2018l). Since 1982, the United Nations has been actively involved in establishing protections for the oceans through the adoption of the Law of Sea Convention, which established (United Nations 2018l):

- Freedom of navigation rights
- Set territorial sea boundaries 19.3 kilometers offshore
- Set exclusive economic zones of up to 321 kilometers offshore
- Set rules for extending continental shelf rights up to 563 kilometers offshore

- Created the International Seabed Authority
- Created other conflict-resolution mechanisms

The United Nations is also involved in protection of the marine environment and biodiversity and reducing pollution from marine shipping, especially shipping in the polar regions, with the adoption of the Polar Code in 2014. As of 2015, there were nearly 90,000 commercial shipping vessels that carry cargo that were registered world-wide that transported nearly 10 billion metric tons of cargo (United Nations 2018l).

7.11   SUMMARY OF ENVIRONMENTAL REGULATIONS AND POLLUTION OF THE WORLD

We have now completed a review of environmental regulations of more than 50 countries and every continent including a brief assessment of Antarctica and the oceans. With respect to pollution, we have learned the most significant points are the following:

- No country, continent, or ocean is pollution free
- Pollution caused by humans is everywhere
- Significant environmental degradation caused by humans has touched every country
- Pollution does not respect country borders
- Pollution will continue to migrate in the air, water, and on land
- The oceans have been treated as the human privy and garbage disposal of the world

With respect to environmental regulations, the most significant information we have learned include the following:

- Each country has environmental regulations
- No two countries are exactly the same
- Each country has unique attributes whether it be related to its climate, geography, geology, social and economic concerns, or politics, which affect the ability of a country to protect human health and the environment that influence many aspects of environmental regulations
- The platform of environmental regulations of each country are similar to that of the United States or the European Union
- Although some countries are doing a better job than others, each country struggles with protecting human health and the environment for various reasons, two of which seem to be universal and include a growing population and urban expansion
- Every country has been significantly affected by pollution of the air, water, and land

The largest polluting activity of land world-wide is agriculture. The largest polluter of air world-wide are motor vehicles that include automobiles, diesel trucks, trains, buses, marine vessels, and aircraft. The largest polluter of water world-wide are biological pollutants largely from human waste and from agricultural activities that include pesticides and herbicides, fertilizers, and erosion (WHO 2013; 2015b; 2016a; WHO and USEPA 2012).

The final resting place for much of the pollution of the world are the oceans.

7.11.1   Evaluating and Ranking Each Country's Regulatory Effectiveness

To evaluate the effectiveness of each country's environmental regulations, we must examine several factors and assigned integers that appropriately reflect where each country is in the process and to provide an enlightened point of view as to where each country can improve.

This evaluation involves assessing several attributes that we have discussed and highlighted throughout this chapter and include whether each country:

1. Has the appropriate environmental regulations for air, water, land
2. Has appropriately implemented and enforced environmental regulations
3. Has the appropriate infrastructure to support the implementation and enforcement of environmental regulations
4. Has the economic and financial resources to support continued pollution controls
5. Has other social issues that impede or override environmental concerns such as poverty, corruption, violence, unstable political issues, population and urban constraints
6. Sustainability initiatives including pollution prevention, recycling, and waste reduction
7. Geographical, geological, climate factors, and other natural conditions such as natural disasters that impede environmental performance

To make this evaluation, we must first consider how each country has chosen to protect the air, the water, and the land under the first criterion listed above. Therefore, the regulations that are most important are associated with protecting the air, water, and land and include:

- Air quality regulations from both stationary and mobile sources
- Water quality regulations for effluent discharges from industrial sources and public-owned treatment works (POTW) and septic tank discharges
- Drinking water quality regulations
- Solid and hazardous waste regulations relating to characterization, generation, transport, and disposal
- Soil, water, and sediment cleanup regulations including investigation and polluters pay principles
- Regulating different land-use designations such as residential, commercial, industrial, and agricultural

The remaining criterion (those listed as 2 through 7 explained above) are straightforward qualitative evaluations. Table 7.6 presents the assigned values for each criterion for each country and are ranked from most effective to least effective. Each criterion in Table 7.6 corresponds to the list of 1 through 7 described above. Each criterion is assigned an integer of 0 through 3 that corresponds to the following qualitative assessment category:

A value of zero (0) indicates that the country lack appropriate controls

- A value of one (1) indicates that the country has poor controls
- A value of two (2) indicates that the country has adequate or average controls
- A value of three (3) indicates that the country has good controls

For comparison purposes, the per capita gross domestic product is listed for each country (United States Central Intelligence Agency 2018) and a ratio of GDP to the total score of each country is presented in the last column.

As you can see from analyzing Table 7.6 there should be very little surprises when each country is ranked. The countries with the most room for improvement include Turkey, Egypt, Malaysia, Russia, Saudi Arabia, and India. These countries have a host of challenges that include lack of regulations or enforcement of existing regulations, lack of infrastructure, financial hardship, population pressures, and may have to deal with corruption issues. The countries in the middle that include Mexico, Indonesia, Brazil, Argentina, South Africa, Kenya, Tanzania, Peru, China, and Chile are developing nations that are struggling to learn and enforce their environmental regulations, have population pressures, lack infrastructure, and are struggling to deal with cleanup of existing sites

TABLE 7.6
Country Comparison of Environmental Regulation Effectiveness

Country	Criteria 1	2	3	4	5	6	7	GDP Total Score	Rank[a]	GDP/Score
Turkey	1	1	0	0	0	0	2	4	69	17.25
Egypt	1	1	1	0	0	1	1	5	105	21
Malaysia	1	1	0	1	1	0	1	5	64	16
Russia	1	1	1	1	1	0	1	6	65	10.8
Saudi Arabia	1	1	1	1	1	0	1	6	20	3.3
India	1	1	0	1	1	1	1	6	132	22
Mexico	1	1	2	1	1	1	1	8	82	10.25
Indonesia	2	1	1	2	1	1	1	9	108	12
Brazil	1	1	1	2	1	1	2	9	95	10.5
Argentina	1	1	1	2	1	1	2	9	95	10.5
South Africa	2	1	2	1	1	2	1	10	103	10.3
Kenya	2	2	1	1	1	1	2	10	154	15.4
Tanzania	2	2	1	1	1	1	2	10	156	15.6
Peru	1	1	1	2	2	1	2	10	104	10.4
China	2	1	2	3	2	1	2	13	94	7.2
Chile	2	2	2	2	3	1	1	13	74	5.7
Japan	3	3	3	3	2	2	1	17	38	2.2
United States	2	3	3	3	2	3	2	18	19	1.05
Canada	2	3	3	3	3	2	2	18	32	1.7
Australia	2	3	3	3	3	2	2	18	27	1.5
New Zealand	3	3	3	3	3	2	2	19	44	2.3
European Union	3	3	3	3	3	3	2	20	43	2.1
Switzerland	3	3	3	3	3	3	2	20	17	0.85
Norway	3	3	3	3	3	3	2	20	12	0.6
South Korea	3	3	3	3	2	3	3	20	42	2.1

[a] United States Central Intelligence Agency. 2018. The World Fact Book. www.cia.gov/library/publi cations/the-world-factbook.gos/NA.html. (accessed October 10, 2018).

of contamination. The countries that rank the highest include Japan, the United States, Canada, Australia, New Zealand, European Union, Switzerland, Norway, and South Korea are all developed nations with financial resources to address environmental issues but still struggle with issues such as population and urbanization, and air and water quality, but have all made significant progress compared to other nations.

Examining the final column in Table 7.6 provides perhaps the most telling information. The countries that rank the highest have a ratio of GDP to total score of less than 2.5. The countries that have the most room for improvement all have a ratio greater than 10, except for Saudi Arabia. Saudi Arabia ranks poorly for environmental controls but high for per capita GDP, which indicates that Saudi Arabia has financial resources to improve its environmental controls but chooses to focus their financial resources elsewhere. Egypt and India are the only countries evaluated that have a total score to GDP ratio greater than 20. China and Chile are unique when examining their ranking and numerical scores. China and Chile rank in between the low performance and high performance. Knowing that China has been improving its environmental controls indicates that China is transitioning from low to high. Chile appears to be currently more stagnant.

7.11.2   Environmental Challenges of Each Country

Some of the challenges that each country face that have significantly influenced the score of each country that have been described in this chapter include:

- Turkey is currently facing financial and economic hardship along with a lack of robust environmental standards and enforcement
- Egypt is experiencing political unrest, a growing population and urbanization, lack of infrastructure, and water shortages
- Malaysia lacks infrastructure, environmental enforcement, financial resources, and has a rapid population growth
- Russia lacks infrastructure, environmental enforcement, financial resources, political framework, and struggles with corruption
- Saudi Arabia lacks robust regulations and enforcement, some lack of infrastructure but does have financial resources
- India has large population pressures, lacks infrastructure and enforcement of regulations
- Mexico lacks infrastructure, especially with water treatment, and struggles with corruption and lack of complete regulations (i.e., stormwater) and enforcement struggles
- Indonesia has pressure from a large population, newer regulations, deforestation, and lack of infrastructure
- Brazil has population pressure, especially in urban areas, infrastructure, lack of progress in cleaning up contaminated sites, water treatment, deforestation, and corruption
- Argentina lacks infrastructure, which places pressure on enforcement, financial resources, water treatment, and cleanup of contaminated sites
- South Africa has a degree of political unrest and financial pressures, climate change, and water issues
- Kenya has financial pressures and lack of infrastructure
- Tanzania has lack of financial and infrastructure resources
- Peru has lack of infrastructure to deal with water contamination issues, especially from mining, and lacks adequate financial resources
- China lacks enforcement measures (but improvements are in the making) and has population challenges and infrastructure to deal with pollution control measures, especially in the more undeveloped western regions
- Chile has urban population challenges and water issues in the desert regions of the northern portion of the country
- Japan has geographical and geologic challenges (i.e., island nation that is prone to earthquakes), population pressure, and lack of landfill space
- United States faces lack of regulations of agricultural areas and has significant issues related to significant chemical use of pesticides and herbicides, erosion, deforestation, pollution from septic systems, and invasive species
- Canada lacks some stormwater controls, and faces habitat destruction from mining, deforestation, and petroleum development
- Australia faces water issues, population pressure, and habitat destruction, especially the Great Barrier Reef, which it can do little at correcting
- New Zealand faces water quality challenges, especially from agriculture and livestock sources
- European Union faces continued challenges from cleaning up historical sites of contamination and population pressure but has done well with regulations and sustainability measures
- Switzerland has erosion challenges and pollution from agriculture and livestock
- Norway faces climate change challenges and pollution from agriculture and livestock
- South Korea faces population pressures and air quality challenges, especially in its capital city of Seoul

A final point to consider is the objectivity of the evaluation conducted through the selection of countries included in this chapter. As stated at the beginning of this chapter, the countries selected include 70% of the human population and 75% of the land area. The countries not selected include countries such as Somalia, Algeria, Libya, Sudan, Democratic Republic of Congo, Angola, Zimbabwe, Botswana, Mozambique, Congo, Syria, Iran, Pakistan, Afghanistan, Venezuela, and many former Soviet bloc countries. These countries were not evaluated because of the more severe social and economic factors, poverty, sickness, and violence they are experiencing, which has degraded the natural environment and directly threaten human health. Therefore, when considering environmental degradation and threats to human health and the environment from pollution, the countries examined in this chapter are biased toward countries with better environmental performance.

7.11.3   SUMMARY AND CONCLUSION

Antarctica and the Oceans have and will play a significant role concerning measuring future sustainability efforts and survival of our species and many other species on Earth. We have learned thus far that Antarctica and the oceans have become polluted from direct discharge and from migration of contaminates through the air, water, and land. We have also learned that pollution migrates following fundamental laws of physics and that fact is evident by examining the impacts observed in Antarctica and the Oceans. We have also learned that pollution does not respect boundaries whether they be political or media-specific.

The oceans were the origin of life on Earth and remain the prime source of nutrients and food that sustains life on our planet but have become polluted by our own actions. If not for any other reason than our own survival, we must be diligent at assessing the risks posed by pollutants, enact appropriate protections, and act on those protections to prevent further degradation on a global scale.

To summarize this chapter is to state that we must now accept the reality that humans have adversely impacted the entire Earth and that our efforts to improve our environment since the enactment of environmental regulations in the United States and world-wide have simply not been enough. Earth scientists are now convinced that out of this fact, we have now moved the needle of geologic time into a new period called the Anthropocene (Smithsonian 2018; Zalasiewicz et al. 2018; International Union of Geological Sciences 2016). This is significant because the definition of a Geologic Age is a time period that affects the entire Earth and will be recorded in Earth history and cannot be reversed.

Now that we have described pollution and its behavior and examined how numerous countries have enacted environmental regulations and have struggled with their own pollution issues, we will now turn our attention to how to maintain compliance with environmental regulations. The next chapter outlines an approach to achieve and maintain environmental compliance anywhere in the world by using a rather simple approach that can be summed up in just a few words, "no matter where you are, it's all the same."

REFERENCES

Agarwal, V.K. 2015. Environmental Law in India: Challenges for Enforcement. *Bulletin of the India National Institute of Ecology.* Vol. 15. pp. 227–238. Delhi, India.

Akkoyuniu, A., Avsar, Y., and Erguven, G.O. 2017. Hazardous Waste Management in Turkey. *Journal of Hazardous, Toxic, and Radioactive Waste.* Vol. 21. No. 4. p. 7.

Al-Omran, A., Al-Barakah, F., Altuquq, A., Aly, A., and Nadeem, M. 2014. Drinking Water Quality Assessment and Water Quality Index of Riyadh, Saudi Arabia. *Water Quality Research Journal of Canada.* Vol. 46. No. 1. pp. 287–296.

Anderson, Sean. 2011. Turkey's Globally Important Biodiversity Crisis. *Journal of Biological Diversity.* Vol. 144. pp. 2752–2769. Elsevier Publishers. New York.

Antarctic and Southern Ocean Coalition. 2018. *Antarctic Environmental Protection.* www.asoc.org/advocacy/antarctic-environmental-protection. (accessed May 11, 2018).

Arctic Center. 2017. *Environmental Impact Assessment Processes in Northwest Russia.* www.articcenter.org/RussianEIA/process. (accessed September 14, 2017).

Argentine Secretariat for the Environment. 2018a. *Ambient Air Standards. Argentina Law 20.284.* www.fmed.uba.ar/depto/toxicol/plaguicidas/20.284. (accessed October 14, 2018).

Argentina Secretariat for the Environment. 2018b. *Argentina Water Management Regulations Law 25.688.* www.mrecic.gov.ar/water. (accessed October 14, 2018).

Argentina Secretariat for the Environment. 2018c. *National Definition of Waste Under Law 24.051.* www.mrecic.gov.ar/waste. (accessed October 14, 2018).

Asia Society. 2018. *Korean History and Geography.* https://asiasociety.org/education/korean-history-and-political-geography. (accessed May 1, 2018).

Australia Department of Environment. 2009. *National Waste Policy. Environmental Protection.* Heritage Council. Canberra, Australia. 22p. www.nepc.gov.au/node/849. (accessed October 13, 2018).

Australia Department of Environment. 2016. *Australian National Waste Report.* Department of Environment. Canberra, Australia. 74p. www.environment.gov.au/national-waste-report-2016. (accessed October 13, 2018).

Australia Department of Environmental and Conservation. 2010. *Assessment Levels for Soil, Sediment and Water.* Department of Environment and Conservation. Canberra, Australia. 56p. www.environment.gov.au/site-cleanup.pdf. (accessed October 13, 2018).

Australia Department of Environment and Energy. 2018a. *Publications and Resources.* www.environment.gov.au/about-us/publications. (accessed October 12, 2018).

Australia Department of Environment and Energy. 2018b. *Air Pollutants in Australia.* www.environment.gov.au/airpollution. (accessed October 12, 2018).

Australia Department of Environment and Energy. 2018c. *Ambient Air Quality Standards.* www.env.gov.au/protection/air-quality-standards. (accessed October 13, 2018).

Australia Department of Environment and Energy. 2018d. *Guidelines for Risk Assessment of Wastewater Discharges to Waterways.* Environmental Protection Agency of Victoria, Australia. 36p. www.epa.vic.gov.au. (accessed October 13, 2018).

Australia Department of Environment and Energy. 2018e. *Solid Waste Classification System.* www.env.gov.au/protection/waste-class.pdf. (accessed October 13, 2018).

Australia Department of Environment and Energy. 2018f. *The Waste and Recycling Industry in Australia.* Report to Australian Government. Canberra, Australia. 152p.

Australia Maritime Safety Authority. 2018. *Discharge Standards.* www.amsa.gov.au. (accessed October 13, 2018).

Australia National Health and Medical Research Council. 2018. *Australian Drinking Water Guidelines.* Canberra, Australia. 1172p. www.nhmrc.gov.au. (accessed October 13, 2018).

Barczewski, B. 2013. *How Well Do Environmental Regulations Work in Kenya? A Case Study.* Center for Sustainable Urban Development. Columbia University. New York. 28p.

Basurto, D. and Soza, R. 2007. Mexico's Federal Waste Regulations. *Journal of the Air and Waste Management Association.* Vol. 57. pp. 7–10.

Baum, R., Luh, J., and Bartman, J. 2013. Sanitation: A Global Estimate of Sewerage Connections without Treatment and the Resulting Impact. *Journal of Environmental Science and Technology.* Vol. 47. No. 4. pp. 1994–2000.

Berkun, M., Aras, E., and Amlan, T. 2011. Solid Waste Management in Turkey. *Journal of Materials and Waste Management.* Vol. 13. No. 1. pp. 305–313.

Blinnikiv, M. 2011. *A Geography of Russia and Its Neighbors.* The Guilford Press. London, UK. 425p.

Bowers, S. 2011. *Swiss Top of the Rich List.* The Guardian. London, UK. www.Theguardian.com/features/2011qualityoflife. (accessed September 16, 2017).

Brain, S. 2016. *Environmental History of Russia.* Oxford Research Encyclopedias. Oxford University Press. Oxford, UK. 237p.

Brazil Ministry of the Environment. 1990. *CONAMA Resolution No. 3. Standards on Air Quality.* www.brazilgovnews.gov.br/standards-on-air-quality.htm. (accessed October 18, 2018).

Brazil Ministry of the Environment. 2018a. Ministry of the Environment. www.brazilgovnews.gov.br/ministry-of-the-environment.htm. (accessed October 18, 2018).

Brazil Ministry of the Environment. 2018b. *National Solid Waste Law of 2010.* www.wiego.gov.br/wast-law.htm. (accessed October 18, 2018).

Broughton, E. 2005. The Bhopal Disaster and Its Aftermath: A Review. *Journal of Environmental Health.* Vol. 4. No. 6. United States Institute of Health. Washington, DC. www.ncbi.nlm.nih.gov/pmc/articles/PMC1142333. (accessed October 1, 2018).

Canada Water. 2019. *Sustainable Stormwater Management in Canada.* www.watercanada.net/feature/sustain able-stormwater-management. (accessed February 10, 2019).

Canadian Council of Ministers of the Environment. 2019. *Canada-wide Strategy for the Management of Wastewater Effluent.* www.ccme.ca/files/Resources/minicipal_wastewater_effluent/cda_wide_Strategy _mwwe_final_e.pdf. (accessed February 10, 2019).

Canadian Environmental Protection Act. 1999. *Canadian Environmental Protection Act.* www.ec.gc.ca/ Lcpe-cepa. (accessed February 10, 2019).

Canadian Government. 2017. *Canada: A Brief Overview.* www.cic.gc.ca/english/newcomers. (accessed April 5, 2017).

Canadian Government. 2019a. *Air Quality Strategy for the Canada-US Border.* https://canada.ca/en/envi ronment-climate-change/issues/quality-stragey-canada-united-states-border.html. (accessed February 10, 2019).

Canadian Government. 2019b. *Canadian Air Quality.* www.canada.ca/en/environment-climate-change/servic es/environmental-indicators/air-quality.html. (accessed February 10, 2019).

Canadian Government. 2019c. *Regulatory Framework for Air Emissions.* www.ec.ge.ca/doc/media_124/rep ort_eng.pdf. (accessed February 10, 2019).

Canadian Government. 2019d. *Overview of Wastewater Regulations in Canada.* http://canada.ca/en/envir onment-climate-change/services/wastewater/regulations.html. (accessed February 10, 2019).

Chakibi, S. 2013. Saudi Arabia Environmental Laws. *Journal of Environmental Health and Safety.* http://ehs journal.org/sanaa-chibi/saudiarabialaws. (accessed October 5, 2018).

Chikanda, A. 2009. *Environmental Degradation in Sub-Saharan Africa. In Environment and Health in Sub-Saharan Africa: Managing an Emerging Crisis.* Springer Publishing. New York. pp. 79–94.

Chile Ministry of the Environment. 2018a. *Overview and USEPA Cooperation.* http://protal.mma.gob.cl/. (accessed October 28, 2018).

Chile Ministry of the Environment. 2018b. *Chile Division of Air and Climate Change.* http://portal.mma.gob. cl.aire/. (accessed October 28, 2018).

Chile Ministry of the Environment. 2018c. *Chile Division of Solid and Hazardous Waste.* http://portal.mma. gob.cl.waste/. (accessed October 28, 2018).

China Ministry of Ecology and Environment (MEE). 2018. *Law of the People's Republic of China on the Prevention and Control of Environmental Pollution by Solid Waste.* Beijing, China, www.china.org.cn/ english/environment/34424.htm. (accessed September 24, 2018).

China Ministry of Environment Protection (MEP). 2014. *Five Standards of Environmental Protection.* MEP Notice No. 14 of 2014. Beijing, China. 98p.

China Ministry of Environmental Protection. 2018. *New Law on Soil Pollution Prevention.* www.china.org.c n/china/2018-09/01/content61417828. (accessed September 25, 2018).

China Research Academy of Environmental Sciences. 2013. *Groundwater Challenges and Environmental Management in China.* Ministry of Environmental Protection. Beijing, China. 28p.

Choudhary, S. 2017. *Drinking Water in Indonesia.* Jakarta, Indonesia. www.indoindians.com/jakartadrinking water. (accessed October 7, 2018).

Clean Air Alliance of China (CAAC). 2015. *China Air Quality Management Assessment Report.* CAAC Secretariat. Beijing, China. 28p. www.iccs.org.cn. (accessed September 3, 2018).

Coble, C.R., Murray, E.G., and Rice, D.R. 1987. *Earth Science.* Prentice-Hall Publishers. Englewood Cliffs, NJ. 502p.

Conservation International. 2018. *Indonesia: A Vast and Beautiful Country – At the Crossroads.* https://co nservation.org/where/pages/indonesia. (accessed October 5, 2018).

Coolaustralia.org. 2018a. *Water Pollution.* www.coolaustralia.org/water-pollution-primary. (accessed October 13, 2018).

Coolaustralia.org. 2018b. *Land Pollution.* www.coolaustralia.org/land-pollution-primary. (accessed October 13, 2018).

Delaware River Watershed Initiative. 2019. *Delaware River Watershed Initiative.* https://4states/source.org/ about. (accessed February 10, 2019).

Diaz-Robles, L. Saavedra, H., Schiappacasse, L., and Cereceda-Balic, F. 2011. The Air Quality in Chile. *Journal of the Air and Waste Management Association.* Vol. 33. pp. 26–33.

Egyptian Environmental Affairs Agency (EEAA). 2018. *Environmental Protection Law.* www.eeaa.gov.eg/ en-us/laws/envlaw.aspx. (accessed August 30, 2018).

Environment Canada. 2016. *Overview of Environmental Act, Regulations and Agreements.* www.ec.gc.ca/ environmental-acts. (accessed April 6, 2017).

Environmental Health. 2017. *Air Pollution and Health in Turkey.* Health and Environment Alliance (HEAL). Brussels, Belgium. www.env-health.org/imf/pdf/150220. (accessed September 19, 2017).

European Environment Agency. 2013. *Turkey Air Pollution Fact Sheet.* www.eea.europe.eu/themes/air/air-po llution-fact-sheets. (accessed December 14, 2017).

European Environment Agency. 2014. *Air Pollution Country Fact Sheet: Turkey.* www.eea.europe.eu/themes/ air/air-pollution-fact-sheets. (accessed September 19, 2017).

European Environment Agency. 2015a. *Environmental Status of Switzerland.* www.eea.europa.eu/switzer-land. (accessed September 16, 2017).

European Environment Agency. 2015b. *Turkey Country Briefing – The European Environment – State and Outlook 2015.* https://eea.europe.eu/soer-2015/countries/turkey. (accessed August 10, 2018).

European Environment Agency (EEA). 2019a. *Overview of Environmental Regulations within the European Union.* www.eea.europa.eu/overview. (accessed February 10, 2019).

EEA. 2019b. *Air Emissions Policy Context.* www.eea.europa.eu/themes/air/policy-context. (accessed February 10, 2019).

EEA. 2019c. *European Union Air Quality Standards.* https://ec.europa/environment/air/quality/standards.htm. (accessed February 10, 2019).

EEA. 2019d. *European Air Permitting Directives and Procedures.* https://ec.europa/air/permitting. (accessed February 10, 2019).

EEA. 2019e. *Drinking Water Standards.* https://ec.europa/drinkingwater. (accessed February 10, 2019).

EEA. 2019f. *Introduction to the European Union Water Framework Directive.* https://ec.europa.eu/environme nt/water/water-framework/information. (accessed February 10, 2019).

EEA. 2019g. *Water Pollution Framework in the European Union.* www.ec.europa/waterframework. (accessed February 10, 2019).

EEA. 2019h. *Resource Efficiency and Waste.* www.eea.europa.eu/themes/waste. (accessed February 10, 2019).

European Environment Agency. 2019i. *Hazardous Waste.* http://ec.europa.eu/environment/waste/hazardo us_index.html. (accessed February 10, 2019).

EEA. 2019j. *Soil Contamination in Europe.* www.eu.europa.soil-contamination. (accessed February 10, 2019).

European Union. 2019. *European Countries in Brief.* www.eu./european-union/about-eu/countries/member-c ountries_eu. (accessed February 10, 2019).

Health Canada. 2017. *Guidelines for Canadian Drinking Water Quality.* Ottawa, Canada. 18p. www.health-canada.gc.ca/waterquality. (accessed February 10, 2019).

Henry, L.A. and Douhovnikoff, V. 2008. Environmental Issues in Russia. *Journal of Annual Review of Environmental Resources.* Vol. 33. No. 1. pp. 437–460.

India Central Pollution Control Board. 2009. *National Ambient Air Quality Standards.* New Delhi. India. 48p. www.cpcb.nic.in/upload/latest_final_air_standard.pdf. (accessed September 28, 2018).

India Ministry of Environment, Forestry, and Climate Change. 2018a. *Environmental Laws. Air Pollution Control Act of 1981.* www.environmentallawsofindia.com/air/pollution/act. (accessed October 1, 2018).

India Ministry of Environment, Forestry, and Climate Change. 2018b. *Environmental Laws.* www.enviro nmentallawsofindia.com/water/pollution/act. (accessed October 1, 2018).

India Ministry of Environment, Forestry, and Climate Change. 2018c. *Effluent Water Discharge Criteria for Industry.* www.envfor.nic.in/decision/water-pollution. (accessed October 1, 2018).

India Ministry of Environment, Forestry, and Climate Change. 2018d. *Solid and Hazardous Waste Regulations.* www.envfor.nic.in/decision/solid-and-hazardous-waste. (accessed October 1, 2018).

India Ministry of Environment, Forestry, and Climate Change. 2018e. *Environmental Impact Assessment.* www.envfor.nic.in/decision/environmental-impact-assessment. (accessed October 1, 2018).

Indonesia Ministry of Environment. 2018a. www.aecon.org/indonesia. (accessed October 5, 2018).

Indonesia Ministry of Environment. 2018b. *Ambient Air Quality Standards.* www.moe.indonesia.org/amb ientairstandards. (accessed October 5, 2018).

Institute for European Environmental Policy. 2019. Institute for European Environmental Policy. https://ieep. eu/. (accessed February 10, 2019).

International Union of Geological Sciences (IUGS). 2016. *The Anthropocene Epoch: Adding Humans to the Chart of Geologic Time.* International Geological Congress. Cape Town, South Africa. www.35igc.org/ verso/5/scientific-programme. (accessed November 11, 2018).

Japan Fact Sheet. 2018. *Environmental Issues: Progress Made but New Challenges Must Be Faced.* www. web-Japan.org/environmental/issues. (accessed September 27, 2018).

Japan Ministry of the Environment. 2018a. *The Basic Environmental Law and Basic Environmental Plan.* www.env.go.jp/en/laws/policy/basic_lp.html. (accessed September 27, 2018).

Japan Ministry of the Environment. 2018b. *Air Pollution Control Act.* Tokyo, Japan. www.env.go.jp/en/laws/ policy/basic_lp.html. (accessed September 27, 2018).

Japan Ministry of the Environment. 2018c. *Environmental Air Quality Standards for Ambient Air in Japan.* http://env.go.jp/en/air/aq/aq.html. (accessed September 27, 2018).

Japan Ministry of the Environment. 2018d. *Japanese Environmental Pollution Experience.* www.env.go.jp/en/coop/experience.html. (accessed September 28, 2018).

Japan Ministry of the Environment. 2018e. *State of Japan's Environment at a Glance: Water Pollution.* www.env.go.jp/en/water/wq/pollution. (accessed September 27, 2018).

Japan Ministry of the Environment. 2018f. *Environmental Issues and Pollution in Japan.* www.env.go.jp/en/pollution/issues. (accessed September 27, 2018).

Japan Ministry of the Environment. 2018g. *Soil Contamination Countermeasures Act of 2002.* www.env.go.jp/en/soilact. (accessed September 27, 2018).

Japan Ministry of the Environment. 2018h. *History and Current State of Waste Management in Japan.* Waste Management and Recycling Department. Tokyo, Japan. 31p.

Japan Ministry of the Environment. 2018i. *Waste and Recycling Standards for Verification.* www.env.go.jp/en/recycle/manage/sy.html. (accessed September 27, 2018).

Japan Ministry of the Environment. 2018j. *Soil Pollution Control Values for Soil Leachate.* www.env.go.jp/en/soil/leachate/html. (accessed September 27, 2018).

Japan Ministry of the Environment. 2018k. *Ambient Water Quality Standard for Groundwater and Criteria for Implementation of Soil Pollution Control Measures.* Water Pollution Control Law. www.env.go.jp/en/spcl/html. (accessed September 27, 2018).

Kenya Air Quality Act. 2009. National Environmental Management Authority. Nairobi, Kenya. www.nema.go.ke/NEMA/Airqualityact. (accessed August 22, 2018).

Kenya Geological Society. 2018. *Summary of the Geology and Minerals Resources of Kenya.* www.apsea.or.ke/index/geological-society-of-kenya. (accessed August 18, 2018).

Kenya Impact Assessment and Audit Act. 2003. National Environmental Management Authority. Nairobi, Kenya. www.nema.go.ke/NEMA/Impactassessmentandauditact. (accessed August 22, 2018).

Kenya National Bureau of Statistics. 2018. *Estimated Population of Kenya in 2018.* www.knbs.or.ke/. (accessed August 21, 2018).

Kenya National Environment Management Authority. 2018. *Kenya State of the Environment and Outlook 2010.* www.nema.go.ke/downloads. (accessed August 22, 2018).

Kenya Waste Management Act. 2006. National Environmental Management Authority. Nairobi, Kenya. www.nema.go.ke/NEMA/Wastemanagementact. (accessed August 22, 2018).

Kenya Water Quality Act. 2006. National Environmental Management Authority. Nairobi, Kenya. www.nema.go.ke/NEMA/Waterqualityact. (accessed August 22, 2018).

Kenya Wildlife Service. 2018. *National Parks and National Wildlife Reserves in Kenya.* www.go.ke/national-parks. (accessed August 22, 2018).

Kichira, S. and Sugai, M. 1986. *The Origins of Environmental Destruction: The Ashio Copper Mine Pollution Case.* United Nations University. Shinyousha, Tokyo. 167p. www.unu.edu.ashiocoppermine. (accessed September 28, 2018).

Kinny, P.L., Gatari, M., Volavka-Close, N., Ngo, N., Ndiba, P.K., Law, A., Gachanja, A., Mwaniki, S., Chillrud, S.N., and Sclar, E. 2011. Traffic Impacts on PM2.5 Air Quality in Nairobi, Kenya. *Journal of Environmental Science and Policy.* Vol. 14. No. 4. pp. 369–378.

Korea Ministry of the Environment. 2018a. *Air Quality Standards and Air Pollution Levels.* www.eng.me.go.kr/eng/web/index.do?menuId+252. (accessed October 2, 2018).

Korea Ministry of the Environment. 2018b. *State of Water Environmental Issue: Republic of Korea.* www.wepa-db.net/policies/state/southkorea/southkorea.htm. (accessed October 4, 2018).

Korea Ministry of the Environment. 2018c. *Drinking Water Management: Drinking Water Quality Standards and Water Quality Monitoring.* www.me.go.kr/eng/web/index.do. (accessed October 4, 2018).

Korea Ministry of the Environment. 2018d. *Expansion of the Total Water Pollution Load Management System (TPLMS).* www.me.go.kr/eng/web/indes.do?menuId=274. (accessed October 4, 2018).

Korea Ministry of the Environment. 2018e. *Enforcement Decree of the Water Quality and Aquatic Ecosystem Conservation Act of 2015.* www.law.go.kr/eng/web/law/waterquality. (accessed October 4, 2018).

Korea Ministry of the Environment. 2018f. *Wastes Control Act of 2016: General Provisions.* www.law.go.kr/DRF/law/wastecontrolact. (accessed October 4, 2018).

Korea Ministry of the Environment. 2018g. *Hazardous Waste Management Flowchart.* www.eng.me.go.kr./eng/web/index.do?menuId=379. (accessed October 4, 2018).

Korea Ministry of the Environment. 2018h. *Korea Waste Prevention and Recycling Policy.* www.eng.me.go.kr/eng/web/index.do?menuId=141. (accessed October 4, 2018).

Korea Ministry of the Environment. 2018i. *Liability, Compensation and Relief System for Damages from Environmental Pollution.* www.eng/me/go.kr/eng/web/law/Actno.166922. (accessed October 4, 2018).

Korea Ministry of the Environment. 2018j. *Soil and Groundwater Contamination Prevention and Restoration.* www.eng.me.go.kr/eng/web/index.do?menuld=311. (accessed May 4, 2018).

Lijzen, J.P., Baars, A.J., Otte, P.F., Rikken, M.G., Swartjes, F.A., Verbruggen, E.M., and van Wezel, A.P. 2001. *Technical Evaluation of the Intervention Values for Soil/Sediment and Groundwater.* Institute of Public Health and the Environment. The Netherlands. 147p.

Lokahita, B. 2017. *Indonesia Municipal Solid Waste: Regulations and Common Practice.* Tokyo Institute of Technology. www.academia.edu/23109258/indonesia/solidwasteregs. (accessed October 7, 2018).

Loomis, D., Castillejos, M., Gold, D.R., McDonnell, W., and Borja-Aburto, V.H. 1999. Air Pollution and Infant Mortality in Mexico City. *Journal of Epidemiology.* Vol. 10. No. 2. pp. 118–123.

Macedo, A.A. and Sant'ana, L.P. 2015. Managing Emerging Hazardous Wastes in Brazil. *Unisanta Science and Technology.* ISSN 2317-1316. pp. 51–54.

Mage, D., Ozolins, G., Peterson, P., Webster, A., Orthofer, R., Vandeweerd, V., and Gwynne, M. 1996. Urban Air Pollution in Megacities of the World. *Journal of Atmospheric Environment.* Vol. 30. No. 5. pp. 681–686.

Malaysia Department of Environment. 2015. *Environmental Issues and Challenges – Malaysian Scenario.* www.doe.gov.my/environmentalissues. (accessed October 7, 2018).

Malaysia Department of Environment. 2018a. *Malaysia Ambient Air Quality Standards.* www.doe.gov.my/portelv1/info-umum/english-ambient-air-standards. (accessed October 7, 2018).

Malaysia Department of Environment. 2018b. *Malaysian Air Pollution Index. Official Portal of Department of State.* www.doe.gov.my/portelv1/info-umum/englisk-api. (accessed October 7, 2018).

Malaysia Department of Environment. 2018c. *Water Pollution and Water Quality Standards.* www.doe.gov.my/waterquality. (accessed October 7, 2018).

Malaysia Department of Environment. 2018d. *Malaysia Sewerage and Industrial Effluent Discharge Standards.* www.doe.gov.my/discharge-standards/malaysia.htm. (accessed October 7, 2018).

Marey, H.S., Gille, J.C., El-Askary, H.M., Shalaby, E.A., and El-Raey, M.E. 2010. Study of the Formation of the "Black Cloud" and Its Dynamics Over Cairo, Egypt Using MODIS and MISR Sensors. *Journal of Geophysical Research.* Vol. 115. No. 10. pp. 1029–1048.

Marsh, W.B. and Kaufman, M.M. 2015. *Physical Geography.* University of Cambridge Press. Cambridge, UK. 647p.

Marsh, W.B. and Grossa, M. 2012. *Environmental Geography.* John Wiley and Sons. New York. 426p.

McConnell, J.R., Maselli, O.J., Sigl, M., Vallelongo, P., Neumann, T., Anschutz, H., Bales, R.C., Curran, M.A., Edwards, R., Das, S.B., Kipfstuhl, S., Layman, L., and Thomas, E.R. 2014. Antarctic-wide Array of High-Resolution Ice Core Records Reveals Pervasive Lead Pollution Began in 1889 and Persist Today. *Scientific Reports.* Vol. 4. No. 5848. p. 5.

Mexico Environmental and Natural Resource Ministry (SAMARNAT). 1997. *Official National Commission on Water.* www.Mexico.nom-003-SEMARNAT-1997. (accessed August 30, 2017).

Mexico Environmental and Natural Resource Ministry (SAMARNAT). 2004. *Official Mexican Remediation Objective Concentrations for Heavy Metals.* www.Mexicana.nom-147-SEMARNAT/ss-2003. (accessed August 10, 2017).

Mexico Environmental and Natural Resource Ministry (SAMARNAT). 2005. *Official Mexican Remediation Objective Concentrations for Hydrocarbons.* www.Mexicana.nom-138-SEMARNAT/ss-2003. (accessed August 10, 2017).

Mexico Environmental and Natural Resource Ministry (SAMARNAT). 2015. *Official Mexican Remediation Objective Concentrations for Hydrocarbons.* www.Mexicana.nom-133-SEMARNAT/ss-2003. (accessed August 10, 2017).

Mexico Environmental and Natural Resource Ministry (SAMARNAT). 2016. *Official National Commission on Water.* www.Mexico.nom-003-SEMARNAT-2016. (accessed August 30, 2017).

Mexico Environmental and Natural Resource Ministry (SAMARNAT). 2017. *Ambient Air Quality Standards for Criteria Pollutants.* www.MexicoAirQualityStandards. (accessed June 17, 2017).

Mexico Environmental and Natural Resource Ministry (SAMARNAT). 2019a. *Mexico's Waste Law.* www.gob.mx/semarnat. (accessed February 10, 2019).

Mexico Environmental and Natural Resource Ministry (SAMARNAT). 2019b. *Remediation of Contaminated Sites in Mexico.* www.gob.mx/semarnat. (accessed February 10, 2019).

Mexico National Water Commission (CONAGUA). 2019. *Water Use in Mexico.* www.gob.mx/CONAGUA. (accessed February 10, 2019).

Mwambazambi, Kalemba. 2010. Environmental Problems in Africa: A Theological Response. *Ethiopian Journal of Environmental Studies and Management.* Vol. 3. No. 2. pp. 54–66. University of South Africa. Pretoria, South Africa.

National Aeronautics and Space Administration (NASA). 2014. *Lead Pollution Beats Explorers to the South Pole.* www.nasa.gov/content/goddard/lead-pollution-in-antarctica. (accessed November 10, 2018).

NASA. 2018. *Map of Antarctica.* https://nasa.gov/images/contnet/600827main_24.jpg. (accessed November 12, 2018).

National Geographic Society. 2017. *China's Solutions to Its Air Pollution.* www.news.nationalgeographic.com/2017/05/china0air-pollution-solutions-environment. (accessed September 3, 2018).

National Geographic Society. 2018a. *Australia and Oceania: Physical Geography.* www.nationalgeographic.org/encyclopedia/oceania-physical-geography. (accessed October 12, 2018).

National Geographic Society. 2018b. *Corals Are Dying on the Great Barrier Reef.* https://news.nationalgeographic.com/2016/03/160321-coral-bleaching-great-barrier-reef.htm. (accessed October 13, 2018).

National Geographic Society. 2018c. *Threats to the Amazon Rainforest.* www.nationalgeographic.com/environment/habitats/rainforest-threats. (accessed October 18, 2018).

National Intelligence Council. 2012. *The Environmental Outlook in Russia.* www.dni.org/nic/special_russianoutlook.html. (accessed August 31, 2017).

National Oceanic and Atmospheric Administration (NOAA). 2017. *DSCOVR: Deep Space Climate Observatory.* www.nesdic.naoo.gov/drcovr-deep-space-climate-observatory. (accessed December 14, 2017).

National Oceanic and Atmospheric Administration (NOAA). 2018a. *Koppen-Geiger Climate Changes.* https://sos.noaa.gov/datasets/koppen-geiger-climate-changes. (accessed February 10, 2019).

NOAA. 2018a. *Koppen-Geiger Climate Changes.* https://sos.noaa.gov/datasets/koppen-geiger-climate-changes. (accessed February 10, 2019).

NOAA. 2018b. *Ocean Pollution.* www.noaa.gov/resource-collections/ocean-pollution. (accessed November 10, 2018).

Natural Resources Defense Council (NRDC). 2018. *Ocean Pollution: The Dirty Facts.* https://nrdc.org/stories/ocean-pollution-dirty-facts. (accessed November 18, 2018).

Netherlands Ministry of Infrastructure and the Environment. 2015. *Dutch Pollutant Standards.* www.government.nl/ministry-of-infrastructure-and-the-environment or at www.sanaterre.com/guidelines/dutch.html. (accessed February 17, 2019).

Newell, J.P. and Henry, L.A. 2017. The State of Environmental Protection in the Russian Federation: A Review of the Post-Soviet Era. *Journal of Eurasian Geography and Economics.* DOI: 10.1080/15387216.2017.1289851. (accessed September 17, 2017).

New Zealand Ministry for the Environment. 2018a. *Environmental Management in New Zealand.* www.mfe.nz/publications/environmental-reporting. (accessed October 14, 2018).

New Zealand Ministry for the Environment. 2018b. *Measuring New Zealand's Progress.* http://archive.stats.govt.nz/sustainability-progress. (accessed October 14, 2018).

New Zealand Ministry for the Environment. 2018c. *New Zealand Air Discharge Permit Application.* www.gw.govt.nz/Form-5a. (accessed October 14, 2018).

New Zealand Ministry for the Environment. 2018d. *Drinking Water for New Zealand.* www.drinkingwater.org.nz. (accessed October 14, 2018).

New Zealand Ministry for the Environment. 2018e. *Drinking Water Standards.* www.mfe.nz/publications/drinking-water. (accessed October 14, 2018).

New Zealand Ministry for the Environment. 2018f. *Wastewater Discharges.* www.mfe.nz/publications/wastewater-discharges. (accessed October 14, 2018).

New Zealand Ministry for the Environment. 2018g. *Hazardous Waste Management in New Zealand.* www.oag.govt.nz/2007/cg-2005-06/part11. (accessed October 14, 2018).

New Zealand Ministry for the Environment. 2018h. *Waste List.* www.mfe.govt.nz/waste/guidance-and-resources. (accessed October 14, 2018).

New Zealand Ministry for the Environment. 2018i. *Waste and Resources.* www.mfe.govt.nz/waste. (accessed October 14, 2018).

New Zealand Ministry for the Environment. 2018j. *Contaminated Sites Remediation Fund.* www.mfe.govt.nz/more/funding/contaminated-stes. (accessed October 14, 2018).

New Zealand Ministry for the Environment. 2018k. *Contaminated Land Management Guideline. Revised 2016.* www.mfe.govy.nz/sites/land. (accessed October 14, 2018).

New Zealand Ministry for the Environment. 2018l. *Contaminated Land Management Risk Screening System.* www.mfe.govt.nz/sites/screening-system. (accessed October 15, 2018).

Norwegian Environment Agency. 2019a. Ministry of Climate and Environment. https://miljodirektoratet.no/no/Om-Miljodirektoratet?Norwegian-Environment-Agency. (accessed September 15, 2019).

Norwegian Environment Agency. 2019b. *Solid and Hazardous Waste.* www.miljodirektoratet.no/en/legislation1/regulations/waste-regulations/chapter11/. (accessed February 10, 2019).

Norwegian Environment Agency. 2019c. *Guidelines for Risk Assessment of Contaminated Sites.* www.miljodirektoratet.no/old/klif/publikasjoner/andre/1691/ya1691.pdf. (accessed February 10, 2019).

Norwegian Polar Institute. 2018. *Regulations for Activities in Antarctica.* www.nplar.no/en/regulations/the-antarctic. (accessed November 11, 2918).

Organization for Economic Co-operation and Development (OECD). 2015. *Summary Report of Mexico's Environmental Laws and Status.* Paris, France. 27p. www.OECD.org/Mexico/environmental. (accessed September 16, 2017).

OECD. 2016. *Summary Report of Switzerland's Environmental Laws.* Paris, France. 21p. www.OECD.org/Switzerland/environmental. (accessed September 16, 2017).

OECD. 2019. *Summary Reports of Russia.* www.oecd.org/russia/publicationsdocuments. (accessed February 10, 2019).

Peace Corps Peru. 2016. *Trash Talking: Solid Waste Management in Peru.* https://psu2.wordpress.com/2016/11/03/trash-talking-Peru. (accessed October 18, 2018).

Peru Ministry of the Environment. 2018a. *Ambient Air Quality Standards.* www.temasactuales.com/assets/pdf/gratis/PERairqual.htm. (accessed October 28, 2018).

Peru Ministry of the Environment. 2018b. *Solid Waste.* www.gob.pe/ambiento. (accessed October 28, 2018).

Peru Ministry of Housing, Construction, and Sanitation. 2018. www.gob.pe/vivienda. (accessed October 28, 2018).

Porter, M.E. 2009. *Turkey's Competitiveness: National Economic Strategy and the Role of Business.* Institute for Strategy and Competitiveness. Harvard Business School. Cambridge, MA. www.ics.hbs.edu. (accessed September 19, 2017).

Province of Alberta Canada. 2017. *Environmental Laws and Regulations.* www.environment.gov.ab.ca/legislation. (accessed April 6, 2017).

Province of British Columbia Canada. 2017. *Environmental Laws and Regulations.* www.environment.gov.bc.ca/legislation. (accessed April 6, 2017).

Province of Manitoba Canada. 2017. *Environmental Laws and Regulations.* www.environment.gov.mb.ca/legislation. (accessed April 6, 2017).

Province of Newfoundland and Labrador Canada. 2017. *Environmental Laws and Regulations.* www.environment.gov.nf.ca/legislation. (accessed April 6, 2017).

Province of New Brunswick. 2017. *Environmental Laws and Regulations.* www2.gnb.ca. (accessed April 6, 2017).

Province Northwest Territories Canada. 2017. *Environmental Laws and Regulations.* www.oag-bvg.ca. (accessed April 6, 2017).

Province of Nova Scotia Canada. 2017. *Environmental Laws and Regulations.* www.environment.gov.ns.ca/legislation. (accessed April 6, 2017).

Province of Ontario Canada. Ministry of Environment, Conservation and Parks. 2011. www.ontario.ca/page/soil-ground-water-and-sediment-standards-use-under-part-xvl-environmental-protection-act. (accessed April 6, 2017).

Province of Ontario Canada. 2017. *Environmental Laws and Regulations.* www.ontario.ca/environmental-laws. (accessed April 6, 2017).

Province of Quebec Canada. 2017. *Environmental Laws and Regulations.* https://legisquebec.gouv.qc.ca. (accessed April 6, 2017).

Province of Saskatchewan Canada. 2017. *Environmental Laws and Regulations.* www.saskatchewan.ca/legislation/environmental-protection. (accessed April 6, 2017).

Province Yukon Territory. 2017. *Environmental Laws and Regulations.* www.env.gov.yk.ca/environment. (accessed April 7, 2017).

Quinn, L.P., de Vos, B.J., Fernandes-Whaley, M., Roos, C., Bouwman, H., Pieters, K.R., and Berg, J. 2011. *Pesticide Use in Africa.* www.environment.gov.sa/docs/pesticides.pdf. (accessed August 18, 2018).

Ramadam, A.R. and Nadim, A.H. 2014. Hazardous Waste in Egypt: Status and Challenges. In: *Waste Management and the Environment.* Popov, V., Itoh, H., Brebbia, C.A., and Kungolos, S. (editors). WIT Press. Southhampton, UK. pp. 125–129.

Rich, K.H., Huang, W.Z., Zhu, T., Wang, H., Lu Se, O., Ohman-Strickland, P., Zhu, P., and Zhang, J.I. 2010. Measurement of Inflammation and Oxidative Stress Following Drastic Changes in Air Pollution During the 2008 Beijing Olympics: A Panel Study Approach. *Annals of New York Academy of Sciences.* August 2010. Vol. 1203. pp. 160–167.

Rodriguez, A., Castrejon-Godinez, M.L., Ortiz-Hernandez, M.L., and Sanchez-Salinas, E. 2015. Management of Solid Waste in Mexico. Fifteenth International Waste Management and Landfill Symposium. CISA Publisher. Cagliari, Italy. 7p.

Russian Federal Law on Environmental Protection. 2002. www.rospotrebnadzor.ru. (accessed September 13, 2017).

Russian Ministry of Economic Development of the Russian Federation. 2017. *Industrial Safety of Major Hazard Installations.* http://en.smb.gov.ru/regulation/116fz. (accessed September 5, 2017).

Russian Ministry of Natural Resources (MNR). 2019. *Ministry of Natural Resources and Environment of the Russian Federation.* www.mnr.ru/english. (accessed September 14, 2017).

Saudi Arabia Presidency of Meteorology and Environment. 2012. *General Environmental Regulations.* www.pme.gov.sa. (accessed February 10, 2019).

Siddiqi, T.A. and Chong-Xain, Z. 1984. Ambient Air Quality Standards in China. *Journal of Environmental Management.* Vol. 8. No. 6. pp. 473–479.

Smithsonian Museum of Natural History. 2017. *Introduction to Human Evolution. Smithsonian Institution's Human Origins Program.* Smithsonian Institution. Washington, DC. http://humanorigins.si.edu/education/introduction-human-evolution. (accessed December 14, 2017).

Smithsonian. 2018. *The Age of Humans: Living in the Anthropocene. Smithsonian Statement on Climate Change.* www.si.edu/newsdesk/releases/smithsonian-statement-climate-change. (accessed November 11, 2018).

South African Department of Environment. 1996. *South African Water Quality Guidelines. Domestic Water Use.* Second Edition. Government Gazette. Pretoria, South Africa. 197p.

South African Department of Environmental Affairs. 2009a. *South Africa Ambient Air Quality Standards.* Government Gazette. No. 32816. Pretoria, South Africa. 12p.

South African Department of Environmental Affairs. 2009b. *National Environmental Waste Act No. 59 of 2009.* Pretoria, South Africa. 309p.

South African Department of Environmental Affairs. 2012. *National Waste Management Strategy.* South Africa Government Gazette. Pretoria, South Africa. 77p.

South African Department of Environmental Affairs. 2013. *National Norms and Standards for the Remediation of Contaminated Land and Soil Quality in the Republic of South Africa.* Government Gazette. Pretoria, South Africa. 15p.

South African Department of Environmental Affairs. 2018a. *Outline of Environmental Regulations.* https://environment.gov.sa/legislative/actsregulations. (accessed August 18, 2018).

South African Department of Environmental Affairs. 2018b. *Cape Town Drought; Current Status.* www.capetowndrought.com. (accessed August 21, 2018).

South Africa Government Gazette. 2018. *South Africa National Environmental Management Act (NEMA). 1998. National Environmental Management Act. Act No. 107.* Pretoria, South Africa. November 27, 1998. 37p.

Statistics South Africa. 2018. *Mid-Year 2018 Population Estimate of South Africa.* www.statsa.gov.sa/?=15. (accessed August 16, 2018).

Switzerland Environmental Protection Act. 2019. *Federal Act on the Protection of the Environment.* www.admin.ch/opc/en/classified-compilation/19830267/index.html. (accessed February 10, 2019).

Switzerland Office for the Environment. 2019. *Reporting for Switzerland on the Protocol for Water and Health.* www.unece.org/fileadmin/DAM/env/water/Protocol/reports/pdf. (accessed February 10, 2019).

Tanzania Air Quality Act. 2007. *The Environmental Management. Air Quality Regulations.* www.parliament.go.tz/acts-list-air-quality. (accessed August 28, 2018).

Tanzania Environmental Management Act. 2007. *Environmental Management Act.* Soil Quality Regulations. Dar es Salaam. Tanzania. 25p.

Tanzania Environmental Management Act. 2008. Hazardous Waste Control and Management Regulations. Dar es Salaam. Tanzania. 49p.

Tanzania Ministry of Water and Irrigation. 2015. *Rural Water Supply and Sanitation Program.* www,tanzania.go.tz/egov_uploads/documents/tanzania. (accessed August 30, 2018).

Tanzania National Bureau of Statistics. 2018. *Current Population of Demographics of Tanzania.* www.nbs.go.tz. (accessed August 28, 2018).

Tanzania National Environmental Standards Compendium (NESC). 2018. *Drinking Water Specifications. TZS 789:2003.* Dar es Salaam, Tanzania. 78p.

Tanzania Water Resources Act. 2009. *Tanzania Water Resources Management.* www.tanzania.go.tz/egov_iploads/documents/water-resources_en_sw.pdf. (accessed August 30, 2018).

Tanzania Water Supply and Sanitation Act. 2009. *The Republic of Tanzania Water Supply and Sanitation Act.* Dar es Salaam, Tanzania. 47p.

Turkey Ministry of Environment and Urbanization. 2018a. *Turkey Environmental Law.* https://csb.gov.tr/. (accessed August 10, 2018).

Turkey Ministry of Environment and Urbanization. 2018b. *Directorate of Environmental Management.* Department of Marine and Coastal Management.

Turkish Statistical Institute. 2017. *Census of the Population of Turkey 2017.* /www.turkstat.gov.tr. (accessed September 19, 2017). http://mavikart.cevre.gov.tr/en/Haberler.aspx. (accessed February 10, 2019).

United Nations. 1997. *Australia Country Profile.* Department of Policy Coordination. New York. www.un.org/esa/earthsummit/astra-cp.htm. (accessed October 12, 2018).

United Nations. 2002. *Saudi Arabia Country Profile.* Johannesburg Summit. United Nations. New York. 59p.

United Nations. 2004. *Kingdom of Saudi Arabia. Public Administration and Country Profile.* Department of Economic and Social Affairs (UNDESA). New York. 16p.

United Nations. 2005. *Republic of Peru: Country Profile.* Department of Economic and Social Affairs. New York. 18p.

United Nations. 2012. *Country Profile Series: Republic of Korea.* New York. 92p. www.un.org.esa/agenda21/natlinfo/rekorea.pdf. (accessed October 2, 2108).

United Nations. 2013a. *Water Pollution in China.* Food and Agriculture Organization of the United Nations. Rome, Italy. 173p.

United Nations. 2013b. *Climate Change Impacts on Chile Water Systems.* www.oecd.org/env/resources/Chile.pdf. (accessed October 28, 2018).

United Nations. 2015a. *International Decade for Action "Water for Life".* United Nations Department of Economic and Social Affairs (UNDESA). New York. www.un.org/waterforlifedecade/quality.shtml. (accessed October 14, 2018).

United Nations. 2015b. *Environmental Performance Review of Chile.* www.oecd.org/environmental/country-review/chile.htm. (accessed October 21, 2018).

United Nations. 2016a. *Global Drinking Water Quality Index and Development and Sensitivity Analysis Report.* UNEP Water Programme Office. Burlington, Ontario, Canada. 58p.

United Nations. 2016b. *World Air Pollution Status.* United Nations News Center. New York. www.un.org/sustainabledevelopment/2016/09. (accessed August 29, 2017).

United Nations. 2016c. *Human Development Report.* United Nations Development Programme. www.hdr.undr.org/sites/2017. (accessed September 16, 2017).

United Nations. 2017a. *World Population Prospects. United Nations Department of Economic and Social Affairs. Population Division.* https://esa.un.org/unpd/wpp/data. (accessed December 14, 2017).

United Nations. 2017b. *Population Trends of Russia.* Department of Economic and Social Affairs. Division of Population. www.UN.org.unpd. (accessed June 17, 2017).

United Nations. 2017c. *Outer Limits of the Continental Shelf Beyond 200 Nautical Miles (370 km) from the Baselines.* www.un.org/depts/los/clcs.htm. (accessed September 17, 2017).

United Nations. 2018a. *Air Pollution Index: Real Time Air Quality Index Visual Map.* www.aqicn.org/map. (accessed October 2018).

United Nations. 2018b. *United Nations Population Programme For Egypt.* www.egypt.unfpa.org/em/united-nations-population-egypt-2. (accessed August 30, 2018).

United Nations. 2018c. *Country Profile for China.* www.china.unfpa.org/em/united-nations-population-china-2. (accessed February 10, 2019).

United Nations. 2018d. *The Demographic Profile of Saudi Arabia with Population Trends.* www.unescwa.org/sites/saudiarbiaprofile. (accessed October 5, 2018).

United Nations. 2018e. *Country Profile and Review of Indonesia.* www.un.org/esa/earthsummit/indon-cp.htm. (accessed October 6, 2018).

United Nations. 2018f. *Country Profile and Review of Malaysia.* www.un.org/esa/earthsummit/mal.htm. (accessed October 7, 2018).

United Nations. 2018g. *Great Barrier Reef: UNESO World Heritage Site.* www.whc.unesco,org/en/list/154. (accessed October 12, 2018).

United Nations. 2018h. *Country Profile of New Zealand.* www.un.org/esa/earthsummit/NZ.htm. (accessed October 14, 2018).

United Nations. 2018i. *Country Profile for Argentina.* www.un.org/esa/earthsummit/AR.htm. (accessed October 14, 2018).

United Nations. 2018j. *Country Profile for Brazil.* www.un.org/esa/earthsummit/brzl-cp.htm. (accessed October 14, 2018).

United Nations. 2018k. *Country Profile for Chile.* www.un.org/esa/earthsummit.ch-cp.htm. (accessed October 21, 2018).

United Nations. 2018l. *Oceans and the Law of the Sea.* www.un.org/en/sections/issues-depth/oceans-and-law-sea. (accessed November 11, 2018).

United Nations. 2019. *Geospatial Information Section. Map of Africa.* www.un.org/Depts/Cartographic/english/htmain.htm. (accessed April 15, 2019).

United States Census Bureau. 2018. *United States Trade in Goods with China 2017.* www.census.gov/foreign-trade/balance (accessed August 31, 2018).

United States Central Intelligence Agency. 2018. *The World Fact Book.* www.cia.gov/library/publications/the-world-factbook.gos/NA.html. (accessed October 10, 2018).

United States Department of Commerce. 2018. *Foreign Trade with China 2017.* www.commerce.gov/tags/china. (accessed August 31, 2018).

US Department of State. 2017. Office of the United States Trade Representative. https://ustr.gov/nafta. (accessed August 31, 2017).

US Department of State. 2018a. *United States Department of State Air Quality Monitoring Program.* www.stateair.net/web/post/1/5.html. (accessed October 1, 2018).

US Department of State. 2018b. *United States Relations with Japan.* www.state.gov/p/eap/ci/ja/. (accessed September 27, 2018).

US Department of State. 2018c. *Profile and Statistics of Saudi Arabia.* www.state.gov/saudiarabia. (accessed October 5, 2018).

United States Environmental Protection Agency (USEPA). 2013. *Air Quality in Africa.* http://epa.gov/international/air/africa.htm. (accessed December 14, 2017).

USEPA. 2017. *EPA Collaboration with Chile.* www.epa.gov.international-cooperation/epa-collaboration-chile. (accessed October 28, 2018).

USEPA. 2018a. *EPA Collaboration with Sub-Saharan Africa.* www.epa.gov.international-cooporation/epa-collaboration-sub-saharan. (accessed August 16, 2018).

USEPA. 2018b. *EPA Collaboration with China.* www.epa.gov.international-cooperation/epa-collaboration-china. (accessed September 24, 2018).

USEPA. 2018c. *EPA Collaboration with the Republic of Korea.* www.epa.gov.international-cooporation-korea. (accessed October 2, 2018.

USEPA. 2018d. *EPA Collaboration with Indonesia.* www.epa.gov.nternational-cooporation-Indonesia. (accessed October 6, 2018).

USEPA. 2018e. *EPA Collaboration with Brazil.* www.epa.gov.nternational-cooporation-Brazil. (accessed October 17, 2018).

USEPA. 2018f. *EPA Collaboration with Peru.* www.epa.gov.nternational-cooporation-Peru. (accessed October 28, 2018).

USEPA. 2018g. *Laws That Protect Our Oceans.* www.epa.gov/beach-tech/laws-protect-our-oceans. (accessed May 11, 2018).

United States Geological Survey (USGS). 2017. *Examples of Land Subsidence in Mexico City.* www.USGS.gov/edu/gallery/landsubsidence-learning.html. (accessed August 30, 2017).

USGS. 2018. *Earthquakes.* https://earthquakes.usgs.gov/earthquakes/map. (accessed October 21, 2018).

USGS. 2019. *Colorado River Basin Census.* https://water.usgs.gov/watercensus/crb-fas/index.html. (accessed February 10, 2019).

United States Library of Congress. 2010. Chile: New Law Creates Ministry for the Environment. www.loc.gov/law/foreign-news/article/chile-new-law. (accessed October 28, 2018).

University of Michigan. 2017. *Environmental Issues of Africa.* www.globalchange.umich.edu.globalchange1. (accessed December 14, 2017).

Water Environment Partnership in Asia. 2018. *State of Water Environmental Issues in Indonesia.* www.wepa-db.net/policies/state/indonesiad/indonesia.htm. (accessed October 7, 2018).

Welch, C. 2018. *Why Cape Town Is Running Out of Water and Who's Next.* www.news.nationalgeographic.com/2018/02/cape-town-running-out-of-water. (accessed August 16, 2018).

Wencong, W. 2012. *China Unveils Plan to Control Hazardous Waste.* China Daily. Beijing, China. www.chinadaily.com.cn/china/2012-11/02content15867364.htm. (accessed September 24, 2018).

World Bank. 2000. *Argentina: Water Resources Management, Policy Issues and Notes: Thematic Annexes. Volume III. World Bank.* Latin America and the Caribbean Office. 190p.

World Bank. 2011. *Environmental Assessment, Country Data: India.* https://data.worldbank.org/countrydata.india.html. (accessed September 28, 2018).

World Bank. 2013a. *The Arab Republic of Egypt, Air Pollution in Greater Cairo.* Sustainable Development Department. Middle East and North Africa Region. Washington, DC. 150p.

World Bank. 2013b. *Matanza-Riachuelo River Cleanup, Argentina.* Blacksmith Institute. http://worldbank.org/projects/argentine/matanza-riachuela-river. (accessed October 14, 2018).

World Bank. 2015. *Economy Profile for Chile.* https://data.worldbank/country/chile. (accessed October 21. 2018).

World Bank. 2016. *India in Profile: Economy and Standard of Living.* https://data.worldbank.org/countrydata. india.html. (accessed October 1, 2018).

World Bank. 2018a. *Australia in Profile. Gross Domestic Product, Economy and Standard of Living.* www.data.worldbank.org/countries/australia. (accessed October 12, 2018).

World Bank. 2018b. *The World Bank in Argentina.* www.worldbank.org/en.country/argentina. (accessed October 14, 2018).

World Bank. 2018c. *Significant Advances in the Recovery of the Matanza-Riachuelo River Basin.* World Bank. Argentina. 6p.

World Bank. 2018d. *Argentina: Policy for Sustainable Development in the 21st Century.* http://openknowledge.wolrdbank.org/handle/10986/14980. (accessed October 14, 2018).

World Bank. 2018e. *Global Economic Prospects: The Turning of the Tide?* World Bank Corp. Washington, DC. USA. 184p.

World Bank. 2018f. *Overview of World Bank in Peru.* www.worldbank.org/en/country/peru/overview. (accessed October 28, 2018).

World Health Organization and United States Environmental Protection Agency. 2012. *Animal Waste, Water Quality and Human Health.* IWA Publishing. London. UK. 489p.

World Health Organization (WHO). 2013. *Water Quality and Health Strategy 2013–2020.* www.who.int/water_sanitation_health/dwq/en/. (accessed October 18, 2018).

WHO. 2014. *Ambient Air Pollution Database May 2014.* www.who.org/ambient-air-database. (accessed September 19, 2017).

WHO. 2015a. *Guidelines for Drinking-water Quality.* Geneva, Switzerland. 516p.

WHO. 2015b. *Progress on Drinking Water and Sanitation.* World Health Organization. New York. 90p.

WHO. 2016a. *World Health Organization Global Urban Ambient Air Pollution Database.* www.who.int/phe/health_topics/outdoorair/database.htm. (accessed October 28, 2018).

WHO. 2016b. *Air Pollution in China: Sources and Amounts.* www.who.org/china-air-pollution. (accessed September 3, 2018).

World Health Organization (WHO). 2018a. *Nine Out of Ten People Breathe Unhealthy Air.* Geneva. Switzerland. www.who.int/news-room/detail/02-02-2018. (accessed May 30, 2018).

WHO. 2018b. *World Health Organization Country Profile – Egypt.* www.who.int/countries/egy/en/. (accessed August 31, 2018).

WHO. 2018c. *Drinking-Water Quality Standards in the Eastern Mediterranean Region.* World Health Organization. WHO-EM/CEH/143/E. Cairo, Egypt. 44p.

WHO. 2018d. *World Health Organization Profile of India.* www.who.int/countries/ind/en. (accessed October 1, 2018).

WHO. 2018e. *United Nations Global Analysis and Assessment of Sanitation and Drinking Water: Indonesia.* www.who.org/drinkingwater/Indonesia. (accessed October 6, 2018).

WHO. 2018f. *Peru: Country Profile.* http://who.int/gho/countries/per/country_profiles/per.htm (accessed October 28, 2018).

World Population Review. 2017. *Population of Mexico City.* http://worldpopulationreview.com/MexicoCity/. (accessed August 29, 2017).

World Water. 2017. *Russian Water Industry Remains at Crossroads.* www.waterworld.com/articles/wwi/print/volume-25. (accessed September 14, 2017).

Yablokov, A. 2016. *Environmental Problems and Projections in Russia.* Institute for Global Communications. Moscow, Russia. www.igc.org/russia. (accessed May 1, 2018).

Yip, M. and Madl, P. 2002. Air Pollution in Mexico City. *Journal of Biophysics.* Salsburg, Austria. www.biophysics.sbg.ac.al/mexico/air.htm. (accessed August 28, 2017).

Zalasiewicz, J., Waters, C., Summerhayes, C., and Williams, M. 2018. The Anthropocene. *Geology Today.* Vol. 34. No. 5. pp. 182–187.

Part Three

Compliance and Sustainability

8 Fundamental Concepts of Environmental Compliance

At this point we now have an appreciation for environmental regulations in the United States and the world in our journey through the environmental regulatory landscape of our planet. We know that the primary purpose of environmental regulations is to protect human health and the environment. We should also have an appreciation of the sheer immensity and often times overwhelming complexity of environmental regulations.

In this chapter, we will boil things down even more than we have in the previous chapter to a few points that will provide assistance and focus our efforts to guide us in achieving compliance with environmental regulations and beyond.

The significant lessons in this chapter include:

1. **It all begins with conducting an environmental audit.**

    The first step to compliance is to conduct an environmental audit. An environmental audit is an assessment of which a facility or organization is observing practices to minimize harm to the environment. United States Environmental Protection Agency (USEPA) defines an **environmental audit** as a systematic evaluation to determine the conformance to quantitative specifications to environmental laws, regulations, standards, permits, or other legally required documents (USEPA 2018a).

2. **Being in compliance with environmental regulations has little to do with sustainability.**

    In this chapter, we will begin to explore what it means to look toward environmental stewardship and sustainability. Environmental stewardship and sustainability require going beyond environmental compliance and addressing deeper questions about how we may impact the environment. Being in compliance means that you are conducting activities consistent with environmental regulations. It does not ask the question whether you should be doing something else that will enhance your organizational goals. In other words, you have to ask yourself, "Just because you can doesn't always mean you should."

3. **Work yourself out of environmental regulations.**

    This is often difficult and challenging but is well worth the trouble and helps pave the way to environmental stewardship and sustainability.

4. **There is little we can control but those things we do control are significant and can make a big difference.**

    Environmental professionals for the most part do not control the location where manufacturing takes place or the type of manufacturing or products that are made. The only thing that environmental professionals typically have input in are the chemicals that are used and what management actions and engineering controls to put in-place to prevent mismanagement and potentially causing harm to the environment.

5. **Limit chemical use when possible.**

    Remember, if there were no hazardous substances there would be little need for environmental regulations or environmental professionals.

6. **Lastly, we must remember that "no matter where you are, it's all the same."**

    We are learning that we live on one planet with no environmental boundaries and that contamination does not respect borders. Another point is that environmental regulations are built on the USEPA or European Union (EU) platforms and those platforms rely on three basic principles:

1. Protect the air
2. Protect the water
3. Protect the land

The central focus of this chapter will be conducting compliance audits. Conducting a compliance audit also provides an opportunity to evaluate a facility's environmental health and whether it's a candidate for further study to identify sustainability opportunities.

8.1 INTRODUCTION

Environmental compliance is sometimes perceived as just another set of rules and is simple. Just follow the rules. However, the Earth is dynamic. There is change, ever getting more complex as more information and science is discovered, learned, and shared. The weather changes constantly and is perhaps the best example of how to imagine environmental management and risk. It also changes, as does everything. Everything obeys the second law of thermodynamics, namely entropy, in that nature tends to become more complex and disordered as time marches on.

Understanding the natural setting of our urban and industrial areas, knowing the chemicals that are used and how they might cause harm to the environment and humans if they are released, what alternative chemicals are available, how to reduce chemical and energy use, preventing chemical releases to the environment, quantifying the costs involved with cleaning up the environment once a release occurs, and developing environmental stewardship are all subjects that should be taken into account when evaluating the environmental health of a facility and it all begins with first conducting an environmental audit.

8.2 USEPA AUDIT POLICY

In 1986, the USEPA published the Environmental Auditing Policy Statement (USEPA 1986). USEPA published the policy statement to encourage the use of conducting "Self Assessments" by the regulated community to help, achieve, and maintain compliance with environmental laws and regulations, and to identify and correct unregulated hazards. In addition, USEPA defined environmental audits as (USEPA 1986):

> a systematic, documented, periodic, and objective review of facility operations and practices related to meeting environmental requirements

The policy also identified several objectives for environmental audits that included (USEPA 1986):

1. Verifying compliance with environmental requirements
2. Evaluating the effectiveness of in-place environmental management systems
3. Assessing risks from regulated and unregulated materials and practices

USEPA has published guidelines or protocols for conducting environmental audits in subject areas that include (USEPA 2018a):

- Comprehensive Environmental Response, Compensation and Liability Act (CERCLA)
- Clean Water Act (CWA)
- Stormwater under the CWA
- Wastewater under the CWA
- Federal Insecticide, Fungicide, and Rodenticide Act (FIFRA)
- Emergency Planning and Community Right-to-Know Act (EPCRA)
- Resource Conservation and Recovery Act (RCRA)

- Safe Drinking Water Act (SDWA)
  - Toxic Substance and Control Act (TSCA)

The audit protocols listed above are a checklist for conducting environmental audits on each subject listed and are 1,387 pages in length. Audit protocols are intended to assist the regulated community in developing programs at individual facilities to evaluate their compliance with environmental requirements under federal law. USEPA states that the protocols are intended solely as guidance in this effort. In addition, USEPA states that the regulated community's legal obligations are determined by the terms of applicable environmental facility-specific permits, as well as underlying statutes, and applicable federal, state, and local law.

Therefore, when examining the audit protocols, the facility must also take into account state, county, municipal, or local regulations that apply in order to fully understand requirements that the facility must comply with before determining whether the facility is in full compliance. This again may require the services of in-house counsel, a qualified and experienced environmental attorney, an experienced and qualified environmental consultant, and/or all of the above. Additionally, contacting the regulatory agency may be justified to clarify certain issues that may arise.

In 1995, USEPA published "Incentives for Self-Policing: Discovery, Disclosure, Correction, and Prevention of Violations," which both reaffirmed and expanded the 1986 policy (USEPA 2018b). USEPA again revised the policy in 2000 and in 2005 (USEPA 2018c). Under USEPA's environmental audit policy, gravity-based penalties for violations of USEPA-administered statutes are reduced or completely eliminated if the violations are voluntarily discovered, promptly disclosed to USEPA, and meet a number of other specified conditions. Those other specified conditions include the following (USEPA 2018c):

1. The violation must be systematically discovered, either through (a) an environmental audit or (b) a compliance management system reflecting the company's actions and methods in preventing, detecting, and correcting violations.
2. The violation must be discovered voluntarily and not by legally mandated sampling or monitoring required by statute, regulation, permit, judicial or administrative order, or consent agreement.
3. The company must fully disclose the specific violation in writing to the appropriate regulatory authority or USEPA within 21 days or within a shorter period of time if necessary (i.e., such as a spill exceeding the corresponding reportable quality in which case the reporting requirement is within 24 hours) after the company discovered that the violation has, or may have occurred.
4. The company must discover and disclose the violation to USEPA prior to (a) the commencement of a federal, state, or local agency inspection or investigation, or the issuance by the agency of an information request to the company, (b) notice of a citizen suit, (c) the filing of a compliant by a third party, (d) the reporting of the violation to USEPA (or other governmental agency) by a "whistleblower" employee, or (e) imminent discovery of the violation by a regulatory agency.
5. The company must correct the violation within 60 calendar days from the date of discovery, certify as such in writing, and take appropriate action or measures as determined by USEPA to remedy any environmental or human harm due to the violation.
6. The company must agree in writing to take steps to prevent a recurrence of the violation.
7. The specific violation (or closely related violation) cannot have occurred previously within the past three years at the same facility, or within the past five years as part of a pattern of multiple facilities owned or operated by the same entity.
8. The violation cannot be one that (a) resulted in serious harm, or may have presented an imminent and substantial endangerment, to human health or the environment or (b) violates the specific terms of any judicial or administrative order, or consent agreement.
9. The company must cooperate with USEPA and provide such information as is necessary and requested by USEPA to determine applicability of the policy.

USEPA has broad statutory authority to request relevant information on the environmental compliance status of regulated entities. However, USEPA believes routine requests for audit reports by the agency could inhibit auditing in the long run, decreasing both the quantity and quality of audits conducted. Therefore, as a matter of policy USEPA will not routinely request environmental audit reports (USEPA 1986).

USEPA's authority to request an audit report, or relevant portions thereof, will be exercised on a case-by-case basis where the agency determines it is needed to accomplish a statutory mission, or where the government deems it to be material to a criminal investigation. USEPA expects such requests to be limited, most likely focused on particular information needs rather than the entire report, and usually made where the information needed cannot be obtained from monitoring, reporting, or other data otherwise available to the agency. Examples would likely include situations where a company has placed its management practices at issue by raising them as a defense; or state of mind or intent are a relevant element of inquiry, such as during a criminal investigation. This list is illustrative rather than exhaustive, since there doubtless will be other situations, not subject to prediction, in which audit reports rather than information may be required (USEPA 1986).

USEPA acknowledges regulated entities need to self-evaluate environmental performance with some measure of privacy and encourages such activity. However, audit reports may not shield monitoring, compliance, or other information that would otherwise be reportable and/or accessible to USEPA, even if there is no explicit "requirement" to generate the data. Thus, this policy does not alter regulated entities' existing or future obligations to monitor, record, or report information required under environmental statutes, regulations, or permits, or to allow USEPA access to that information. Nor does this policy alter USEPA's authority to request and receive any relevant information, including that contained in audit reports, under various environmental statutes, such as Clean Water Act Section 308, and Clean Air Act Sections 114 and 208, or in other administrative or judicial proceedings (USEPA 1986).

Regulated entities also should be aware that certain audit findings by law may have to be reported to governmental agencies. However, in addition to any such requirements, USEPA encourages regulated entities to notify appropriate state or federal officials of findings which suggest significant environmental or public health risks, even when not specifically required to do so (USEPA 1986).

Questions and reporting requirements related to this policy and the conducting of environmental audits should be directed to in-house counsel or to a qualified and experienced environmental attorney.

8.3 STARTING THE PROCESS

The first step in achieving environmental compliance is to conduct a compliance audit. Conducting a compliance audit will answer some very important questions concerning any operation. There are two basic questions that must be answered right at the beginning to determine if a compliance audit is advisable and include:

- Is the facility subject to any environmental regulations?
- Does the facility require an environmentally related permit for any of its operations?

Today, it is rare that a facility doesn't know the answer to those two basic questions, but for the sake of argument, let's assume that they do not know and want to do the right thing and evaluate whether they are subject to environmental regulations. Under this scenario, the following questions apply:

- Does the facility store or use any hazardous substances as defined by USEPA in Chapter 6 If the answer is no, then does the facility store or use any hazardous substances as defined by state or local regulations?
- Does the facility operate any equipment that generates any exhaust or has any air emissions?

- Does the facility use or discharge any water other than sanitary?
- Does the facility generate any solid waste?

If the answer is yes to any of the above questions, the facility is likely subject to environmental regulations and may require one or more environmental permits. If questions persist, involvement of company management and in-house counsel, a qualified and experienced environmental attorney, an experienced and qualified environmental consultant, and/or all of the above may be necessary. Additionally, contacting the regulatory agency may be justified to clarify specific questions.

8.4 CONDUCTING AN ENVIRONMENTAL AUDIT

Conducting an environmental audit is the first step on a long and winding road toward environmental compliance and ultimately can lead to the path of sustainability and environmental stewardship. Most large companies with several manufacturing locations conduct environmental audits on a regular basis. Some choose to conduct environmental audits more often than others, but most seem to choose to conduct an environmental audit every year or every two years. Some companies conduct environmental audits with in-house staff to retain institutional knowledge. Some choose to retain an environmental consulting firm to conduct environmental audits because it's conducted by an independent third party. Some environmental audits are conducted without any prior notice and some are conducted with prior notice. Whichever path is chosen, environmental audits are usually always conducted at the request of counsel.

8.4.1 CATEGORIES OF FINDINGS

When conducting an environmental audit, findings are typically weighted by priority, severity, or category and typically include:

1. Non-compliance
2. Best management practice (BMP)
3. Observation

An issue of non-compliance is the highest priority of any finding and would be considered a regulatory violation. USEPA (2018c) defines an issue of **environmental non-compliance** as not conforming with environmental law, regulations, standards, or other requirements such as an environmental permit. An example of a non-compliance issue would be an improper or unlabeled drum of hazardous waste.

A **best management practice** was first defined by USEPA in 1974 (USEPA 2018d) in the Clean Water Act and has since been adopted by the regulated community. Best management practices are defined for environmental purposes as specific practices that are capable of improving, preventing, and/or minimizing the potential of an occurrence or situation that can lead to environmental non-compliance and/or a release of a hazardous substance to the environment. A best management practice issue, when identified, does not mean that a non-compliant situation exists, but if left unattended, could result in a non-compliance issue in the future. Examples of environmental best management practices include (Connecticut Department of Environmental Protection 2018):

- Storing products and wastes indoors
- Clean catch basins on a regular basis
- Ensuring that lids on dumpsters remain closed

An **observation** is an issue that could be improved but does not reach the threshold of a best management practice. For instance, a better method for keeping track of monitoring results or improving housekeeping are examples of an observation.

8.4.2   Document List

If the environmental audit is announced prior to the audit date, a request for applicable documents to be available to be reviewed in advance of the audit will save time. As applicable to the facility being reviewed, the following list of documents should be available for review during the site visit.

1. Current site map
2. Current organization chart
3. Current process flow diagrams depicting inputs, process units, waste streams, and other outputs
4. List of processes that have been shut down since the last audit
5. List of processes that are expected to start up within the next 12 to 24 months
6. Spill Prevention Control and Countermeasures (SPCC) Plan
7. Spill reports for the last three years
8. All RCRA manifests since the last audit
9. All special waste notifications and shipping documents since the last audit
10. All analytical test results and determination for special and hazardous waste
11. All analytical tests for each non-hazardous waste stream
12. Contingency plan
13. Personnel training records since the last audit
14. List of arrangements with local agencies
15. National Pollutant Discharge Elimination System (NPDES) permit and most recent application
16. Discharge monitoring reports and any associated letters of explanation of permit limit exceedances since the last audit
17. Stormwater plan, intent to comply documents, and all monitoring reports and sample data
18. All notices of violation, non-compliance letters, or consent orders including facility's non-compliance explanations and, if in significant non-compliance, the facility's plan to return to compliance
19. Water flow diagrams and water balances
20. Sketches of all wastewater treatment systems within the facility
21. Name of wastewater analytical laboratory, a list of all analytical methods used for wastewater analysis, and the quality assurance/quality control procedures used for wastewater sample collection and analysis
22. Air permit application modifications
23. All construction and operating permits for air sources
24. Copies of all visible emission observations taken by facility personnel or contractors for the past three years
25. Environmental management plan
26. Any other environmental permit to operate (associated with air, land, or water)
27. Waste shipment summary for the past two years
28. Toxic Release Inventory Reports (Form R) for the past two years

For facilities not in the United States, and as applicable to the specific country, province, or state, additional documents to be made available to the reviewers include:

1. Noise studies and noise compliance (e.g., Mexico, China, etc.)
2. Environmental Impact Assessment (EIA)
3. Operating permits
4. EU Registration, Evaluation, Authorization, and Restriction of Chemicals (REACH) certification status for those locations within the EU

8.4.3 OPENING MEETING

The following sections describe the environmental review process in greater detail to ensure that a more complete review of the facility compliance status is conducted. In addition, this section also provides guidance for identifying and establishing pollution prevention opportunities to further minimize a potential environmental impact in the future.

To improve the efficiency of conducting the environmental review, prior notice to the facility should be made so that all necessary information can be made readily available and appropriate plant personnel are in attendance for the review. Some companies may choose to conduct unannounced audits to emulate an agency inspection, which often are unannounced. Whichever method is used, it's recommended that the facility be reminded that an agency inspection can occur any time on any day and that proper preparation prevents poor performance. Therefore, organization is key and backup systems should be in-place in case the facility person in charge of compliance is not present.

A list of documents is provided in Section 8.4.2 that should be available for review, as applicable to the facility.

An opening meeting should be scheduled in advance of the environmental audit. The purpose of the opening meeting is to:

(1) Inform appropriate facility management of the purpose and objectives of the audit
(2) Coordinate logistics and review procedures
(3) Remind facility management with respect to:
   - Internal policy pertaining to environmental audits
   - Report privilege and distribution
   - Procedures to follow during agency inspections and emergency and spill events
   - Periodically review environmental permits
   - Pollution prevention initiatives
   - Preparation for environmental compliance requires an awareness and daily attention to such matters
   - Discovery of any non-compliance issue must be corrected immediately, if possible
   - Discovery of a non-compliance issue which cannot be corrected immediately should be brought to the attention of counsel
   - Reporting requirements and corrective actions for all non-compliance issues should be reviewed with counsel
(4) Ensure that priority items are presented to facility management quickly so that appropriate corrective actions can be initiated promptly
(5) Remind management and all personnel that compliance with environmental regulations is not optional
(6) Remind management that the process for identified non-compliance items is that they must be addressed immediately and with the utmost priority
(7) Remind management that controls must be evaluated and updated as necessary if non-compliance issues are identified to minimize the possibility of a re-occurrence
(8) Remind facility personnel that compliance is not an option

8.4.4 ENVIRONMENTAL AUDIT CHECKLIST

It is always very helpful to review and become familiarized with the overall facility layout and manufacturing process of the facility before the actual audit begins. This is usually accomplished by reviewing a diagram or figure of the general facility layout and manufacturing process flow and then reviewing a diagram or figure of the facility that identifies (1) locations of all air emission sources, (2) solid and hazardous waste stream sources and points of generation, (3) water and wastewater flow through the facility, and (4) areas where hazardous substances enter the facility, where they are stored and used within the facility, and where they are stored as wastes before offsite transportation for disposal.

The items listed in the checklist in Sections 8.4.4.1 through Section 8.4.4.10 are intended to be used as a guide in conducting the environmental audit. Some of the sections or listed items included in the checklist may not be applicable to a particular facility. Therefore, the checklist should be facility specific and should be used as applicable.

8.4.5 AIR

Regulatory requirements for air vary significantly from facility to facility. Each facility's air emissions are not standardized under similar regulations. Therefore, a country's federal, state, and local guidelines that affect each facility should be reviewed independently. In addition, there are specific guidelines under Maximum Achievable Control Technology or MACT that should also be evaluated.

8.4.5.1 General Background Information

The following items should be considered first before conducting the detailed review of air to become familiarized with the current status of the facility with respect to air status and compliance:

- Evaluate whether the facility is located in an attainment or non-attainment area for specific compounds.
- Evaluate whether the facility has any air emissions from industrial sources.
- Review whether the facility is a minor, synthetic minor, or major emission source of criteria pollutants and of hazardous air pollutants (HAPS).
- Evaluate whether the facility has emission units that are registered by the state or local municipality.
- Evaluate whether any changes or modifications to existing equipment have been made or new equipment has been installed that may have increased production and hence may have increased air emissions from any source. This may indicate that a permit or permit modification may have been required.
- Each facility should also have a keen awareness that any process or manufacturing changes or installation of new equipment may have a profound effect on the types, amounts, and chemical and physical characteristics of emissions that are generated which may require a permit or permit modification.
- Review any recent correspondence from or to any regulatory agency.
- Review the nature and focus of any regulatory inspections that have been conducted since the last environmental audit.
- Review and evaluate whether the facility has received any Notice of Violations (NOVs) since the last environmental review.
- Review and evaluate any corrective actions that have been undertaken to address any NOVs.

8.4.5.2 Air Operating Permit

- Review air operating permit, such as Title V (major source)
- Review permit terms
- Review emission source list and flow diagram
- Review emission limits
- Review testing requirements
- Review monitoring requirements
- Review air pollution control (APC) operation & maintenance (O&M) plans
- Review compliance assurance monitoring (CAM) plans
- Inspect examples of APC maintenance records

- Evaluate any changes in regulatory status since operating permit issued
- Review reporting requirements

8.4.5.3 Construction Permits (Issued Since Operating Permit Last Issued)
- Review permit of any affected emission units
- Discuss any new or modified APC units
- Review permit limits
- Review testing requirements
- Review monitoring requirements
- Review request to modify operating permit
- If Prevention of Significant Deterioration (PSD) was triggered; review best available control technology (BACT) analysis & air modeling
- Review any recent changes to facility which were not covered by a construction permit

8.4.5.4 Air Compliance Documentation
- Discuss recent stack testing with respect to frequency, analytical parameters, analytical methods, results, trends
- Review annual emission inventory
- Review monitoring reports
- Review annual compliance certification and derivation reports
- Obtain copies of any NOVs since last review
- Review any agency inspection reports
- Discuss any other significant agency correspondence

8.4.5.5 Air Toxics/Hazardous Air Pollutant (HAP)
- Review potential-to-emit of HAP emissions
- Determine Iron & Steel Foundry MACT Rule Applicability
- Discuss Iron & Steel Foundry Area Source Rule Applicability (rule under development)
- Review state air toxic program applicability

8.4.5.6 Ozone Depleted Substance (ODS) Requirements
- Review facility's procedures for repair and replacement of ODS-containing equipment
- Review list of all ODS-containing equipment
- Inspect copies of technician certifications (in-house and/or outside contractor)
- Inspect documentation associated with units containing more than 50 pounds of ODS
- Ensure all ODS-containing units are properly evacuated prior to disposal

8.4.5.7 New Equipment
- Review any new construction
- Evaluate potential for new equipment or construction
- Evaluate whether any new equipment has been installed
- Evaluate whether any new equipment installed has increased production
- Evaluate whether any new equipment increases emissions

8.4.5.8 Asbestos
- Review asbestos management plan, if available
- Review list of asbestos contained and presumed asbestos contained materials (PACM)
- Review internal asbestos inspections
- Review asbestos training records
- Review asbestos abatement activities and documentation since last review, if any
- Review provisions for managing and sampling of PACM

8.4.5.9   Agency Inspections
- Review whether there have been any agency inspections
- Evaluate items reviewed by agency
- Record items of special interest

8.4.6   WATER

Similar to air regulations, the regulatory requirements for water vary significantly from facility to facility. Each facility's water discharge requirements and permits are not standardized under similar regulations. Therefore, federal, state, and local guidelines that affect each facility should be reviewed independently.

8.4.6.1   General Background Information

The following items should be considered first before conducting the detailed review of water to become familiarized with the current status of the facility with respect to water status and compliance:

- Evaluate the volume of water that the facility uses and the source of the water (i.e., onsite water well, city, or other)
- Evaluate the type, volume, and location of the facility's water discharges
- Evaluate whether the facility currently uses or historically operated a septic system
- Evaluate whether the facility has processed wastewater discharges
- Evaluate the overall operation of any wastewater treatment system, especially maintenance concerns and operational concerns
- Check to make sure that the wastewater system operator is licensed, if necessary
- Evaluate whether the facility is a categorical discharger and if so, which one, and evaluate any special requirements
- Evaluate whether the facility has a wastewater discharge permit
- Review any recent correspondence with the regulatory agency
- Review the nature and focus of any regulatory inspections that have been conducted since the last environmental review
- Review and evaluate whether the facility received any NOVs since the last environmental review
- Review and evaluate any corrective actions that have been undertaken to address any NOVs

8.4.6.2   Water Source(s)
- Evaluate the volume of water that the facility uses on a daily basis
- Record the sources of water that the facility receives (i.e., onsite water well, city, surface water, or other)
- Evaluate whether the facility has or had an onsite water well
- Evaluate and review the facility water well permit
- Evaluate whether the facility has a water usage permit
- Evaluate whether the facility is located in an area where water rights and consumption is a regulated resource
- Evaluate closure requirements if the facility has an onsite water well that is not currently in use
- Obtain and review records of any onsite water wells installed since the last audit or for historical wells if not previously conducted (i.e., materials of construction, diameter, installation date, depth, strata, pumping rates and capacity, onsite storage capacity, treatment requirements, and monitoring)

8.4.6.3 Wastewater Streams

- Review list of processed and non-processed wastewater discharges
- Review water/wastewater flow diagrams
- Review all wastewater sources, volume of discharges, and history of discharges
- Review all available charts, graphs, and drawings of wastewater discharges

8.4.6.4 Direct Discharges

- Review list of direct discharges
- Evaluate whether the facility currently uses or historically operated a septic system
- Review septic system permit, if applicable
  - Permit number
  - Discharge limits
  - Monitoring requirements
- Review NPDES permit:
  - Permit number
  - Discharge limits
  - Monitoring requirements
- Review the most recent (usually one year) DMRs (discharge monitoring reports):
  - Facility's compliance status
  - Reports of exceedances and violations
  - Timeliness
  - Fees
- Review any agency inspections
- Review any Notice of Violations

8.4.6.5 Indirect Discharges (to Sanitary Sewer)

- Review discharge status of facility
  - Industrial user discharge
  - Categorical discharger, for example:
    - 40CFR 464 Subpart A – Aluminum Casting Subcategory
    - 40CFR 464 Subpart C – Ferrous Casting Subcategory
    - 40CFR 433 – Metal finishing
    - Other
- Review sewer permit:
  - Permit number
  - Discharge limits
  - Monitoring requirements
- Review the sewer monitoring reports covering the most recent 12 months:
  - Identify party responsible for conducting sampling
  - Facility's compliance status
  - Reports of exceedances and violations
  - Timeliness
  - Fees
- Review any agency inspections
- Review any Notice of Violations

8.4.6.6 Wastewater Monitoring

- Review list of sampling points
- Review sampling procedures
- Review analytical methods

- Review flow meter and sampling equipment calibration records
- Evaluate outside laboratory credentials
- Evaluate whether qualified and authorized personnel are signing regulatory reports
- Review record-keeping procedures and organization

8.4.6.7   Stormwater Discharge Associated with Industrial Activity

- Review stormwater permit
  - Permit number
  - Type of permit (i.e., general permit, individual permit)
  - Discharge limits
- Review stormwater pollution prevention plan
  - Date of the most recent revision
  - Identification and implementation of best management practices
  - Periodic training
- Review stormwater monitoring, if required
  - Compliance status
  - Sampling reports (i.e., DMR)
  - NOVs
- Review sampling techniques
- Review laboratory methods
- Review record-keeping procedures and organization
- Evaluate whether training records are up to date
- Evaluate whether qualified and authorized personnel are signing regulatory reports
- Review non-stormwater discharge assessment forms
- Review if there have been any agency inspections
- Review any Notice of Violations

8.4.6.8   Potable Water

- Evaluate potable water sources
- Evaluate whether there are any onsite wells
- Evaluate sampling requirements
- Review compliance requirements
- Review analytical results
- Evaluate sampling location for appropriateness
- Review certification of sampling and laboratory

8.4.6.9   Agency Inspections

- Review any agency inspections
- Review any Notice of Violations

8.4.7   Solid and Hazardous Waste

Solid and hazardous waste regulations are generally considered the most complex and also vary significantly from facility to facility. To assist in conducting the solid and hazardous waste portion of the environmental review process, this section is divided into several subsections that include (1) hazardous waste, (2) solid waste, (3) universal waste, (4) used oil, (5) company owned or operated landfills, (6) recycling, and (7) pollution prevention plans.

As an additional item to be noted, each facility's solid and hazardous waste requirements and permits may not be standardized under similar regulations. Therefore, federal, state, and local guidelines that affect each facility should be reviewed independently.

8.4.7.1 General Background Information

The following items should be considered first before conducting the detailed review of solid and hazardous waste to become familiarized with the current status of the facility with respect to status and compliance:

- Evaluate the types of solid waste that the facility generates
- Evaluate whether the facility generates hazardous waste
- Evaluate and record the facility EPA ID number
- Evaluate the current and previous facility generator status (e.g., permit, large quantity generator [LQG], small quantity generator [SQG], etc.)
- Evaluate whether the facility has a permit to store waste onsite
- Review requirements (40 CFR, 264 and 265 [interim requirements]) and any state and local requirement if facility is permitted to store wastes onsite
- Evaluate whether the Boiler & Furnace Rule, 40 CFR 266 Subpart H, applies
- Verify that the correct EPA ID number is recorded on all documentation (i.e., manifests)
- Evaluate whether any processes have changed since the last audit that could influence the chemical or physical characteristics of any waste stream. If so, has the associated generated waste been re-characterized. If not, why?
- Review any correspondence from or to any regulatory agency
- Review whether any solid waste inspections have been conducted since the last environmental audit
- Review and evaluate whether the facility has received any solid waste NOVs since the last environmental audit
- Review and evaluate any corrective actions that have been undertaken to address any NOVs

8.4.7.2 Hazardous Waste

8.4.7.2.1 Hazardous Waste Determination

- Evaluate and verify that each hazardous waste stream is hazardous by either (1) listing in regulations, (2) laboratory analysis, or (3) knowledge of materials and processes used
- Evaluate generator status
- Evaluate whether waste needs to be reclassified or characterized
  a. Evaluate whether any process or manufacturing changes occurred that may impact the classification or chemical characterization of the waste (i.e., installation of new machinery)
  b. Evaluate whether any other changes have occurred that may impact the characterization of the waste (i.e., differences in parent material or re-formulation of bulk product inputs)
  c. Check to ensure that each hazardous waste stream is re-characterized at least every three years or sooner if there is any process changes that could influence the chemistry of the waste material
  d. Evaluate appropriateness of analytical list
  e. Evaluate whether point of generation is appropriate
- Review waste stream source diagram

8.4.7.2.2 90-Day Storage

- Review inspection documentation
- Verify that no container has accumulated waste for more than 90 days, unless a 30-day extension was granted
- Verify each container and tank is labeled with the words HAZARDOUS WASTE and the start accumulation date and applicable waste codes

8.4.7.2.3   Treatment and Disposal

- Verify wastes are hauled by transporters with valid EPA identification numbers
- Verify wastes are taken to disposal facilities with valid hazardous waste permits
- Review annual waste shipment summaries
- Check to make sure the facility has a tracking procedure to ensure that appropriate actions are undertaken in case a waste manifest is not returned to the facility within the required time frame

8.4.7.2.4   Biennial Reports

- Verify that the biennial report was completed and submitted by March 1
- Verify that copies of the report are kept for three years

8.4.7.2.5   Manifests

- Inspect for documentation accuracy
- Verify that manifests are used when shipping waste offsite
- Verify that exception reports are filed when a copy of the manifest is not received within the required time frame of the waste being accepted by the initial transporter (45 days for SQG and 30 days for LQG)
- Check to make sure the facility has a tracking procedure to ensure that appropriate actions are undertaken if a waste manifest is not returned to the facility within the required time frame
- Verify that manifests are kept for three years
- Verify that manifests are signed and dated correctly and clearly

8.4.7.2.6   Storage Areas

- Ensure internal communications system available
- Ensure telephone or two-way radio to summon emergency assistance
- Check for portable fire extinguishers
- Inspect spill control equipment
- Inspect decontamination equipment
- Check for fire hydrants
- Some of the above may not be necessary depending on the waste
- Determine if equipment is tested and maintained
- Verify sufficient aisle space is maintained
- Verify arrangements with local fire, police, and emergency response teams if necessary

8.4.7.2.7   Personnel Training

- Verify that personnel complete classroom instruction
- Training must include contingency plan implementation, if a LQG
- Training must be completed within six months of employment/assignment
- Verify that annual review training is provided
- Verify that employees do not work unsupervised until training is completed
- Verify specifically that waste storage area managers and hazardous waste handlers have been trained

8.4.7.2.8   Training Records

- Include job title and description for each employee by name
- Written description of how much training each position will receive
- Documentation of training received by name
- Determine if records on former employees are retained for three years
- Determine if records on current employees are maintained

8.4.7.2.9   Contingency Plans and Emergency Coordinators
- Verify that the contingency plan is designed to minimize hazards from fires, explosions, or releases, if a LQG
- Must describe actions to be taken in an emergency
- Must describe arrangements made with local agencies as appropriate
- Must include the name, address, and phone number of the emergency coordinator and any alternates
- Must include a list of any emergency equipment, its location, and description
- Must include an evacuation plan if required
- Ensure that communications systems are satisfactory and are defined in case of an emergency
- Verify that revisions are maintained and submitted to local emergency services where required
- Verify that adequate aisle space is available for unobstructed movement of emergency equipment in case of an emergency
- Verify that the plan is routinely reviewed and updated when regulations change, the plan fails, the facility changes, the emergency coordinators change, or the emergency equipment changes
- Review personnel training records to ensure that facility personnel are able to respond effectively during an emergency. This should include but is not limited to (1) inspecting and maintaining equipment, (2) shut off systems, (3) communication and alarms, (4) fire and explosions, (5) contamination, and (6) shut down procedures.

8.4.7.2.10   Emergency Coordinators
- Verify that at all times there is at least one employee at the facility or on call with responsibility for coordinating emergency response measures
- Verify that he is familiar with the facility and the contingency plan
- Verify that he has the authority to commit resources to carry out the contingency plan

8.4.7.2.11   Containers
- Verify that containers are not leaking, bulging, rusting, damaged, or dented
- Verify that the storage area is inspected at least weekly
- Verify that empty containers have less than the regulated material left in them
- Verify that containers are compatible with the waste
- Verify that containers are kept closed except when adding or removing waste
- Verify that containers are properly labeled and legible

8.4.7.2.12   Satellite Accumulation
- Verify that the satellite accumulation point is at or near the point of generation
- Verify that the containers are in good condition, compatible with the waste, and kept closed except when adding or removing waste
- Verify that the containers are marked HAZARDOUS WASTE or other identified marking, as required
- Verify that when waste is accumulated in excess of quantity limitations that the date is marked on the container and within three days the container is moved to the 90-day storage area
- Verify proper containment is provided

8.4.7.2.13   Restricted Wastes
- Determine by analysis or process knowledge if wastes are restricted from land disposal
- Verify that all notifications and certifications has been made
- Verify that records are kept for three years

8.4.7.2.14   Transporter Inspection
- Evaluate each transporter compliance history
- Evaluate distance to each disposal site
- Evaluate transportation route to each disposal site

8.4.7.2.15   Disposal Site Evaluation
- Evaluate each disposal site location
- Review compliance history
- Evaluate company compliance history

8.4.7.2.16   Record Keeping and Analytics
- Evaluate that records are organized appropriately
- Evaluate frequency of shipments
- Evaluate frequency of shipments for each transporter and disposal site
- Evaluate generation of wastes per year by type and volume

8.4.7.2.17   Agency Inspections and Notice of Violations
- Review agency review records
- Review areas of emphasis
- Review any Notice of Violations

8.4.7.3   Solid Waste

8.4.7.3.1   Inventory of Non-Hazardous Waste
- Evaluate each non-hazardous waste stream
- Ensure record keeping is up to date
- Evaluate volumes of wastes generated

8.4.7.3.2   Point of Generation Analysis
- Evaluate the appropriateness of the point of generation
- Evaluate waste volumes

8.4.7.3.3   Non-Hazardous Waste Determination
- Verify that proper waste determinations are performed (at a minimum every three years or more every time the process changes that could result in input changes of waste or chemical or physical characteristics)
- Verify that non-hazardous wastes are actually non-hazardous. Ask yourself the question, How do I know its non-hazardous? If you don't know, test it
- Review solid waste flow diagram
- Review list of solid waste streams

8.4.7.3.4   Storage
- Verify that materials are properly stored and labeled
- Verify that no applicable storage time frames are exceeded
- Ensure that storage containers are in good condition

8.4.7.3.5   Transportation
- Verify that materials are hauled according to any applicable state and local requirements
- Evaluate each transporter's compliance records
- Evaluate distance to each disposal locations
- Evaluate transportation routes' appropriateness

8.4.7.3.6   Disposal

- Verify that the materials are disposed at an approved site
- Review special waste permits or waste disposal authorizations if applicable
- Review current waste profiles
- Ensure proper and organized record-keeping procedures

8.4.7.3.7   Agency Inspections

- Review agency review records
- Review areas of emphasis
- Review any Notice of Violations

8.4.7.4   Universal Waste

8.4.7.4.1   Inventory Universal Wastes

- Evaluate each universal waste type
- Evaluate whether there are any special handling or storage requirements
- Evaluate storage locations

8.4.7.4.2   Review Storage Requirements and Locations

- Verify batteries, pesticides, mercury-containing equipment, and lamps and electronic equipment are being segregated and properly disposed at a licensed facility
- Verify that necessary employees receive required training
- Verify that waste is not accumulated for more than the allowed time period
- Review storage locations
- Review the appropriateness of each location
- Ensure wastes are being accumulated in a timely manner
- Evaluate to ensure each universal waste is appropriately separated and stored properly
- Review that universal wastes are stored in appropriate containers

8.4.7.4.3   Review Labeling Requirements

- Review that labels are clear and not obstructed
- Review any specific labeling requirement

8.4.7.4.4   Review Disposal Records and Manifests

- Review manifests for each waste type of universal waste
- Ensure tracking of waste volumes and types

8.4.7.4.5   Transportation

- Verify that materials are hauled according to any applicable state and local requirements
- Verify that the materials are shipped properly and arrive at the final destination
- Evaluate each transporter's compliance records
- Evaluate distance to each disposal location
- Evaluate transportation routes' appropriateness

8.4.7.4.6   Disposal

- Verify that the materials are disposed at an approved site
- Review special waste permits or waste disposal authorizations if applicable
- Review current waste profiles
- Ensure proper and organized record-keeping procedures

8.4.7.4.7   Agency Inspections
- Review agency review records
- Review areas of emphasis
- Review any Notice of Violations

8.4.7.5   Used Oil

8.4.7.5.1   Storage
- Verify that the tanks, containers, and fill pipes are labeled USED OIL
- Verify that the tanks or containers are in satisfactory condition
- Verify that containers are appropriate and not damaged

8.4.7.5.2   Transportation
- Verify that the used oil is transported only by transporters with a valid U.S. EPA identification number
- Verify that materials are hauled according to any applicable state and local requirements
- Evaluate each transporter's compliance records
- Evaluate distance to each disposal location
- Evaluate transportation routes' appropriateness

8.4.7.5.3   Recycle
- Verify the final destination of the used oil
- Verify that the materials are disposed at an approved site
- Inspect certificate of recycling

8.4.7.5.4   Agency Inspections
- Review agency review records
- Review areas of emphasis
- Review any Notice of Violations

8.4.7.6   Company Owned or Operated Landfills

8.4.7.6.1   Landfill Permit
- Review operational plan
- Review closure/post-closure plan
- Review water quality monitoring plan
- Discuss most recent permit renewal application
- Review list of acceptable/current waste streams

8.4.7.6.2   Compliance Documentation
- Inspect landfill operator certificates, if required
- Review latest engineering reports
- Review water monitoring reports
- Review closure/post-closure financial assurance
- Obtain copy of any NOV since last review
- Review any agency inspection reports
- Discuss any other significant agency correspondence

8.4.7.6.3   Other Landfill Related Documents
- Discuss any other significant agency correspondence
- Review closure/post-closure cost estimates

- Review remaining volume/life estimates
- Review recent topographic surveys

8.4.7.6.4   Beneficial Reuse Efforts

- Discuss facility's efforts towards beneficial reuse of landfill materials
- Discuss any state regulations, policies, guidance regarding foundry waste streams
- Discuss any state regulations, policies, guidance regarding beneficial reuse

8.4.7.6.5   Agency Inspections

- Review agency review records
- Review areas of emphasis
- Review any Notice of Violations

8.4.7.7   Recycling and Pollution Prevention

- Evaluate whether the facility has a recycling plan or initiative
- Evaluate what items, products, and materials are or can be recycled
- Evaluate whether items, products, and materials are recycled internally or externally
- Evaluate whether the facility could benefit from a regulatory-sponsored recycling or pollution prevention program (i.e., USEPA's Climate Leaders Program)
- Evaluate the status of the facility's pollution prevention plan and awareness
- Evaluate whether the facility uses chlorinated solvents and enact measures to eliminate the continued use if chlorinated solvents are used or stored at the facility
- Evaluate whether the facility uses other targeted compounds such as cadmium, mercury, chromium, and lead and evaluate measures to eliminate or reduce use of these compounds
- Evaluate whether the facility uses large quantities of volatile organic compounds (VOCs) other than chlorinated solvents such as xylenes, toluene, methyl ethyl ketones (MEKs), etc. and evaluate initiating a pollution prevention strategy
- Evaluate the effectiveness of the facility's chemical ordering procedures as a method of pollution prevention and awareness

8.4.8   PCBs

8.4.8.1   Review and Record Keeping

- Record the type (liquid, dry, pole-mounted, pad, polychlorinated biphenyl [PCB], non-PCB), contents, capacity, age, and location of each transformer at the facility.
- Review list of PCB equipment including:
- Transformers (>50 ppm, > 500 ppm, made before July 1, 1979)
- Evaluate ownership records of all transformers on facility property
- Capacitors (large, high voltages, made before July 1, 1979)
- Hydraulic presses (made before July 1, 1979)
- Oil-filled electrical switches (made 1935–1979)
- Oil-filled electrical motors (made before July 1, 1979)
- Oils, paints, inks, sealants (made 1935–1971)

8.4.8.2   Compliance Documentation

- Review analysis of oils for above units
- Inspect PCB labeling and non-PCB labeling
- Review documents regarding storage and disposal of PCB equipment
- Review inspection reports of PCB equipment in use
- Review registration of PCB transformers

8.4.8.3   PCB Spill Review

- Evaluate whether any recent or historical PCB spills have occurred at the facility
- Obtain copy of any PCB or non-PCB spill reports

8.4.8.4   Agency Inspections

- Review agency review records
- Review areas of emphasis
- Review any Notice of Violations

8.4.9   SPILLS

8.4.9.1   Liquid Inventory and SPCC Plan

- Review SPCC applicability and current total quantity of regulated liquids stored onsite
- Review SPCC plan
- Review list of potential spill materials
- Review waste and material storage areas with diagram
- Review contingency plans
- Review training records
- Ensure that SPCC plan is signed by a professional engineer (PE)
- Ensure that a Certification of Substantial Harm Determination is completed
- Review pre-calculations for reportable quantities

8.4.9.2   Spill History

- Review any spills since last review
- Review incident reports
- Review NRC or state spill reports
- Inspect spill areas
- Review corrective action
- Discuss historical spills

8.4.9.3   Agency Inspections

- Review agency review records
- Review areas of emphasis
- Review any Notice of Violations

8.4.10   STORAGE TANKS

8.4.10.1   Review SPCC and Contingency Plans

- Review spill/release prevention and response procedures
- Review reporting procedures

8.4.10.2   Type and Location of Tanks

- Review record keeping on the design, construction, installation, alterations, repairs
- Record material of construction, capacity, contents, and location of each aboveground storage tank (AST) and underground storage tank (UST) at the facility
- Review contents and identification (labeling)
- Review documented unloading procedures and training records
- Procedures – Are employees present during the unloading operations

8.4.10.3 Containment Areas
- Review containment for integrity, size, and impermeable nature, etc.

8.4.10.4 Spill/Leak History
- Documentation, open issues

8.4.10.5 Tank Fail-Safe Measures
- Review calibration and testing of alarms and gauges
- Documented procedures and preventative maintenance (PM)
- Record keeping

8.4.10.6 Inspections – PM and Record Keeping
- Routine (weekly) housekeeping, levels, leaks, etc.
- Formal (quarterly) – same as above to include labeling, spill kits, and personal protective equipment (PPE) adequacy and location, piping, tanks, containment, repairs
- Integrity (internal and external) and leak testing (not to exceed every five years and only applies to USTs)

8.4.10.7 Training Records
- Spill team, employees responsible for handling, loading, unloading, inspections

8.4.10.8 Agency Inspections
- Review agency review records
- Review areas of emphasis
- Review any Notice of Violations

8.4.11 Hazardous Materials (HAZMAT)

8.4.11.1 Review DOT Requirements for Hazardous Material Shipments
- Registration requirements
- Hazardous Waste Handler Training documentation (every three years)
- Management of hazardous materials for shipment
- Record keeping
- DOT security plan

8.4.11.2 Agency Inspections
- Review agency review records
- Review areas of emphasis
- Review any Notice of Violations

8.4.12 Toxic Substance and Control (TSCA)

8.4.12.1 Chemical Ordering Procedure
- Review written procedure
- Discuss environmental personnel involvement
- Discuss customer requirements on chemical use limitations
- Ensure permitting issues and applicability are addressed
- Identify efforts to eliminate chlorinated solvents

8.4.12.2   New Chemical Evaluation

- Review any new chemicals which are produced or imported into the United States since last review
- Review training records for shipping and receiving personnel
- Evaluate whether a security plan for the facility has been developed, if required

8.4.13   EMERGENCY PLANNING AND COMMUNITY RIGHT-TO-KNOW-ACT (EPCRA)

8.4.13.1   Hazard Communication Plan

- Employee training on workplace chemicals, procedures, and record keeping
- Review material safety data sheet (MSDS) management, updating, chemical listing, tracking, employee access/availability procedures

8.4.13.2   Emergency Planning Notification

- Extremely hazardous substances threshold planning quantities
- Submission of emergency response and/or contingency plans to Local Emergency Planning Committee (LEPC) (verify if plans are up to date and current)

8.4.13.3   Tier I and II Report Submissions

- Review other state requirements for supplemental reporting
- Hazardous chemical listing submissions (local fire department or fire marshal)

8.4.13.4   Toxic Release Inventory (TRI) Form R Report Submission

- Review applicability determination (used, manufactured, or processed) and submission of reports

8.4.14   SITE INSPECTION

Following the review of written materials, a site inspection of the entire facility should always be conducted. It is recommended that the site inspection be structured in such a way that it follows the manufacturing process as much as possible. Photographs are recommended to document the current environmental condition of the facility, especially those items that may require a corrective action. It is also recommended that a photograph be taken after any corrective action to document task completion and to demonstrate compliance.

It will be helpful to have a map of the facility while conducting the site inspection that shows the entire property, buildings, and operations that are clearly marked. In addition, it will be helpful to have waste storage areas, air emission points, points of generation of solid and hazardous waste, points of generation of wastewater and wastewater discharge points, and stormwater outfalls labeled on the map.

During the site inspection, the following items should be observed, evaluated, and noted:

- If there is an item observed during the site inspection or at any time that may present an immediate threat or imminent danger to human health or the environment, follow the organization's emergency notification procedures. Examples may include but would not be limited to the following:
  - Evidence of existing release of hazardous substances or petroleum products
  - Evidence of material threat of a release of hazardous substances or petroleum products
- Other areas to inspect, review, and/or observe:
  - General housekeeping
  - Any new equipment?

- Any new processes?
- Any new waste streams?
- Review history of neighborhood complaints (documented)
- Vehicle maintenance areas
- Spill kit's location, contents listing, and adequacy
- Emergency response phone list – postings, updated, controlled document, etc.
- Chemical and waste storage areas
- Labeling, storage, and housekeeping
- Waste storage areas, piles, drop boxes, leaks, proper contents
- New product storage areas (oil and chemicals)
- Universal waste storage practices and labeling
- Drum management
- Drum labeling, management, residue issues, secure when not in immediate use, empty drum storage
- Secondary containment areas
- Plant and property security – alarm system, guard service, fencing, gates, lighting, access control, etc.
- Roof inspection
- Drainage
- Emission deposits
- Exhaust stack(s) deposits
- Roof vent(s) deposits
- Floor drains identified on plant drawings, purpose, discharge, potential discharge, drain blocker in the event of a spill
- Stormwater conveyance systems' preventative maintenance
- Erosion issues
- Catch basins
- Detention ponds
- Outfalls
- Outside processes, if any
- Equipment and product storage potential for contamination
- Utility supply and shutoff valves, identification, security, management, emergency procedures, and training
- Compressor discharge management
- Exhaust fan and vent deposits to the exterior of the building
- Baghouse inspection, record keeping, and preventative maintenance
- Review the facility's control of chemicals for chlorinated solvent content
- Secondary container labeling
- Neighboring properties identified and note (1) the potential for hazardous substance use or storage such as ASTs and drum storage areas, (2) stained soil, (3) topography, (4) drainage patterns, and (5) types of operations
- Property boundaries inspected, non-facility contributions, observations/activity, any issues

8.4.15 Closing Meeting

A closing meeting should be conducted at the end of the environmental audit. The purpose of the closing meeting is to:

(1) Inform facility management of the results of the environmental audit
(2) Set a timeline for when the draft report will be completed

(3) Establish a timeline and assign appropriate personnel for implementing corrective actions as a result of the findings of the environmental audit, if any

(4) Remind facility management with respect to:
- Internal company policy pertaining to environmental audits
- Report privilege
- Report distribution
- Procedures to follow during agency inspections
- Procedures to follow during an emergency or spill event
- Periodically review environmental permits
- Review company policy for non-routine correspondence with a regulatory agency
- Review company policy for permit applications
- Remind all employees that environmental compliance requires an awareness and daily attention
- Remind all employees that compliance is not an option

8.4.16 REPORT PREPARATION

A draft report should be prepared soon after the audit has been conducted, certainly no more than 15 to 30 days following the audit. The report should be draft and marked as such when submitted for review. For items that require a corrective action that are discovered during the environmental review, it is recommended that they be separated into either best management practices (BMPs)-related issues, compliance-related issues, or observations at the end of each section of the report.

An example environmental audit report outline is presented below:

- Fundamentals
  - Air
  - Water
  - Land
- Conducting the audit
  - The opening meeting
    - Air
    - Water
    - Hazardous waste
    - Universal waste
    - Non-hazardous waste
    - Spills
    - PCBs
    - Hazardous materials (HAZMAT)
    - Toxic Substance and Control Act (TSCA)
    - Emergency Planning and Community Right-To-Know-Act (EPCRA)
  - Site inspection
  - The closing meeting
  - Summary and conclusions
  - Photographs

After the report has been reviewed and finalized, a system should be put in-place to ensure that all the recommended items are properly addressed and correct, especially any identified non-compliance items. As a reminder, any identified non-compliance items are of the highest priority and must be addressed immediately.

Typically, a spreadsheet is prepared for each item identified in the report as a compliance, BMP, or observation that typically includes the following:

- Name of person responsible for supervising or conducting corrective action
- Time period allowed to complete corrective action
- Actions to be taken to complete corrective action
- Completion sign off
- Completion date
- Management sign off

8.5   AGENCY INSPECTIONS

We first discussed the importance of environmental enforcement in Chapter 4. We learned that the United States has an established and effective enforcement policy and program. The United States has also consulted with many developing nations in providing assistance in establishing environmental regulations and enforcement programs. Much of the success that the United States has achieved has been, in part, through its environmental enforcement program.

The USEPA has also helped themselves by promoting "Self Audits". Self Audits have benefited the regulated community in many ways including education and conducting corrective actions when issues are discovered. Self Audits have benefited USEPA as well by shifting much of the responsibility to the regulated community. However, this is predicated on conducting inspections to ensure compliance.

Inspections by regulatory authorities vary from location to location and state to state. Some inspections may target one particular media such as air, water, or solid and hazardous waste, while others may cover several all at once, but that is rare. Usually agency inspections will focus on one particular set of regulations.

Typically, an agency inspection will consist of the following:

1. Opening meeting. This is to discuss the reason behind the inspection (e.g., routine inspection or compliance), the facility layout, production processes, plant history, wastes generated and storage areas, air emission points and stacks, air emission controls, water distribution, and water use, water treatment and discharge, and other environmental aspects.
2. Document review. The document review usually follows the opening meeting. Document review usually focuses on demonstrating compliance with applicable environmental permits such as air and water discharge. Document reviews may also include inspection of manifests and waste characterization. Waste characterization may also focus on non-hazardous waste streams to ensure that a non-hazardous waste has been properly characterized. Training records, logs, plans, reports, forms, and correspondence records are also routinely inspected.
3. Site inspection. From the information obtained from the document review, the agency inspector will have a much more educated perspective on operations and what regulations apply to the facility. The inspector will usually request that a site inspection be conducted to visually inspect the area or areas that the inspection has focused on (air, water, solid and hazardous waste, etc.). The regulatory inspector may also ask if photographs can be taken during the site inspection. The regulatory inspector is typically looking to evaluate the consistency between the document review with site observations.
4. Closing meeting. The closing meeting will usually involve whether any violations have been identified and may also involve additional requests for information and documentation that was not readily available at the time of the inspection.

To ensure that the facility and individuals properly respond to an agency inspection and that the highest professional and ethical standards are maintained, the following should be considered:

- Detailed notes should be taken and every issue identified by the inspector(s) should be listed and confirmed with an explanation as to the reason the identified issue is either not in compliance or requires further evaluation.
- As important as an inspection is concerning environmental compliance, it is even more important and critical to tell the truth when responding to an agency inspector who asks a question. However, it is equally important and critical that an appropriate answer to a question be accurate. Therefore, if you don't know that answer to a question, it's acceptable to state that you don't know and that if necessary, you will look into the issue and provide a response at a later time.
- Address and correct as many identified issues as possible before the inspector(s) leaves. Depending on the location of the facility, the person conducting the inspection, the type of inspection, and the reason for the inspection, if an issue is addressed immediately and is considered a minor infraction, a formal Notice of Violation (NOV) may not be issued for that specific identified item. Furthermore, correcting other identified items, with proper documentation, before a formal written response to the inspection is released may also result in the same or similar response by the agency.
- In most all instances, notification to appropriate higher management and in-house or outside counsel should be made as soon as possible to assist in responding and managing an agency inspection.
- Remember that perfection may be the goal but will likely never be achieved in a dynamic system such as a complex manufacturing facility. However, if perfection is the goal, excellence may be achieved along the way.
- Maintaining compliance is an everyday challenge.

As we have discovered, most environmental regulation is based on self-disclosure. This means that a facility monitors itself and usually discovers an issue before the regulatory agency. An example would be when compliance sampling is required to document that a facility is in compliance with a permit term. Often the permit holder would notify the appropriate agency and provide documentation as to the facts surrounding the non-compliance, provide a cause, and then also provide actions undertaken to ensure future compliance. The regulatory agency may or may not issue a Notice of Violation (NOV) depending on the facts surrounding the issue. Some factors that are considered include:

1. How was the agency notified
2. Did the issue cause harm to human health or the environment
3. Was the issue that caused the non-compliance a purposeful act
4. Compliance history

8.6 ENVIRONMENTAL AUDITS AND SUSTAINABILITY

Environmental audits should not be confused with sustainability. An environmental audit is a measure of compliance with environmental laws, regulations, standards, and other requirements. An environmental audit is not a measure of sustainability. However, an environmental audit is a good resource to begin evaluating sustainability. Sustainability is based on a simple principle that states that (USEPA 2018e):

Everything we need for our survival and well-being depends, either directly or indirectly, on our natural environment

USEPA defines **sustainability** as creating and maintaining conditions under which humans can exist in productive harmony to support present and future generations (USEPA 2018e). Note that the definition states "productive harmony." One may ask, in productive harmony with what? The answer is with the environment. Therefore, in order to achieve some level of sustainability we must understand our environment and also understand the aspects of how facility operations impact the environment. Sustainability is then the outcome of analyzing the aspects of facility operations with the natural environment and developing and engineering methods to either minimize or eliminate potential harmful impacts.

A good start for understanding facility operations that may negatively impact the environment is with a comprehensive environmental audit. The other key element in the definition of sustainability is understanding the natural environment. This is often a difficult task as we shall see in the next chapter where we will explore the pursuit of sustainability.

8.7   SUMMARY OF FUNDAMENTAL CONCEPTS OF ENVIRONMENTAL COMPLIANCE

The primary purpose of environmental regulations is to protect human health and the environment. We should also have an appreciation of the sheer immensity and often times overwhelming complexity of environmental regulations.

An important aspect of environmental regulations and conducting environmental audits is that environmental regulations focus on what is called "end of the pipe" or wastes that are generated. For instance, looking back over this chapter on how to conduct an environmental audit, the regulations focused on the type and amount of air emissions, water discharges, and solid and hazardous wastes that were generated.

Each facility should also have a keen awareness that any process or manufacturing changes or installation of new equipment may have a profound effect on the types, amounts, and chemical and physical characteristics of emissions, discharges, and wastes that are generated that may require a permit or permit modification and may render existing waste profiles obsolete and require associated wastes be re-characterized.

In this chapter, we have boiled things down to a few points that will provide assistance and focus our efforts to guide us in achieving compliance with environmental regulations and beyond.

The significant lessons in this chapter include:

1. **It all begins with conducting an environmental audit.**

    The first step ito compliance is to conduct an environmental audit. An environmental audit is an assessment in which a facility or organization is observing practices to minimize harm to the environment. USEPA defines environmental audit as a systematic evaluation to determine the conformance to quantitative specifications to environmental laws, regulations, standards, permits, or other legally required documents (USEPA 2018a).

2. **Being in compliance with environmental regulations has little to do with sustainability.**

    In this chapter, we have begun to explore what it means to look toward environmental stewardship and sustainability. Environmental stewardship and sustainability requires going beyond environmental compliance and addressing deeper questions about how we may impact the environment. Being in compliance means that you conduct activities consistent with environmental regulations. It does not ask the question whether you should be doing something else that will enhance your organizational goals. In other words, you have to ask yourself, "Just because you can doesn't always mean you should."

3. **Work yourself out of environmental regulations.**

    This is often difficult and challenging but is well worth the effort and will assist in paving the way to environmental stewardship.

4. **There is little we can control but things that are within our control are significant and can make a big difference**

   Environmental professionals for the most part do not control the location where manufacturing takes place or the type of manufacturing or products that are made. The only thing that environmental professionals typically have input in are the chemicals that are used and what management actions and engineering controls to put in-place to prevent mismanagement and potentially causing a release and harming the environment.

5. **Limit chemical use when possible.**

   Remember, if there were no hazardous substances there would be little need for environmental regulations or environmental professionals.

6. **Lastly, we must remember that "no matter where you are, it's all the same."**

   We are learning that we live on one planet with no environmental boundaries and that contamination does not respect borders. Another point is that environmental regulations are built on the USEPA or EU platforms and those platforms rely on three basic principles:

1. Protect the air
2. Protect the water, and
3. Protect the land

The focus of the next chapter is the pursuit of sustainability.

REFERENCES

Connecticut Department of Environmental Protection. 2018. *Environmental Best Management Practices Guide.* Hartford, CT. 4p. https://www.ct.gov/dep/compliance-assistance. (accessed December 1, 2018.)

United States Environmental Protections Agency (USEPA). 1986. Environmental Auditing Policy Statement. *Federal Register.* Vol. 51. No. 131. pp. 25004–25010. Washington, DC. Wednesday, July 9.

USEPA. 2018a. *United States Environmental Protection Agency QA Glossary.* https://www.epa.gov/emap/archive-emap/web/html. (accessed December 1, 2018.)

USEPA. 2018b. *Audit Protocols.* https://www.epa.gov/compliance/audit-protocols.

USEPA. 2018c. *Compliance.* https://www.epa.gov/compliance. (accessed December 1, 2018.)

USEPA. 2018d. *National Menu of Best Management Practices (BMPs) for Stormwater.* https://www.epa.gov/npdes/national-menu-best-management-practices-bmps-stormwater. (accessed December 1, 2018.)

USEPA. 2018e. *What Is Sustainability?* https://www.epa.gov/sustainability. (accessed December 1, 2018.)

9 Fundamental Concepts of Sustainability

9.1 INTRODUCTION

This chapter is dedicated to the concept of sustainability. United States Environmental Protection Agency (USEPA) defines **sustainability** as creating and maintaining conditions under which humans can exist in productive harmony to support present and future generations (USEPA 2018a). Note that the definition states "productive harmony." One may ask, in productive harmony with what? The answer is with the environment. Therefore, in order to achieve some level of sustainability we must understand the environment and also understand the aspects of how facility operations impact the environment. Sustainability is then the outcome of analyzing the aspects of facility operations with the natural environment and developing and engineering methods to either minimize or eliminate harmful potential impacts.

Sustainability is based on a simple principle that states that (USEPA 2018b):

Everything we need for our survival and well-being depends, either directly or indirectly, on our natural environment

The United Nations defines sustainability as (United Nations 2018a):

development that meets the needs of the present without compromising the ability of future generations to meet their own needs

Sustainability had its origin in 1969 in the incorporation of the United States National Environmental Policy Act (NEPA):

to foster and promote the general welfare, to create and maintain conditions under which humans and nature can exist in productive harmony and fulfill the social, economic, and other requirements of our present and future

The term "sustainable development" was first used by the United Nations in 1987 (United Nations 2018a).

9.2 SUSTAINABILITY POINTS OF CONTENTION

Points of contention exist between countries concerning sustainability and include differences in power and responsibility, and environmental and economic concerns that must be solved before sustainability can proceed on a path to become a reality on a global level (United Nations 2018a). Additional points of contention are religious conviction and population, which are difficult and sensitive to many and are often not openly discussed.

9.2.1 POWER AND RESPONSIBILITY CONTENTION

These points of contention are realized between developing and developed countries. It is of concern for developed nations because they are perceived to be imposing their new-found sustainable values upon developing countries. Developed nations have already greatly benefited from exploitation of environmental resources, whereas developing nations have not had the chance or enough time to exploit their natural resources and feel resentment toward developed nations. A significant focus in the sustainable development movement has been on freeing people in developing nations from the bonds of poverty and starvation. Freeing nations from poverty and starvation would mean consuming enormously more energy and resources in order to raise the standard of living of these individuals which number in the billions. One of the most important arguments of sustainability states that the rights provided by governments, such as the right to vote and freedom of speech, are not of much use unless the society provides its citizens with an opportunity of an education, access to food, and employment (United Nations 2018a).

9.2.2 Economic Contention

This brings us to perhaps one of the two most problematic hurdles in achieving a sustainable world and that is sustainability does not yet appear to be compatible with a capitalistic economic system (United Nations 2018a). An economic system of capitalism relies on an ever-growing economy. How can humans achieve sustainability with a growing population, increasing energy needs, lifting vast human populations (e.g., more than 2 billion people) in undeveloped nations' standard of living and achieve a sustainable world where we all live in harmony with our environment? This all must take place in a fragile biosphere called Earth. So far, our economic system continues to grow and gather momentum and as it does it threatens to destabilize the global ecological balance in many significant ways. How can humans modify a capitalistic system that has ecological destructive outcomes and achieve a sustainable planet?

There are specific issues, such as reversing much of the damage to the ozone layer, that have been successful and lends hope and a call to action that not all is lost and that balance can be restored. After all, humans are an adaptable species and we are just recently putting our collective intellectual and cooperative skills to work to confront this challenge which we must win.

9.2.3 Religious Contention

The last hurdle relates to religious conviction. In May 2015, Pope Francis of the Roman Catholic Church released an Environmental Encyclical from the Vatican that addressed climate change, global warming, pollution, economics, and population control. The Environmental Encyclical, as it is now referred to, states that humans must forcefully reject the notion that we were given dominion over the Earth and absolute dominion over other creatures (Pope Francis 2015). This is yet another example of hope and resolve that we all have a part to play in, as Pope Francis further states that there is an urgent need for a radical change in the conduct of humanity (Pope Francis 2015).

9.3 GLOBAL SUSTAINABILITY GOALS

The United Nations has framed sustainability goals termed "Global Goals," which include (United Nations 2018b):

• No Poverty	• Zero Hunger	• Good Health
• Well Being	• Quality Education	• Gender Equality
• Clean Water	• Effective Sanitation	• Affordable Energy
• Clean Energy	• Economic Growth	• Satisfactory Employment
• Industry	• Innovation	• Infrastructure
• Reduced Inequalities	• Sustainable Cities	• Responsible Consumption
• Climate Action	• Life Below Water	• Life on Land
• Peace and Justice	• Partnerships	• Life in the Air

Much of what we have discussed up to this point in this chapter may sound like a bleak outlook from a world view, but let us not ponder and debate the political, social, economic, and religious challenges on a global scale before we examine what can be accomplished at the local level, which as you will see in this chapter, is significant.

For instance, sustainability has been a priority in the United States since the USEPA was first formed nearly 50 years ago. This has been achieved through improving air quality, water quality, and land impact through the Clean Air Act (CAA), Clean Water Act (CWA), Resource, Conservation, and Recovery Act (RCRA), and Comprehensive Environmental Response, Compensation and Liability Act (CERCLA) and a multitude of other environmental regulations, many of which we have examined. In addition, actions by the United Nations, the World Bank, private nonprofits, thousands of businesses, and many other nations have had positive results.

A good start for understanding and identifying facility operations that may negatively impact the environment is a comprehensive environmental audit. The other key element in the definition of sustainability is understanding the natural environment. This second key element is often the most difficult task.

We have covered several subjects within this book that now can be pulled together into a cohesive framework that can be used to implement a comprehensive sustainability program at any level with the end goal of environmental stewardship. The pursuit of sustainability involves many subject areas and levels. We will discuss what must be enacted at the global level, national level, and at the community and individual level.

Those subject areas that will be addressed in the following sections include:

- Cohesive action from the global community
  - Global environmental regulations
  - Obtaining a sustainable human population
  - Modifying the human diet
  - Financial fairness and equity distribution
- Improving environmental regulation at national levels
- Understanding the natural setting of our urban and industrial areas
  - Protecting sensitive ecological areas
  - Smart urban development
  - Improving urban transportation
  - Improving building design
  - Incentivizing multi-family housing
- Knowing the chemicals that are used and how they might cause harm to the environment and humans if they are released
- Knowing what alternative chemicals are available
- How to reduce chemical use
- Preventing chemical releases
- Reducing energy use
- Using alternative sources of energy
- Minimizing solid waste, air emissions, and water discharges
- Reducing water use
- Minimizing stormwater runoff
- Repairing environmental damage to the extent possible
- Attaining environmental stewardship

9.4   COHESIVE ACTION FROM THE GLOBAL COMMUNITY

At least five actions at the global level are required if sustainability is to have a chance and include:

- Cohesive environmental policy and enforcement
- Controlling the human population

- Modifying the human diet
- Financial fairness and equity distribution
- Modifying business models

9.4.1 GLOBAL ENVIRONMENTAL REGULATIONS

Pollution does not respect boundaries. Like all matter, pollution follows the laws of physics and will disperse from a point of origin and given enough time will be detected world-wide if it does not degrade. One country can adopt and implement effective sustainability measures but, in the end, will eventually become polluted at similar levels as other countries who do not implement the same measures regardless of financial resources or technology. Therefore, we must develop a global community and enact cohesive environmental policies and enforcement at the international level. Climate change is a good example that action must be taken at a global level since greenhouse gases and a plethora of other pollutants travel the globe and become ubiquitous. Just in the last few decades there is much less pack ice in the artic and glaciers are shrinking at a significant pace throughout the world, as shown in Figures 9.1, 9.2, and 9.3.

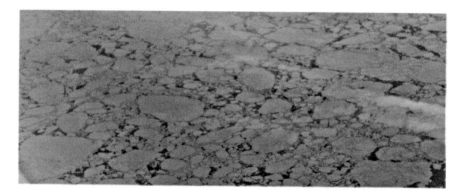

FIGURE 9.1   Melting pack ice near the North Pole. (Photograph taken in September 2018 by Daniel T. Rogers.)

FIGURE 9.2   Receding glaciers in Greenland. (Photograph taken in August 2008 by Daniel T. Rogers.)

Significant recent retreat of glacial ice, especially the Nisqually Glacier

FIGURE 9.3 Retreating glacial ice on Mount Rainier, USA. (Photograph by Daniel T. Rogers, August 2018.)

Building a global community is possible and is achieved periodically. An example is the Olympic Games, which bring the world together to celebrate sport every 2 years. Building a global community will require significant cooperation and also must include removing nationalistic political, social, and religious boundaries (Harari 2018). Patriotism for a nation, which is common, must graduate to a global level.

We have discussed a significant potential impediment in achieving sustainability, which is the current capitalistic economic system. Every country of the world is unique but also shares commonalities with respect to basic economic systems. The current capitalistic economic system must change from growth-minded to sustainability-minded. Sustainability incentives must be integrated at the international, national, community, and individual economic levels. We will discuss this aspect in Section 9.4.5 where businesses are incorporating sustainability into their business models.

9.4.2 THE HUMAN POPULATION

In Chapter 7, we examined environmental regulations of many countries of the world and discovered that there are at least three factors common to countries that have significant pollution issues. Those include (1) lack of financial resources, (2) over population, and (3) lack of effective environmental regulations.

We have discussed in the previous section that capitalism currently relies on growth for itself to be sustainable. Much of the economic growth that fuels capitalism is achieved through an endless and continued increase in the human population and a growing standard of living, especially in underdeveloped countries. In 1900, the human population was approximately 1 billion and currently is estimated at more than 7 billion (United Nations 2017; Harari 2015). A continued growing human population is clearly not sustainable and has grabbed the attention of world and religious leaders (Pope Francis 2015).

The question is what is a sustainable population of humans on Earth? Clearly, we have exceeded that balance with our current practices and disregard for our planet. The answer lies in reducing adverse impacts and balancing those achievements with a human population that does not adversely impact future generations and other life on Earth.

Many human technological innovations have been developed throughout our history that include (1) the Agricultural Revolution approximately 12,000 years ago, (2) the Industrial Revolution in the last 300 years, and (3) the information and computer revolution just recently. Each of these technological marvels have enabled humans to establish large cities, have time for recreation, and greatly increase our population (Harari 2015). If humans can solve complex technological unknowns, humans can also solve the human population issue before nature does it for us.

9.4.3   MODIFYING THE HUMAN DIET

Much of the farming in the United States is concentrated in the production of meat (USDA 2011). Meat production in the United States also consumes huge volumes of water and creates significant biological pollution and is also a significant contributor of methane, which is a powerful greenhouse gas (USEPA 2018c). Slight modifications in the diet of Americans that include reducing meat and animal fat consumption combined with modifying farming techniques through the addition of urban farms and other measures can have a significant positive impact on:

- Improving the health of Americans by improving our diet and reducing exposure to contaminants such as pesticides and herbicides
- Reducing greenhouse gas emissions
- Reducing fugitive emissions
- Significantly increasing the amount of natural land
- Decreasing water use
- Decreasing fertilizer, pesticide, and herbicide use
- Decreasing exposure to harmful chemicals
- Decreasing biological pollution
- Improving surface water quality
- Improving water quality of the oceans

An additional factor is the amount of food that is wasted. According to the USEPA (2017), as much as 40% of food produced in the United States is thrown away. According to the United Nations (2019a), one-third of food produced world-wide is wasted.

Evaluating ocean fisheries is much more difficult because the oceans are large, have many different jurisdictions, and counting the number of fish is difficult even with advanced technologies. However, there is evidence that suggests that ocean fisheries are under stress from overexploitation. This is evidenced by the fact that the number of fish species caught is declining even though the number of fishing vessels has increased and fishing technologies have improved. For example, in 1974, only 10% of ocean fisheries were considered stressed whereas nearly 30% were stressed in 2010 (World Ocean Review 2019). However, the National Oceanic and Atmospheric Administration (NOAA) of the United States released a report in 2017 that concluded that only 15% of fisheries were stressed, which represents a decrease over previous evaluations (NOAA 2018).

In 2010, China caught more than double the amount of fish (nearly 14 million metric tons) than any other nation. Indonesia and the United States rank second and third, respectively, at just over 4 million tons each (World Ocean Review 2019).

9.4.4   FINANCIAL FAIRNESS AND EQUITY DISTRIBUTION

As we identified in Chapter 7, there is a correlation between effective environmental regulations and standard of living. Those countries without financial resources have more significant environmental uses than countries with deeper financial resources. Fortunately, the United Nations and The World Bank are very aware of this situation and have been working diligently to improve the environment of undeveloped nations (United Nations 2019b; World Bank 2019). In addition, the USEPA is also aware and has been assisting numerous other nations at improving the environmental quality by implementing several environmental initiatives, which we have discussed in Chapter 7 (USEPA 2019a).

9.4.5   MODIFYING BUSINESS MODELS

Thousands of businesses world-wide are adjusting their business models by incorporating sustainability measures and incentives into their culture and using these modifications to promote themselves by meeting the highest verified standards of social and environmental performance, public transparency, and legal accountability to balance profit and purpose. Two widely used methods include ISO 14001 and Certified B Corporations.

The International Organization for Standardization (ISO) 14000 is a family of standards originating in 1992 and has now grown into detailed international standards that specifies requirements for an effective environmental management system (EMS) (American Society for Quality 2019). ISO 14001 is a voluntary standard that examines every aspect of a particular business and evaluates how each aspect potentially impacts the environment. ISO 14001 then examines how to reduce environmental impact by modifying aspects. ISO 14001 examines every step of the manufacturing process from each raw material, energy consumption and type, chemical use, packaging, space, location, transportation, and each type of waste generated (American Society for Quality 2019). ISO 14001 essentially states that everything that is not included in the manufactured product eventually become a waste.

ISO 14001 is widely used throughout the world, especially in the automobile manufacturing industry. Automobile manufacturers required suppliers to be ISO certified as early as 2002 (Thornton 2001).

ISO 14001 is an environmental management system that has also been used as a tool to evaluate the status of environmental compliance but has found more purpose in reducing waste and generating manufacturing efficiencies that translate into business advantages by lowering costs and liabilities and other factors.

Certified B Corporations is an additional method that can steer business toward sustainability. Certified B Corporations believe (B Corporations 2019):

- That they must be the change that people seek in the world
- That people and places matter
- That businesses should aspire to do no harm and benefit all through their products, practices, and profits
- That businesses should act with the understanding that each person depends upon another, including future generations

Currently, there are hundreds of Certified B Corporations throughout the world, with 470 just in the United States (B Corporations 2019).

9.4.6 Managing Natural Resources

Many of the countries we have evaluated where there has been an environmental incident that has caused enormous harm to human health and the environment were or are related to natural resource exploitation. Examples include:

- Exxon Valdez oil spill in Alaska
- Deepwater Horizon oil spill in the Gulf of Mexico
- Poor mining techniques throughout the world
- Poor forest management techniques throughout the world
- Hydroelectric power generation throughout the world
- Decreasing erosion by improving farming techniques

These examples highlight the need for common regulatory treatment on a global scale for resource exploitation.

9.5 IMPROVING ENVIRONMENTAL REGULATION AT NATIONAL LEVELS

We have spent much of this book examining the environmental regulations of the United States and numerous other countries of the world. We have concluded that no country is perfect; some countries are better than others when it comes to effective environmental regulation, and all countries can improve and must if humans desire to inhabit Earth sustainably. We have also discovered that there is a direct correlation between effective environmental regulations and a high standard of living and financial resources, and relatively stable population growth.

As discussed in Chapter 6, the majority of subject areas where environmental regulations in the United States are either in need of improvement or where new regulations are needed include the following:

- Modify environmental enforcement emphasis and policy
- Science-based improvements to environmental regulations to account for regional risks and climate change
- Regulate the farming industry and farmland no different than other industries and regulate the farm industry through the USEPA
- Pesticides and herbicides
- Fertilizers
- Septic tanks
- Invasive species
- Urban planning and land use
- Residential and household waste
- Urban air
- Further or new restrictions on harmful chemicals, especially those that are persistent and mobile in the environment
- Emerging contaminants, such as those described in Chapter 2
- Sustainability
- Noise
- Plastic

Many of the items listed above fall into the exceptions scenario that we first introduced in Chapter 1 and explained in the introduction of Chapter 6 and mentioned again in this chapter. The main theme with the exceptions is that environmental regulations should address all identified risks for a given practice, situation, or exposure scenario. It makes little sense if human health and the environment are protected from nine out of ten exposure risks when the remaining risk factor causes harm. This example is realized when considering the lack of regulations of the items listed above.

Another item worthy of discussion is that there are political factors and areas where there are overlapping jurisdictions that may have negative outcomes that must be addressed when discussing the need for further environmental regulations in the United States. For instance, the United States Department of Agriculture would almost certainly interject an interest in environmental regulation of farmland. In addition, the Department of Interior would also be interested in sustainability regulations as it pertains to national forests and parks and environmental regulations on rangeland would be of interest for the Bureau of Land Management (BLM).

9.6   NATURAL SETTING OF OUR URBAN AREAS

The natural setting of our urban and industrial areas perform a crucial role in how contaminants behave and how they affect us when they are released into the environment. As we discussed in Chapter 4, all of us walk on a history book in each urban area of the world but yet most of us go unaware of its significance and the potential impact it may have on all our lives and its just centimeters beneath our feet.

9.6.1   GEOLOGIC VULNERABILITY

We do not have control over the natural environment. Therefore, we must understand the natural environment where our urban areas are located and develop methods to minimize or eliminate the potential harmful effects of contaminants upon human health and the environment. A logical first step to this end is through an *understanding* of urban geology, followed by an *evaluation* of the extent that a given urban area's geology influences the migration of contaminants. And, since water plays a critical role in assessing a region's vulnerability to contamination, the analyses performed during the evaluation step require an understanding of water occurring at the Earth's surface and beneath (Rogers 2014). These factors control the severity of the damage and are (1) the geologic and hydrogeologic environment, (2) the physical chemistry of the contaminants and amounts released, and (3) the mechanism in which the release occurs (Rogers 1996; Murray and Rogers 1999; Rogers et al. 2006, Kaufman et al. 2013; Rogers et al. 2016; Rogers 2018).

Despite the availability of specific methods and procedures, the environmental assessment of many urban areas can become a daunting task. This situation arises because the near-surface geologic deposits in urban areas are poorly understood, difficult to study, complex, have been anthropogenically disturbed, and exhibit high variability over short distances. Therefore, to achieve any level of success in mitigating environmental contamination, it becomes a prerequisite to understand how contaminants migrate in any given urban area which makes it necessary to understand its geology and hydrology (Rogers et al. 2016).

9.6.2 PHYSICAL CHEMISTRY OF POLLUTANTS

We learned in Chapter 2 that if it were not for hazardous substances there would be little need for the majority of environmental regulations. However, that is certainly not the case. There are currently more than 84,000 chemicals in use today and more and more every year. Contaminants are everywhere – in the air, soil, water, inside buildings, and in our homes. Most households contain chemicals that would be considered contaminants if they were released into the environment or disposed of improperly. Therefore, we must understand which chemicals are harmful and we must not use or we must limit the use of those chemicals that are most harmful. We have learned that USEPA has not banned any chemicals since 1984, so it's up to industry and the consumer to evaluate the risk and take appropriate action.

In addition, chemicals or substances become pollution when they are released into the environment either inadvertently or improperly – at the wrong place or in the wrong amounts. For example, milk becomes a contaminant when large quantities are released into a stream. Therefore, it's not just what chemicals are used, but preventing releases of most any chemical, especially to sensitive areas (Rogers 2011; Kaufman et al. 2011).

9.6.3 SMART URBAN DEVELOPMENT

The human population is increasing and requires more and more space to live, grow food, and for building roads. The rapid physical growth of urban areas leads to complex processes of landscape transformation which alters the structure and function of ecosystems. Urban invasion into natural areas decreases productivity, decreases infiltration of water, increases impervious surfaces, increases flooding, increases erosion, increases pollutant load, increases energy demand, and decreases species biodiversity. The size of a single-family home in the United States has more than doubled in the last 70 years (United States Department of Housing and Urban Development 2019; United States Census Bureau 2019a).

Many opportunities for smart urban development are available within many cities of the world. Some of the numerous opportunities include the following:

- Less single-family homes and more multi-family units. The average size of a condominium is approximately 130 square meters (1,200 square feet) versus 300 square meters (2,700 square feet) for a single-family home. The average land required for a single-family home in the United States is 815 square meters or 0.25 acres per home. This includes the land required for easements and roads for a typical subdivision (National Association of Home Builders 2018).

• Building large skyscrapers that have hundreds to sometimes over a thousand condominium units have enormous sustainability improvements over the single-family home. Take for example a large skyscraper currently being constructed in Chicago in the United States (see Figure 9.4) that will have an estimated 1,200 condominium units on a footprint of approximately 5,000 square meters (2 acres). This represents a 99.5% reduction in occupied land if those 1,200 condominium units were single-family homes. There are also significant reductions in the amount of construction material that would be used if each were a single-family home that includes forest materials, insulation, roofing, paint, metal piping, sewers, stormwater, roads and curbs, and sidewalks. There are also significant long-term energy savings including heat and electrical.

Reducing the size of housing units. The size of a single-family home in the United States has nearly doubled since 1970 and nearly tripled since 1940. This is in contrast with the fact that the children

FIGURE 9.4 Residential building under construction in Chicago, USA. (Photograph by Daniel T. Rogers.)

per family in the United States has decreased by more than 60% from 3.76 children per family in 1940 to 1.58 children per family in 2018 (United States Census Bureau 2019b).

- Living closer to work. The average commute to work in the United States is 20 miles one way and takes nearly 30 minutes (United States Census Bureau 2019c).
- Increase mass transit availability. Mass transit is much more efficient and emits much less pollution than an automobile per person per kilometer travel (see Figure 9.5). Estimates are as high as 95% more efficient when traveling by commuter train compared to an average automobile when considering fuel use alone and not taking into consideration the cost and inefficiencies of building and maintaining highways (United States Department of Energy 2019; National Geographic Society 2019).
- Increase other forms of transport that include using scooters, bicycles, and walking. Many cities in the United States and throughout the world are improving bike-friendly and foot transport routes to increase these forms of transportation by promoting health and enjoyment benefits (see Figure 9.5 and 9.6).

FIGURE 9.5 Example of bicycle use and mass transit station in London. (Photograph by Daniel T. Rogers.)

FIGURE 9.6 Use of electric scooters in China. (Photograph by Daniel T. Rogers.)

Integrating more mixed land use into urban areas. The United States has much to learn from other countries when it comes to this urban development technique. Historically, the United States has not taken an integrated approach to land use. Many European and Asian countries, namely South Korea (see Figure 7.27, have embraced integration of farmland, industry, and housing.

- Increasing clean energy production such as solar and wind. Both solar and wind energy have become more efficient and less costly. A key point for solar efficiency is cleaning and preventing dirt buildup on the solar panels. Different types of solar energy methods available include direct and indirect electrical generation. Indirect commonly uses an array of mirrors to focus sunlight to a central location and is used to heat a medium such as aluminum that creates enough heat to melt the aluminum, which then is used to generate electricity through a steam turbine. Figure 9.7 is an example of a solar array in the desert of southwest United States. Wind turbines are available in different engineering configurations that can be applied in small areas and minimize bird deaths (United States Department of Energy 2019).

Increasing permeable surfaces. To decrease flooding, increase infiltration of precipitation through several available and simple techniques. Figure 9.8 shows an example of paving bricks which are more permeable than asphalt or concrete. Figure 9.9 is as example of a rain garden that includes a

FIGURE 9.7   Solar array in the southwestern United States. (Photograph by Daniel T. Rogers.)

FIGURE 9.8   Example of paving brick to increase permeability. (Photograph by Daniel T. Rogers.)

FIGURE 9.9 Example of a rain garden. (Photograph by Sarah Anderson. With permission.)

combination of techniques including materials and vegetation that increases surface infiltration and decreases stormwater runoff.

Currently, examples of sustainable and smart urban development using some or all of the items listed above are mainly confined to Europe, Asia, and some parts of the United States. Two impediments that have been slow in becoming a reality in the United States concerning smart urban development is the American love affair with the automobile and achieving the American dream in which owning a single-family home has often been utilized as a degree of measure that defines making that dream come true.

9.7 SUSTAINABILITY AND POLLUTION PREVENTION AT THE LOCAL AND PARCEL LEVEL

We learned in Chapter 4 that it is difficult and costly to remediate a contaminant after it has been released into the environment. The contaminant may spread into soil, water, and air, and often causes harm to the environment before cleanup can occur. These reasons underscore why preventing the release of contaminants is a prerequisite for creating a sustainable environment.

FIGURE 9.10 Pollution prevention hierarchy. (From USEPA. Pollution Prevention. http://www.epa.gov/p2.) (Accessed April 6, 2019), 2019b.)

Preventing contaminants from entering the environment is called pollution prevention. To be successful, this process includes reducing or eliminating waste at the source by modifying production processes; promoting the use of non-toxic or less toxic substances; implementing conservation techniques; and reusing materials instead of putting them into the waste stream (USEPA 2010a). Simply put, the USEPA defines **pollution prevention** as any practice that reduces, eliminates, or prevents pollution at its source (see Figure 9.10) (USEPA 2019b).

The legal framework for this pollution prevention effort is embodied in the Pollution Prevention Act of 1990, which focused industry on these measures (USEPA 1990):

- Pollution should be prevented or stopped at the source whenever feasible
- Pollution that cannot be prevented should be recycled
- Pollution that cannot be prevented or recycled should be treated in an environmentally safe manner
- Releases of pollution into the environment should be conducted only as a last resort and should be conducted safely

Sources of pollution can be designated as point sources or nonpoint sources (USEPA 2003). **Point-source pollution** originates from identifiable sources, such as smokestacks or sewage outfall pipes. **Nonpoint source pollution** emanates from diffuse or unknown sources, and is the leading cause of water pollution in the United States (USEPA 2002). Examples of nonpoint source pollution include:

- Contaminated groundwater from an unknown source
- Contaminated stormwater from runoff originating from parking lots, roads, and lawns
- Air deposition of contaminants and particulates
- Erosion
- Runoff from agricultural areas

The road to sustainability must incorporate effective pollution prevention that yields observable results. In the United States, the observation of major pollution events has been a catalyst for significant levels of response (USEPA 2018a). As a corollary to this, observable progress in preventing pollution, we postulate, will provide additional incentives to continue those efforts. At the onset and throughout, science must guide the planning process, and the results should be published and open for critical review by scientists, professionals, and the public. Other components of this process include: 1) the maximization of resource efficiency; 2) implementation of existing and developing technical innovations; 3) minimization of use of toxic chemicals; and 4) education (USEPA 2019b).

The following sections will concentrate on sustainability measures that can be employed at the facility and local level. We will focus our discussion on pollution prevention. As you will see, pollution prevent is the key component on the road to sustainability at the facility and local level. We will discuss a framework for preventing pollution at industrial point sources. Since point and nonpoint sources are characterized by common transport media and transport processes, portions of this framework are then applied to the source reduction efforts for the nonpoint pollution variants of stormwater and erosion.

9.7.1 POLLUTION PREVENTION IN THE UNITED STATES

USEPA pollution prevention efforts have resulted in a reduction of hazardous waste by approximately 300 million kilograms, reduction of carbon dioxide emissions by 7 million metric tons, and reduction of water used by approximately 30 million liters (USEPA 2019b). As calculated by USEPA, the resulting pollution prevention efforts have resulted in a cost savings of over $1.2 billion dollars (USEPA 2019b). These numbers alone, however, do not tell the complete story of pollution prevention – there are additional quantitative and qualitative considerations. From a quantitative perspective, the quantities of pollution reduction must be weighed against the quantities of pollution produced. If the annual rates of reduction are consistently surpassing the annual rates of pollution released into the environment, then continued improvements must be considered. Progress, however, depends not only on consistent quantitative reductions, but also on the *qualitative* nature of the pollution released, including the types, toxicity, persistence, and mobility of contaminants released, and an accounting of the impacts on the geologic, ecologic, hydrologic, and atmospheric environments where these contaminants enter (Rogers 2011). A full characterization of progress, therefore, would need to document the changing status of ecosystems, receiving water quality (surface water and groundwater), soil conditions, and air pollution levels realized through pollution prevention efforts. Moreover, due to the human pathways present with many contaminant releases, there would need to be evidence of fewer emergency hospital visits and a trend towards less chronic occupational-related diseases, such as the skin and lung problems associated with the use of chromium.

It is beyond the scope of this book to carry out this type of analysis. What can be managed here is the presentation of a procedure for pollution prevention that has had success with industrial point-source applications, and an extension of this procedure to nonpoint sources. These experiences can then be applied to help achieve the broader quantitative and qualitative goals of pollution prevention outlined above.

9.7.2 IMPLEMENTING POLLUTION PREVENTION TECHNIQUES – POINT SOURCES

Successful implementation of pollution prevention involves careful planning. Within the broad array of planning venues and forms (e.g., urban, environmental, strategic, business), there are common threads to the planning process:

- Identification of what you want to do (goals/objectives)
- Collection of data
- Specification of methods for achieving the goals/objectives
- Implementation using the selected methods
- Assessment of the results

The planning process shown here is cyclical. Assessment may lead back to more data collection, or if implementation fails with the methods selected, new methods can be developed and implemented until success is achieved. Sometimes the outcome changes the entire goal of the project, especially in cases where you bit off more than you could chew. Science should be infused into the planning process wherever appropriate. Accurate measurement is a foundation of good science, so to ensure scientific standards for data collection are met, the procedures should include a statistically sound specification of the sample size.

Another area where science must be incorporated into the execution of a plan is the **experimental design**. Science is fundamentally about identifying and explaining variation, and the experimental design – which is the assignment of subjects to experimental groups – provides the roadmap. Although it may sound obvious, at contaminated sites there are two types of locations: contaminated and uncontaminated. Assigning these locations into two groups allows investigators to study the similarities and differences between them. This separation is how we learned that contamination tends to occur more frequently near low points in buildings, and will be described in greater detail later in this chapter. We then use this knowledge to help design the most effective measures for pollution prevention – the "where" of intervention.

Successful source control also requires an understanding of the process producing the pollution. Processes occur over time, so the specification of where to intervene should be accompanied by the proper timing of our pollution prevention efforts – the "when" of intervention. If loading docks are areas in a facility more prone to a contaminant release, then busy times at these locations require special diligence. The initiation of pollution prevention efforts occurs within organizations, and represents a form of change or innovation. To succeed it is important not only to get the science right, the innovation must 1) be testable and implementable at small-scales; 2) represent an observable improvement over existing conditions; 3) be culturally acceptable (in the corporate/organizational and social senses); 4) be economically feasible; and 5) be convenient to implement. These five conditions characterize successful innovations (Rogers 1995).

What follows is a planning process based on the experiences of implementing a successful pollution prevention effort at a major manufacturing company (Rogers et al. 2006). This process contains the basic elements of plans, incorporates scientific aspects of the geologic environment and contaminant properties, and recognizes the social context for implementation.

9.7.2.1　Step 1: Establishing Objectives and Gathering Background Data

Establishing objectives or goals for pollution prevention provides a baseline for measuring success. Making decisions with better information can avoid the specification of arbitrary objectives and goals. Information can be obtained with targeted data collection and evaluation, and this will help with the attainment of goals and focus limited resources where they will produce maximum benefit. It is recommended these data are collected:

- Mass and volumes of each type of solid wastes generated, including:
  - Stormwater volumes and contents
  - Sanitary discharge volumes
  - Industrial wastewater volumes and contents
  - Regulated liquid wastes in containers
  - Regulated liquid hazardous wastes in containers
- Mass and volumes of air emissions generated, including:
  - Volatile organic compound (VOC) emissions
  - Particulate emissions
  - Heavy metal emissions
  - All other identifiable air emissions
- Energy
  - Sources, types, consumption, and rates through time
  - Energy loss

- Purchasing habits
  - Bulk containers compared to smaller containers and amounts
  - Types of containers (e.g., gallon containers vs. spray cans)
  - Packaging (plastic vs. cardboard or biodegradable material)
- Production efficiencies
- Product packaging
- Shipping (rail vs. truck)

Where possible, a mass balance should be calculated to ensure the accuracy of the data. Once the data have been collected and evaluated, Steps 2, 3, and 4 should be completed before firmly establishing goals and objectives for any pollution prevention program.

9.7.2.2 Step 2: Inventory of Hazardous Substances

The next step in evaluating the need for developing a pollution prevention plan at any location – whether a manufacturing facility or a household – is to inventory and assess hazardous substance use. Chapter 2 covered many common contaminants present in urban areas, including households. As they are often very close to industrial sites, households can assist with any inventory since the average American household stores 3–10 gallons of hazardous materials (Smolinske and Kaufman 2007). To inventory chemicals, CAS Registration Numbers should be used since many products display either trade names or synonyms. If available, Safety Data Sheets (SDS) often provide valuable information.

Inspecting the label on chemical containers is required, especially when the container is a mixture of chemical products. In these cases, the name of the product typically is a trade name and is not very helpful in identifying the specific chemicals contained. Inspection of the label is the only effective method. An example of a label on a chemical product is shown in Figure 9.11.

Chemicals should be inventoried by chemical group, such as VOCs, polycyclic aromatic hydrocarbons (PAHs), semi-volatile organic compounds (SVOCs), polychlorinated biphenyls (PCBs), metals, acids, bases, and whether they are present in gas, liquid, or solid form. Transformers containing PCBs should also have appropriate labels, as shown in Figure 9.12. After the hazardous substance inventory has been completed, a map should be created showing the following: location of where each hazardous substance enters the property; where the substances are stored before use; where they are used; and where they are stored after being used and before any residuals are discarded. Since many contaminant release locations occur from low points in buildings, highlighting pits, sumps, trenches, underground storage tanks, and floor drains on the map can also help provide valuable information for pollution prevention efforts (Rogers et al. 2006).

9.7.2.3 Step 3: Assessing CRFs

The hazardous substances having the highest contaminant or pollution risk factors (CRFs) should now be evaluated. This evaluation helps to prioritize the pollution prevention efforts. Many facilities

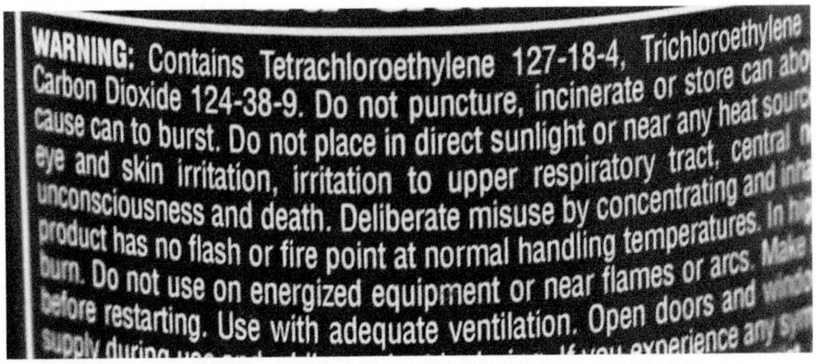

WARNING: Contains Tetrachloroethylene 127-18-4, Trichloroethylene Carbon Dioxide 124-38-9. Do not puncture, incinerate or store can above cause can to burst. Do not place in direct sunlight or near any heat source eye and skin irritation, irritation to upper respiratory tract, central unconsciousness and death. Deliberate misuse by concentrating and product has no flash or fire point at normal handling temperatures. In burn. Do not use on energized equipment or near flames or arcs. Make before restarting. Use with adequate ventilation. Open doors and

FIGURE 9.11 Inspecting labels. (Photograph by Daniel T. Rogers.)

FIGURE 9.12   Inspecting labels. (Photograph by Daniel T. Rogers.)

and households may have more than 100 different hazardous substances. Therefore, prioritizing is critical to achieve maximum benefit. As noted in Chapter 5, chromium VI and dense nonaqueous phase liquid (DNAPL) VOCs have the highest CRFs for groundwater, with PCBs, mercury, chlordane, and PAHs having the highest CRFs for soil. Concentrating pollution prevention efforts on those contaminants with the potential to significantly impact groundwater such as chromium VI and DNAPL VOCs will have the most long-term benefit.

9.7.2.4   Step 4: Preliminary Assessment of Geologic Vulnerability

Most urban areas have not been geologically mapped, so the detail necessary to accurately evaluate the vulnerability at any location may not be available (Rogers et al. 2016; Kaufman et al. 2011). The suggestions offered here can help determine if any given area presents enough potential risk to warrant a more detailed examination, and whether an aggressive pollution prevention initiative should be pursued:

- Source, location, and type of potable water. Contact the local municipality to evaluate the source of potable water for the area in question. In addition, inquire whether there are any groundwater extraction wells of any type within at least a one-mile radius of the location being evaluated. If any wells exist, request a copy of the installation records for further examination.
- Nearest surface water body. Determine where the nearest surface water body is located with respect to the location being evaluated. Topographic maps (7.5 minute) can help with this procedure.
- Stormwater collection and discharge. The local municipality should have information about stormwater collection, its treatment, and the discharge locations for the area being evaluated.
- General geological conditions. Examining well records could provide valuable information on soil type, stratigraphy, and depth to groundwater if well records can be obtained in the vicinity of the location being evaluated.

If the source of potable water in the area being evaluated is obtained from groundwater and extraction wells are nearby, environmental risk should be considered high. In this case the geologic vulnerability rating will likely exceed a score of 50, and pollution prevention efforts should become the highest of priorities. The other factors listed above will likely require examination and evaluation by a qualified professional before an appropriate geologic vulnerability rating can be determined.

9.7.2.5 Step 5: Preventing Pollution through Elimination, Substitution, Prevention, and Minimization (ESPM) Methods

Pollution prevention can be implemented with a stepwise evaluation process that proceeds from the most preventative measure to the least preventative measure. This process is referred to as ESPM, and consists of these steps:

- **E**limination: Not using potentially harmful chemicals
- **S**ubstitution: Using a less potentially harmful chemical instead of a harmful one
- **P**revention: Using engineering controls and other measures to minimize the potential for a release; employed if eliminating or substituting a potentially harmful chemical is not possible
- **M**inimization: Reducing usage of harmful chemicals or a reduction in generated wastes through process changes, recycling, or other methods

After steps 1 through 4 have been completed, a focused and achievable plan for pollution prevention can be developed and implemented. Facilities where a synergistic effect may be present should receive the highest initial effort characterized by aggressive pollution prevention initiatives for reducing the potential risks (Rogers et al. 2006; Rogers 2011).

9.7.2.5.1 Elimination

The most aggressive form of pollution prevention is *elimination* of hazardous chemical use. Where possible, elimination of hazardous substance use is the preferred pollution prevention method because it is the easiest to manage and has the greatest benefit to the environment. As noted in Chapters 2 and 3, if there is no hazardous chemical use – there is minimal risk. In most cases, total elimination of hazardous chemical use is usually not possible, so the elimination of chemicals should focus on contaminants with high CRFs. For example, chromium VI and DNAPL VOCs have high CRFs for groundwater and PAHs, chlordane, PCBs, and mercury have high CRFs in soil.

A crucial step in the process of eliminating hazardous substances is to develop a chemical ordering procedure. This procedure protects against unauthorized hazardous substances making their way into operations at the facility without prior knowledge. All proposed chemicals or substances should undergo a review process to evaluate whether they are acceptable for use.

9.7.2.5.2 Substitution

The next most aggressive form of pollution prevention is *substitution*. Substitution involves using alternative chemicals with the goal of greatly reducing risk. For instance, if a facility uses DNAPL VOCs for cleaning, an effective substitute may be citrus-based cleaners. The Solvent Alternatives Guide provides options and guidance for evaluating available chemical substitutes for common solvents (USEPA 2019c).

Other alternatives exist for chemical substitution. For instance, mercury-containing devices such as switches, thermometers, and monometers can be substituted by digital devices. Liquid transformers containing PCBs can be substituted with dry transformers or with transformers not containing any detectable concentration of PCBs. Other examples of substitution include:

- Using paints without VOCs and certain heavy metals
- Using biodegradable oils
- Using paraffin as a lubricant instead of oil

9.7.2.5.3 Prevention

Prevention of contamination entering the environment can be accomplished through engineering controls. Engineering control methods include release prevention, release detection, release containment, and release cleanup immediately after a spill.

The following examples (denoted by bullets) highlight some ways activities and operations can be modified to minimize the potential for the release of hazardous substances to the environment:

- Evaluating and eliminating potential points of release. Sumps, pits, trenches, floor drains, and chemical storage and usage areas are common points for hazardous substance release. Conducting an inventory of these locations and locations of chemicals present at a facility will assistance in identifying areas where releases may occur.
- Figure 9.13 provides an example of a potential release area, and illustrates several practices that should be avoided, including:
- Storing liquid wastes in improperly labeled containers. The drum on the center on the left side has a non-hazardous label, yet it is also labeled as containing a waste solvent.
  - Storing liquid wastes in upside-down containers.
  - Storing liquid wastes on bare ground.
  - Storing liquid wastes outside without a roof to prevent contact with storm water and corrosion.
  - Storing liquid wastes in drums where the structural integrity has been compromised.
- Modifying liquid waste storage areas. Figure 9.14 demonstrates one way to properly store liquid chemicals and substances prior to disposal. This storage method for liquid wastes has multiple and redundant engineered systems to prevent a release and contain a release if one does occur. These engineered systems include:
- Storing liquids wastes inside and under a roof.
  - Coating the floor with epoxy.
  - Providing secondary containment if a release were to occur. In this example, drip pans are located beneath each outer container.
  - Locating the liquid waste storage area at a location without floors drains, sumps, trenches, and pits.

FIGURE 9.13 Improper storage of drums containing liquid waste. (Photograph by Daniel. T. Rogers.)

- Using redundant storage containment. Sealed drums containing liquid wastes are located inside the outer containers pictured in this example.
- Properly labeling the contents and potential hazards.
- Sealing sumps, pits, trenches, floor drains, and liquid storage areas with an impervious surface. In most instances, leaks will occur from structures composed of concrete storing or conveying liquid wastes. Leaks can occur from seams, cracks, or in some cases from the direct migration of liquid through the concrete surface itself. These surfaces should be sealed to prevent a release to the subsurface. Figure 9.15 is an example of a sealed floor drain.
- Providing secondary containment. Engineering redundant systems can effectively prevent an uncontrolled release to the environment. Figure 9.16 is an example of a liquid storage area inside a building with a sealed floor with secondary containment in a restricted area. Figure 9.17 is an example of secondary containment for a machine that uses oil. Not the redundant systems, the machine itself is self-contained, beneath the machine is secondary containment and the machine is located inside located on a sealed floor.
- Spill containment and cleanup. Despite the existence of engineering controls, accidental releases do occur. Therefore, proper response is necessary to prevent the uncontrolled release of a hazardous substance and to protect human health and the environment. Spill stations outfitted with an assortment of tools, containers, personal protective gear, and instructions can ensure small spills of liquids not presenting an immediate threat to human health are addressed quickly and safely. These stations should be located near hazardous substances. Figure 9.18 shows a typical spill station; refer to Figure 9.14 for an example of a spill kit located near a liquid waste storage area.

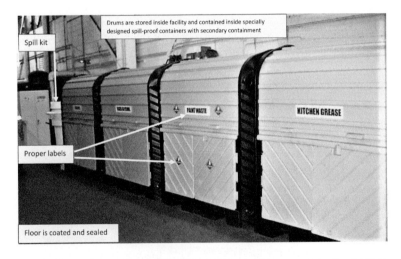

FIGURE 9.14 Proper storage of drums containing liquid waste. (Photograph by Daniel T. Rogers.)

FIGURE 9.15 Example of a sealed floor drain. (Photograph by Daniel T. Rogers.)

FIGURE 9.16   Liquid storage inside a restricted area. (Photograph by Daniel T. Rogers.)

- Education and training. Education and training are critical for the prevention, response, and cleanup of spills. These efforts can also prevent the response to a spill or accidental release of a hazardous substance when evacuation and immediate notification are the necessary courses of action, and qualified emergency personnel are required on site. Figure 9.19 shows a spill response and training exercise.

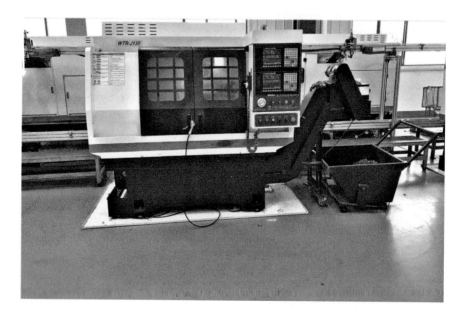

FIGURE 9.17   Machine inside a building with secondary containment. (Photograph by Daniel T. Rogers.)

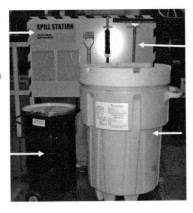

Markings noting location of spill station

Tools to assist in spill response and cleanup

Empty drum available for spill cleanup

Drum containing absorbents and other materials

FIGURE 9.18 Example of a spill station. (Photograph by Daniel T. Rogers.)

9.7.2.5.4 Minimization

Waste minimization involves using less hazardous substances through conservation efforts or process changes. The result is a reduction in the amount of wastes requiring disposal. In some instances, minimization also includes recycling. For instance, reducing the discharge or generation of liquids through process changes, recycling of water, or the evaporation/recycling of liquid or solid waste can greatly lower the volumes and mass of waste streams. An example of waste minimization is shown in Figure 9.20. Here, the additional treatment of wastewater using reverse osmosis reduces the amount of heavy metals in the discharged wastewater.

Recycling of metals, plastic, glass, wood products, and many other materials can greatly minimize the amount of solid waste generated and disposed of in a landfill (see Figure 9.21). Often, these activities result in significant cost savings. The future in this area is promising, as the discovery of new options for the beneficial reuse of waste materials parallels the appearance of new waste sources (USEPA 2019b).

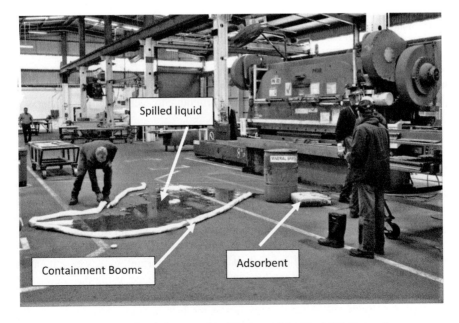

Spilled liquid

Containment Booms

Adsorbent

FIGURE 9.19 Spill response and training exercise. (Photograph by Daniel T. Rogers.).

Removal of heavy metals from waste water through reverse osmosis

FIGURE 9.20   Reducing heavy metals in wastewater. (Photograph by Daniel T. Rogers.).

FIGURE 9.21   Glass recycling in Scotland. (Photograph by Daniel T. Rogers.)

Minimization is also achieved through awareness, tracking, and accountability. An example is shown in Figure 9.22, which is a dispensing machine for supplies that often includes cleaners and other liquids. The machine requires a tracking number unique to each employee so that work habits and precise amounts and types of chemical can be tracked.

Energy reduction strategies evaluate the energy types consumed, and their usage rates by location over time. In addition, energy loss and potential recovery/recycling may also come into play.

Purchasing habits can greatly assist minimization efforts. When supplies are purchased in bulk containers, the number of containers requiring disposal is reduced. It is also possible to realize reductions in energy usage and air emissions through production efficiencies. For instance, switching to more efficient gas turbines in certain production processes may save energy, while conducting energy-intensive activities at night when energy demand is low can lower air emissions. At off-peak times power companies are more likely to substitute a non-fossil fuel such as hydroelectric for coal.

Product packaging and transportation methods are also important areas where many improvements can be made from an environmental perspective. For instance, using packaging made from biodegradable material instead of plastic can have a significant positive impact on the environment. In transportation, the use of rail or barges to transport products consumes much less fossil fuel and reduces air emissions.

FIGURE 9.22 Supply dispensing machine. (Photograph by Daniel T. Rogers.)

9.7.2.6 Step 6: Assessing Results

Practicing pollution prevention has qualitative and quantitative benefits that can result in significant financial gains and cost avoidance. The evidence of quantitative pollution prevention benefits can be found by tracking reductions, especially if data have been collected before a pollution prevention program was initiated. Quantitative evaluation is reflected by the reductions of (USEPA 2019b):

- Waste volumes
- Raw materials
- Energy usage

On-site qualitative benefits are realized by (USEPA 2019b):

- Release avoidance
- Additional protection of human health and the environment
- Lowering regulatory reporting requirements
- Lowering environmental liability

Across an urbanized watershed, qualitative improvements will be seen in ecosystem health, receiving water quality (surface water and groundwater), soil conditions, and air pollution levels.

Continuous improvement, evaluation, and inspection should be developed as part of program assessment, because the lack of an effective pollution prevention program can result in significant liability and cost.

One final note: pollution prevention represents an innovation within an organizational context. With this in mind, the key "in-house" actions for achieving environmental project objectives, realizing cost savings, and reducing environmental risk are: (a) establish easily identifiable objectives with the involvement of senior management; (b) perform an effective accounting of cost savings and tracking of other related benefits; and (c) communicate the results across all levels of the organization. In addition, it is beneficial to have compliance personnel trained in environmental science, with these personnel positioned at a high level within the organization (Rogers et al. 2006).

9.7.2.7 Financial Incentives for Pollution Prevention

The major financial incentives for implementing pollution prevention at most industrial facilities are the direct and indirect cost savings realized from a successful program. Direct cost savings accrue from reducing the amounts of solid waste generated and water consumed. Indirect cost savings occur when the long-term liabilities associated with the disposed waste at a licensed facility are lowered because the amount of waste decreases. Other long-term liability reductions such as savings on litigation costs and reduced contaminant releases can also significantly reduce the costs associated with cleanup. Finally, waste reduction often reduces the amount of regulatory reporting requirements, and this outcome can significantly cut costs.

Implementation of an effective pollution prevention plan at any manufacturing facility requires the commitment, input, and cooperation of every employee. Representatives from purchasing, production, maintenance, human resources, and environmental must work together in close cooperation to identify, implement, and measure every aspect of a pollution prevention program. Tracking progress is necessary to sustain a program and create more involvement with its execution and outcomes. Many pollution prevention programs fail because they do not fully quantify their benefits.

9.7.3 Implementing Pollution Prevention Techniques – Nonpoint Sources

Stormwater and erosion are related, as both involve the transport of materials by fluids.

Stormwater originates from precipitation, and is the polluted overland flow of water in urban areas. As precipitation flows over roads, rooftops, parking lots, construction sites, and lawns, it becomes contaminated with oil and grease, pesticides, fertilizers, litter, and pollutants from vehicles. The EPA estimates over 10 trillion gallons of untreated stormwater make their way into US surface waters each year (USEPA 2019d). Dense urbanization exacerbates the problem, since the amount of pollution present in stormwater runoff is correlated with the amount of impervious cover (Schueler 1994). As noted in Chapter 6, pollutants from farmland are responsible for much of the pollutant load of pesticides, herbicides, and fertilizers that enter the Gulf of Mexico and many other river systems that discharge to the oceans.

The Clean Water Act (CWA) of 1972 gave the USEPA the authority to regulate point-source discharges through the National Pollution Discharge Elimination System (NPDES) Program. In 1987, a survey of the nation's waters indicated point-source control alone was not sufficient to achieve the "fishable and swimmable" goal of the CWA because nonpoint sources, especially agricultural and urban runoff, were contributing substantial amounts of pollution (Humenik et al 1987). Also in 1987, the USEPA initiated the NPDES Stormwater Program, requiring municipalities with separate storm sewer systems located in incorporated areas with populations of 100,000 or more to obtain NPDES permits for stormwater discharges (USEPA 2019d). In 1999, Phase II of this program was extended to smaller municipalities and required permit holders to implement post-construction stormwater management programs using Best Management Practices (BMPs). Examples of BMPs for stormwater management include education, road salt management, street cleaning, and erosion control measures such as silt fences and covering exposed soil (USEPA 2019e).

9.7.3.1 Applying the Source Control Framework to Nonpoint Sources

Since the overland flow of water (stormwater) forms the transport component of erosion, stormwater becomes the focal point of pollution prevention methods. The goal here is not to develop a list of stormwater BMPs; the EPA has a "menu of BMPs" designed to help communities with their implementation of the Phase II stormwater rules (USEPA 2019e). Instead, our focus is this; given the limitations of the Phase II Stormwater controls (they only apply to new development or redevelopment), how can we achieve effective pollution prevention methods for stormwater in older urban areas? To help answer this question, we now apply the pollution prevention framework used for point-source control to nonpoint sources.

9.7.3.1.1   Step 1: Establishing Objectives and Gathering Background Data

Objectives for nonpoint pollution control will vary according to specific watershed conditions and needs. To place this effort on a firm foundation and to obtain the best information, these general objectives are recommended:

- *Integrate nonpoint pollution control fully with watershed management/watershed restoration efforts.* There is a growing movement to implement non-structural source control measures for stormwater (e.g., bioretention, swales, porous pavement) – collectively referred to as "low-impact development." Unfortunately, the adoption of these procedures at larger scales across watersheds is being hampered by: (1) uncertainties in performance and cost, (2) insufficient engineering standards and guidelines, (3) fragmented responsibilities, (4) lack of institutional capacity, (5) lack of legislative mandate, (6) lack of funding and effective market incentives, and (7) resistance to change (Roy et al. 2008). Integration of these efforts with watershed-wide initiatives can help overcome some of these obstacles and increase their extent of implementation.
- *Manage the land to achieve better water quality.* Most watershed surface areas are 95% land, so proper land management is necessary to improve water quality. For example, a study of 27 water suppliers conducted by the Trust for Public Land and the American Water Works Association in 2002 found that the more forest cover in a watershed the lower the treatment costs. According to the study: For every 10% increase in forest cover in the source area, treatment and chemical costs decreased approximately 20%, up to about 60% forest cover. Approximately 50–55% of the variation in treatment costs was explained by the percent of forest cover in the source area (Ernst et al. 2004).
- *Integrate quantity and quality.* Reducing stormwater quantity will improve the quality of receiving waters.
- *Recognize the linkages between and within point and nonpoint sources, and optimize the control measures based on these linkages.* Point and nonpoint sources are linked through surface runoff, groundwater flow, and particulate transport. For example, using phytoremediation at brownfield sites provides vegetative cover and minimizes erosion and surface runoff.
- *Opt for non-structural (low-cost and non-permanent) controls first.* Many urban areas built expensive and permanent combined sewer overflow (CSO) control basins before adequately characterizing the potential of runoff reduction from non-structural controls within the watershed (Kaufman and Marsh 1995).
- *Try to maximize the vertical movement of non-polluted or minimally degraded water.* Many stormwater strategies advocate moving stormwater vertically, but this action can degrade groundwater and ultimately surface waters if the water is highly contaminated. Locations for infiltration should be chosen and prepared carefully so they have adequate organic content to assist with biodegradation and the adsorption of organics and heavy metals.
- *Pay close attention to erosion control in the headwaters region of the watershed.* The highest energy gradients are in the first-order stream segments found in the headwaters, so lowering erosion here lowers the kinetic energy of the stream and minimizes channel erosion. Reducing the erosion capacity of the streams will reduce the deposition of materials downstream, that include lakes and impounds that function as settling basins downstream of headwaters.
- To support the general objectives above, these data can be collected and generated:
- *Current measures of water quality at several locations throughout the watershed.* Universities and watershed organizations may conduct regular-stream sampling and benthic monitoring, so check with them first. Benthic monitoring collects samples of materials dislodged from stream beds and assesses the species diversity. The presence of certain

insects, for instance stoneflies, may indicate high water quality. Remember to also collect samples from wells within the watershed. This sampling is often free, with collection vials provided by the county public health department. At a minimum, the parameters tested for in streams and groundwater should include total suspended solids, pH, temperature, phosphorus, nitrate, fecal coliform, lead, and chromium.

- *A stream order map for the watershed.* As noted earlier, lower-order streams are more vulnerable to contamination, so identifying their locations can assist with prioritizing source control efforts. Some universities have done this already.
- *Hydrographs for all gauged streams within the watershed.* A hydrograph is a plot of discharge (Q) over a specified time interval. These are available in different time increments (day, month, year) from the US Geological Survey. Hydrographs will help characterize the flood patterns within the watershed. Flood waters often carry massive amounts of contamination, so knowing where floods occur more frequently can help direct efforts to minimize stormwater runoff, which contributes large volumes of water to streams during wet-weather events.
- *Aerial photographs of the watershed taken during "leaf on" conditions.* Try to get the most recent images available. The Natural Resource Conservation Service has aerial imagery for many urban areas. Land cover and tree patterns can assist with strategies to provide more interception storage and more infiltration capacity. Knowing the locations of brownfields and their proximity to surface water and groundwater recharge zones is also important for minimizing the impacts of contaminated runoff. In addition, aerial photos can identify locations along urban streams where public access is limited. In heavily urbanized watersheds, rail lines and industrial facilities often block access to the water. If the properties are abandoned, these areas can be restored to improve access.
- *A map of current water drainage infrastructure.* Many communities have digital maps of their sanitary and storm sewer lines, surface drains, and the locations of detention basins and access structures. These maps can help you identify where runoff volumes are large and natural drainage patterns have been altered.
- *A map showing the locations of brownfields in the watershed, and the results of the soil and groundwater testing from these sites.* The brownfield map can help identify areas where there are potential storage areas for stormwater runoff and for increasing the local density of vegetation. Soil and groundwater data can be used to characterize the extent and pattern of contamination, and the erosion potential at these abandoned sites.

9.7.3.1.2   Step 2: Inventory of Hazardous Substances

With point sources, reading product labels was the important activity. For nonpoint sources, determining the locations within the watershed where hazardous substances may be released and transported by erosion and stormwater are the essential actions. During wet weather, lawns release nitrogen, phosphorus, and pesticide residues, feedlots contribute bacteria, parking lot runoff has oil and antifreeze, construction sites release soil, and industrial facilities may have on hand barrels of hazardous material capable of being swept away in a flood. It is important to know the locations of the hazardous substances and their relationship to the ephemeral drainage network capable of transporting their releases. For example, some erosion may reach a catch basin 100 feet away, yet in other cases the eroded material will never make it to a storm drain because it reaches a natural depression first. Aerial photos and topographic maps can help identify potential routes of overland flow; field checking, however, is required for verification. In addition, dye testing is often used to help identify infiltration/inflow of precipitation into sewers and to reveal the paths of stormwater and wastewater. During this process, non-toxic dyed water is introduced into roof drain leaders, driveway drains, or area drains. In some instances, dyed water is injected into the ground around foundations to check for the illegal connection of foundation drains. After introducing the dyed water, the downstream sanitary sewer access structure is checked for its presence to ascertain the path.

9.7.3.1.3   Step 3: Assessing CRFs

The locations where the highest CRFs are present should now be evaluated to help with the prioritization of the pollution prevention efforts. The results of the water sampling should guide the selection of locations. For example, if aquatic ecosystems have already been damaged and the groundwater samples produce high levels of chromium, this contaminant may need to be addressed as a priority.

9.7.3.1.4   Step 4: Preliminary Assessment of Geologic Vulnerability

Stormwater is surface water, and erosion is a surface process. Due to their placement, geologic vulnerability to these two pollution sources is realized through infiltration. Since soil is the medium for infiltration, efforts should be focused here. Some suggestions for assessing the soil vulnerability in an urbanized watershed include:

- Do not rely upon the county soil surveys – use them for general orientation to the soils in the watershed. Urban soils are highly disturbed, and frequently the soil surveys do not capture their variation because the survey may be 10–20 years old.
- Instead, conduct a soil sampling program that enables a correlation between soil type, infiltration rates, and routes of overland flow during wet-weather events. The objective is to reveal those locations in the watershed where larger-grained vulnerable soils and less-vulnerable fine-grained soils exist. Use the soil data available from environmental site investigations to help determine the locations for additional sampling. Once the sites for sampling have been selected, compare their locations to the major pathways of overland flow in the watershed, such as steep unvegetated slopes, large parking lots, and major streets. Then sample those sites within or along the drainage paths. At the sampled sites, use an infiltrometer to measure the infiltration capacity of the soil, and have the soil samples analyzed for type, texture, and organic content by the local agricultural extensions of the nearest state university.
- Delineate floodplains and other areas where groundwater levels are likely to be closer to the surface. Try to avoid the placement of infiltration activities in these zones, and concentrate on areas where the depth to groundwater is higher, thus giving more time for organisms and structures in the soil to reduce the contamination.
- If groundwater is used as a source of potable water in the area being considered for enhanced infiltration, and extraction wells are nearby, environmental risk should be considered high.

9.7.3.1.5   Step 5: Preventing Pollution through ESPM Methods

Elimination of stormwater and erosion are impossible since they are triggered by natural processes. From the perspective of creating a sustainable watershed, elimination translates into reducing the amounts of stormwater volume and eroded material transported, as well as lowering the use of contaminants entrained by stormwater. These actions must obtain results that work towards returning to dynamic equilibrium in streams, reducing the sedimentation within lakes and reservoirs, and preserving soil for future generations.

Stormwater management in older urban areas presents special problems. House lots and vegetated spaces are typically smaller, downspout discharges are often routed directly onto paved surfaces (e.g., driveways), and sidewalks add to the amount of impervious surface. Compounding the problem is the presence of combined sewers and combined sewer overflows (CSOs). A CSO occurs when a pipe designed to carry sanitary and storm flows is surcharged, and a special sewer, called an interceptor, is used to transport the untreated overflow to the nearest water body. Over 45 million people in 746 US communities within 32 states are affected by CSOs. EPA estimates that about 850 billion gallons of untreated wastewater and storm water are released as CSOs each year in the United States (USEPA 2019d).

In urban areas, overland flow originates from impervious surfaces such as rooftops, roads, and sloped residential lots. The size of these areas is frequently small – often less than 0.10 ha (1/4 acre), with flows greatly influenced by the micro-topography. For instance, at various locations stormwater and sediment sinks exist with the micro-topography of a watershed; certain soils may infiltrate the entire storm volume, surface depressions store and infiltrate runoff, and locally dense vegetation canopies can intercept as much as 30% of the incoming precipitation (Cape et al. 1991).

The incorporation of micro-topography in the urban landscape has immense practical significance for preventing stormwater pollution. If the capabilities of the local micro-topography are exploited to store and/or redirect stormwater, large volumes of stormwater in urban areas can be kept out of the human-built drainage network. Since 95% of the total flow volume within combined sewers is stormwater, source reduction applied to stormwater management can help developed urban areas avoid substantial investment costs in structural stormwater management infrastructure (e.g., regional detention basins to contain CSOs or sewer separation) (USEPA 2019e).

Table 9.1 shows the water balance of an urban parcel (e.g., house, business, or industrial site) during a precipitation event. Inflows, outflows, and storage components of the water balance are routed through the four spheres of the geosphere (atmosphere, biosphere, hydrosphere, lithosphere) at a small geographic scale (0.1–0.3 ha). The micro-topographic elements of the landscape affecting the pathways of water are identified in the third column. For instance, precipitation moves along two paths: the first path is from the atmosphere to the biosphere, where vegetation affects the flows; the second path is from the atmosphere to the lithosphere, where soil and pavement influence the flows. Strategies to reduce stormwater runoff volumes must include the modification of those landscape features capable of rerouting stormwater back to its natural downward path. Structural (engineering-based) approaches to this problem result in building more pipe capacity, or providing basins for storage. An earth science approach looks at the scale of the processes generating the runoff, the source location, and direction of the flows, and the landscape components affecting the transmission system at the relevant scales.

TABLE 9.1

Urban Parcel Water Balance during Precipitation[a]

Component	Paths	Micro-Topography
Inflows		
Precipitation	Atmosphere → biosphere	Vegetation soil, pavement (driveways,
	Atmosphere → lithosphere	sidewalks, road surfaces)
Runoff	Lithosphere → lithosphere	Soil, grass surfaces, pavement, roofs, gutters
Outflows		
Runoff	Lithosphere → lithosphere → hydrosphere	Soil, grass surfaces, pavement, storm sewers
Evaporation	Lithosphere → atmosphere	Ground surface
Evapotranspiration	Biosphere → atmosphere	Vegetation
Storage		
Interception	Atmosphere → biosphere	Vegetation
Infiltration	Atmosphere → lithosphere → hydrosphere[b]	Soil, vadose zone water
Surface depression storage	Atmosphere → lithosphere	Ground surface

[a] including the road frontage

[b] interflow

Source: Kaufman, M.M., Rogers, D.T. and Murray, K S. 2011. Urban Watershed: Geology, Contamination, and Sustainable Development. CRC Press. Boca Raton, FL. 583 p.

TABLE 9.2

Opportunities for Stormwater Storage and Diversion at the Micro-Topographic Scale

Micro-Scale Landscape Feature	Opportunities for Storage/Diversion
Branches	Certain tree canopy shapes (such as beech) act to funnel water downward to zones containing more permeable soil near the main root zone (Mosley 1982)
Leaves	Conifers have generally higher interception capacity (Cape et al 1991)
Soil	Increasing the organic content of soil increases its water retention capacity (Hudson 1994)
Median strips	Concave or flat median strips can increase stormwater storage
Surface depressions	Downspouts can be routed to slight depressions to increase infiltration, instead of discharging to paved surfaces. These areas can also delay stormwater from entering the pipe network
Root buttresses	Roots from larger trees and shrubs create friable soil more capable of infiltrating water (Bartens et al. 2008)
Flat rooftops	Can become storage areas with the addition of vegetation or rooftop cisterns
Swales	Store overland flow and prevent entry into street drains
Downspouts	Should be disconnected from footing drains, and routed to cisterns or natural depression storage areas for infiltration or used as graywater; "in-line" storage bins can store up to 100 gallons per downspout
Driveways	Use porous pavement to reduce runoff
Sidewalks	Use porous pavement

Source: Kaufman, M.M., Rogers, D T. and Murray, K.S. 2011. Urban Watershed: Geology, Contamination, and Sustainable Development. CRC Press. Boca Raton, FL. 583 p.

In Table 9.2, opportunities for stormwater storage or diversion at the micro-topographic scale are shown for each landscape feature influencing stormwater flows. The objective is to maximize storage at the site closest to the origin of runoff. At the micro-topographic scale of the house lot or single parcel, this translates into keeping stormwater on the lot and out of the storm sewers (Kaufman 1999). This is accomplished by directing water downward, instead of horizontally.

As Table 9.2 shows, increasing interception and infiltration and minimizing overland flow from impervious surfaces form the core components of a stormwater retrofitting strategy in developed urban areas. Increasing the organic content of soils (by topfilling with humus and cutting with a mulching mower), using available storage in surface depressions, and the use of swales can provide large amounts of infiltration if the lot layout permits their construction.

Flat roofs (such as those on parking lots) converted into vegetated green roof systems can reduce stormwater volumes and delay its release, and also help reduce the urban heat island effect. One study showed that green roofs not only reduced the amount of stormwater runoff, they also extended its duration over a period of time beyond the actual rain event (van Woert et al. 2005). In another study, rain and roof runoff data collected from seven rains during October and November 2002 showed that the green roofs delayed the start of runoff an average of 5.7 hours. The green roofs retained an average of 45% of the rain from the seven storms evaluated and delayed the peak runoff by 2 hours. Roof temperature data collected between April 2002 and February 2003 showed that the green roof maximum surface temperatures averaged 6°C higher in the winter and more than 19°C lower in the summer (DeNardo et al. 2005).

Downspout diversion provides excellent potential for stormwater flow reduction in highly urbanized communities lacking available space for large-scale detention facilities. In the Beecher Water District, a highly urbanized area near Flint, Michigan, a downspout diversion program was implemented from 1996–1997. Here, downspouts were connected to footing drains and into sanitary sewer pipes, resulting in frequent sanitary sewer pipe overflows during wet weather. Downspout diversion contributed to a reduction of over 35% in the mean flow volumes within the sanitary sewer collection network, and reduced overtime costs associated with overflow maintenance (Kaufman and Wurtz 1997). In Portland, Oregon, the Downspout Disconnection Program has disconnected 50,000 downspouts, and these disconnections are removing more than 1 billion gallons of stormwater annually from the combined sewer system (City of Portland 2019). Where downspouts are already disconnected, many can be rerouted to prevent direct discharge onto paved surfaces. It may not be practical or affordable to convert driveways and sidewalks to porous pavement. There are, however, opportunities for some conversions whenever homeowners and commercial establishments repave their driveways and parking lots.

As with point sources, the elimination of chemicals should focus on contaminants with high CRFs. Many lake associations ban the use of fertilizers and pesticides due to concerns over cultural eutrophication. It would be good practice if every homeowner considering the use of a timed fertilization program offered by "lawn care experts" would conduct an inexpensive soil sample to assess the level of nutrients already present in the soil. Many lawns do not need four to six fertilization/herbicide applications per season. Contaminants released into stormwater from lawns can be significantly reduced through improved lawn management. Most state universities have agricultural extension services, and these organizations offer excellent advice on how to reduce the amount of fertilizer used, and the least environmentally offensive types.

Road salt is a significant contaminant of surface waters and groundwater. A survey of 23 springs in the Greater Toronto Area of southern Ontario recorded chloride contamination levels, resulting from the winter application of road de-icing salt, ranging from <2 to >1200 mg l^{-1} (Williams et al. 2000). Many communities, especially those with lakes present, substitute sand for road salt.

Soil stabilization – especially at construction sites – can help prevent erosion and the transport of contaminated soil by stormwater. One activity not used enough for preventing soil erosion is the inspection of erosion control measures after rainstorms. Another frequent problem is the failure of communities to require a soil sample before development occurs at a site (Kaufman 2000). Basic information about soil type and texture can help direct specific erosion control measures. For example, the performance of certain geotextiles varies due to the difference in the characteristics of suspended solids associated with different soils (Barrett et al. 1998).

Special consideration should also be given to road placement in developing areas and wellhead protection. Roads should be designed so as to not join partial areas of runoff, and stormwater runoff should be considered when creating protection zones around wells. Since homes store an average of 3–10 gallons of hazardous waste, all of the substitution recommendations made in the point-source section apply here as well. Recycling paper, plastic, and metal cans is also an effective way to reduce inputs of undesired substances into streams. In addition, individual households can advance pollution minimization through the purchase of environmentally friendly products and packaging, increasing their efficiencies of energy and water use, and selecting less polluting transportation forms such as carpooling and mass transit.

9.7.3.1.6   Step 6: Assessing Results

Efforts to reduce erosion and stormwater runoff can be evaluated on several levels. Visually, improvements across the watershed will be seen in ecosystem health, receiving water quality (surface water and groundwater), and soil conditions. Hydrologically, there would be a lower frequency of floods, and a less "flashy" stream response; that is, a slowing down of the time required to reach peak runoff typical of urban watersheds. Testing of the stormwater runoff would reveal consistently lower amounts of suspended solids, nitrogen, phosphorus, heavy metals, bacteria, and other contaminants responsible for degrading water quality. Groundwater quality would improve based on systematic well testing, and base flow rates would increase as more water was infiltrated. The increased base flow would also help to keep stream channels at higher levels throughout the year, and make aquatic ecosystems less vulnerable to the effects of higher water temperatures.

The results of any sampling/monitoring efforts undertaken to reduce erosion and stormwater within watersheds should be published. This helps educate the public about these efforts and may increase public involvement. It also makes the organizations conducting the work more accountable.

9.8 SUSTAINABILITY AT THE INDIVIDUAL LEVEL

There is almost an endless list of sustainability measures that each of us should practice in our daily life. As stated earlier in this book, individual efforts do make a difference. As with governments and large organizations, practicing ESPM methods is advisable even at the individual level. As stated in Chapter 2 and then again in Chapter 6, when examining pollution sources and environmental regulations of the United States, households in the United States are exempt from many environmental regulations.

This is more significant now because as other sectors are regulated, they have improved to the point that households now pollute a much greater share compared to levels emitted a few decades ago. Figure 9.23 shows a large industrial manufacturing facility in the United States that successfully implemented numerous sustainability measures to the point that it produces approximately twice that of the average household per week (see Figure 9.24). This highlights the fact that households in the United States can improve sustainability efforts.

Below is a partial list of individual behavior modifications that each of us can undertake with little effort that together will make a significant impact on moving toward a sustainable future. These items include (USEPA 2018b):

- Bike to work, to school
- Walk
- Carpool
- Check daily air quality forecasts
- Consider living closer to work
- Install solar house panels
- Evaluate the volume and types of food wasted in your household
- Compost yard waste and food waste
- Plant a vegetable garden
- Collect rain water from roof drains to water plants and for other outdoor needs
- Drive a more fuel-efficient automobile, hybrid, or electric automobile
- Consume less meat
- Consume more plant food
- Use mass transit
- Avoid drive-thru lines
- Fill fuel tanks in the evening or at night during summer months
- Get regular engine tune ups
- Request flexible work hours
- Dry clothes by hanging them outside
- Consider an electric lawn mower and snow blower
- Lower water heater temperature
- Keep wood stoves and fireplaces well maintained and limit use, if possible
- Do not smoke indoors
- Do not consume water from plastic bottles
- Recycle plastic, paper, glass, and aluminum (contact for municipality for additional information and recycling opportunities)
- Properly dispose of electronic equipment (contact your local municipality for more information concerning other items such as light bulbs, batteries, mercury switches)
- Properly dispose of other household items including cleaners, solvents, paints, and other items (contact your local municipality for a comprehensive local list)
- Utilize reusable shopping bags
- End the use of plastic bags and plastic bottles
- Choose to purchase items with minimal packaging material (remember if you don't consume or use it, it becomes a waste)

FIGURE 9.23   Manufacturing facility in the United States. (Photograph by Daniel T. Rogers.)

- Upgrade appliances with improved energy efficiency
- Install a heat blanket on the hot water heater and lower its temperature 10 to 20%
- Use low-flow shower heads
- Take shorter showers
- Wash clothes in cold water
- Purchase organically grown and eco-friendly food
- Become more educated
- Encourage and inspire others

FIGURE 9.24   Waste generated weekly at industrial manufacturing facility. (Photograph by Daniel T. Rogers.)

- Turn your computer off rather than using sleep mode or screen saver
- Print documents only when required
- Unplug electrical devices such as toasters, printers, chargers, computers, and other items when not in use
- Use recycled paper
- Drink more tap water
- Don't drink from disposable cups
- Replace your vehicle's air filter every 3,000 miles
- Keep vehicle tires at recommended air pressure
- Do full loads of laundry
- Run the dishwasher only when full
- Limit dry cleaning clothes and other items
- Turn off the water while brushing your teeth
- Receive and pay bills online
- Call companies that participate in junk mail advertisements
- Avoid purchasing food in individual serving sizes
- Avoid purchasing items that are considered single-use
- Turn down the heat or turn air conditioning up when not at home
- Lower the thermostat during winter and especially at night and increase the thermostat during the summer
- Contact your local and state-elected officials and encourage them to support sustainable legislation
- Participate in volunteer outdoor cleanup efforts in your community
- Plant trees, especially deciduous, near your home for shade in the summer
- Consider planting native vegetation rather than grass
- Limit fertilizer and pesticide and herbicide use
- Use citrus-based cleaners rather than volatile organic compounds
- Consider downsizing your home
- Improve insulation of your home
- Purchase more items from garage sales, thrift stores, resale, and consignment shops
- Donate more items
- Sell items at a garage sale, resale, or consignment shop
- Purchase bleach-free paper products
- Reuse wrapping paper and envelopes
- Purchase only items that you need
- Collect and donate packing peanuts to local shipping store
- Purchase rechargeable batteries
- Recycle motor oil, ink cartridges, and tires (contact your local municipality for more information)
- Conserve water (contact your local municipality for more information)

Embracing some or all of the above-listed items into the everyday life of individuals will have a positive impact on the environment. In addition, incorporating the sustainability measures listed above will result in significant cost savings for each individual.

9.9   SUMMARY AND CONCLUSION

Sustainability is complex and will require significant effort at the global, national, local, and individual level. This chapter has outlined numerous sustainability measures and has presented evidence to support why these measures are important. In summary, in order to support a human population on Earth we must not consume and pollute at current levels. In order to achieve a sustainable living, vibrant, and diverse world instead of a world without humans, we need to diligently work cooperatively towards living in productive harmony with nature instead of trying to change nature.

We have dedicated a significant portion of this chapter to pollution prevention in a framework of ESPM techniques. Pollution prevention benefits everyone, especially the environment, and accomplishes much on the road to sustainability. Perhaps most importantly, practicing ESPM techniques in pollution prevention efforts require human action at the individual level, which must be a focus of priority if sustainability is to be achieved. The next chapter will present a model for achieving sustainability at the local level.

REFERENCES

American Society for Quality (AQS). 2019. ISO 14001: *Environmental Management Systems Standard*. https ://asq.org/quality-resources/iso-14001. (accessed January 12, 2019).

Barrett, M.E., Malina, J.F., Jr., and Charbeneau, R.J. 1998. An Evaluation of Geotextiles for Temporary Sediment Control. *Water Environment Research*. Vol. 70. pp. 283–290.

Bartens, J., Day, S.D., Harris, J.R., et al. 2008. Can Urban Tree Roots Improve Infiltration through Compacted Subsoils for Stormwater Management? *Journal of Environmental Quality*. Vol. 37. pp. 2048–2057.

B Corporations. 2019. *About B Corporations*. https://bcorporation.net/about-b-corps. (accessed January 12, 2019).

Cape, J.N., Brown, A.H.F., Robertson, S.M.C., et al. 1991. Interspecies Comparisons of Throughfall and Stemflow at Three Sites in Northern Britain. *Forest Ecology and Management*. Vol. 46. pp. 165–177.

City of Portland, Bureau of Environmental Services. 2019. http://www.portlandonline.com/bes/index.cfm?c=43081&a=177712 (accessed April 6, 2019).

Denardo, J.C., Jarrett, A.R., Manbeck, H.B., et al. 2005. Stormwater Mitigation and Surface Temperature Reduction by Green Roofs. *Transactions of the ASABE*. Vol. 48. No. 4. pp. 1491–1496.

Ernst, C., Gullick, R., and Nixon, K. 2004. Protecting the Source: Conserving Forests to Protect Water. *Opflow*. Vol. 30. No. 5. pp. 1, 4–7.

Harari, Y.N. 2015. *Sapiens*. Harper Collins Publishers. New York. 443p.

Harari, Y.N. 2018. *21 Lessons for the 21st Century*. Random House Books. New York. 372p.

Hudson, B.D. 1994. Soil Organic Matter and Water Quality. *Journal of Soil and Water Conservation*. Vol. 49. pp. 189–193.

Humenik, F., Smolen, M., and Dressing, S. 1987. Pollution from Nonpoint Sources, *Journal of Environmental Science and Technology*. Vol. 23. pp. 737–742.

Kaufman, M.M. and Marsh, W.M. 1995. CSO Control Is No Longer Mere Engineering. *American City and County*. Vol. 110. No. 2. p. 10.

Kaufman, M.M. and Wurtz, M. 1997. Hydraulic and Economic Benefits of Downspout Diversion. *Journal of the American Water Resources Association*. Vol. 33. No. 2. pp. 491–497.

Kaufman, M.M. 1999. Micro-topographic Opportunities for Stormwater Management in Urban Landscapes. *Papers and Proceedings of the Applied Geography Conferences* 22: pp. 86–95.

Kaufman, M.M. 2000. Erosion Control at Construction Sites: The Science-Policy Gap. *Journal of Environmental Management*. Vol. 26. No. 1. pp. 89–97.

Kaufman, M.M., Rogers, D.T., and Murray, K.S. 2011. *Urban Watersheds: Geology, Contamination, and Sustainable Development*. CRC Press. Boca Raton, FL. 583p.

Kaufman, M.M., Rogers, D.T., and Murray, K.S. 2013. Water Pollution. *Encyclopedia of Global Social Issues*. C.G. Bates and J. Ciment, eds. M.E. Sharpe Publishers. New York. Vol. 2. Chapter 45. 1450p.

Mosley, M.P. 1982. The effect of a New Zealand Beech Forest Canopy on the Kinetic Energy of Water Drops and on Surface Erosion. *Journal of Earth Surface Processes and Landforms*. Vol. 7. pp. 103–107.

Murray, K.S. and Rogers, D.T. 1999. Groundwater Vulnerability, Brownfield Redevelopment and Land Use Planning. *Journal of Environmental Planning and Management*. Vol. 42. No. 6. pp. 801–810.

National Association of Home Builders. 2018. *Typical America Subdivisions*. https://www.nahbclassic.org. (accessed January 6, 2019).

National Geographic Society. 2019. *Energy Efficiency of Modes of Transportation*. https://www.nationalgeographic.com/carbon-footprint-transportation-efficiency. (accessed January 6, 2019).

National Oceanic and Atmospheric Administration (NOAA). 2018. *Status of Fish Stocks 2017*. https://www.fisheries.noaa.gov/status-stocks-2017. (accessed January 12, 2019).

Pope Francis. 2015. *Environmental Encyclical and Care for Our Common Home.* The Vatican. 184p. https://www.w2.vatican.va/content/francesco/en/encyclicals/documents/papa-francesco-2015024-enciclica-laudato-si.html. (accessed December 25, 2018).

Rogers, E.M. 1995. *Diffusion of Innovations*, 4th ed. The Free Press. New York.

Rogers, D.T. 1996. *Environmental Geology of Metropolitan Detroit.* Clayton Environmental. Novi, MI. 152p.

Rogers, D.T., Kaufman, M.M., and Murray, K.S. 2006. Improving Environmental Risk Management through Historical Impact Assessments, *Journal of Air and Waste Management.* Vol. 56. pp. 816–823.

Rogers, D.T. 2011. Why Geology and Chemistry Matter in Forming the Foundation in Sustaining Urban Areas, American Institute of Professional Geologists 49(2). Denver, Colorado.

Rogers, D.T. 2014. Scientists Call for a Renewed Emphasis on Urban Geology, American Geophysical Union 95(47): 431–432. Earth and Space News. Eos.

Rogers, D.T., Kaufman, M.M., and Murray, K.S. 2016. Development of an Environmental Risk Management and Sustainability Model Using Chemistry, Geology, and Human Activity, and Response Inputs. *35th International Geological Congress.* Paper number 2090. Cape Town. South Africa.

Rogers, D.T. 2018. Derivation of a Comprehensive Environmental Risk Model for Urban Groundwater Protection, International Association of Hydrogeologists Congress 1: 879. Daejeon, Korea.

Roy, A.H., Wenger, S.J., Fletcher, T.D., et al. 2008. Impediments and Solutions to Sustainable, Watershed-Scale Urban Stormwater Management: Lessons from Australia and the United States. *Journal of Environmental Management.* Vol. 42. pp. 344–359.

Schueler, T.R. 1994. The Importance of Imperviousness. *Journal of Watershed Protection Techniques.* Vol. 1. pp. 1–11.

Smolinske, S. and Kaufman, M.M. 2007. Consumer Perception of Household Hazardous Materials. *Clinical Toxicology.* Vol. 45. pp. 1–4.

Thornton, R.V. 2001. ISO 14001 Certification Mandate Reaches the Automobile Industry. *Journal of Environmental Quality Management.* Vol. 10. No. 1. pp. 89–96.

United Nations. 2017. *World Population Prospects. United Nations Department of Economic and Social Affairs. Population Division.* https://esa.un.org/unpd/wpp/data. (accessed December 14, 2017).

United Nations. 2018a. *What Is Sustainable Development.* https://www.un.org/sustainabledevelopment. (accessed December 24, 2018).

United Nations. 2018b. *Global Sustainability Goals.* https://www.undp.org/content/undp/en/home/sustainable-development-goals.html. (accessed December 25, 2018).

United Nations. 2019a. *UN Announces First-Ever Global Standard to Measure Food Loss Waste.* https://www.un.org/sustainabledevelopment/blog/2016/06/un-announces-first-ever-global-standard-to-measure-food-loss-and-waste/. (accessed May 24, 2019).

United Nations. 2019b. *Funds, Programmes, Specialized Agencies and Others.* https://www.un.org/en/sections/about-un/funds-programmes-specialized-agencies-and-others/. (accessed April 6, 2019).

United States Census Bureau. 2019a. *Census Bureau Average Population Per Household.* https://www.census.gov/population/socdemo/hh-fam/tabHH-6.pdf. (accessed January 6, 2019).

United States Census Bureau. 2019b. *Median and Average Square Feet of Floor Area in New Single-family Homes.* https://www.census.gov/const/c25ann/sftotalmedavsqft.pdf. (accessed January 6, 2019).

United States Census Bureau. 2019c. *Average One-Way Commuting Time and Distance.* https://www.census.gov/library/visualization/interactive/travel-time.html. (accessed January 6, 2019).

United States Department of Agriculture. 2011. *Land Use in the United States.* https://www.ers.usda.gov/data-products/major-land-use. (accessed December 8, 2018).

United States Department of Energy. 2019. *Energy Efficiency and Renewable Energy. Alternative Fuels Data Center.* https://www.afdc.energy.gov.data. (accessed January 6, 2019).

United States Department of Housing and Urban Development. 2019. *Median and Average Square Feet of Floor Area in New Single-family Homes.* United States Census Bureau. https://www.census.gov/const/c25ann/sftotalmedavsqft.pdf. (accessed January 6, 2019).

United States Environmental Protection Agency (USEPA). 1990. *Pollution Prevention Act of 1990.* Code of Federal Regulation 42 CFR, Chapter 133. Washington, DC.

USEPA. 2002. *National Water Quality Inventory: Report to Congress: 2002 Recycling Report.* USEPA. EPA-841-R-07-001. Washington, DC.

USEPA. 2003. *National Management Measures to Control Nonpoint Source Pollution from Agriculture.* USEPA. EPA 841-B-03-004. Washington, DC.

USEPA. 2017. *Reducing Wasted Food at Home.* https://www.epa.gov/recycle/reducing-wasted-food-home. (accessed May 24, 2019).

USEPA. 2018a. *United States Environmental Protection Agency QA Glossary.* https://www.epa.gov/emap/archive-emap/web/html. (accessed December 1, 2018).

USEPA. 2018b. *What Is Sustainability?* https://www.epa.gov/sustainability. (accessed December 1, 2018).

USEPA. 2018c. *Regulations for Greenhouse Gas (GHG) Emissions.* https://www.epa.gov/regulations-emissions-vehicles-and-engines/regulations-greenhouse-gas-ghg-emissions. (accessed December 24, 2018).

USEPA. 2019a. *International Cooperation.* https://www.epa.gov/international-cooperation. (accessed April 6, 2019).

USEPA. 2019b. *Pollution Prevention.* http://www.epa.gov/p2. (accessed April 6, 2019).

USEPA. 2019c. *SAGE 2.1: Solvent Alternatives Guide.* https://cfpub.epa.gov/si/si_public_record_report.cfm?Lab=NRMRL&dirEntryId=115684. (accessed April 7, 2019).

USEPA. 2019d. *Nonpoint Source Pollution.* https://19january2017snapshot.epa.gov/nps_.html. (accessed April 7, 2019).

USEPA. 2019e. *Stormwater Best Management Practices.* http://www.epa.gov/npdes/stormwater/menuofbmps. (accessed April 7, 2019).

Van Woert, N.D., Rowe, D.B., Andresen, J.A., et al. 2005. Green Roof Stormwater Retention. *Journal of Environmental Quality.* Vol. 34. pp. 1036–1044.

Williams, D.D., Williams, N.E., and Cao, Y. 2000. Road Salt Contamination of Groundwater in a Major Metropolitan Area and Development of a Biological Index to Monitor Its Impact. *Journal of Water Research.* Vol. 34. pp. 127–138.

World Bank. 2019. *The Fight to Improve Lives of the World's Poor.* http://www.worldbank.org/en/news/press-release/2013/07/23/improve-lives-world-poor-world-bank-group-delivers-nearly-53-billion-support-developing-countries-fy13. (accessed April 6, 2019).

World Ocean Review. 2019. *The Global Hunt for Fish.* https://worldoceanreview.com. (accessed January 12, 2019).

10 Building and Implementing a Model for Sustainability

10.1 INTRODUCTION

A sustainability model for any location in the world may seem impossible because of the sheer number of potential variables that greatly increase the complexity and because of the many differences between countries. However, this is just not the case. In this chapter, we shall build a model for sustainability.

We defined sustainability in the previous chapter as creating and maintaining conditions under which humans can exist in productive harmony to support present and future generations (United Nations 2018; USEPA 2018). We also examined what is meant by "productive harmony" in the definition of sustainability and concluded that productive harmony applies to the environment. Therefore, two attributes make up the sustainability universe that are (1) the natural environment and (2) our interaction with the natural environment. It's with these two basic fundamentals of sustainability that we can build our model.

10.2 BUILDING THE ELEMENTS OF A SUSTAINABILITY MODEL

As stated above, in order to achieve a model for sustainability we must understand the environment and also understand the aspects of how humans impact the environment. At the local or site level, sustainability is the outcome of analyzing the aspects of facility operations with the natural environment and developing engineering methods to eliminate or reduce the potential for realization of harmful impacts.

Therefore, simplifying sustainability into its basic elements we arrive at the following:

- The natural environment
- Aspects of operations
- Risk reduction actions or lack of risk reduction actions

We have been introduced to two important elements of building a model of sustainability already (Chapter 3 – the contaminant risk factor [(CRF]) and Chapter 4 – geologic vulnerability).

Geologic vulnerability is a characterization of the natural environment. Essentially, geologic vulnerability is a measure of nature's response when it becomes polluted (Rogers 1996; Rogers 2011; Rogers et al. 2012).

The contaminant risk factor is an aspect of operations. We learned in Chapter 2 that if it weren't for chemicals, environmental risk would be greatly reduced but would not be eliminated. We have learned about chemicals and their behavior when released into the environment that includes each chemical's toxicity, persistence, and mobility. Each of these factors are critically important when combined with geologic vulnerability because it can be used as a predictor to evaluate whether a release of any particular chemical will cause harm and at what scale (Rogers et al. 2016). As you will realize, the contaminant risk factor is integrated into the aspect risk of operations in the following sections.

We learned how the natural environment reacts to different chemicals and identified certain geologic and hydrologic areas that are especially sensitive to contamination where a synergistic effect is realized that greatly amplifies potential harmful effects. Each chemical has a different associated risk factor for each environmental media –air, water, and soil – that is dependent upon its toxicity and physical chemistry (mobility and persistence) (Rogers 2014; 2018).

Now we need to combine what we have learned about pollution and the environment with aspects of facility operations and risk reduction actions. A convenient place to start this portion of our model development is with the environmental audit. The environmental audit is convenient because it examines facility operations that have the potential to emit pollutants into the air, water, and land. In addition, an environmental audit also examines management practices and engineering controls undertaken to eliminate or manage the risks involved with minimizing the potential for a release of contaminants to the environment. It is also a convenient place to begin because most every industrial facility on the planet should conduct an environmental audit at some scale. Therefore, much of the required information already exists.

Facility risks can then be characterized with two measured variables that include aspect risks of operations and risk reduction actions. Much of this information can be obtained and scored from information in an environmental audit. The following section evaluates the aspect risks of operations. There are also integers that are assigned to each of the many aspects that are measured. For instance, let's look at three examples that include our three most important avenues for protecting human health and the environment which include the air, water, and land. We will examine each separately below (Rogers 2018).

1. Air permits. If a facility has an air emission source that emits a pollutant, a permit is almost always required. The permit will outline pollutants to be emitted along with concentrations and numerous other operational items to ensure that the air-emitting source limits the discharge of pollutants into the air. If a facility has many emitting sources, a Title V permit may be necessary. Title V permits are generally the most complex for any facility in the United States. If a facility has a Title V permit, it means that the facility emits pollutants from many sources. Therefore, it seems logical to assign a higher aspect of operations value for those facilities that have a Title V permit as compared to a facility with little or no air emission sources.

2. Water discharge permits. Industrial plants that use water for anything other than sanitary discharge usually are required to obtain a wastewater discharge permit. Wastewater from an industrial facility is either discharged to a publicly owned treatment works (POTW) directly to surface water, or to a septic system. From an environmental perspective, discharge of wastewater directly to surface water or to a septic system carries more risk because it's a direct discharge to the environment compared to wastewater discharged to a POTW as the water gets additional treatment at the POTW before it's discharged. Other factors that create higher environmental risk to the environment is based on what chemicals are contained in the water and the amount of water discharged and whether any pretreatment is required. Therefore, a facility wastewater that discharges high volumes of water (greater than 50 million liters a month) which contains compounds with high relative CRFs, requires pretreatment, and discharges directly to surface water is assigned a much higher aspects of operations value compared to a facility with low CRF compounds, low water volume, no pretreatment, and discharges to a POTW.

3. Solid waste. A facility that does not generate hazardous waste generally has lower risk than a facility that is a large quantity generator (LQG) of hazardous waste. However, it not that easy because there are mitigating factors that must be examined which include the number of waste streams, volumes, processes generating the waste, CRFs, whether wastes are solids or liquids, and other factors.

Now that we have presented three examples, we turn our attention to examining an environmental audit in detail so that we can gather enough information on the aspects of operations to arrive at an integer value that reflects the environmental risk aspects from operations.

An aspect checklist is provided below that lists each aspect that is evaluated with a corresponding integer assignment range based on risk that follows the same methodology we have discussed above. Note that occasionally there is the possibility for a negative integer to be assigned to a particular attribute. This does not occur often but is possible in situations where risk is significantly or totally eliminated through engineering controls.

10.3 CHECKLIST FOR ENVIRONMENTAL ASPECT RISK ATTRIBUTES AND SCORING

As we have discussed as one of the themes of this book, "no matter where you are, it's all the same." Protecting human health and the environment should be the same everywhere and focuses on protecting the air, water, and land. Therefore, as you proceed through the checklist you will notice that more weight is applied to the air, water, and land. Most countries of the world have relied on this foundation of air, water, and land protection, which is the United States Environmental Protection Agency (USEPA) framework for environmental regulations.

The recommended point values for each attribute is given with a range. Higher point values indicate higher relative risk. Categories that are included in the checklist following the environmental audit guidelines are presented in the previous chapter and are (Rogers 2018; 2019):

- Air
- Solid and hazardous waste and landfills
- Water
- Spills
- Polychlorinated biphenyls (PCBs)
- Toxic Substance and Control Act (TSCA)
- Emergency Planning and Community Right-to-Know Act (EPCRA)
- Tanks
- Asbestos
- Regulatory inspections
- Site inspection

10.3.1 Air

Three scenarios are included in the checklist for air, which include a Title V permit, an air permit but not a Title V, and no air permit. To evaluate an appropriate assignment of risk in points, the user should carefully consider each of the secondary bullets listed below:

- Title V permit or equivalent = 5 to 7 points
- Air permit but no Title V = 1 to 5 points
- No air permit but some air emissions = 1 to 2 points
- No air permit and no emissions = 0 points

Sub-factors to evaluate include:

- Number of air emission points
- PM10/PM2.5 limits, if applicable
- Chemicals emitted
- Volumes (cubic meters per second)
- Emission rates
- Mass (kilograms per hour or day) for each compound

- Chemistry of compounds
- Contaminant risk factor for each compound
- Air pollution control efficiency (baghouse vs. wet scrubber)
- Opacity requirements
- Stack heights
- Distance to property boundary
- Offsite property use (i.e., residential, commercial, or industrial)
- Facility located in non-attainment location
- State located
- Potential sensitive receptors
- Governing regulatory authority
- Permit limits
- Difficulty in documenting permit compliance

10.3.2 Solid and Hazardous Waste

Solid and hazardous waste is separated into four separate subcategories that include:

- Hazardous waste
- Solid waste
- Onsite or offsite landfill
- Universal waste

10.3.2.1 Hazardous Waste

Hazardous waste is separated into three subcategories that include large quantity generator (LQG), small quantity generator (SMQ), and a very small quantity generator (VSQG). To evaluate an appropriate assignment of risk in points, the user should carefully consider each of the secondary bullets listed below:

- LQG = 3 to 7 points
- SQG = 1 to 4 points
- VSQG = 1 to 2 points
- No hazardous waste = 0 points

Sub-factors to evaluate include:

- Number of waste streams
- Volume
- Mass
- Content
- Degree of waste characterization
- History of waste characterization (e.g., when was the last time each waste stream was re-classified)
- Chemistry of compounds
- Differentiation between liquids and solids
- Distance
- Stabilization required for any wastes
- Number of compounds
- Contaminant risk factor for each compound
- Volatile organic compound content
- Highly toxic compounds (i.e., chlorinated volatile organic compounds, etc.)

- Compliance history
- Storage inside or outside or separate building
- Compliance history

10.3.2.2 Solid Waste

Solid waste are materials generally considered to be common trash that can't be readily recycled and may contain residue contaminants such as oil or metal. In the case of solid wastes, point values are generally less than those applied to hazardous wastes and are dependent upon several factors listed below under sub-factors.

- Solid wastes = 1 to 5 points

Sub-factors to evaluate include:

- Number of waste streams
- Degree of waste characterization
- History of waste characterization (e.g., when was the last time each waste stream was re-classified)
- Processes generating each waste stream
- Chemical content
- Volume generated
- Mass
- Differentiation between liquids and solids
- Chemistry of compounds
- Contaminant risk factor for each compound
- Compliance history
- Disposal frequency

10.3.2.3 Onsite or Offsite Landfill

Hazardous and non-hazardous solid wastes are typically disposed of in landfills. Landfills are known to be sources of contamination when any number of unanticipated issues may occur that include liner failure, excessive erosion, or a natural disaster. Since landfills are highly variable in construction, location, and compliance history, a rather wide point range is considered. Evaluation of sub-factors may require conducting an environmental audit of the landfill.

- Onsite or offsite landfill = 1 to 7 points and depend upon the following factors:
  - Liner construction
  - Leachate recovery construction, if applicable
  - Age
  - Distance
  - Travel routes
  - Transporter compliance history
  - Drainage
  - Nightly cap
  - Security
  - Monitoring results and trends
  - Location to sensitive ecological receptors
  - Compliance history
  - Classification
  - State located
  - Compliance history

- Regulatory inspection frequency
- Agency findings
- Geologic vulnerability
- Geology and hydrogeology
- Geography
- Distance to nearest surface water body

10.3.2.4   Universal Waste

Universal wastes are a special category of hazardous wastes and include items such as (USEPA 2018:

- Batteries
- Mercury switches and other mercury-containing devices such as thermometers and monometers
- Electronic equipment such as computers and many other consumer items
- Fluorescent bulbs
- Recalled pesticides
- Some state-specific universal waste regulations also include:
  - Aerosol cans
  - Anitfreeze
  - Ballasts
  - Barometers
  - Oil-based finishes
  - Paint waste
  - Pharmaceuticals

The point range recommended for universal waste is from 1 to 3 points. Every facility should generate many types of universal wastes and must follow state and federal requirements that include at the minimum:

- Labeling requirements
- Designating a specific area for universal waste accumulation
- Follow disposal requirements
- Secure universal wastes in appropriate containers
- Follow manifesting requirements

10.3.2.5   Used Oil

Used oil is usually a recycled material. However, there are storage and handling requirements. This is usually to prevent a spill which can significantly affect the environment.

10.3.3   Water

Water is separated into several subcategories that include:

1. Source of incoming water (i.e., onsite water well, municipal supply)
2. Water usage
3. Onsite wastewater treatment
4. Discharge to surface water (National Pollutant Discharge Elimination System [NPDES] permit)
5. Discharge to municipal sewer (POTW)

6. Discharge to septic system
7. Stormwater

To evaluate an appropriate assignment of risk in points, the user should carefully consider each of the secondary bullets listed below:

- Onsite water well or other direct or onsite water source (i.e., river or stream) = 1 to 7 points
  - Depth
  - Age
  - Rate of withdraw (liters per day/month/year)
  - Location
  - Water quality
  - Sampling requirements
  - Does onsite well supply potable water for human ingestion? If so, higher point total necessary
  - Any pretreatment of water necessary?
  - Regional groundwater flow direction
  - Geologic and hydrogeologic conditions
  - Evaluation of potential upgradient contaminant sources
  - Compliance history
- Municipal water supply = 0 to 3 points
  - Source of municipal water supply
  - Distance from treatment plant
  - Age of infrastructure piping
  - Volume used
  - Compliance history
  - Monitoring trends
  - Number of users
- Water usage = 0 to 5 points
  - >10M gallons/year = 5 points
  - <10M gallons/year = 3 points
  - <1M gallons/year = 1 point
  - Source
  - Water quality
  - Source
  - Pretreated?
  - Discharge volume
- Onsite wastewater treatment operation = 1 to 7 points
  - Complexity
  - Volumes
  - Permit limits or mass (kilograms per day) for each compound
  - Treatment requirements
  - Discharge limits, etc.
  - Compliance history
  - Contaminant risk factor for each chemical compound
- Surface water discharge (NPDES permit = 1 to 7 points)
  - Monitoring frequency
  - Volume of discharge
  - Permit limits or mass (kilograms per day) for each compound
  - Discharge body and watershed
  - Chemistry of compounds
  - Discharge limits, etc.

- Compliance history
- Contaminant risk factor for each compound
- Evaluate that each compound is treated appropriately
- Ecological risks
- POTW permit = 0 to 4 points (evaluate whether the facility discharges to the POTW instead of to surface water under an NPDES permit)
  - Volume
  - Chemistry of each compound
  - Permit limits or mass (kilograms per day) for each compound
  - Contaminant risk factor for each compound
  - Confirm POTW treatment for each compound present
  - Compliance history
- Stormwater permit = 1 to 5 points (also evaluate other requirements if no permit required)
  - Volume
  - Chemistry of each compound
  - Discharge body and watershed
  - Distance to point of discharge
  - Discharge limits
  - Discharge limits or mass (kilograms per day) for each compound
  - Compliance history
  - Frequency of sampling required
  - Contaminant risk factor for each compound
- Onsite septic system = 1 to 7 points
  - Types
  - Age
  - Volumes
  - Discharge limits or mass (kilograms per day) of each compound
  - Content of discharges
  - Chemistry of compounds
  - Contaminant risk factor for each compound
  - Distance to nearest ecological or human receptor
  - Depth to groundwater
  - Is groundwater used as potable water supply?
  - Regional groundwater flow direction?
  - Geology and hydrogeologic conditions
- No industrial wastewater discharge = 0 points
  - Confirm zero discharge
  - Confirm that there are no cross connections

10.3.4 POLYCHLORINATED BIPHENYLS (PCBs)

In the United States and most other countries, polychlorinated biphenyls (PCBs) may still be present in transformers and capacitors, especially those that were manufactured before 1978. In addition, transformers in the United States that have a non-PCB label are permitted to have a concentration of PCBs at less than 50 milligrams per kilogram. Therefore, to evaluate the potential environmental risks for PCBs, the following can be used as a guide:

- Facility-owned PCB transformers (greater than 50 milligrams per kilogram) = 5 to 7 points for each depending on the following:
  - Capacity (liters of fluid)

- Location
  - Inside or outdoors
  - Distance to sensitive receptor
  - Physical condition
  - Age
  - Distance to nearest migration pathway (i.e., floor drain, storm drain, etc.)
  - Engineering controls
  - Inspection frequency
- Facility-owned transformers containing PCBs (less than 50 milligrams per kilogram) = 1 to 3 points for each depending on the following:
  - Capacity (liters of fluid)
  - Location
    - Inside or outdoors
    - Distance to sensitive receptor
    - Physical condition
    - Age
    - Distance to nearest migration pathway (i.e., floor drain, storm drain, etc.)
    - Engineering controls
    - Inspection frequency
- Utility-owned transformers containing PCBs on facility property = 1 to 3 points for each depending on the following:
  - Capacity (liters of fluid)
  - Location
    - Inside or outdoors
    - Distance to sensitive receptor
    - Physical condition
    - Age
    - Distance to nearest migration pathway (i.e., floor drain, storm drain, etc.)
    - Engineering controls
    - Inspection frequency
- Capacitors potentially containing PCBs = 1 to 5 points
  - Capacity (liters of fluid)
  - Location
    - Inside or outdoors
    - Distance to sensitive receptor
    - Physical condition
    - Age
    - Distance to nearest migration pathway (i.e., floor drain, storm drain, etc.)
    - Engineering controls
    - Inspection frequency
- No fluid-containing transformers = No points
- Transformers containing mineral oil or other similar fluid with no PCBs = No points

10.3.5 TANKS

Aboveground storage tanks (ASTs) and underground storage tanks (USTs) are two types of tanks that are evaluated. USTs will generally have a higher point assignment than ASTs because they are more difficult to monitor and inspect. The following is presented as a guide for tanks:

- ASTs = 0.5 to 1 point each dependent upon the following:
  - Age

- Contents
- Physical condition of tank
- Capacity
- Usage rate
- Location (indoor or outdoor)
- Secondary containment
- Compliance history
- Distance to nearest receptor
- Potential transport pathways (i.e., distance to floor drains, storm drains, bare ground, etc.)
- Contaminant risk factor for each compound stored
- Other engineering controls
- USTs = 1 to 3 points for each UST dependent upon the following:
  - Age
  - Contents
  - Physical condition of tank
  - Capacity
  - Usage rate
  - Location (indoor or outdoor)
  - Secondary containment
  - Compliance history
  - Distance to nearest receptor
  - Potential transport pathways (i.e., distance to floor drains, storm drains, bare ground, etc.)
  - Contaminant risk factor for each compound stored
  - Other engineering controls

10.3.6 SPILLS

The two most important factors in evaluating the potential environmental risk posed by spills at any facility are what chemicals are present in volumes sufficient enough to cause a release and what is the compliance history of the facility. If the facility is required to have a spill plan that means that the facility stores a volume greater than 5,300 liters (1,320 gallons) in the United States. The following is suggested to be used as a guide to evaluate the environmental risk from spills:

- Spill plan required = 3 points
- Spill plan not required but chemicals present = 1 to 3 points
  - Type of chemical(s)
  - Contaminant risk factor for each chemical
  - Location(s) stored
  - Distance to nearest receptor
  - Potential transport pathways (i.e., distance to floor drains, storm drains, bare ground, etc.)
- Reportable spill occurrence within past 24 months = 1 to 7 points
  - Volume
  - Mass
  - Potential environmental damage
  - Sensitive receptor evaluation, offsite?
  - Training
  - Lessons learned
  - Prevention measures and engineering controls employed

10.3.7 Emergency Planning and Community Right-to-Know Act (EPCRA)

EPCRA is important since the associated documents required by EPCRA list chemicals stored, used, and released from each specific facility. The main goal of EPCRA as we have explained in Chapter 5 is informing and protecting the public. The recommended point total for EPCRA is 1 to 3 points and involves:

- Review Tier I and Tier II forms
  - Ensure list is up to date
  - Ensure onsite storage, use, and waste areas are up to date and clearly identified
  - Ensure volumes are accurate
- Review Toxics Release Inventory (TRI) Form R
  - Evaluate methods of calculating volumes or mass are accurate
  - Evaluate points of generation
  - Identify media (air, water, or solid)

10.3.8 Toxic Substance and Control Act (TSCA)

TSCA is where the evaluation is conducted on whether a facility has a chemical ordering procedure and whether that procedure is effective. The recommended point total for TSCA is 0 to 7 points and is dependent upon the following:

- The amount and different types of chemicals stored, used, and generated
- Does the facility have a written chemical ordering procedure or program
- Does the facility environmental engineering sign off on new chemicals
- Check chemical use during site inspection to ensure compliance with chemical ordering program

10.3.9 Asbestos

Asbestos remains a significant environmental issue world-wide. Buildings constructed before 1980 may contain asbestos. The recommended point total if a specific facility has asbestos-containing materials is from 1 to 3 points and is dependent upon the following:

- Amount of asbestos-containing materials
- Type (pipe wrap, floor tile, mastic, caulking, roofing, siding, insulation, etc.)
- History
- Condition or amount potentially friable
- Accuracy and completeness of the asbestos management plant

10.3.10 Regulatory Inspection History

Regulatory inspection history is an important item when evaluating environmental risk. If a specific facility has a record of non-compliance or is frequently inspected by regulatory agencies, it generally indicates a higher level of risk associated with environmental risk to cause harm from the regulatory agency perspective. The recommended point total for each regulatory inspection is 1 to 3 points within the last 3 years and is dependent upon the following:

- Type of inspection. If inspection is air, water, or solid and hazardous waste, a higher point value should be considered.
- Routine inspection or complaint. An inspection due to a complaint should carry a higher point total.

10.3.11　Summary of Environmental Risk Parameters

Under environmental risk parameters, we have included the main risks are air, water, and solid and hazardous wastes and then have examined other attributes which include PCBs, tanks, spills, EPCRA, TSCA, asbestos, and history of regulatory inspections. One we have left out is site inspection. We will address site inspection under the next section where we evaluate risk reduction actions and performance, which essentially is a measure of what actions a facility undertakes to eliminate or reduce environmental risk.

10.4　RISK REDUCTION ACTIONS AND PERFORMANCE

Risk reductions actions and performance are activities a facility undertakes to eliminate or reduce environmental risk identified by the environmental aspects of each facility. As you can probably guess by now, actions a facility undertakes to reduce environmental aspect risks related to air, water, or land (solid and hazardous waste) carry more weight than many of the other attributes we will evaluate. The reason we concentrate so heavily on air, water, and solid and hazardous waste is because that is how pollution is transported offsite and can potentially cause harm to humans and the environment.

Items we will evaluate as part of risk reduction actions and performance include (Rogers 2018; 2019):

- Environmental audits
- Site inspection
- Regulatory agency inspections
- Level of environmental professional attention
- Integration with leadership of management
- Chemical ordering procedure
- Elimination, substitution, prevention, and minimization (ESPM) program and achievements

Each of the above-listed items will be separately evaluated and scored in the following sections.

10.4.1　Environmental Audits

If you remember in the previous two chapters, we highlighted the importance of environmental audits in building a sustainability model. Well, here is where we put the environmental audit to another good use other than as just a measure of compliance status.

The recommended scoring for items identified in a compliance audit are tiered according to importance according to the following:

- Compliance issues related to air, water, or solid and hazardous waste have a recommended point value of from 1 to 5 points for each compliance issue discovered.

  The range in scoring is related to several factors that can involve an almost endless list of possibilities. However, we can provide some guidance on a few of the most commonly observed items to assist the user in developing a representative score that includes:
  - Is the compliance issue identified currently causing immediate harm? If so, then immediate attention is required and a higher point value should be assigned.
  - Is the compliance issue related to air, water, or solid and hazardous waste? If so, then a higher point value should be assigned.
  - Is the compliance issue identified related to the discovery of an extremely dangerous chemical? If so, then a higher point total should be assigned.
  - Waste characterization issues. This can be a significant issue if the waste was previously characterized as non-hazardous but due to other factors such as process changes

would now be characterized as hazardous because the chemistry of the waste has changed.
- Labeling. This is a difficult issue to score. The nightmare scenario is a container that is not labeled and is an extremely hazardous waste, stored outside, in a large container that is inappropriate on bare ground in extreme weather conditions (see Figure 9.13 in Chapter 9 for an example).
- Compliance issues related to other items that are not associated with air, water, or solid and hazardous waste. The recommended point score for these items ranges from 1 to 3 points and again is evaluated on the level at which the compliance issue can result in causing immediate harm to human health of the environment.
- Best management practice (BMP) recommendations range in point values from 0.25 to 1 point for each identified issue. Following the same methodology as compliance issues, items that may result in a higher point total for BMPs include:
  - Air
  - Water
  - Waste
  - Outside storage
  - Maintenance
  - Chemical ordering
  - Housekeeping
  - Repeat BMP

As a reminder, the following is a quick reference to those items that are related to scoring risk reduction actions:

- Air
  - Review of emission sources with diagram
  - Review annual emissions inventory
  - Review permits including all permit conditions
  - Ask if there are any concerns or exceedances
  - Review maintenance records and any concerns
  - Any new equipment? Any new permits needed?
  - Frequency of agency inspections, areas of concentration, and notice of violations (NOVs)
- Water
  - Review wastewater sources with diagram
  - Review permits and POTW or NPDES status
  - Ask if there are any concerns or exceedances
  - Review stormwater pollution prevention plan (SWPPP) or equivalent
  - Review maintenance records and any concerns
  - Frequency of agency inspections, area of concentration, and NOVs
- Solid and hazardous waste
  - Review of wastes and sources with diagram
  - Review waste stream characterization profiles for each waste stream
  - Evaluate points of generation
  - Evaluate whether any wastes are comingled before transport for disposal
  - Review permits and permit conditions
  - Review contingency plans
  - Review manifests
  - Ask if there are any concerns or tracking issues
  - Review written logs and other records not already mentioned

- Frequency of agency inspections, area of concentration, and NOVs
- Spills
  - Review Spill Prevention Control and Countermeasures (SPCC) plan
  - Review the need for a plan if not required
  - Review waste and material storage areas with a diagram
  - Review contingency plan
  - Review if the facility has had any recent spills
  - Review if the facility has had any recent reportable spills
  - Frequency of agency inspections, areas of concentrations, and NOVs
- PCBs
  - Any transformers?
  - Liquid filled? (check quantity, ownership, maintenance records, age, inspection)
  - Verify that no PCBs present through analytical or other type of certification
  - Any capacitors? (age, place of manufacture, etc.)
  - Other PCB-containing materials (caulking, light ballasts, etc.)
- Tanks
  - Review type, location, age, inspection records, dispensing, engineering, contents, volumes, and calculate CRF for each chemical of concern
  - Review location on SPCC or contingency plan
- HAZMAT
  - Review hazardous material shipments
  - Review general awareness and response training and logs
- TSCA
  - Review chemical and new chemical ordering procedure and sign off
  - Review environmental engineer involvement
- EPCRA
  - Review Tier I and Tier II forms
  - Review TRI Form R

10.4.2 SITE INSPECTION

A detained site inspection is part of conducting an environmental audit. However, the site inspection is a crucial part of the environmental audit and as such, we highlight the site inspection with its own section. As we discussed in the previous chapter, following the review of written materials, a site inspection of the entire facility should always be conducted.

It is recommended to use the point value system described above to assign points for items discovered during the site inspection. As an example, if a compliance issue is discovered and it's related to air, water, or solid and hazardous waste, it should receive a higher score. Score ranges for the site inspection range from 0.25 points for a BMP issue to 5 points for an air, water, or solid and hazardous waste compliance issue.

It is recommended that the site inspection be structured in such a way that it follows the manufacturing process as much as possible. Photographs are recommended to document the current environmental condition of the facility, especially those items that may require a corrective action. It is also recommended that a photograph be taken after any corrective action to document task completion and to demonstrate compliance.

It will be helpful to have a map of the facility while conducting the site inspection that shows the entire property, buildings, and operations that are clearly marked. In addition, it will be helpful to have waste storage areas, air emission points, points of generation of solid and hazardous waste, point of generation of wastewater and wastewater discharge points, and stormwater outfalls labeled on the map.

During the site inspection, the following items should be observed, evaluated, and noted:

- If there is an item observed during the site inspection or at any time that may present an immediate threat or imminent danger to human health or the environment, follow the organizations emergency notification procedures. Examples may include but would not be limited to the following:
  - Evidence of existing release of hazardous substances or petroleum products
  - Evidence of material threat of a release of hazardous substance or petroleum products
- Other areas to inspect, review, and/or observe:
  - General housekeeping
  - Any new equipment?
  - Any new processes?
  - Any new waste streams?
  - Any chemicals with high CRFs?
  - Review history of neighborhood complaints (documented)
  - Vehicle maintenance areas
  - Spill kit's location, contents listing, and adequacy
  - Emergency Response Phone List – postings, updated, controlled document, etc.
  - Chemical and waste storage areas
  - Labeling, storage, and housekeeping
  - Waste storage areas, piles, drop boxes, covered, leaks, proper contents
  - New product storage areas (oil and chemicals)
  - Universal waste storage practices and labeling
  - Drum management
  - Drum labeling, management, residue issues, and empty drum storage
  - Secondary containment areas
  - Plant and property security – alarm system, guard service, fencing, gates, lighting, access control, etc.
  - Roof inspection
  - Drainage
  - Emission deposits
  - Exhaust stack(s) deposits
  - Roof vent(s) deposits
  - Floor drains identified on plant drawings, purpose, discharge, potential discharge, drain blocker in the event of a spill
  - Stormwater conveyance system's preventative maintenance
  - Erosion issues
  - Catch basins
  - Detention ponds
  - Outfalls
  - Outside processes, if any
  - Equipment and product storage potential for contamination
  - Utility supply and shutoff valves, identification, security, management, emergency procedures, and training
  - Compressor discharge management
  - Exhaust fan and vent deposits to exterior of the building
  - Baghouse inspection, recordkeeping, and preventative maintenance
  - Review facility's control of chemicals for chlorinated solvent content
  - Secondary container labeling
  - Neighboring properties identified and note (1) the potential for hazardous substance use or storage such as ASTs and drum storage areas, (2) stained soil, (3) topography, (4) drainage patterns and (5) types of operations

- Property boundaries inspected, non-facility contributions, observations/activity, any issues

As a note of importance and to become familiar with how contaminants behave in relation to a specific facility, a few items that should be of emphasis and are deserving of mentioning again and again include:

- Housekeeping
- Location of floor drains, storm drains, catch basins, ponds, streams, lagoons, pits, sumps, outfalls, storage areas, trenches, maintenance areas, laboratories, waste storage, receiving, shipping, production, open areas, back doors, pipes, floor tile, parts washers, storage containers, secondary containment, property line, roof, stacks, and fugitive emissions
- Garbage, stained pavement, recent grading, recent paving
- Any new pieces of equipment
- Any new permits, updated or renewed permits

10.4.3 LEVEL OF PROFESSIONAL ENVIRONMENTAL ATTENTION

Level of professional environmental attention is a human resource evaluation for those directly responsible for environmental matters. To score this section will likely include consultation with plant management and plant human resources and direct interaction with those who are being evaluated. Point values assigned for this attribute range from -5 to 5 points. Notice that a negative value is possible. This is because of rare circumstances where a facility has an exceptionally qualified and effective environmental professional who has demonstrated the leadership and technical abilities to engage facility personnel in not just achieving facility compliance goals but those also associated with sustainability goals.

Recommend items to evaluate include:

- Part time or full time
- Other potential duties
- Work ethic
- Attitude
- Regulatory knowledge
- Education
- Applicable experience
- Length of employment
- Communication skills with regulators
- Leadership and management skills
- Written and verbal communication skills
- Organization of files
- Effectiveness of employment
- Degree to which outside consultants are used and managed

10.4.4 REGULATORY INSPECTIONS

Regulatory inspections are a benchmark and an excellent measurement of the effectiveness of a facility to comply with environmental regulations. Every facility should consider a compliance audit as a practice exam and the regulatory inspection as the actual exam.

The recommended point values for regulatory inspections do have the possibility of a negative value, which are related to those findings of compliance which is a good and certainly independent verification of the level of performance at any facility. The recommended point values for this section are outlined below and include:

- Each regulatory inspection non-finding by media (air, water, or solid and hazardous waste) within the previous 24 months = -0.25 to -5 points
  - Depends on type of inspection and level of detail of inspection
  - Comprehensive multi-media inspections should carry more weight
  - Air, water, and solid and hazardous waste inspections should carry more weight
- Evaluate number and type of inspections and the level of detail of inspections
  - Inspection for air, hazardous waste, and water (including stormwater) are more of a priority than others (e.g., tire inspection)
  - Evaluate ratio of inspections to violations
  - How often is facility inspected and why
- Evaluate whether inspection was routine or employee compliant or other
- Each confirmed inspection finding of non-compliance = 1 to 5 points (based on type of non-compliance issue). Note that the same appears under complexity. Therefore, evaluation should be conducted as to the most appropriate location to assign points and ensure that duplication is minimized.
  - Were NOVs referred to attorney general
  - Was facility under a consent decree
  - Will NOV result in a consent decree

10.4.5 INTEGRATION WITH LEADERSHIP AND MANAGEMENT

Integration with leadership and management is an evaluation of senior management at a facility. Specifically, it is a measure of each department and the plant manager and above on commitment through actions to protect human health and the environment. Recommended point values for this section range from -3 to 3 points. Note that again, a negative score is possible if management demonstrates exceptional commitment to protecting human health and the environment. Criteria to evaluate include:

- Management support demonstrated and documented by:
  - Requested capital projects
  - Approved capital projects
  - Regular meetings with management and entire workforce
  - Regular site inspections
  - Regular interview and development sessions
  - Root cause analysis outcomes on any corrective actions
- Departmental support demonstrated and documented by:
  - Knowledge of environmental matters of other departments such as maintenance, production, HR, and purchasing
  - Regular communications
  - Frequency and depth of environmental analysis
  - Frequency of inspections and meetings

10.4.6 CHEMICAL ORDERING PROCEDURE

Chemical ordering procedures have become a significant issue in the last three decades. Especially since there have been no new chemicals banned by USEPA since 1984, industry has had to take the lead on learning what they should and should not use. This is often learned the hard way when a company experiences a release and then is hit with the cleanup bill which all too often is in the millions of dollars. So it's crucial that a company know what chemicals it uses, why and how much is used, and what is the potential harm they can inflict if they are released into the environment. Although this is also discussed under EPCRA, we will discuss how a negative situation could be avoided here. Chemical ordering is an obvious place to start since eliminating the use of a specific chemical should always be the first choice at reducing risk.

The recommended point range is -3 to 3 points. Note again that a negative point value is achievable and should be reserved to those situations where exceptional performance is realized. Criteria for evaluating chemical ordering procedures include:

- Documented chemical ordering procedure
- Reviewing recent activities
- Evaluating if any legacy chemicals have been discovered
- Examining forms and sign offs for new chemicals
- Results of site inspection and spot-checking chemicals and products within the plant

10.4.7 ESPM Performance

One of the best measurements for evaluating risk reduction action is performance at reducing risk through elimination, substitution, prevention, and minimization actions. The recommend point range is -5 to 5 points. Note again that negative points are possible and the point range is more significant than the other criterion because of the relative importance and significance of ESPM performance. The criteria to evaluate ESPM performance includes:

- Does the facility have a written ESPM or other pollution prevention program?
- Evaluate whether there are any VOCs
- Any CVOCs
- Does the facility use Cr^{+6}
- Any Hg, Cd, PCBs?
- Evaluate commitment for improvement
- Evaluate sustainability program and progress (e.g., reducing water use, recycling programs, etc.)
- Check chemical storage, use, and waste areas
- Is anything stored outdoors?
- Drains sealed?
- Engineering plans adequate?
- How often are goals reviewed?
- Any recent capital improvements with an environmental focus
- Oil management initiatives
- Improved engineering controls
- Employee involvement
- Tracking
- Improved housekeeping

10.5 SUSTAINABILITY MODEL EQUATION

We have all the elements to build a model for sustainability, which includes calculating the three basic elements we have discussed in this chapter and include the following:

- Understanding the natural environment
- A detailed analysis of the aspects of operations
- Evaluating the effectiveness or ineffectiveness of risk reduction actions

Combining these three elements, we arrive at an equation with an output that is termed the sustainability index and is presented as Equation 10.1 (Rogers 2018).

Geologic Vulnerability × Operational Aspects ×Risk Reduction Measures=Sustainability Index
Equation 10.1

Examining the method for calculating the sustainability index, a higher sustainability index value indicates higher risk.

One factor that we have spent much time developing and discussing is the contaminant risk factor. At first glance at Equation 10.1, it would appear that the CRF is missing but that is just not the case. The contaminant risk factor is a critical value and appears in many places within the operational aspects variable. The CRF appears at several locations that include air, water, solid and hazardous waste, EPCRA, TSCA, and ESPM.

We learned in Chapter 2 that if it weren't for chemicals, the environmental risk would be greatly reduced but would not be eliminated. We have learned about chemicals and their behavior when released into the environment, which includes each chemical's toxicity, persistence, and mobility. Each of these factors are critically important especially when combined with geologic vulnerability because it can be used as a predictor to evaluate whether a release of any particular chemical will cause harm and at what scale. In addition, imbedding the chemical risk factor into the aspects of operations is a logical and sensible location since the aspects of operations includes many opportunities where contaminants can be released into the air, water, and land. To account for these many chances for a chemical to pollute the environment, evaluating CRFs for each media (air, water, and land) and for each operational aspect that may use chemicals is again a logical and sensible location to embed the CRF to ensure that nothing was overlooked.

Contaminant risk factors are also part of the evaluation for the risk reduction measures. Since CRFs are part of two of the three elements of the sustainability model including operational aspects and risk reduction measures, it ensures that the sustainability index that is calculated results in a higher risk to the environment if chemicals are used at a specific facility that have high CRFs. On the other hand, by eliminating chemicals with high CRFs from operations can have a significant positive effect at lowering risks and greatly lowering the sustainability index. Hence, here we have embedded in the equation an additional benefit and incentive to eliminate chemicals that are especially harmful to human health and the environment.

Eliminating chemicals that are harmful from operations is where environmental professionals have input and where companies have a choice that can have a significant impact on achieving sustainability and environmental stewardship.

The third value is a measure of risk reduction measures and is essentially presented as an opportunity to lower risk through proactive steps to eliminate or reduce the potential to cause environmental harm.

We learned how the natural environment reacts to different chemicals and identified certain geologic and hydrologic areas that are especially sensitive to contamination where a synergistic effect is realized that greatly amplifies potential harmful effects. Each chemical has a different associated risk factor for each environmental media – air, water, and soil – that is dependent upon its toxicity and physical chemistry (mobility and persistence).

10.6 SUSTAINABILITY CASE STUDY

To evaluate whether the sustainability index can actually be effective at lowering environmental risk, increasing sustainability, and achieving environmental stewardship, it must be implemented at operating manufacturing facilities and put through rigorous testing. For a period of 10 years that included five 2-year evaluation periods, as many as 67 manufacturing facilities in 12 countries were tracked using the sustainability model presented in the previous section. Table 10.1 lists each facility, what evaluation period the facility participated, country, and type of operations. Some facilities were not evaluated during all five periods because of mergers, acquisitions, divestitures, new construction, or were shut down.

An environmental compliance audit was conducted at each facility every 2 years. The environmental compliance audit was used to gather the necessary information to score each facility's

TABLE 10.1
Industrial Manufacturing Facilities Evaluated

Facility	Location (Country)	Type of Manufacturing	2009 to 2011	2011 to 2013	2013 to 2015	2015 to 2017	2017 to 2019
			Evaluation Period				
1	Australia	Commercial/industrial metal products					
2	Belgium	Commercial/industrial metal products					
3	Brazil	Steel foundry (Electric Arc Furnace)					
4	Brazil	Fabrication and assembly plant					
5	Brazil	Fabrication and assembly plant					
6	Canada	Steel foundry (Electric Arc Furnace)					
7	Canada	Automobile metal parts manufacturing					
8	Chili	Industrial metal parts manufacturing					
9	China	Commercial/industrial metal products					
10	China	Commercial/industrial metal products					
11	China	Automobile parts manufacturing					
12	China	Automobile parts manufacturing					
13	China	Automobile parts manufacturing					
14	China	Heavy truck parts manufacturing					
15	China	Heavy truck parts manufacturing					
16	China	Heavy truck parts manufacturing					
17	China	Industrial metal parts manufacturing					
18	China	Automobile plastic parts manufacturing					
19	China	Automobile metal parts manufacturing					
20	China	Automobile metal parts manufacturing					
21	China	Automobile metal parts manufacturing					
22	China	Steel foundry (Electric Arc Furnace)					
23	France	Automobile metal parts manufacturing					
24	Italy	Commercial/industrial metal products					
25	Italy	Automobile metal parts manufacturing					
26	Mexico	Heavy truck parts manufacturing					

(Continued)

TABLE 10.1 (CONTINUED)
Industrial Manufacturing Facilities Evaluated

Facility	Location (Country)	Type of Manufacturing	2009 to 2011	2011 to 2013	2013 to 2015	2015 to 2017	2017 to 2019
			Evaluation Period				
27	Mexico	Automobile metal parts manufacturing					
28	Mexico	Steel foundry (Electric Arc Furnace)					
29	Russia	Industrial metal parts manufacturing					
30	South Africa	Commercial/industrial metal products					
31	South Africa	Industrial metal parts manufacturing					
32	United States	Iron foundry (cupula)					
33	United States	Iron foundry (cupula)					
34	United States	Iron foundry (cupula)					
35	United States	Steel foundry (Electric Arc Furnace)					
36	United States	Steel foundry (Electric Arc Furnace)					
37	United States	Steel foundry (Electric Arc Furnace)					
38	United States	Steel foundry (Electric Arc Furnace)					
39	United States	Industrial/commercial R&D					
40	United States	Industrial/commercial R&D					
41	United States	Automobile metal parts manufacturing					
42	United States	Automobile metal parts manufacturing					
43	United States	Automobile metal parts manufacturing					
44	United States	Heavy truck assembly plant					
45	United States	Heavy truck assembly plant					
46	United States	Heavy truck assembly plant					
47	United States	Heavy truck assembly plant					
48	United States	Heavy truck assembly plant					

(Continued)

TABLE 10.1   (CONTINUED)
Industrial Manufacturing Facilities Evaluated

Facility	Location (Country)	Type of Manufacturing	Evaluation Period 2009 to 2011	2011 to 2013	2013 to 2015	2015 to 2017	2017 to 2019
49	United States	Heavy truck parts manufacturing					
50	United States	Heavy truck parts manufacturing					
51	United States	Heavy truck parts manufacturing					
52	United States	Automobile metal parts manufacturing					
53	United States	Automobile metal parts manufacturing					
54	United States	Automobile metal parts manufacturing					
55	United States	Automobile metal parts manufacturing					
56	United States	Automobile metal parts manufacturing					
57	United States	Electronic parts manufacturing					
58	United States	Industrial metal parts manufacturing					
59	United States	Industrial metal parts manufacturing					
60	United States	Industrial metal parts manufacturing					
61	United States	Industrial metal parts manufacturing					
62	United States	Industrial metal parts manufacturing					
63	United States	Industrial metal parts manufacturing					
64	United States	Industrial metal parts manufacturing					
65	United States	Industrial metal parts manufacturing					
66	United States	Industrial metal parts manufacturing					
67	USA	Industrial metal parts manufacturing					

operational aspects and risk reduction measures. Table 10.2 presents the environmental risk, aspects of operations, and risk reduction measures score for each facility for each evaluation period. Figure 10.1 through 10.6 shows a graphical representation of aspects of operations and risk reduction measures for each facility during each of the 2-year evaluation periods.

Upon examination of Figures 10.2 through 10.6, it becomes immediately clear that aspects of operations did not vary and only decreased slightly during the 10-year evaluation period. However,

TABLE 10.2
Facility Sustainability Scoring

	Scores				
	Environmental Risk × Operational Aspects × Risk Reduction Measures = Sustainability Index/100				
Facility	Years 2009–2010	Years 2011–2012	Years 2013–2014	Years 2015–2016	Years 2017–2018
1	48×20×13=125	48×19×7=64	48×20×7=67	48×18×10=86	48×18×9=77
2	15×16×1=2.4	15×16×3=7.2	15×16×2=4.8	15×16×5=12	15×16×4=9.6
3	Not Scored	42×36×9=136	42×36×3=45	42×36×2=30	42×36×2=30
4	Not Scored	Not Scored	15×20×7=21	15×20×5=15	15×20×4=12
5	Not Scored	Not Scored	15×18×8=26	15×18×5=13	18×18×3=9.7
6	11×33×28=101	11×30×7=23	11×27×3=9	11×27×2=6	11×25×2=5.5
7	11×15×5.5=9	11×13×5=7	11×15×4=7	11×15×3=5	11×15×2=5
8	Not Scored	Not Scored	Not Scored	11×8×5=4.4	11×8×4=3.5
9	Not Scored	11×16×15=26	11×16×10=18	11×16×7=12	11×16×5=8.8
10	Not Scored	11×14×14=21	11×14×14=21	11×14×6=9.2	11×14×4=6
11	Not Scored	Not Scored	Not Scored	11×16×6=10	11×18×5=10
12	Not Scored	Not Scored	11×14×11=17	11×16×7=12	11×16×5=8.8
13	Not Scored	Not Scored	11×13×114=20	11×17×6=11	11×17×4=5.5
14	Not Scored	11×27×14=42	11×27×10=30	11×27×5=15	11×29×4=13
15	Not Scored	11×21×8=18	11×21×6=14	11×21×5=12	11×21×4=9.2
16	Not Scored	11×20×11=24	11×20×8=18	11×21×4=9.2	11×21×4=9.2
17	Not Scored	11×24×4=11	11×24×6=16	11×24×5=13	11×24×3=8
18	Not Scored	11×18×8=16	11×18×6=12	11×18×5=10	11×18×4=8
19	42×25×83=871	42×30×66=831	42×24×38=383	42×21×4=35	42×20×2=17
20	11×12×3=4	11×12×3=4	11×12×4=5	11×12×3=4	11×15×4=6.6
21	Not Scored	Not Scored	11×15×10=16	11×15×5=8.2	11×15×4=6.6
22	Not Scored	Not Scored	11×36×9=36	11×36×7=28	Not Scored
23	15×20×20=60	15×20×15=45	Not Scored	Not Scored	Not Scored
24	48×26×16=200	48×21×3=30	48×20×2=19	48×21×2=20	48×21×2=20
25	48×20×8=77	48×18×11=95	48×18×7=60	48×19×2=18	48×19×2=18
26	15×27×15=61	15×29×8=35	15×31×15=70	15×31×9=42	15×32×5=24
27	15×22×15=50	15×22×8=26	15×22×7=23	15×20×5=15	15×20×3=9
28	15×36×41=221	15×36×7=38	15×36×4=22	15×36×2=11	15×36×10=54
29	Not Scored	15×15×5=11	15×15×2=4.5	15×20×4=12	15×21×5=16
30	48×20×18=173	48×20×6=58	48×19×3=27	48×19×3=27	48×19×9=82
31	Not Scored	15×14×5=10	15×14×5=10	15×15×2=4.5	15×15×3=6.7
32	45×37×13=216	45×37×14=233	45×37×11=183	45×37×8=133	Not Scored
33	45×42×37=700	45×42×15=283	Not Scored	Not Scored	Not Scored
34	45×39×28=491	45×49×41=904	45×44×42=831	45×44×40=792	Not Scored
35	15×42×62=390	15×44×10=66	15×42×6=38	15×42×2=13	15×42×2=13
36	Not Scored	15×40×5=30	15×40×3=18	15×40×2=12	Not Scored
37	15×40×8=48	15×40×3=18	15×40×1=6	15×40×6=36	15×40×10=60
38	15×38×7=40	15×36×1=5.5	15×36×4=21	15×36×3=16	15×36×2=16
39	30×40×15=180	30×40×3=36	30×40×3=36	30×40×4=48	30×40×3=32
40	15×15×5=11	15×15×4=9	15×8×4=4.8	15×8×2=4.8	15×8×2=4.8
41	15×30×9=40	15×25×6=23	15×23×6=21	15×23×2=7	15×23×1=3.5
42	15×22×3=10	15×22×4=13	15×23×2=6.9	15×19×3=8.5	15×19×2=5.7

(Continued)

TABLE 10.2   (CONTINUED)
Facility Sustainability Scoring

	Scores				
	Environmental Risk × Operational Aspects × Risk Reduction Measures = Sustainability Index/100				
Facility	Years 2009–2010	Years 2011–2012	Years 2013–2014	Years 2015–2016	Years 2017–2018
43	$15\times13\times3=6$	$15\times13\times1=2$	$15\times11\times2=3.3$	$15\times15\times4=9$	$15\times12\times2=3.6$
44	Not Scored	Not Scored	Not Scored	$15\times10\times6=9$	$15\times10\times6=7.7$
45	Not Scored	Not Scored	Not Scored	$15\times10\times4=6$	$15\times10\times4=6$
46	Not Scored	Not Scored	Not Scored	$15\times12\times6=11$	$15\times12\times6=9$
47	Not Scored	Not Scored	Not Scored	$15\times9\times4=5.4$	$15\times9\times2=2.7$
48	Not Scored	$15\times14\times3=6.3$	$15\times14\times3=6.3$	$15\times11\times2=3.3$	$15\times11\times2=3.3$
49	$15\times25\times13=49$	$15\times25\times6=22$	$15\times25\times8=30$	$15\times25\times5=19$	$15\times25\times4=15$
50	$15\times20\times12=36$	$15\times20\times4=12$	$15\times21\times8=25$	$15\times21\times6=19$	$12\times21\times2=5$
51	Not Scored	Not Scored	Not Scored	$11\times13\times6=8.5$	$11\times13\times5=7$
52	$11\times15\times2=3.3$	$11\times15\times4=6.6$	$11\times15\times7=12$	$11\times15\times2=3.3$	$11\times15\times2=3.3$
53	$11\times12\times6=8$	$11\times14\times10=15$	$11\times14\times9=14$	$11\times14\times5=7.7$	$11\times14\times3=4.6$
54	$11\times14\times7=11$	$11\times12\times9=12$	$11\times12\times12=16$	$11\times12\times5=6.6$	$11\times12\times4=5.2$
55	$11\times15\times5=8$	$11\times15\times6=10$	$11\times15\times5=8.2$	$11\times15\times2=3.3$	$11\times15\times3=5$
56	$11\times12\times3=4$	$11\times12\times3=4$	$11\times12\times2=2.6$	$11\times12\times2=2.6$	$11\times12\times2=2.6$
57	$11\times10\times5=5.5$	$11\times10\times5=5.5$	$11\times10\times3=3.3$	$11\times10\times2=2.2$	$11\times10\times2=2.2$
58	Not Scored	Not Scored	Not Scored	$11\times17\times6=11$	$11\times17\times3=5.5$
59	Not Scored	Not Scored	$11\times12\times2.5=3.3$	$11\times12\times2.5=3.3$	$11\times12\times2=2.6$
60	$15\times32\times35=168$	$15\times31\times7=33$	$15\times32\times5=24$	$15\times32\times4=19$	$15\times32\times7=34$
61	$59\times22\times10=130$	$59\times21\times5=62$	$59\times21\times3=37$	$59\times21\times2=25$	$59\times21\times2=25$
62	$15\times10\times7=10$	$15\times10\times5=7.5$	$15\times10\times2=3$	$15\times10\times2=3$	$59\times10\times2=3$
63	$59\times36\times27=573$	$59\times35\times20=413$	$59\times35\times10=206$	$59\times35\times4=83$	$59\times32\times2=38$
64	$11\times22\times16=39$	$11\times19\times6=13$	$11\times19\times4=8.3$	$11\times19\times3=6.3$	$11\times19\times1=2$
65	$42\times34\times23=309$	$42\times29\times8=97$	$42\times31\times10=130$	$42\times31\times9=118$	$42\times31\times4=26$
66	$42\times28\times9=106$	$42\times30\times4=50$	$42\times30\times2=25$	Not Scored	Not Scored
67	Not Scored	$11\times18\times8=16$	$11\times18\times7=14$	$11\times18\times4=8$	$11\times18\times4=4$

increased performance and attention to risk reduction measures were significant. A 40% improvement in risk reduction was realized during the first evaluation and continued, albeit at a less significant percentage, through each of the five evaluation periods and is steady during the 2019 scoring compared to 2017. The main reason that operational aspects did not appear to improve significantly is because many facility's operational aspect improvements were offset by expanding operations which tended to increase the complexity of operations.

Another measure of risk reduction measures is to compare the number of regulatory inspections with the number of notice of violations (NOVs) issued by the regulatory authorities. Table 11.4 and Figure 11.9 show the number of inspections and NOVs.

From examining Table 10.3 and Figure 10.6, the number of inspections has remained relatively constant but the number of NOVs has decreased by approximately 75%. This is due to a significant increase in risk reduction measures that may also reduce operational aspects.

Over the 10-year period from 2009 through 2018, some of the most significant risk reduction actions have included the following:

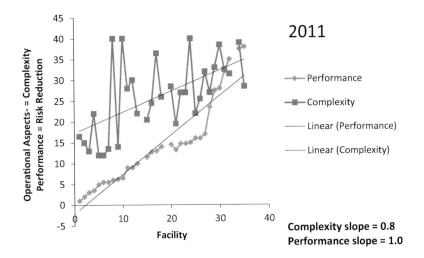

FIGURE 10.1 Evaluation period 2009 through 2010.

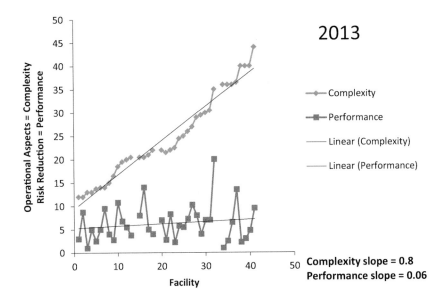

FIGURE 10.2 Evaluation period 2011 through 2012.

- Improved housekeeping
- Improved training and education
- Improved environmental compliance
- Elimination of chlorinated volatile organic compound use
- Elimination of extremely hazardous compounds such as hydrofluoric acid, picric acid, cyanide, and arsenic
- Reduction in the use of other volatile organic compounds (i.e., benzene, xylenes, toluene)
- Elimination of hexavalent chrome use
- Elimination of any detectable concentration of PCBs in transformers
- Appropriate disposal of capacitors that may have contained PCBs
- Decrease in oil use

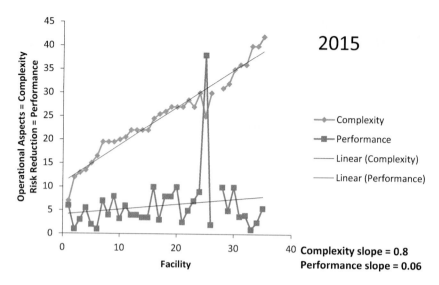

FIGURE 10.3 Evaluation period 2013 through 2014.

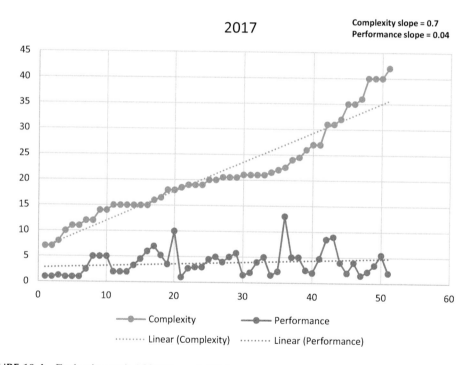

FIGURE 10.4 Evaluation period 2015 through 2016.

- Reduction in the amount of solid and hazardous wastes generated
- No outdoor storage of wastes
- Reducing water use
- Significant reduction in spills and reportable releases of hazardous substances to the environment
- Increased recycling efforts
- Upgrading wastewater treatment with new technologies
- Decreased process water discharges through water recirculating and recycling systems
- Sealing floor drains

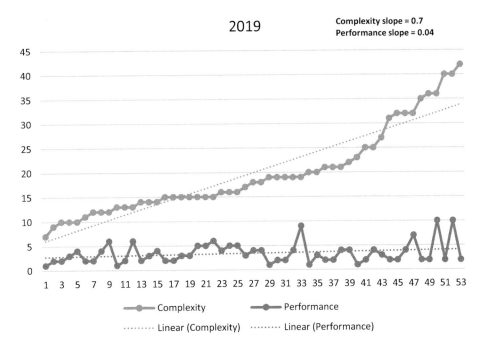

FIGURE 10.5 Evaluation period 2017 through 2018.

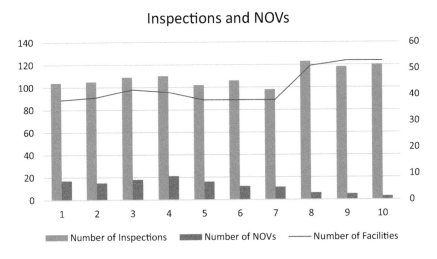

FIGURE 10.6 Inspections and NOVs from 2009 through 2018.

- Elimination of sumps, pits, and trenches
- Installation of rain gardens and other methods to decrease stormwater runoff
- Upgrading engineering controls in waste storage areas such as construction of dedicated buildings, improved secondary containment, epoxy coating floors, color-coded waste containers, and installation of electronic controls for continuous monitoring and warning monitoring
- Improved lighting

The risk reduction action listed above also represents initial efforts to create and improve sustainability initiatives. In fact, some manufacturing locations have achieved nearly a zero waste-generating status of solid or hazardous waste and have also achieved decreases in water use by more than 95%.

TABLE 10.3
Regulatory Inspections and Notice of Violations for Each Evaluation Year

Year	Number of Inspections	Number of NOVs	Inspection/NOV Ratio	Number of Facilities
2009	104	17	6	38
2010	105	15	7	39
2011	109	18	6	42
2012	110	21	5.5	41
2013	102	16	6.5	38
2014	106	12	8.5	38
2015	98	11	9	38
2016	123	6	12.5	51
2017	118	5	24.5	53
2018	120	4	30	53

10.7   SUMMARY AND CONCLUSIONS

The sustainability model presented in this chapter has undergone 10 years of testing at as many as 67 manufacturing locations in 12 different countries of the world. Over the 10-year period, improvements in risk reduction measures have been realized by an average of 80%, resulting in a significant reduction in environmental non-compliance and releases of hazardous substances to the environment and has greatly improved the sustainability of each facility (Rogers 2019).

Efforts to reduce operational aspects were largely offset by new or more strict environmental regulations, facility expansion, new equipment, or increased production that required additional permitting and regulatory limitations. However, even with the offsets, slightly more than 50% of facilities reduced operational aspects. These opportunities were most often realized through ESPM efforts, especially through the following:

- Eliminating chemical use with high CRFs (i.e., chlorinated VOC, PCBs, other VOCs, and hexavalent chrome)
- Reducing hazardous waste generation that resulted in a lower hazardous waste classification (i.e., large-quality generator to small quantity generator)
- Eliminating the requirement for a stormwater permit by increasing onsite surface water infiltration and eliminating any offsite discharge of stormwater
- Changing wastewater discharge from surface water to a POTW
- Upgrading wastewater treatment systems to reduce the amount of pollutant load and reducing the amount of water discharged
- Reducing wastewater discharges through operational modifications
- Reducing wastewater discharge through recycling efforts
- Reclassification of solid waste to beneficial reuse
- Assisting other companies in networking and increasing recycling opportunities
- Upgrading operations with new technologies that produce less waste
- Working with suppliers and purchasing to reduce packaging wastes and increase recycling efforts

The most significant challenge over the 10-year implementation and evaluation period was changing cultural attitude and behavior. Up until the 1970s, there generally was a disregard for the environment that perhaps had persisted for centuries. The 1970s saw the birth of the environmental movement. The 1980s and 1990s saw the passage and implementation of most of the major environmental laws of the United States and in many other countries of the world. The United States and many other developed countries have made significant progress at cleaning up the environment and making many places a better place to live for us and all the other animals and plants on Earth. However, much remains to be done, even in the United States. The most difficult actions that have yet to be taken but are required relate to changing our cultural attitude and behavior toward Earth (Rogers 2019).

The success of implementing the sustainability program outlined in this chapter could not have happened without acknowledging and overcoming a cultural attitude and behavior ingrained in most capitalist organizations, which is a constant drive for growth and profit first and everything else second. Hopefully by now each of you have come to the conclusion after reading this book that if humans do not change their behavior there might not be a place for us on Earth in the future. The issues and challenges surrounding and weaved within the solutions for overcoming pollution will take commitment and sacrifice from each of us. One person may not make much of a difference but collectively we can and must (Rogers 2019).

Changing attitude and culture required several steps and involved all organization levels. Those steps included (Rogers 2019):

1. A corporate culture of respect and commitment of establishing and maintaining environmental stewardship for the Earth
2. Education and training
3. A clear plan with stated purpose and goals
4. Incentives, rewards, and penalties
5. Measuring and recording performance
6. Data analysis
7. Plan adaptations including additions or subtractions and adjusting timelines

REFERENCES

Rogers, D.T. 1996. *Environmental Geology of Metropolitan Detroit*. Clayton Environmental. Novi, MI. 152p.

Rogers, D.T. 2011. Why Geology and Chemistry Matter in Forming the Foundation in Sustaining Urban Areas. *American Institute of Professional Geologists*. Vol. 49. No. 2. Denver, CO. p. 26–31.

Rogers, D.T., Kaufman, M.M., and Murray, K.S. 2012. Key Environmental Changes Required to Reverse Urban Groundwater Degradation. International Association of Hydrogeologists 38th Congress. Vol. 1. p. 198. Niagara Falls, Ontario, Canada.

Rogers, D.T. 2014. Scientists Call for a Renewed Emphasis on Urban Geology. *American Geophysical Union*. Vol. 95. No. 47. pp. 431–432. Earth and Space News. Eos.

Rogers, D.T., Kaufman, M.M., and Murray, K.S. 2016. Development of an Environmental Risk Management and Sustainability Model Using Chemistry, Geology, and Human Activity, and Response Inputs. *35th International Geological Congress*. Paper No. 2090. Cape Town, South Africa.

Rogers, D.T. 2018. Derivation of a Comprehensive Environmental Risk Model for Urban Groundwater Protection. International Association of Hydrogeologists 45th Congress. Vol. 1. p. 879. Daejeon, Korea.

Rogers, D.T. 2019. Key Policy Changes Required to Reduce Environmental Degradation in the United States and Beyond. International Association of Hydrogeologists 46th Congress. Vol. 1. p. 529. Malaga, Spain.

United Nations. 2018. *What Is Sustainable Development*. https://www.un.org/sustainabledevelopment. (accessed December 24, 2018).

United States Environmental Protection Agency (USEPA). 2018. *What Is Sustainability?* https://www.epa.gov/sustainability. (accessed December 1, 2018).

Glossary of Terms

Abiotic degradation: degradation of a substance by processes not involving microorganisms or fungi.

Absorption: a process by which a particle of gas or liquid enters a substance.

Acid: any substance that when dissolved in water increases the concentration of the hydrogen ion H^+ and lowers the pH of the solution.

Acid rain: precipitation with a pH of less than 7, resulting in large part from the reaction between sulfur dioxide and water vapor in the atmosphere that produces sulfuric acid.

Acute response: characterized as a single high dose with rapid onset and disappearance of symptoms.

Adsorption: the adhesion of molecules of gas, liquid, or dissolved solids to a surface.

Advection: bulk transport of a liquid.

Advection: when applied to groundwater, is the movement of contaminants by the bulk motion of groundwater flow.

Adverse health effect: a change in body function or cell structure potentially leading to disease or health problems.

Air pollution: degradation of any layer of the atmosphere by the introduction of pollutants from direct or indirect anthropogenic activity.

Amines: a group of organic compounds that contain nitrogen and are basic.

Anthropogenic: effects, processes, or materials derived from human activities.

Anthrosphere: human-built or developed world.

Area source: any stationary source that is not a major source.

Aroclor: mixture of PCB compounds distinguished by a four-digit numbering system that denotes characteristics of composition.

Asbestos: a name applied to a group of six different fibrous minerals that include amosite, chrysotile, crocidolite, and the fibrous forms of tremolite, actinolite, and anthophyllite.

Asthenosphere: upper portion of the mantle.

Atomic mass: the mass of an atom. Also referred to as atomic weight.

Atomic weight: the mass of an atom. Also referred to as atomic mass.

Bacteria: single-celled organisms.

Base: any substance that when dissolved in water increases the concentration of the hydroxide ion OH^- and raises the pH of the solution.

Base-neutral-acid compounds: a group of organic compounds much less volatile than VOCs. Also referred to as semi-volatile organic compounds.

Basic: a compound with a pH over 7.

Benzene ring: A hexagonal ring arrangement found in benzene and other aromatic compounds, consisting of six carbon atoms with alternating single and double bonds between them, and with each carbon atom bonded to a hydrogen atom, or to other atoms or groups of atoms in derivatives of benzene.

Best management practice: specific practices that are capable of improving, preventing, and/or minimizing the potential of an occurrence or situation that can lead to environmental noncompliance and/or a release of a hazardous substance to the environment.

Bioaccumulation: contaminant accumulation in the body of an organism at a concentration greater than what is generally defined as background or naturally occurring.

Biotic degradation: degradation of a substance by microorganisms or fungi.

Biphenyl: organic compound composed of two benzene rings.

Brownfield: an abandoned, idled, or underutilized industrial or commercial facility reuse of which may be complicated by the presence or potential presence of a hazardous substance, pollutant, or contaminant.

Cancer: when cells in the body become abnormal and grow or multiply out of control.

Capture zone: in reference to groundwater, the areal extent where groundwater will be captured by a pumping well.

Carbon dioxide: CO_2, a common air pollutant and greenhouse gas. Anthropogenically produced through the combustion of fossil fuels.

Carbon monoxide: CO, a common pollutant toxic to humans and animal life. Anthropogenically produced through the combustion of fossil fuels.

Cementation: the precipitation of minerals at grain to grain contacts and within the pore spaces of sedimentary deposits; after this process is completed, a sedimentary deposit is termed a sedimentary rock.

Chemical: a substance with a specific atomic composition.

Chemical dehalogenation: a remedial method that removes halogens from contaminants.

Chemical oxidation: the loss of electrons.

Chemical reduction: the addition of electrons.

Chlorinated solvents: a group of organic solvents containing one or more chlorine atoms within their atomic structure.

Chronic response: a stimulus lingering for a period of time after exposure to a chemical.

Combined Sewer Overflow: occurs when the wastewater volume in a combined sewer system exceeds the capacity of the sewer system or treatment plant; the excess wastewater is discharged directly to nearby streams, rivers, or other water bodies.

Compaction: occurs over time within a sedimentary deposit where individual grains are rearranged to form a more tightly packed sediment; the mass stays the same but the volume decreases.

Congener: refers to a specific PCB chemical family.

Conjunctive use: the planned and interchangeable use of groundwater and surface water.

Contaminant: the human introduction into the environment of substances harmful to human health or ecosystems. Also referred to as pollutant.

Contaminant fate and transport: a description of what happens to a contaminant after it has been released into the environment.

Contaminant risk factor: a method to calculate the potential risk of a contaminant in air, soil, or water by combining the elements of toxicity, mobility, and persistence.

Contaminant Risk Factor for Groundwater: the environmental risk posed by specific contaminants to contaminant groundwater.

Contamination: the introduction of harmful substances into an ecosystem, typically anthropocentric-manufactured chemicals.

Convection: vertical advection of air, water, or other fluid as a result of thermal differences.

Corrosivity: materials, including solids, that are acids or bases, or that produces acidic and alkaline solutions.

Covalent bond: a form of chemical bonding characterized by sharing pairs of electrons between atoms.

Cultural eutrophication: when anthropogenic activities such as fertilization or sewage discharges bring excess nutrients into water bodies.

Cultural landscape: the product of human interaction with the natural landscape, with culture acting as the agent, and the natural area as the medium.

Cyanide: any chemical compound containing the cyano group consisting of a carbon atom triple-bonded to a nitrogen atom.

Dense nonaqueous phase liquids: organic compounds that do not readily mix or dissolve in water and are heavier than water.

Diffusion: the movement of a chemical from an area of higher concentration to an area of lower concentration due to the random motion of the chemical molecules. Also referred to as molecular diffusion.

Dioxins: a diverse set of halogenated substances; includes other compounds called furans.

Dispersion: the tendency for contaminants to spread out from the path normally expected from advective flow. Also referred to as hydrodynamic dispersion.

Dissolve: to cause a substance to pass into solution.

Ecosystem: a self-regulating association of living plants, animals, and their non-living physical and chemical environments.

Eddies: turbulent swirls in a fluid.

Edge effects: processes located at the margins of physical entities, such as the receipt of more light at the edge of forests.

Electrokinetics: a remedial method applied to heavy metals involving the imposition of an electrical field via electro-osmosis to induce migration of heavy metals to a desired location where they are removed or encapsulated.

Elimination: a pollution prevention term used to describe the removal or storage of a certain hazardous substance.

Environment: a broad term encompassing all living and non-living things on the Earth or at smaller geographic scales. The environment encompasses the natural world existing within the atmosphere, hydrosphere, lithosphere, and biosphere, and also includes the built or developed world.

Environmental audit: a systematic, documented, periodic, and objective review of facility operations and practices related to meeting environmental requirements.

Environmental compliance: conforming with environmental law.

Environmental concern: a condition or situation that generally does not present a threat to human health or the environment and is typically not the subject of an enforcement action.

Environmental impact statement: a study to identify and evaluate the positive and negative biophysical, social, and other environmental effects that a proposed development action may have on the environment.

Environmental non-compliance: not conforming with environmental law, regulations, standards, or other requirements such as an environmental permit.

Environmentalism: a value system that seeks to redefine humankind's relationship with nature.

Environmental regulation: a set of laws, sometimes referenced as an amalgam of principles, treaties, laws, regulations, directives, ordinances, policies, at the local, state or federal level that are enacted to protect the environment in some shape or form.

Environmental risk: the probability of an event resulting in an adverse impact on the environment or humans.

Environmental risk assessment: a qualitative or quantitative assessment or investigation of the risks posed by the presence of contamination to human health or the environment.

ESPM: pollution prevention implemented with a stepwise evaluation process that proceeds from the most preventative measure to the least preventative measure; e=elimination, s=substitution, p=prevention, m=minimization.

Esters: a group of chemical compounds containing a modified carboxylic acid group; the acidic hydrogen atom has been replaced by a different organic functional group.

Eutrophication: the natural process of enrichment of surface waters with plant nutrients.

Experimental design: the assignment of subjects to experimental groups.

F-Listed wastes: wastes from many common manufacturing and industrial processes, such as solvents that have been used for cleaning or degreasing.

Feasibility study: an evaluation of the potential remediation methods after the nature and extent of contamination has been completed.

Feedback: an outcome of a process within a system that affects the overall function of the system.

Fertilizers: chemical compounds designed to promote plant and fruit growth when applied.

Fick's law of diffusion: a relationship stating the flux of a diffusing species is proportional to the concentration gradient.

Flux density: the mass of a chemical transported across an imaginary surface of unit area per unit of time.

Fragmentation: the breaking of ecosystems into spatially separated units or fragments.

Geographic scale: areal extent defined through the observation and/or measurement of events and processes.

Geologic formation: a fundamental unit of lithostratigraphy that consists of geologic strata with similar features in lithology, origin, or other criteria.

Geologic vulnerability: the natural properties of geologic materials acting to increase or decrease potential exposure risk to a contaminant.

Geology: the science dedicated to the study of the history, structure, and composition of the Earth.

Greenhouse gases: gases in the atmosphere capable of absorbing and emitting radiation within the thermal infrared range.

Groundwater: water beneath the surface of the ground present in pore space, fractures, or void spaces.

Groundwater recharge: a hydrologic process where water migrates downward from the surface and recharges groundwater.

Half life: the average amount of time required to degrade half or 50% of a specific contaminant population.

Half reaction: the loss or gain of an electron.

Halogenated volatile organic compounds: a group of organic compounds with a halogen atom as part of its atomic structure.

Halogens: a group of elements composed of fluorine (F), chlorine (Cl), bromine (Br), and iodine (I).

Hazardous chemical: any chemical which is a physical hazard or health hazard.

Hazardous waste: any discarded material considered a listed waste under RCRA or waste that is corrosive, ignitable, reactive, or infectious, explosive, radioactive or a solid waste, or combination of solid waste, which because of its quantity, concentration, or physical, chemical, or infectious or has characteristics that may (a) cause, or significantly contribute to, an increase in mortality or an increase in serious irreversible, or incapacitating reversible, illness; or (b) pose a substantial present or potential hazard to human health or the environment when improperly treated, stored, transported, or disposed of, or otherwise managed.

Heavy metals: a group of naturally-occurring metals that generally consist of arsenic, barium, chromium III, chromium VI, cadmium, copper, lead, mercury, nickel, selenium, silver, and zinc.

Henry's Law: a measure of the tendency for substances to volatilize.

Heptotoxin: a chemical posing a risk of liver damage.

Herbicide: a chemical compound used to kill unwanted plants.

Heterogeneity: highly variable and poorly sorted geologic materials.

Hexavalent chromium: a group of chemical substances containing the metallic element chromium in its positive-6 valence (hexavalent) state; all Cr(VI) compounds are potential occupational carcinogens.

Humid microthermal climate: continental climate characterized by strong seasonal variations and highly variable weather with ample precipitation throughout the year.

Hydraulic conductivity: the ability of saturated geologic media to conduct water under an induced hydraulic or pressure gradient.

Hydrodynamic dispersion: the tendency for contaminants to spread out from the path normally expected from advective flow. Also referred to as dispersion.

Hydrograph: a plot of stream discharge over a specified time interval.

Hydrolysis: cleavage of a molecule into two parts by the addition of a molecule of water.

Hygroscopic water: water tightly held onto soil particles through adsorption and not available for flow.

Ignitibility: wastes that can create fires under certain conditions, undergo spontaneous combustions, or have a flashpoint of less than 60 C° (140° F).

Insecticide: a chemical compound with the function of preventing, destroying, repelling, or mitigating insects.

Invasive species: any kind of organism that is not native to an ecosystem and causes harm that is directly or indirectly introduced by humans or as a result of human activity.

Isomer: chemicals with the same molecular formula but a different molecular structure.

K-Listed wastes: wastes from specific industries, such as petroleum refining or pesticide manufacturing. Also, certain sludges and wastewaters from treatment and production in these specific industries are examples of source-specific wastes.

Land pollution: defined as a degradation or even destruction of the Earth's surface, soil, rock or subsurface as a result of anthropogenic activities.

Land use restrictions: a remedial method of placing restrictions on land use to protect human health and environment so any contamination left in place is not disturbed.

Latency period: the duration of time without an observable effect following exposure to a chemical.

Law: the written statute passed by an appropriate and recognized governing body.

Light nonaqueous phase liquids: organic compounds that do not readily mix or dissolve in water and are lighter than water.

Lithification: a complex process where unconsolidated geologic materials become rock mainly through compaction and cementation.

Lowest-observed-adverse-effect level: the lowest tested dose of a chemical or substance causing a harmful or adverse health effect. Also referred to as threshold effect value.

M-Listed wastes: wastes known to contain mercury, such as fluorescent lamps, mercury switches, and the products that house mercury switches, and mercury-containing novelties.

Major source: a stationary or group of stationary sources that emit or have the potential to emit 10 tons or more of hazardous air pollutants per year.

Maximum achievable control technology: air pollution control technology capable of the maximum degree of reduction in emissions of hazardous air pollutants.

Maximum contaminant level: the maximum allowable amount of a contaminant in drinking water.

Mesothelioma: a type of cancer attributed to the exposure to asbestos.

Minimization: pollution prevention term used to describe a reduction in use of harmful chemicals or production of waste through process changes, recycling, or other methods.

Mobile source: all on-road and off-road vehicles that emit air pollutants including automobiles, trucks, trains, ships, aircraft, and farm equipment.

Mobility: a measure of a substance's potential to migrate in the environment.

Molecular attraction: a force pulling molecules together.

Molecular diffusion: the movement of a chemical from an area of higher concentration to an area of lower concentration due to the random motion of the chemical molecules. Also referred to as diffusion.

Molecular mass: the sum of the atomic weights of all the atoms in a molecule. Also referred to as molecular weight.

Molecular weight: the sum of the atomic weights of all the atoms in a molecule. Also referred to as molecular mass.

Monitoring well: a device inserted into the subsurface that penetrates the groundwater surface for the purposes of collecting hydrologic, chemical, or other information with respect to groundwater. Also referred to as a piezometer.

Mutagen: a chemical capable of causing genetic changes which could affect future generations through reproduction.

National Primary Drinking Water Standards: drinking water standards that are based on health-related criteria established by USEPA.

Natural landscape: earth surface formations resulting from the interactions within and between the physical systems comprising the four spheres of the geosphere: atmosphere, biosphere, hydrosphere, and lithosphere.

Natural resource: air, land, water, fish, biota, wildlife, groundwater, rock, soil, aquifers, ice, or other identifiable entities on Earth considered a resource and occurring naturally.

Natural resource damage assessment: a study evaluating damages or injuries to the environment.

Navigable waters: waters that are under the influence of tidal fluctuations and are used, or have been used in the past or potentially in the future, to transport interstate or foreign commerce, and includes coastal and inland water, lakes, rivers, streams that are navigable or are connected to a navigable waterway, and territorial seas.

Negative feedback: when an output from a process in a system slows down or dampens the overall operation of the system.

Nonpoint source pollution: pollution from a diffuse or unknown source.

Observation: an issue that could be improved but does not reach the threshold of a best management practice.

Operator: defined as any person, company, or organization in control of, or having responsibility for, the daily operation of a manufacturing or industrial facility.

Organochlorines: an organic compound containing at least one covalently bonded chlorine atom.

Overgrowth: the importation of excess nutrients into an ecosystem.

Oxidant: an atom donating an electron.

Owner: any person, company, or organization that owns a manufacturing or industrial facility.

Ozone: O_3, a gas present in the upper and lower atmosphere. Ozone in the upper atmosphere acts as a filter of ultraviolet radiation produced by the sun and is beneficial. In the lower atmosphere, ozone is considered a pollutant and is a major component of smog.

P-Listed wastes: specific commercial chemical products that have not been used, but that will be (or have been) discarded. Industrial chemicals, pesticides, and pharmaceuticals are examples of commercial chemical products that appear on the P-List and become hazardous waste when discarded.

Parasite: an organism living on or within a different organism or host at the expense of the host organism.

Particle pollution: complex mixture of very small particles and liquid droplets emitted into the atmosphere. Also referred to as particulate matter.

Particulate matter: complex mixture of very small particles and liquid droplets emitted into the atmosphere. Also referred to as particle pollution.

Permeability: a measure of a geologic material's ability to be penetrated by water.

Persistence: a measure of a substance's ability to remain in the environment before being degraded, transformed, or destroyed.

Pest: living organisms occurring where they are not wanted or causing damage to crops or humans or other animals.

Pesticide: a chemical compound with the function of preventing, destroying, repelling, or mitigating any pest.

Phase I environmental site assessment: a qualitative study of the current environmental condition of a property or site.

Phase II investigation: quantitative investigation conducted at a property or site after a Phase I environmental site assessment with the objective of confirming or refuting the presence of contamination, or defining the nature and extent of known contamination.

Phenol: a group of organic compounds composed of one or more hydroxyl groups attached to a carbon atom in a benzene ring.

Photochemical degradation: a chemical reaction where a substance is broken down or degraded by photons. Commonly occurs in the presence of sunlight. Also referred to as photodegradation or photolysis.

Photochemical smog: a form of air pollution formed through a reaction of air contaminants and sunlight.

Photodegradation: a chemical reaction where a substance is broken down or degraded by photons. Commonly occurs in the presence of sunlight. Also referred to as photolysis or photochemical degradation.

Photolysis: a chemical reaction where a substance is broken down or degraded by photons. Commonly occurs in the presence of sunlight. Also referred to as photodegradation or photochemical degradation.

Phthalate: a group of approximately 25 organic compounds consisting of esters of phthalic acid.

Piezometric surface: an imaginary surface representing the static level groundwater would rise to. Also referred to as potentiometric surface, isopotential level, and pressure level.

Point source pollution: pollution originating from an identifiable source such as air pollution emitted by an exhaust pipe of stack.

Polarity: directional charge.

Polluters pay principle: the party responsible for the pollution pays for cleaning up the pollution.

Pollution: the human introduction into the environment of substances harmful to human health or ecosystems. Also referred to as contamination.

Pollution prevention: preventing contaminants from entering the environment by reducing or eliminating waste at the source by modifying production processes, promoting the use of non-toxic or less toxic substances, implementing conservation techniques, and re-using materials rather than putting them into a waste stream.

Polychlorinated biphenyl: a group of synthetically produced chemicals with 1 to 10 chlorine atoms attached to a biphenyl.

Polycyclic aromatic hydrocarbons: a group of more than 100 organic compounds with multiple benzene rings. Also referred to as polynuclear aromatic hydrocarbons.

Polyfluoroalkyl substances: a group of synthetic mobile and persistent chemicals containing fluorene that have been manufactured and used in a variety of industries world-wide since the 1940s. They are commonly referred to as PFOA or PFAS.

Polynuclear aromatic hydrocarbons: a group of more than 100 organic compounds with multiple benzene rings. Also referred to as polycyclic aromatic hydrocarbons.

Porosity: the volume of open space within sediment, usually expressed as a percent.

Positive feedback: when the outcome of a process within a system speeds up or magnifies the system's activity or work output.

Potable water: water used for any human purpose.

Potency: the degree to which a substance or agent can cause harm to an organism when exposed. Also referred to as toxicity.

Potential to emit: the maximum capacity of a facility to emit any air pollutant under its physical and operational design.

Potentiometric surface: an imaginary surface representing the static level groundwater would rise to. Also referred to as piezometric surface, isopotential level, and pressure level.

Prevention: a pollution prevention term used to describe actions undertaken – usually engineering controls – to minimize the potential of a contaminant entering the environment.

Primary standards: air standards provided for public health.

Property: a parcel of land with a specific and unique legal description.

Radioactive decay: a spontaneous loss of energy through the emission of ionizing particles and radiation by an unstable atomic nucleus.

Reactivity: wastes that are unstable under normal conditions. They can cause explosions or release toxic fumes, gases, vapors when heated, compressed, or mixed with water.

Receptor: the organism or ecological habitat where exposure to a substance or agent may occur.

Recognized environmental condition: the presence or likely presence of any hazardous substance or petroleum product on a property or site under conditions that may materially affect or threaten the environmental condition of the property or site, human health, or the environment.

Reductant: an atom receiving an electron.

Reduction: the loss of areal coverage of an ecosystem or community.

Regulation: standards and rules adopted by administrative agencies that govern how laws will be enforced.

Remedial action: a step or steps undertaken to remedy or lower the risks posed by the presence of contamination at a property or site to human health or the environment.

Remedial action plan: a description of the steps to be undertaken to remedy or lower the risks posed by the presence of contamination at a property or site to human health or the environment.

Remedial investigation: a comprehensive qualitative and quantitative environmental investigation conducted at large and complex contaminated properties or sites.

Remediation: meaning "remedy" – the process of cleaning up or reducing the risks posed by the presence of contamination.

Retardation: the slowing of the migration of contaminants.

Risk assessment: a procedure used to evaluate whether there is an unacceptable risk posed to humans or the environment from natural events, human activity, or specific substances.

Rock: a solid object composed of minerals.

Rodenticide: a chemical compound with the function of preventing, destroying, repelling, or mitigating rodents.

Routes of exposure: the path chemicals and substances enter the human body; includes inhalation, ingestion, and dermal adsorption.

Sand: clastic sediment of a size between 1/16 and 2 millimeters.

Science-based landscape planning: the process of understanding the physical processes inherent to a natural landscape's formation and its long-term sustainability, and transforming selected components of these processes into planning principles and actions to preserve or enhance the cultural landscape.

Secondary standards: air standards provided for public welfare protection

Sediment: settled matter at the bottom of a liquid.

Sedimentation: the deposition of eroded material in a sink.

Semi-volatile organic compounds: a group of organic compounds much less volatile than VOCs. Also referred to as base-neutral-acid compounds.

Silt: a clastic sediment with a size range between 1/16 and 1/256 of a millimeter.

Simplification: the reduction in the number of species in an ecosystem or community.

Sink: location of deposition.

Site: a parcel of land including one or more than one property or easement; a specific location.

Soil: the top layer of the Earth's surface consisting of rock and mineral particles mixed with organic matter.

Soil excavation: a remedial method involving the removal of contaminated soil and its disposal at an offsite location, typically a regulated landfill.

Solidification: a remedial method involving the addition and mixing of a substance designed to immobilize or entomb contamination. Also referred to as stabilization.

Solid waste: discarded material, solids, sludges, liquid, semisolids, or gaseous materials.

Solubility: a measured property of a solid, liquid, or gaseous chemical substance termed a solute to dissolve in a liquid solvent to form a homogeneous solution.

Solute: a solid, liquid, or gaseous chemical substance dissolved in another substance.

Solvent: a solid, liquid, or gaseous chemical substance that dissolves another substance.

Sorption: refers to the action of both absorption and adsorption.

Source control: stopping or abating contamination at its source.

Spatial resolution: the smallest identifiable element in a sequence.

Speciation: a process over time where one species evolves into a different species or one species diverges to become two or more species.

Specific heat capacity: a measure of the heat energy required to increase the temperature of a unit quantity of a substance by a unit of temperature.

Specific retention: the volume of water retained after a saturated geologic material has been drained under the force of gravity.

Stormwater: nonpoint pollution occurring in urban areas initiated by wet-weather events.

Stratification: the layering of sediments as they are formed and deposited – also referred to as bedding.

Stratigraphy: the study of rock layers or unconsolidated sediment and strata; particularly their ages, composition, and relationship with other layers.

Substitution: a pollution prevention term used to describe the use of a less toxic or potentially harmful substance instead of a more toxic or harmful substance; in ecology, when one or more organisms/species are replaced by others.

Sulfur dioxide: SO_2; occurs naturally through volcanic eruptions and anthropogenically through the combustion of fossil fuels and some industrial processes. Sulfur dioxide is a pollutant and a component of smog.

Surface risk: the probability any given site will contaminate the environment given the best available data from public sources.

Surface water: water at the surface of the Earth.

Sustainability: human activities that do not inflict excessive levels of damage upon the environment; the environment is given adequate time to repair itself from prior damages; these efforts will help return physical systems to dynamic equilibrium. USEPA defines sustainability as creating and maintaining conditions under which humans can exist in productive harmony to support present and future generations. United Nations defines sustainability as development that meets the needs of the present without compromising the ability of future generations to meet their own needs.

Sustainable development: in urban watersheds, this concept encompasses those activities designed to achieve and maintain dynamic equilibrium within and between the major physical systems performing their work in the atmosphere, biosphere, hydrosphere, and lithosphere.

Teratogen: a chemical capable of causing an adverse effect on a developing fetus.

Threshold effect value: the lowest tested dose of a chemical or substance causing a harmful or adverse health effect. Also referred to as lowest-observed-adverse-effect level.

Toxicity: the degree to which a substance or agent can cause harm to an organism when exposed. Also referred to as potency.

Toxicity: RCRA definitions are wastes that are harmful or fatal when ingested or absorbed (e.g., wastes containing mercury, lead, DDT, PCBs, etc.).

Trihalomethanes: a group of volatile organic compounds where three of the four atoms of methane are replaced by halogen atoms.

Trophic level: the position an organism occupies in a food chain.

Unsaturated zone: that portion of the Earth between the surface and the water table or zone of saturation. Also referred to as the vadose zone.

Urban geologic map: geologic map of an urban area.

Urban heat island: results from the excess energy input into the atmosphere from anthropogenic activities. The primary source of the additional energy comes from the replacement of vegetation with structures and pavement.

Urbanization: the processes contributing to urban growth; specifically: an increased number of people coming to the cities to live, and the subsequent city expansion through the annexation of surrounding land and adjacent communities.

Used oil: any oil that has been refined from crude oil, or any synthetic oil that has been used and, as a result of use, is contaminated with physical or chemical impurities.

Vadose zone: that portion of the Earth between the surface and the water table or zone of saturation. Also referred to as the unsaturated zone.

Vapor pressure: the pressure of a vapor in thermodynamic equilibrium with its condensed phases in a closed container.

Virus: a subcellular agent capable of replicating itself inside the cells of another organism.

Volatile organic compounds: organic compounds tending to volatilize or evaporate readily under normal atmospheric pressure and temperature.

Volatilization: conversion into a vapor or gas without chemical change.

Water pollution: degradation of any water body, pond, lake, stream, river, ocean, or groundwater through the direct or indirect anthropogenic introduction of a pollutant.

Watershed: the extent of land drained by a given water feature, such as a river system, lake, swamp, estuary, reservoir, wetland, bay, sea, or ocean.

Watershed management: the minimization of the human impacts within and between different land use systems, with the objective of achieving a sustainable landscape.

Wellhead protection zone: a capture zone for a protected water supply well.

Wetland: lands where saturation with water is the dominant factor determining the nature of soil development and the types of plant and animal communities living in the soil and on its surface.

Zone of dispersion: a diffuse boundary in the saturated subsurface where freshwater and saline water mix. Also referred to as transition zone.

Zone of saturation: the region within the subsurface where all available pore space is saturated with water. Also termed phreatic zone.

List of Important Environmental Websites Alphabetically by Country or Region

Antarctic – www.asoc.org/antarctic-environmental-protection.
Argentina – www.mrecic.gov.ar.
Australia – www.nepc.gov.au.
Austria – www.umweltbundesamt.at/en/
Belgium – www.en.vmm.be/
Brazil – www.brazilgovnews.gov.br/standards-on-air-quality.htm.
Bulgaria – www.eea.europa.eusoer/countires/bg
Canada – www.ec.gc.ca/environmental-acts
    Alberta – www.environment.gov.ab.ca
    British Columbia – www.environment.gov.bc.ca
    Manitoba – www.environment.gov.mb.ca
    Newfoundland – www.environment.gov.nf.ca
    New Brunswick – www2.gnb.ca
    Northwest Territories – www.oag-bvg.ca
    Nova Scotia – www.environment.gov.ns.ca
    Ontario – www.ontario.gov.ca
    Quebec – www.legisquebec.gouv.qc.ca
    Saskatchewan – www.saskatchwan.ca
    Yukon – www.env.gov.yk.ca
Chile – http://protal.mma.gob.cl/.
China – www.china.org.cn/english.
Croatia – www.corine.azo.hr/en
Cyprus www.moa.gov.cy/moa/environment
Czech Republic – www.mzp.cz/en
Denmark – www.eng.mst.bl/
Egypt – www.eeaa.gov.eg/en-us/laws/envlaw.aspx.
Estonia – www.keskkonnaagentuur.ee/en
Europe – www.eea.europa.eu/overview.
Finland – www.environment.fi/
France – wwwademe.fr/en
Germany – www.umweltbunesamt.de/en
Greece – www.eea.europa/soer/countires/gr
Hungary – www.eea.europa.eu/soer/countries/hu
India – www.environmentallawsofindia.com/air/pollution/act.
Indonesia – www.aecon.org/indonesia.
Ireland – www.epa.ie/irelandsenvironment.
Italy – www.isprambiente.gov.it/en
Japan – www.env.go.jp/en/laws/policy/basic_lp.html.
Kenya – www.nema.go.ke.
Korea – www.eng.me.go.kr.

Latvia – www.vvd.gov.lv/eng/
Lithuania – www.eea.eropa.eu/soer/countires/lithuania
Luxemburg – www.aev.gouvernement.lu/en
Malaysia – www.doe.gov.my.
Malta – www.eea.europa.er/counties/malta
Mexico – www.Mexico.nom-003-SEMARNAT-1997.
National Geographic Society – www.nationalgeographic.com.
Netherlands – www.government.nl/ministry-of-infrastructure-and-the-environment.
New Zealand – www.mfe.nz.
Norway – www.miljodiretoratet.no.
Peru – www.gob.pe.
Poland – www.mos.go.pl/en
Portugal – wwwapambienta.pt/en/
Romania – www.gppbest.eu/?page_id=46&lang=en
Russia – www.mnr.r/english.
Saudi Arabia – www.pme.gov.sa.
Scotland – www.sepaorg.uk/
Slovakia – www.sazp.sk/en/slovak-environment-agency.
Smithsonian – www.si.edu.
South Africa – www.environment.gov.sa.
Spain – www.eea.europa.eu/countries-and-regions/spain
Sweden – www.swedishepa.se/
Switzerland – www.enece.org.
Tanzania – www.tanzania.go.tz.
Turkey – www.csb.gov.tr/.
Union of Concerned Scientists – www.ucsusa.org
United Kingdom – www.gov.uk/government/rganisations/environment-agency/
United Nations – www.un.org.
United Nations – Laws of the Oceans – www.un.org/en/sections/issues-depth/oceans.
United States Agency of Toxic Substances and Disease Registry – www.atsdr.cdc.gov
United States Center for Disease Control – www.cdc.gov
United States Census Bureau – www.census.gov.
United States Central Intelligence Agency – /www.cia.gov.
United States Department of Commerce – www.commerce.gov.
United States Department of State – www.state.gov.
United States Environmental Protection Agency– www.epagov.
   Alabama – www.adem.state.al.us
   Alaska – https://dec.alaska.gov/
   Arizona – www.azdeq.gov
   Arkansas – www.adeq.state.ar.us
   California – www.calepa.gov
   Colorado – www.colorado.gov/cdphe
   Connecticut – www.ct.gov
   Delaware – www.dnrec.delaware.gov
   Florida – https://floridadep.gov
   Georgia – https://epd.georgia.gov
   Hawaii – www.health.hawaii.gov/oeqc/
   Idaho – www.deq.daho.gov/
   Illinois – www.iepa.gov
   Indiana – www.in.gov/idem/
   Iowa – www.iowadnr.gov

    Kansas – www.kdheks.gov/
    Kentucky – https://dep/ky.gov
    Louisiana – www.deq.louisiana.gov/
    Maine – www.maine.gov/dep/
    Maryland – www.mde.maryland.gov/
    Massachusetts – www.mass.gov/mdep
    Michigan – www.mdeq.gov
    Minnesota – www.pca.state.mn.us/
    Mississippi – www.mdeq.ms.gov/
    Missouri – www.dnr.mo.gov/env/
    Montana – www.deq.mt.gov
    Nebraska – www.deq.state.ne.us/
    Nevada – www.ndep.nv.gov/
    New Hampshire – www.des.nh.gov/
    New Jersey – www.njdep.gov
    New Mexico – www.env.nm.gov/
    New York – www.dec.ny.gov/
    North Carolina – www.deq.nc.gov/
    North Dakota – www.deq.nd.gov/
    Ohio – www.oepa.gov
    Oklahoma – www.deq.state.ok.us/
    Oregon – www.regon.gov/DEQ
    Pennsylvania – www.dep.pa.gov/
    Rhode Island – www.de.ri.gov/
    South Carolina – www.scdhec.gov/
    South Dakota – www.denrsd.gov/
    Tennessee – www.tn.gov/environment/
    Texas – www.tceq.gov
    Utah – www.deq.utah.gov/
    Vermont – www.dec.vermont.gov/
    Virginia – www.deq.virginia.gov/
    Washington – www.ecology.wa.gov/
    West Virginia – www.dep.wv.gov/
    Wisconsin – www.dnr.wi
    Wyoming – www.deq.wyoming.gov/
United States Geological Survey – www.usgs.org.
United States National Aeronautics and Space Administration – www.NASA.org.
United States National Oceanic and Atmospheric Administration – www.noaa.org.
World Bank – www.worldbank.org.
World Health Organization – www.who.org.

Index